SOIL MECHANICS AND TRANSPORT IN POROUS MEDIA

Theory and Applications of Transport in Porous Media

Series Editor:
Jacob Bear, *Technion – Israel Institute of Technology, Haifa, Israel*

Volume 19

The titles published in this series are listed at the end of this volume.

Soil Mechanics and Transport in Porous Media

Selected Works of G. de Josselin de Jong

Edited by

Ruud J. Schotting
Utrecht University, The Netherlands

Hans (C.J.) van Duijn
Eindhoven University of Technology,
The Netherlands

and

Arnold Verruijt
Formerly at Delft University of Technology,
The Netherlands

 Springer

A C.I.P. Catalogue record for this book is available from the Library of Congress.

ISBN-10 1-4020-3536-5 (HB)
ISBN-13 978-1-4020-3536-4 (HB)
ISBN-10 1-4020-3629-9 (e-book)
ISBN-13 978-1-4020-3629-3 (e-book)

Published by Springer,
P.O. Box 17, 3300 AA Dordrecht, The Netherlands.

www.springer.com

Printed on acid-free paper

Every effort has been made to contact the copyright holders of the articles which have been
reproduced from other sources. Anyone who has not been properly credited is requested to
contact the publishers, so that due acknowledgement may be made in subsequent editions.

Printed in the Netherlands.

1

Preface

Each topical area in science has its own pioneers. Pioneers in science are typically people with unorthodox and original ideas, ideas that change our way of thinking about the world that surrounds us. In the fields of geomechanics and geohydrology, Gerard De Josselin De Jong is a typical example of such a pioneer. His scientific career started in the late fifties of the previous century, and from that time he produced a number of highly significant papers that contributed to the basic understanding of the aforementioned topical areas. He could achieve these results because of his rather unusual and unorthodox way of solving scientific problems. First, he "visualized" the problem in his mind. He always said: "I need to see a "picture" of what's going on". Then he translated this virtual "picture" into a mathematical model, and subsequently tried to solve the resulting mathematical problem. In many cases, his strategy was successful.

Visualization is maybe the key-word in De Josselin De Jong's life. Not only visualization of complex scientific problems, but also visualization of the world surrounding him: as an graphical artist. He is able to capture the real world in beautiful paintings, drawings, litho's, etchings, etc. The real world brought back to its basics: beautiful, exciting, and maybe most important: recognizable and understandable. Abstract art is not his game, neither abstract science! 'If I am not able "see" what's going on, I am not interested'.

Almost all graphs in his scientific papers were hand-drawn. No ruler was ever used. Looking at these graphs is a pleasure, almost works of art. No computer graphics tool is or will ever be able to produce such eye catching and beautiful scientific graphs. Remarkable, but true.

In this volume we present a selection of Gerard De Josselin De Jong's scientific papers. The papers are reproduced in their original form: in the original format (as they appeared in the journals or reports), including typo's, errors, and misprints. The volume consists of two parts. The first part is devoted to his scientific contributions to the topical field of soil mechanics, his main field of interest as full-professor of Soil Mechanics (Geo-technics) at Delft University of Technology. Although the subject of subsurface flow and transport processes did not belong to the chair he hold as a full-professor, he was very interested in these subjects. This interest resulted in a series of highly original papers, which are still relevant for our basic understanding of flow and transport processes in porous media. A selection of these papers can be found in the second part of this volume.

The editors,
Ruud J. Schotting, Hans (C.J.) van Duijn and Arnold Verruijt

2

Short Curriculum Vitae of G. de Josselin de Jong

1915 Born in Amsterdam

1934 Gymnasium-β in Haarlem

1941 Civil Engineering degree at Delft University of Technology

May 1941 - Sept. 1942 Engineer at Delft Soil Mechanics Laboratory (currently Geo Delft)

Sept. 1942 Arrested by the German Navy during an attempt to escape to England

Nov 1942 Sentenced to 15 years imprisonment in Germany

May 1945 Liberated by English troups in the northern part of Germany

1945 - 1947 Lived in Amsterdam, main activities drawing and painting

1947 - 1949 Lived in Paris, worked with different architects and for Bureau d'Etude de Béton Précontraint

Nov. 1947 Marriage with Cara Waller

1949 - 1959 Researcher at Delft Soil Mechanics Laboratory

Febr. 1959 Ph.D. degree at Delft University of Technology

1959 - 1960 Visiting Research Assistant, University of California, Berkeley, USA

1960 - 1980 Full Professor of Soil Mechanics, Delft University of Technology

1980 Retirement

73

"Jugendstil house at the Hooistraat seen from the Nieuwe Uitleg, The Hague",
by G. de Josselin de Jong, 1982. *Washed pen, 38cm x 27,5 cm.*
Property of Mrs. Lagaay-Govers.

viii

Contents

PART I

SOIL MECHANICS

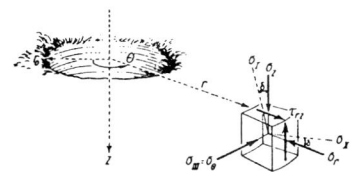

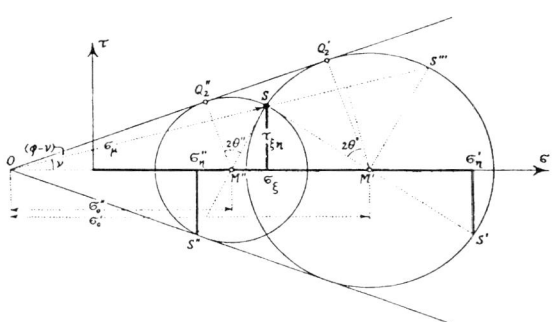

3

Soil Mechanics

3.1 Introduction to Soil Mechanics

The most important papers by Professor G. de Josselin de Jong on soil mechanics can be subdivided into three main topics: consolidation of soils, the stability of a vertical cut off, and the kinematics of granular soils in the plastic zone. This last topic contains his main contribution to theoretical soil mechanics, and has been rather controversial for some time, before being recognized as an important fundamental frame work for the analysis of soil behaviour. He also made significant contributions to the development of measuring techniques in the laboratory and in the field. Some of these can be found in his theoretical papers, some were published separately.

The first basic assumption of De Josselin de Jong's model for plastic flow is that plastic deformations are generated when the stresses satisfy the Mohr-Coulomb yield criterion, which is a certain condition on the stresses, namely that no plastic flow occurs if in all directions (i.e. on all planes) the shear stress τ and the normal stress σ satisfy the condition $\tau < c + \sigma \tan \phi$, and that plastic flow may occur if on any plane $\tau = c + \sigma \tan \phi$. This is generally accepted as a good basis for the description of the behaviour of isotropic materials with internal friction.

The essential part of De Josselin de Jong's model of plastic flow, the *double sliding free rotating model*, is that the plastic deformation consists of three components: sliding deformations on the two planes on which the Coulomb criterion is satisfied, plus an arbitrary rotation. Essential in the model is that the intensities of the two sliding deformations are unrelated, the only restriction being that their signs must be such that positive amounts of energy are dissipated. This indepence of the two sliding deformations leads to an important consequence of the model, namely that the tensor of plastic strain rates need not be coaxial with the tensor of stress. Another property of the original version of the model is that it assumes that during plastic flow the volume remains constant, which constitutes another form of non-coaxiality. A third essential property of the model is that the two sliding deformations do not completely describe the displacement field, but that the displacements may contain an arbitrary additional (free) rotation.

This model, and in particular some of its consequences, initially met with considerable opposition, although it seems that all of this has now vanished.

3

The constant volume assumption, or, to be more precise, the possibility of constant volume plastic deformation, violates an assumption derived from work of Prager and Drucker that was used with great success in metal plasticity, namely the assumption that the plastic potential, which governs the direction of plastic flow, and the yield surface, are identical. This is now called the assumption of an *associated flow rule*. It took some time before it was realized that this is not a physical necessity, but simply a convenient property of certain materials. It is now generally accepted that for frictional materials, such as soils, a *non-associated flow rule* describes reality much better, and that the constant volume assumption often applies, especially for large strains. In modern models a volumetric component of plastic flow is often incorporated as a possibility, depending upon the density of the material, but always with the constant volume case as the limiting situation for large deformations, or even the default condition.

The independence of the two sliding components of the double sliding free rotating model also met with some opposition, because it means that there may also be a deviation of the principal direction of plastic strain with the principal direction of stress in the plane of shear deformation. This may seem strange, because it may be surmised that the coaxiality of the plastic strains and the stresses is a necessary consequence of the isotropy of the material. The proof of that property presupposes the existence of a unique relation between stresses and strains, however, and this is just what De Josselin de Jong denies, at least for a rigid plastic material. Although it is now widely acknowledged that this type of non-coaxiality may indeed occur, in many modern numerical models that include plastic flow, the coaxiality of stresses and incremental plastic strains is still assumed, for definiteness or for simplicity. That there may indeed be a deviation of these two pricipal directions was proved experimentally by Drescher and De Josselin de Jong in 1971. In a contribution to the discussions at a conference in Oslo De Josselin de Jong presented some interesting results from large scale shear tests on sand, which also seem to indicate non-coaxiality.

Another important property of his model is the free rotation, which states that while the deformations may be determined by the stresses, the displacement field may include an arbitrary additional rotation. This may now seem rather trivial, but at the time of the presentation of the model, which were the days of analytical solutions of elementary problems, it gave rise to considerable controversy. This was particularly evident in the analysis of the results of simple shear tests. The classical interpretation of this type of test is that the critical ratio of shear stress to normal stress is reached on horizontal planes, so that the friction angle can immediately be determined from this ratio. De Josselin de Jong realized that the uniform shear deformation is also consistent with shearing along vertical planes, plus a rotation (the toppling book row mechanism), and that this failure mode is much more likely to occur if the horizontal normal stress is smaller than the vertical normal stress. It gave him great satisfaction when one of the leading English scientists, Peter Wroth, appeared to support his views. De Josselin de Jongs model could be used to explain the highly variable results of shear tests. In modern finite element models that include plastic flow the free rotation usually is automatically ensured, but it seems that the assumption of coaxiality of stresses and strain rates may be an unsafe constraint in many of these models.

Another somewhat controversial topic was the derivation of lower limits for the maximum height of a vertical cut off in a uniform cohesive material, without internal friction. An upper limit, on the basis of a circular slip surface, was obtained by Fellenius in 1927: $h < 3.83c/\gamma$. Simple lower limits can be obatined from equibrium fields as $h > 2c/\gamma$ and $h > 2,82c/\gamma$. Using his graphical technique of constructing stress fields that satisfy the two equilibrium equations and the yield condition De Josselin de Jong succeeded in gradually raising this lower limit, reaching a value $h > 3.39c/\gamma$ in 1978. Unfortunately, in the same year Pastor obtained an even higher lower limit, $h > 3.64c/\gamma$, using a completely different method. It has been conjectured that perhaps the existing upper limit, $h < 3.83c/\gamma$, is also a lower limit, and it may seem that certain variational techniques can be used to prove that. In the early 1980's this lead to considerable controversy in the pages of Géotechnique. De Josselin de Jong (and others) argued, rather convincingly, that it is extremely difficult to avoid certain hidden fallacies in the variational approach, and the hope on a breakthrough seems to have vanished.

Among soil engineers De Josselin de Jong was one of the first to realize that the three dimensional consolidation theory of Biot (and not the much simpler heat conduction analogy) was the proper generalization of Terzaghi's one dimensional theory. The theoretical proof is elementary, as Biots theory incorporates elasticity theory as a special case, in the absence of pore water pressures. Experimental support came from laboratory tests at the Delft University on spherical samples, although his friend Robert Gibson preceded him in that respect by a few months. He published a series of papers on three dimensional consolidation, in Dutch, with some of his collaborators, presenting analytical solutions to a variety of problems. Plans to expand this into a book, together with Gibson and Robert Schiffman never materialized, perhaps because the subject matter expanded faster than solutions could be derived, and perhaps also because the development of numerical methods made analytical solution methods somewhat obsolete. On the subject of consolidation it may also be mentioned that his admiration of the pioneer of Dutch soil mechanics, Professor A.S. Keverling Buisman, led him to try to generalize Buismans theory of secular (or secondary) consolidation to a beautiful model including viscoelastic deformation and an early version of a multiple porosity.

1.6 Lower Bound Collapse Theorem and Lack of Normality of Strainrate to Yield Surface for Soils

By

G. de Josselin de Jong

In soil mechanics practice there is a need for a lower bound collapse theorem, which permits an analysis with a result on the safe side. The usual analysis of slip surfaces may give unsafe results for a purely cohesive soil, since it is based upon a kinematically admissible collapse system and therefore constitutes an upper bound. It is therefore necessary to investigate a great number of slip surfaces and the smallest load is an approximation to the actual load which will produce collapse, but it is never known how much the computed load exceeds the actual one.

Upper bound theorems for a material possessing COULOMB friction have been treated by DRUCKER (1954, 1961), but it is still necessary to establish a lower bound theorem. Indeed a lower bound theorem would seem to be of more practical value since it would lead to a result on the safe side. Unfortunately the virtual work proofs of lower bound theorems break down if the material does not obey the postulate of DRUCKER: that additional loads cannot extract useful net energy from the body and any system of initial stresses.

Now in soils there are two possible ways of extracting work, since soils in general are friction systems. The first possibility was mentioned by DRUCKER [1954] and is obtained by changing the isotropic stress in the body with internal friction. The second way to extract work is a consequence of the possible deviation angle between the principal directions of strain rate and stress tensors. This can be shown by considering the extreme case of deviation corresponding to the sliding of the upperleft block in Fig. 1 along a slip surface at $\left(45° - \frac{1}{2}\,\varphi\right)$ to the direction of the major principal stress. The slip occurs under constant volume conditions. Initially the stress state is represented by the points AA in the stress diagram of Fig. 2, lying just inside the limit circle. The additional forces are the stresses AB which bring

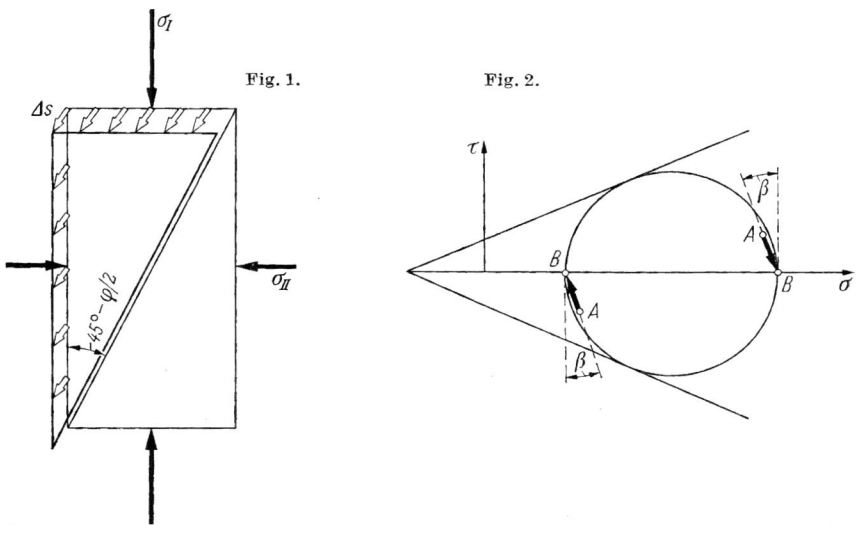

Fig. 1.

Fig. 2.

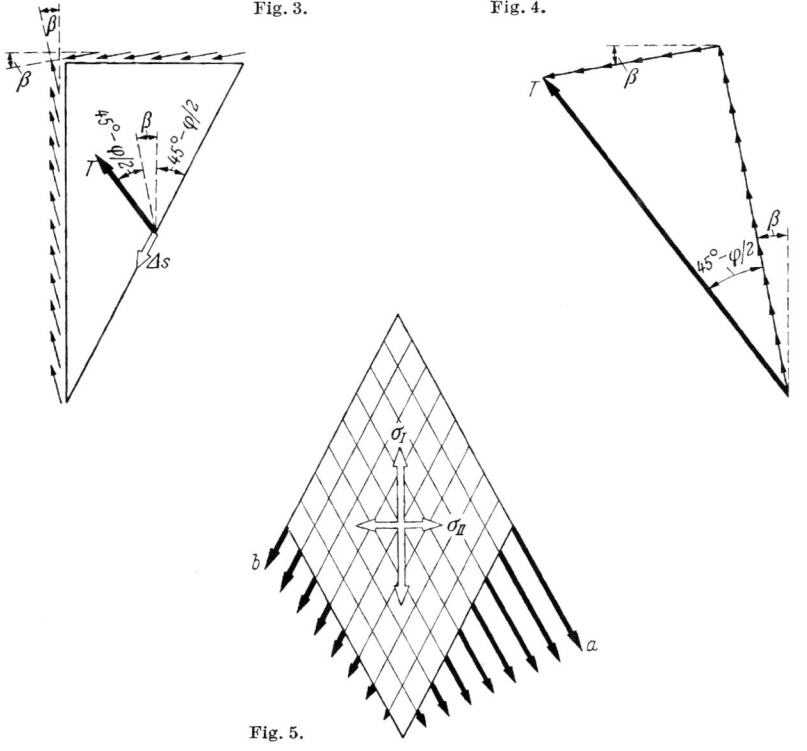

Fig. 3.

Fig. 4.

Fig. 5.

the system to a failure condition at BB. Let us consider the case when the vectors AB make an angle β with the τ-axis. The angle β can be made as small as we please by letting A approach B.

The additional loads on the moving upper left block then consist of stresses uniformly distributed along the vertical and horizontal faces and acting at an inclination β to these faces, Fig. 3. The resultant T of the additional forces on the upper left block is shown in Fig. 4 to make an angle of $\left(45° - \dfrac{1}{2}\varphi + \beta\right)$ with the vertical.

Under the influence of the existing stresses the block slides in a direction, at $\left(45° - \dfrac{1}{2}\varphi\right)$ downwards. If the displacement of the block is ΔS in that direction, then the work done by the body and the system of initial stresses on the added stress resultant T is equal to ΔS times the component of T in the direction opposite to ΔS. The work is therefore.

$$\Delta S \cdot T \cos(90° - \varphi + \beta) = \Delta S \cdot T \sin(\varphi - \beta).$$

This is positive if β is smaller than φ, thus positive work can be extracted.

Work can be extracted from a yielding system if the plastic strain rate tensor plotted as a vector in the corresponding generalised stress space is not normal to the yield surface.

In order to show the lack of normality in the case of soil explicitly, it is convenient to consider a stack of parallel cylinders which form a two dimensional analogy of a grain system with internal friction. Then the generalised stresses are the 4 stresses σ_x, σ_y, τ_{xy}, τ_{yx}, and the generalised stress space is therefore 4-dimensional. Fortunately τ_{xy} is equal to τ_{yx} and only the diagonal of length $\tau\sqrt{2}$ is a relevant coordinate. Therefore the generalised stress space can be reduced to the 3-dimensional space of Fig. 6, with coordinates σ_x, σ_y, $\tau\sqrt{2}$.

Let the material obey a COULOMB friction law, such that the yield criterion is:

$$(\sigma_x - \sigma_y)^2 + 4\tau^2 = [\sin\varphi(\sigma_x + \sigma_y + 2c \cot\varphi)]^2. \qquad (1)$$

To obtain a simpler expression for the yield surface, the coordinates are changed in the orthogonal system p, q, t according to

$$p = \frac{1}{2}\sqrt{2}(\sigma_x - \sigma_y),$$

$$q = \frac{1}{2}\sqrt{2}(\sigma_x + \sigma_y + 2c \cot\varphi),$$

$$t = \tau\sqrt{2}$$

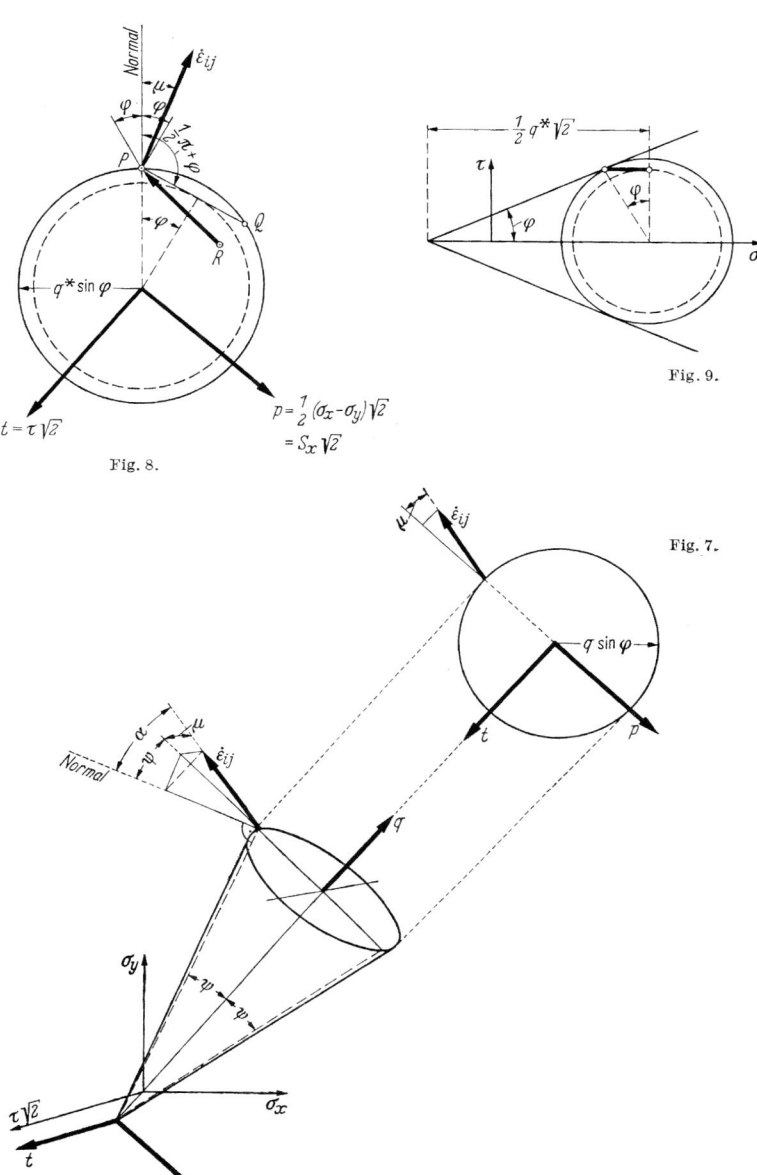

Fig. 8.

Fig. 9.

Fig. 7.

Fig. 6.

Then q is the bisectrix of σ_x and σ_y, and p is a coordinate in the σ_x, σ_y-plane perpendicular to q. In these coordinates the yield criterion is

$$2p^2 + 2t^2 = 2q^2 \sin^2 \varphi. \tag{2}$$

This shows that the yield surface is a cone with q as axis and which intersects the planes for $q = $ constant by a circle with radius $q \sin \varphi$. The angle ψ is then related to φ by

$$\tan \psi = \sin \varphi. \tag{3}$$

If the rod material is assumed to behave as the mechanical model proposed by the author (1958, 1959) plastic shear strain rates consist of volume conserving slip in the directions at $\left(45° - \dfrac{1}{2}\varphi\right)$ with the major principal stress. The two conjugate shear strain rates need not be equal. If they are a and b respectively as shown in Fig. 5, then the deviation angle μ between principal directions of strain rate tensor is given by

$$\tan \mu = \frac{a - b}{a + b} \tan \varphi. \tag{4}$$

Since a and b can only be positive, this relation implies

$$-\varphi \le \mu \le \varphi. \tag{5}$$

It can be shown by a straightforward but somewhat tedious computation that the deviation angle α between the strain rate vector and the normal to the yield surface is then given by:

$$\cos \alpha = \cos \psi \cos \mu. \tag{6}$$

Since the sliding motion is considered to take place at constant volume the strain rate vector $\dot{\varepsilon}_{ij}$, plotted in a coordinate system corresponding to the generalised stresses, lies in the $q = $ constant plane. This plane makes an angle ψ with the normal to the yield surface as shown in Fig. 6. In order that the angle α between $\dot{\varepsilon}_{ij}$ and the normal obeys (6) it is necessary that $\dot{\varepsilon}_{ij}$ is not normal to the circle in the $q = $ constant plane of Fig. 7, but makes an angle μ with the radius of that circle.

According to the first collapse theorem a body is capable of supporting the external loads in any loading program, if it is possible to find a safe statically admissable stress distribution $\sigma_{ij}^{*(s)}$. A stress distribution is called statically admissable if it obeys the equilibrium conditions inside the body, if it satisfies boundary conditions on the part of the boundary where surface tractions are given and if a yield inequality is nowhere violated. For perfectly plastic materials the yield inequality simply requires that $\sigma_{ij}^{*(s)}$ lies inside the yield surface. This requirement is clearly necessary and is also sufficient because convexity of the yield surface and normality of the strain rate vector to that surface ensure that the real collapse stress state σ_{ij} is such that the quantity

$$[\sigma_{ij} - \sigma_{ij}^{*(s)}]\,\mathring{\varepsilon}_{ij}$$

is always positive. The proof of the first collapse theorem follows then by use of virtual work considerations [for a comprehensive description of this theorem and related matter see f.i. KOITER (1960)].

Since there is not always normality in the case of soils the yield inequality condition has to be modified. The modification necessary to take care of the angle μ is only small if by some other means it is possible to prove that q cannot decrease below a certain value q^*.

If the mechanical model of Fig. 5 is applicable, Eqs. (4) and (5) say that the absolute value of μ cannot exceed φ. Now let P represent a real collapse stress state $\sigma_{ij}^{(P)}$, then P lies on the circle with radius $q^* \sin \varphi$ in the plane $q = q^*$, Fig. 8. All stress states $\sigma_{ij}^{(R)}$ represented by a point R lying below PQ, the line at an angle $\left(\frac{1}{2}\pi + \varphi\right)$ to the normal in P, may be called statically admissable with respect to P, because the angle, between any line PR and the vector $\mathring{\varepsilon}_{ij}$ for $\mu = \varphi$, will be larger than $\frac{1}{2}\pi$. Therefore the quantity

$$[\sigma_{ij}^{(P)} - \sigma_{ij}^{(R)}]\,\mathring{\varepsilon}_{ij}$$

will always be positive for $\mu = \varphi$, and clearly this result is generally valid in the interval $0 \le \mu \le \varphi$.

Since the actual collapse stress will be everywhere on the circle, the statically admissable stress state $\sigma_{ij}^{*(s)}$ is limited by all lines PQ drawn from all points of the circumference. This means that the stress states are limited by the dotted circle in Fig. 8, with a radius of length $q^* \sin \varphi \cos \varphi$.

Since the coordinates p and t actually are $\sqrt{2}$ times the deviator-stresses s_x, τ_{xy}, the requirement of the dotted circle can be represented in the usual MOHR-diagram of Fig. 9 by the dotted circle whose radius is equal to the shear stress at the tangent point of MOHR-circle and COULOMB envelope line. This means that a safe statically admissable stress state is limited by the dotted circle (Fig. 9) which is equivalent to reducing the angle of shearing resistance to a value φ^* given by:

$$\sin \varphi^* = \sin \varphi \cos \varphi.$$

Although by this modification of the definition for a statically admissable stress state, the difficulties created by the uncertainty about the deviation angle between principal directions of stress tensor and strain rate tensor are circumvented, it must be emphasized that this only applies if by other means it is established that q cannot decrease below the value q^*. The region limiting the statically admissable stress states is therefore given by a circular cylinder starting

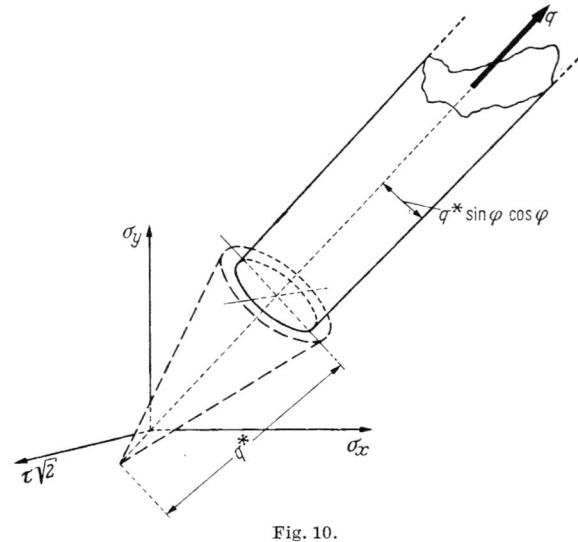

Fig. 10.

on the base of the cone with height q^* and running up to infinity with a radius $q^* \sin \varphi \cos \varphi$.

Literature

DRUCKER, D. C.: Coulomb Friction, Plasticity and Limit Loads. Appl. Mech. **21**, №. 1, 71—74 (1954).

DRUCKER, D. C.: On Stress-Strain Relations for Soils and Load Carrying Capacity. Proc. 1st Int. Conf. Mech. of Soil Vehicle Systems, Turin, 1961.

DE JOSSELIN DE JONG, G.: Indefinitness in Kinematics for Friction Materials. Proc. Conf. Brussels on Earth Pressure Problems, 1958, Vol. I, Brussels 1958, pp. 55—70.

DE JOSSELIN DE JONG, G.: Statics and Kinematics in the Failable Zone of a Granular Material. Doctors Thesis Delft, 1959.

KOITER, W. T.: General Theorems for Elastic-Plastic Solids. Progress in Solid Mechanics, Vol. I, 1960, pp. 165—221.

Discussion

Contribution de K. H. ROSCOE: I would like to question the universal application of Professor DE JONG's statement that the normality condition does not apply to soils. The following remarks are very tentative since I have not had an opportunity to make a proper study of DE JONG's proposals. It does however seem that he is considering soil to be a non-dilatant material possessing constant cohesion and constant internal friction and he is concerned only with states of failure of such a medium. I wish to make two observations regarding these assumptions. Firstly soil is a dilatant medium and as it dilates the apparent cohesion and internal friction will change. Secondly the MOHR-COULOMB envelope is not a true yield surface for soils. If yield is defined as permanent irrecoverable deformation then soils yield, and of course dilate, at stress levels well below those required to satisfy the MOHR-COULOMB criterion of failure.

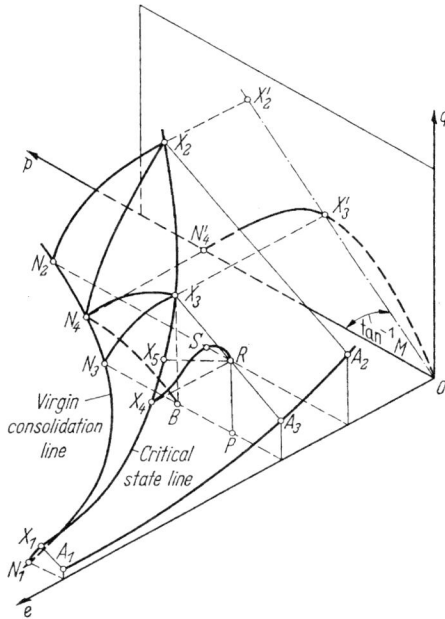

Fig. 1. Isometric view of idealised (p, q, e) yield surface for clays.

The position can be made clearer by referring to Fig. 1 which represents our concepts of the yield surface, obtained from triaxial tests on samples of a saturated remoulded clay, in p, q, e space; where $p = (\sigma'_1 + 2\sigma'_3)$, $q = (\sigma'_1 - \sigma'_3)$, e is the voids ratio and σ'_1 and σ'_3 are the major and minor principal effective compressive stresses respectively. In Fig. 1 the curve $N_1 N_2$ is the isotropic virgin consolidation curve and $X_1 X_2$ is the critical state line. The projection of the critical state line on the (p, q) plane is the straight line $O X'_2$. When a sample reaches a state corresponding to a point on the curve $X_1 X_2$ it will continue to distort in shear without further dilation and without change of stress.

The (p, q, e) yield surface for virgin and lightly over-consolidated clays is represented by the curved surface $N_1 N_2 X_2 X_1$ and in my paper to this symposium I have endeavoured to show that there is some experimental justification for such a surface. Its precise shape is open to some doubt as discussed by ROSCOE, SCHOFIELD and THURAIRAJAH (1963), and ROSCOE and SCHOFIELD (1963). Typical (p, q, e) state paths for undrained tests on normally consolidated samples are represented by curves $N_1 X_1$, $N_2 X_2$ and $N_3 X_3$, while a typical path for a drained test is $N_4 X_2$. It is important to notice that whenever a sample is at a state corresponding to a point on the surface $N_1 N_2 X_2 X_1$, and the deviator stress is increasing, it will be yielding. Consider for example a sample initially at state N_4. If it traverses any state path on the yield surface within the sector $N_2 N_4 X_3$ it will work harden as it yields but it will not fail until the critical state is attained. If the state change corresponds to the path $N_4 X_3$ which lies vertically above the elastic swelling curve $N_4 B$ then the sample will yield and not work harden. The relevant plastic potential curve is then $N'_4 X'_3$. We have called the curve $N_4 X_3$ an elastic limit curve. As a sample work hardens the relevant plastic potential curve continuously grows in size but remains geometrically similar to curve $N'_4 X'_3$. We have proposed that the form of the plastic potential curves is governed by the equation $q = M p \log_e \dfrac{p_0}{p}$ where p_0 is the initial consolidation pressure, and M is as shown in Fig. 1.

Let us now consider more heavily over-consolidated clays. The experimental data that is available for such clays is much less reliable than for lightly over-consolidated clays, hence the following remarks are extremely tentative. We suggest that the (p, q, e) yield surface for undrained tests is $A_1 A_2 X_2 X_1$ in Fig. 1. Consider an over-consolidated sample initially in a state represented by the point P. If it is subjected to an undrained test it will follow a state path which may be idealised by the path $P R X_3$ in Fig. 1. During the portion $P R$ the sample behaves virtually elastically but it begins to yield at R and continues to yield and work harden until it reaches the peak deviator stress, as well as the critical state, at X_3. If the sample was allowed to dilate during a test then present evidence suggests that the state path comes above the undrained surface. For example an ideal representation of a $p = $ constant test is given by the path $P R S X_4$. In such a test yield begins at R but the sample continues to work harden over the range $R S$ and attains the peak deviator stress at S. The sample then becomes unstable and subsequent successive states correspond to $S X_4$. I suggest that some path above a line such as $R X_5$ may be found in which this unstable portion is not present. For such a test the deviator stress would never diminish as the state changed from P to X_5. Hence as a sample, of initial state P, traverses any state path between $R X_5$ and $R X_3$, it will continually work harden until it attains the critical state when it fails. It is possible that a family of plastic potential curves of the type shown by $O X_3'$ apply during all the work hardening processes undergone by over-consolidated samples. The curve $O X_3'$ may have the same equation as $N_1' X_3'$, but adequate exerimental evidence is not available to be able to see how such plastic potentials relate to the yield surfaces for anything other than lightly over-consolidated clays. We have a little indirect evidence on the heavily overcon-solidated or "dense" side from simple shear tests on steel balls. This medium appears, during any work hardening process, to have plastic potential curves of the type shown in Fig. 2. The equation of these curves is $\tau = M \sigma \log_e \dfrac{\sigma_0}{\sigma}$, where τ is the maximum shear stress and σ the mean normal stress under conditions of plane

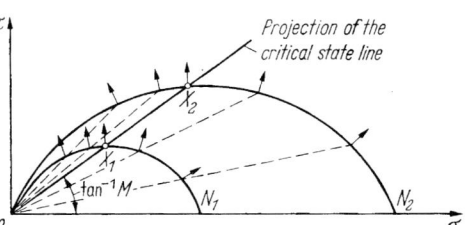

Fig. 2. Plastic potential curves for steel balls.

strain. This equation follows directly from the application of the normality condition to the boundary energy equation which was discussed by POOROOSHASB and ROSCOE (1961) for steel balls. Further work is still required to connect these potential curves with the observed yield surfaces.

Finally I would like to make the point that far too much effort has been made in soil mechanics to study failure conditions. Engineers design, and hope their structures operate, at much lower stress levels. This is the region of yielding that should be studied in detail. The MOHR-COULOMB envelope may, or may not, be shown to be valid for the failure of soils but it is not a yield surface in the true sense of the word since the yielding of a sample cannot be related to a movement on the envelope. With such a theory yield does not occur until failure takes place.

References

POOROOSHASB, H. B., and H. H. ROSCOE (1961): The correlation of the results of shear tests with varying degrees of dilatation. Proc. 5th. Int. Conf. Soil Mech. Vol. 1, pp. 297—304.

ROSCOE, K. H., and A. N. SCHOFIELD (1963): Mechanical behaviour of an idealised "wet-clay". Proc. European Conf. Soil Mech., Wiesbaden, October 1963, Vol. 1, pp. 47—54.

ROSCOE, K. H., A. N. SCHOFIELD and A. THURAIRAJAH (1963): Yielding of clays in state wetter than critical. Géotechnique **13**, No. 3, 211—240.

Réponse de G. DE JOSSELIN DE JONG: It was not my intention to say that for soils there never is normality, but that normality is not necessary. In the cases studied by M. ROSCOE normality may have been observed, but these are special cases, which are not representative for the situation in general.

That M. ROSCOE did not observe the deviation of the principal directions of stress and strain rate tensors, is due to the fact, that the stress coordinates p and q in his diagrams are not the complete set of generalised stresses. The samples were 3-dimensional, so the testresults require a representation in a 9 dimensional stress space. Since shearstresses on perpendicular faces are equal the amount of dimensions can be reduced to 6. The system I talked about this morning, is 2-dimensional and so there are 4 generalised stresses, from which τ_{xy} is τ_{yx}, reducing the system to 3 stress coordinates.

Since M. ROSCOE only considers the stress combinations p and q, his graphs correspond in a way to the σ_x, σ_y plane which intersects the cone enclosed by the yield surface along the axis. The deviation of the principal directions of the tensors is only visible in the plane perpendicular to the axis.

Cf. aussi, p. 46, la citation de D. C. DRUCKER.

PROCEEDINGS OF THE GEOTECHNICAL CONFERENCE OSLO 1968

on Shear Strength Properties
of Natural Soils and Rocks

COMPTES RENDUS DE LA CONFERENCE GEOTECHNIQUE OSLO 1968

sur les propriétés de résistance au
cisaillement des sols naturels
et des roches

VOLUME II

NORWEGIAN GEOTECHNICAL INSTITUTE
OSLO 1968

PROF. G. DE JOSSELIN DE JONG (Netherlands):

In their paper (3/14) Roscoe, Bassett and Cole review concepts pertaining to the coincidence of principal directions of stress and strain. Besides the points mentioned, it must be noted that a case of non-coincidence is to be expected if rupture planes or rupture zones develop erratically throughout the soil mass. Such planes or zones originate if the material yields in these discrete regions before the rest of the soil mass deforms excessively.

Since planes at an angle of $\pm (\pi/4 - \varphi/2)$ with the major principal stress direction have to transmit a stress combination which is most unfavourable to support, it is approximately in these directions that the rupture planes or zones develop. These directions coincide with the stress characteristics. Because the material is yielding in the rupture zones the shear strain rate is undetermined and may be unequal for the two conjugate directions. This inequality is not yet sufficient to create non-coincidence; it is also necessary that the angle φ is unequal to zero.

For an ideally isotropic material it can be expected that in a soil mass of sufficiently large dimensions the average shear strain rate in the two conjugate directions will be the same, thus resulting in coincidence of principal directions of stress and strain. However, even the slightest deviation from isotropy may result in a considerable difference between these principal directions. The mechanism is in a way similar to the instability of a rod under axial compression. If a perfectly straight cylindrical rod is compressed by forces exactly along its axis, then theoretically the rod should only reduce in length, but in reality it will always buckle in some unpredictable sidewise direction.

It may be difficult to visualize the strain rate tensor in this case where the strain rate is concentrated in discrete zones, because a tensor essentially only can be defined for a continuous deformation. The discrete rupture pattern can be replaced, however, by a continuous deformation which averages the discrete jumps and the tensor associated to this representative deformation is the one considered. That this concept represents a physical reality was demonstrated by the following test.

A sandblock ($60 \times 60 \times 15$ cm) schematically represented in Fig. 1 is enclosed on the sides by 12 loading elements provided with teeths, and two thin plastic sheets along the lower and upper plane. The air pressure in the pores was reduced by 0.9 atm. in order to create an allround pressure. Then forces were applied to the loading elements in the direction of the sides, such that a system of pure shear stress was added to the allround pressure. The principal stress directions are then parallel to the diagonals. The sand was deposited in layers parallel to one of the diagonals in order to prevent that anisotropy created by deposition would offer a preference for one of the two conjugate directions of imminent shear.

The difference with Roscoe's simple shear apparatus is that in his apparatus deformation is enforced and stresses are measured, whereas in our test the stresses are applied and the sandblock is left to move in the manner it pleases.

The deformation was measured by photographing the upper sheet, which being transparent showed the grains.

In Fig. 2 two photographs of the block representing a loading from $\sigma_1/\sigma_3 = 1.65$ to 3.15 are superimposed. The photographs are shifted and rotated in such a manner that the particle traces form a family of curves with orthogonal asymptotes. These asymptotes then have the direction of the principal strain rate directions. The smoothness of the curves which are approximately hyperbolas indicates that the deformation was practically uniform. A detailed survey of the deformation by use of a chartographic stereomicrometer at the International Institute for Aerial Survey and Earth Sciences at Delft showed an average deviation angle between principal stress and strain rate direction of $12°$, with a spread of about $3°$.

In order to investigate whether this homogeneous deformation field existed throughout the sample the sand was mixed with a small amount of cement, enough to solidify the sand mass by adding water after the test. The sand was deposited in black and white layers.

After the bloc had solidified it was abrased to show the successive sections at 2.5 cm intervals of the height. A typical view of such a section is given in Fig. 3. The deformation was concentrated in narrow zones which followed the direction of the stress characteristics. These rupture zones were found in the same location for every section, which indicates

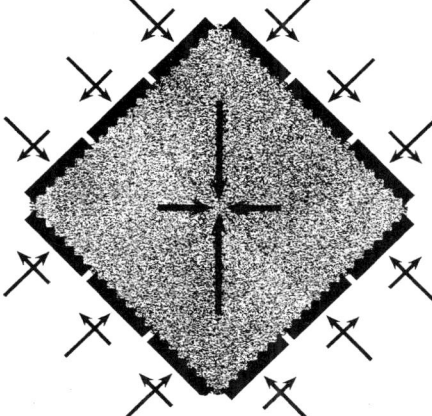

Fig. 1. *Sand block loaded by normal and shear stress.*

Fig. 2. *Superimposed photographs of sand block before and after loading.*

199

that vertical planes are formed throughout the sample. Some originated in the middle of a loading element. The shear-strain-rate differed for all the ruptive zones.

The reason that they were not observed in the photograph was the relative rigidity of the plastic envelopping sheet. The sheet averaged the deformation and showed therefore a homogeneous deformation with deviating principal directions because of the underlying mechanism of unequal strain-rate in the erratic rupture zones.

The mechanism observed in the test was very similar to the one proposed in ref. 1, p. 57, see Fig. 4 taken from that publication.

The indeterminancy of the deformation created by this mechanism needs not to be of too much concern for further use in predictions of soil behaviour. It will anyhow not be possible to predict deformation of soil masses in detail because the initial stress conditions are mostly impossible to ascertain. A more realistic approach to the determination of stability analysis is the use of a lower bound theorem, which gives an answer that is on the safe side and irrespective of previous loading history. It has been shown in ref 2, that the noncoincidence is taken care of by a small reduction of φ to φ^*, such that

$$\sin \varphi^* = \sin \varphi \cos \varphi$$

However, the noncoincidence is only a minor difficulty, the major one is the ignorance with respect to the isotropic stress. How a solution can be constructed which gives higher and therefore more attractive values than the solution mentioned in ref. 2, is beyond the scope of this discussion, but amounts to the construction of a field of stress characteristics taking the solution with the lowest isotropic stress everywhere.

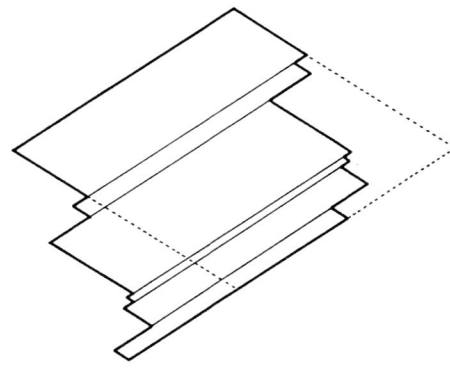

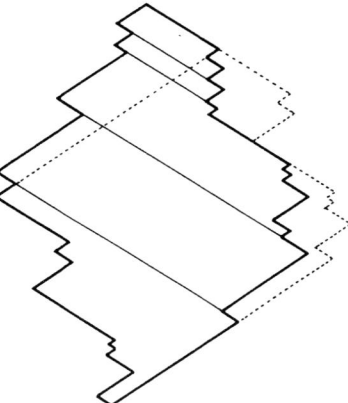

Fig. 4. *Failure mechanism for granular materials.*

References

de Josselin de Jong, G., 1959. *Statics and kinematics in the failable zone of a granular material.* Doctors Thesis Delft.
de Josselin de Jong, G. *Lower bound collapse theorem and lack of normality of strainrate to yield surface for soils.* Proc. IUTAM symp. on Rheology and Soil Mechanics, Grenoble, p. 69–75.

CHAIRMAN:

Thank you Professor de Josselin de Jong – I am sorry we are short of time. The following speaker is Professor Šuklje.

PROF. L. ŠUKLJE (Yugoslavia):

Monsieur le Président,

Je vous demande d'accorder l'hospitalité de la 3ème Section au sujet que j'ai traité dans mon rapport apparu

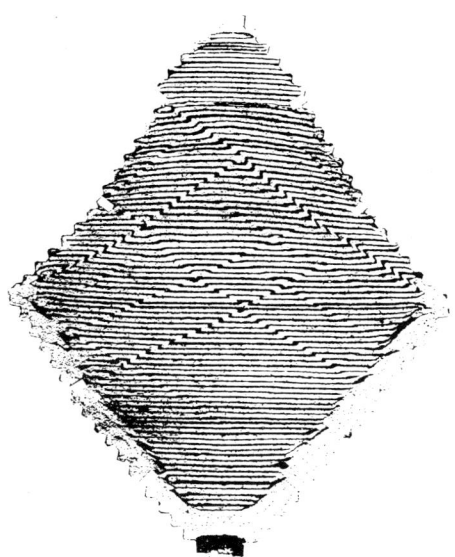

Fig. 3. *Discontinuities in interior of sand block.*

200

THE DOUBLE SLIDING, FREE ROTATING MODEL FOR GRANULAR ASSEMBLIES

G. de Josselin de Jong*

INTRODUCTION

The sliding block model for the mechanism of deformation, in a body composed of grains, is based on the concept that movements of grains with respect to each other occur along planes that coincide preferably with the stress characteristic planes. In the case of plane strain these planes intersect the two-dimensional plane of consideration along two characteristic lines called S_1 and S_2.

The object of this Note is not to consider the probable veracity of such a model, but to establish the flow rule and the constitutive equations which follow from the special character of the model. The properties are taken to be those that were proposed by de Josselin de Jong (1958, 1959). In that model sliding can occur simultaneously in the S_1 and S_2 directions at different shear strain rates, but limited in sense, and in addition the sliding elements are free to rotate.

Geniev (1958) considered such a model, but restricted sliding to one of the two characteristic directions. Most investigators reject this restriction and agree that it is desirable to permit a double sliding motion. Mandl and Fernández Luque (1970) reconsidered the double sliding model and confirmed equations, obtained by Spencer (1964) and Zagainov (1967) for the stationary case, that principal directions of stress remain fixed. However, the equations refer to a model that is restricted in its rotation, as if the sliding elements are forced by an external agency to conserve their orientation in space. Therefore these equations refer to a different model from that of de Josselin de Jong. Their model is not free to rotate and so it cannot execute motions which are commonly accepted to have been observed in reality, e.g.

---

* Professor of Mechanics, Department of Civil Engineering, Delft University of Technology, the Netherlands.

the rotation of the soil mass separated from an embankment by a circular slip plane.   Another kind of rotation was observed in a verification experiment by Drescher (1971).

Mandel (1966, p. 307) directed attention to this lack of freedom of rotation and re-established equations obtained previously (Mandel, 1947).   However, he remarks that the concept of double sliding and rotation combined is void because every deformation without volume change can be decomposed in such a manner.   This remark is correct if sliding is free to occur along each characteristic plane in both senses, i.e. either in the direction of the shear stress on that plane or against it.

By restricting the sliding sense as proposed by de Josselin de Jong the model is made to obey the thermodynamic requirement that energy is dissipated during sliding.   The necessity of this requirement is a consequence of the frictional character of the mechanism.   The grains of, say, a dry sand do not move with respect to each other because friction forces in the contact points between them prevent this.   Sliding can only occur if the friction is surmounted and therefore shear strain will develop only in the direction of the shear stress in the plane o sliding and never against it.   De Josselin de Jong (1959, p. 57) called this the requirement off direction and formulated it as

$$a \geqslant 0$$
$$b \geqslant 0$$

By this restricting requirement the concept of double sliding and rotation combined is no longer meaningless because, when introduced mathematically, a system of hyperbolic differential equations is obtained with a limited range of solutions.   This hyperbolic system is unusual, because its coefficients, instead of being fixed for every point in the field, obey inequalities which determine, instead of unique characteristic directions at every point, a fan of possible directions for the characteristics.   This has been shown graphically by de Josselin de Jong (1959, pp. 72–80).   The pertinent differential equation (de Josselin de Jong, 1959, p. 92) was given in terms of the undetermined characteristics and their curvatures and is unattractive.

The object of this Note is to re-establish the constitutive equations as referred to a cartesian x, y co-ordinate system.   These co-ordinates are straight and so the curvatures of the characteristics disappear from the equations, which simplifies their form.   Since the constitutive equation contains coefficients that obey inequalities, it can be presented as an inequality.

A common objection against the double sliding mechanism is that the principal directions of stress and strain rate tensors can deviate.   It is often proposed that such a deviation can occur only if the material is not isotropic.   However, the reasoning to substantiate this starts with the assumption that an analytic functional relationship exists between the invariants of the two tensors (see e.g. Eringen, 1962, p. 158).

Since for the double sliding model the constitutive law contains an inequality, no such analytic function exists and therefore there is no need for coincidence of principal directions. Nevertheless the requirements of isotropy (see Eringen, 1962, p. 139) are fulfilled because the inequality is invariant for the full orthogonal group of co-ordinate transformations.   This is also true for three dimensions.

Mandl and Fernández Luque (1970) tried to remove the objection to non-coaxiality in isotropic materials by mentioning that in two dimensions the co-ordinate transformations for reflexion cannot be obtained from those for rotation simply by taking the negative of all matrix components, as can be done in three dimensions.   However, this only proves that a proof based on such a sign inversion cannot be applied in two dimensions; it does not mean that another proof might not exist.   Another proof exists if there is a functional relationship between the two tensors.   The functional relationship does not exist in this case and therefore non-coaxiality is acceptable in three as well as in two dimensions.

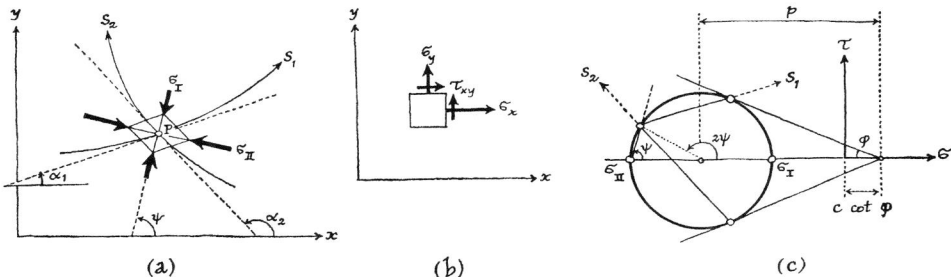

Fig. 1(a). Stress characteristics S_1 and S_2 in the x, y plane (bold arrows indicate directions of principal stresses σ_I and σ_{II}), (b) arrows showing directions in which stresses are taken as positive, (c) limiting stress circle in Mohr diagram

STRESS CHARACTERISTICS AND RELATIVE VELOCITIES ALONG THEM

The directions of the stress characteristics S_1 and S_2 at a point P are given by the angles α_1 and α_2 of their tangents with respect to the x axis such that (see Fig. 1(a))

$$\left. \begin{array}{l} \alpha_1 = \psi - \tfrac{1}{4}\pi - \tfrac{1}{2}\phi \\ \alpha_2 = \psi + \tfrac{1}{4}\pi + \tfrac{1}{2}\phi \end{array} \right\} \quad \cdots \cdots \cdots \quad (1)$$

In these expressions ψ is the angle between the algebraically larger principal stress and the x axis. Stresses are taken as positive in the direction of the arrows of Fig. 1(b) and so larger algebraically means smaller compression stress. The angle of internal friction is ϕ.

The limiting stress condition, supposed to be fulfilled in P, can then be written as

$$\left. \begin{array}{l} \sigma_x = -p + p\sin\phi\cos 2\psi + c\cot\phi \\ \tau_{xy} = \tau_{yx} = p\sin\phi\sin 2\psi \\ \sigma_y = -p - p\sin\phi\cos 2\psi + c\cot\phi \end{array} \right\} \quad \cdots \cdots \quad (2)$$

In these expressions p is the distance between the centre of the Mohr circle and the intersection point of the Coulomb envelope lines (Fig. 1(c)). According to the Coulomb theory stress circles can exist only to the left of the intersection point of the envelope lines so that the requirement on p is

$$p \geqslant 0 \quad \cdots \cdots \cdots \cdots \quad (3)$$

In order to avoid the complication of differentiations along curvilinear co-ordinates the constitutive inequalities are developed here for V_x and V_y, the x, y components of the velocities. The velocities themselves are not of interest, but the relative velocities are, because the physical properties of the model only provide considerations concerning the relative velocities of points on stress characteristics. This leads to differential equations in the velocity components.

The infinitesimal distance vector $\mathrm{d}\boldsymbol{l}_1$ of length $\mathrm{d}l_1$ from point P to point P_1 (see Fig. 2(a)) on the S_1 stress characteristic has x, y components

$$\left. \begin{array}{l} \mathrm{d}x_1 = \cos\alpha_1\,\mathrm{d}l_1 \\ \mathrm{d}y_1 = \sin\alpha_1\,\mathrm{d}l_1 \end{array} \right\} \quad \cdots \cdots \cdots \quad (4)$$

The x, y components $\mathrm{d}V_x^1$, $\mathrm{d}V_y^1$ of the relative velocity vector $\mathrm{d}\boldsymbol{V}^1$ of point P_1 with respect to P can be written as a total differential

$$\left.\begin{array}{l} dV_x^1 = V_{x,x}\,dx_1 + V_{x,y}\,dy_1 \\ dV_y^1 = V_{y,x}\,dx_1 + V_{y,y}\,dy_1 \end{array}\right\} \quad\cdots\cdots\quad (5)$$

where a comma indicates differentiation with respect to the variable following it.

Elimination of dx_1 and dy_1 from these equations gives

$$\left.\begin{array}{l} dV_x^1 = (V_{x,x}\cos\alpha_1 + V_{x,y}\sin\alpha_1)\,dl_1 \\ dV_y^1 = (V_{y,x}\cos\alpha_1 + V_{y,y}\sin\alpha_1)\,dl_1 \end{array}\right\} \quad\cdots\cdots\quad (6)$$

for the x, y components of the relative velocity of two points on an S_1 stress characteristic. By changing 1 into 2 the relative velocity components of two points at an infinitesimal distance dl_2 on an S_2 line are obtained.

The curvatures of the S_1 and S_2 lines give only second order terms in these expressions that can be disregarded.

SEPARATION OF DOUBLE SLIDING AND ROTATION

In order to introduce the physical properties of the double sliding, free rotating model, it is necessary to decompose the relative velocity vector dV^1 into two components: dV_b^1 parallel to S_2, the conjugate of S_1 and dV_p^1 perpendicular to S_1. These vector components are taken as positive if they are in the direction of the arrows in Fig. 2(a).

The special manner of decomposition results in the following relations between the magnitudes dV_b^1 and dV_p^1 of these vectors and the x, y components of the relative velocity between P_1 and P

$$dV_b^1 = [-dV_x^1\cos\alpha_1 - dV_y^1\sin\alpha_1]/\sin\phi \quad\cdots\cdots\quad (7)$$

$$dV_p^1 = [+dV_x^1\sin\alpha_2 - dV_y^1\cos\alpha_2]/\sin\phi \quad\cdots\cdots\quad (8)$$

The separation is such that dV_b^1 is due exclusively to the sliding mechanism and dV_p^1 is created by the rotation.

A similar decomposition of the vector dV^2, representing the relative velocity with respect to P of a point P_2 on the S_2 stress characteristic through P, gives a vector dV_a^2 parallel to S_1 and dV_q^2 perpendicular to S_2 (see Fig. 2(b)). The special manner of decomposition gives

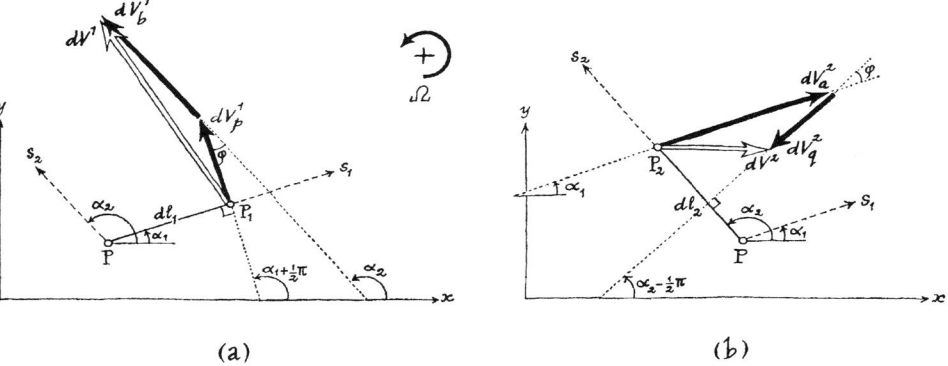

(a)   (b)

Fig. 2(a). **Decomposition of relative velocity dV^1 of point P_1 with respect to P into the components dV_b^1 parallel to S_2 and dV_p^1 perpendicular to S_1, (b) decomposition of relative velocity dV^2 of point P_2 with respect to P into the components dV_a^2 parallel to S_1 and dV_q^2 perpendicular to S_2**

for the magnitude of these vectors

$$dV_a^2 = [-dV_x^2 \cos \alpha_2 - dV_y^2 \sin \alpha_2]/\sin \phi \quad \ldots \ldots \quad (9)$$

$$dV_q^2 = [+dV_x^2 \sin \alpha_1 - dV_y^2 \cos \alpha_1]/\sin \phi \quad \ldots \ldots \quad (10)$$

The vectors are taken as positive if they are in the direction of the arrows in Fig. 2(b).

DOUBLE SLIDING

The components dV_a^1 and dV_a^2 are due to the character of double sliding of the model.

Considering first dV_b^1, because sliding occurs along the stress characteristics a line element on an S_1 stress characteristic is not affected by a sliding along the S_1 family. Therefore the double sliding contributes only a component dV_b^1 parallel to S_2. Curvature of the stress characteristics does not affect the decomposition of the vectors or the sliding directions at point P.

In accordance with previous work, the quantity b is introduced to represent the magnitude of the shear strain rate along the S_2 family by the definition

$$dV_b^1 = +b \, dl_1 \cos \phi \quad \ldots \quad \ldots \ldots \quad (11)$$

Substituting this in equation (7) and using equations (6) gives

$$-V_{x,x} \cos^2 \alpha_1 - (V_{x,y} + V_{y,x}) \cos \alpha_1 \sin \alpha_1 - V_{y,y} \sin^2 \alpha_1 = +b \sin \phi \cos \phi$$

Elimination of α_1 with equations (1) gives

$$-(V_{x,x} + V_{y,y}) - (V_{x,x} - V_{y,y}) \sin (2\psi - \phi) + (V_{x,y} + V_{y,x}) \cos (2\psi - \phi) = +b \sin 2\phi \quad (12)$$

Considering second dV_a^2, the double sliding model infers that a shear strain rate along the S_1 family can exist, with a magnitude dV_a^2, independent of slidings along the S_2 family. A quantity a for the shear strain rate along the S_1 family was introduced in previous work, which is defined by

$$dV_a^2 = +a \, dl_2 \cos \phi \quad \ldots \quad \ldots \ldots \quad (13)$$

Substituting equation (9) and using equations (6) with 1 replaced by 2, and eliminating α_2 from equation (1) gives

$$-(V_{x,x} + V_{y,y}) + (V_{x,x} - V_{y,y}) \sin (2\psi + \phi) - (V_{x,y} + V_{y,x}) \cos (2\psi + \phi) = +a \sin 2\phi \quad (14)$$

Equations (12) and (14) were not mentioned by Mandel (1966) or Spencer (1964) although they can be derived directly from their analyses. In Mandl and Fernández Luque's (1970) notation $c_1 = -b \cos \phi$, $c_2 = -a \cos \phi$, $\epsilon_{xx} = -V_{x,x}$ and so on and equations (12) and (14) can be deduced from the first two of their equations (83).

ROTATION

The components dV_p^1 and dV_q^2 are due to rotation of the sliding elements. The representation of reality by the model is such that these elements, which can be visualized as infinitesimal curvilinear rhomboids, slide and rotate but remain rigid during motion conserving their shape. This means that every line of such an element rotates at the same rate, with an angular velocity of magnitude Ω anticlockwise. Then the relative velocity components (8) and (10) perpendicular to the stress characteristics have a magnitude

$$\left. \begin{array}{l} dV_p^1 = \Omega \, dl_1 \\ dV_q^2 = \Omega \, dl_2 \end{array} \right\} \quad \ldots \ldots \ldots \ldots \quad (15)$$

Rigid body rotation of the elements can be generated by several situations. Spencer (1964) mentioned two causes: sliding of the elements along stress characteristics that are curved, and rotation of the principal stresses in a point, in the non-stationary case. These rotations are due to local circumstances, but are not the only reason for the occurrence of rotation.

If the model is free to rotate, the local sliding elements will follow every rotation without resistance, and also the rotations imposed by the surrounding elements up to the boundaries of the body. These additional rotations can be different in every other situation and are introduced as an unknown variable and arbitrary function of x and y. The locally generated rotations mentioned by Spencer are submerged in the unknown magnitude of all those rotations together.

The quantity Ω introduced by equations (15) is the total rotation that includes all these effects and whose magnitude, being an unknown variable of x and y, cannot be specified from local conditions only. Therefore flow equations containing Ω cannot be considered as constitutive equations.

Introducing equations (8) and (10) and combining them with equations (6) gives

$$+ V_{x,x} \cos \alpha_1 \sin \alpha_2 + V_{x,y} \sin \alpha_1 \sin \alpha_2 - V_{y,x} \cos \alpha_1 \cos \alpha_2 - V_{y,y} \sin \alpha_1 \cos \alpha_2 = \Omega \sin \phi \qquad (16)$$

$$+ V_{x,x} \cos \alpha_2 \sin \alpha_1 + V_{x,y} \sin \alpha_2 \sin \alpha_1 - V_{y,x} \cos \alpha_2 \cos \alpha_1 - V_{y,y} \sin \alpha_2 \cos \alpha_1 = \Omega \sin \phi \qquad (17)$$

Elimination of Ω from these equation gives
$$(V_{x,x} + V_{y,y}) \sin (\alpha_2 - \alpha_1) = 0$$
Since $\alpha_2 - \alpha_1 = \tfrac{1}{2}\pi + \phi$ this reduces to
$$V_{x,x} + V_{y,y} = 0 \qquad \qquad (18)$$

This is the relation for volume incompressibility, a property known to be exhibited by the double sliding, free rotating model.

Adding equations (16) and (17) and substituting equations (1) gives

$$(V_{x,x} - V_{y,y}) \sin 2\psi - (V_{x,y} + V_{y,x}) \cos 2\psi + (-V_{x,y} + V_{y,x}) \sin \phi = 2\Omega \sin \phi \quad (19)$$

Equations (18) and (19) are identical to Spencer's (1964) equations (3.20) and (3.21).

CONSTITUTIVE INEQUALITIES

In the previous sections flow rules have been derived that describe the behaviour of the double sliding, free rotating model. In order to compute a velocity field from boundary conditions, it is sufficient to know two relations concerning V_x and V_y or their derivatives in every point. However, the flow rules (equations (12), (14), (18) and (19)) also contain the unknown sliding rates a and b, and the unknown rotation Ω.

Spencer (1964), Zagainov (1967) and Mandl and Fernández Luque (1970) use only equations (18) and (19) because in their opinion Ω is a known quantity. Mandel (1966) realized that Ω is unknown and concluded that the equations (18) and (19) are insufficient.

In this Note equation (19) is discarded because it contains the unknown Ω. This leaves equations (12), (14) and (18) and the unknowns V_x, V_y, a and b. These three equations are apparently insufficient for four unknowns. However, so far the thermodynamic requirement of energy dissipation has not been used and when introduced it produces a treatable system, although it consists of inequalities.

Adding equations (12) and (14) and substitution of equation (18) gives

$$(V_{x,x} - V_{y,y}) \cos 2\psi \sin \phi + (V_{x,y} + V_{y,x}) \sin 2\psi \sin \phi = (a+b) \cos \phi \sin \phi \quad (20)$$

Multiplication by p and substituting equations (2) gives

$$\tfrac{1}{2}(V_{x,x} - V_{y,y})(\sigma_x - \sigma_y) + (V_{x,y} + V_{y,x})\tau_{xy} = p(a+b) \cos \phi \sin \phi$$

Using equation (18) this gives

$$V_{x,x}\sigma_x + V_{y,y}\sigma_y + V_{x,y}\tau_{xy} + V_{y,x}\tau_{yx} = p(a+b)\cos\phi\sin\phi \qquad . \quad . \quad (21)$$

The terms on the left-hand side of equation (21) together form the energy produced by the stresses on the strain rates. In order that no work is extracted from the system, this quantity should always be positive, and this means that the term on the right-hand side of equation (21) must be positive. The angle of internal friction lies between zero and $\frac{1}{2}\pi$ and $p \geqslant 0$, so $a+b$ must be positive. Since the sliding rates a and b are independent it follows from thermodynamic considerations that a and b are both positive, so that equation (21) dictates the requirements

$$\left. \begin{array}{l} a \geqslant 0 \\ b \geqslant 0 \end{array} \right\} \qquad . \qquad . \qquad . \qquad . \qquad . \qquad . \qquad . \qquad (22)$$

These requirements can be written in terms of V_x and V_y and their derivatives by use of equations (12) and (14). Using equation (14) and $a \geqslant 0$ gives the first part of the consecutive inequality (23) and using equation (12) with $b \geqslant 0$ gives the second part

$$\left. \begin{array}{l} [-(V_{x,x}-V_{y,y})\cos 2\psi - (V_{x,y}+V_{y,x})\sin 2\psi]\sin\phi \\ \leqslant [-(V_{x,x}-V_{y,y})\sin 2\psi + (V_{x,y}+V_{y,x})\cos 2\psi]\cos\phi \\ \leqslant [+(V_{x,x}-V_{y,y})\cos 2\psi + (V_{x,y}+V_{y,x})\sin 2\psi]\sin\phi \end{array} \right\} \qquad . \quad . \quad . \quad (23)$$

This inequality together with equation (18) gives the constitutive relations for the double sliding, free rotating model of a granular assembly. It can be verified that (23) is invariant for the full orthogonal group of co-ordinate transformations. This justifies the use of the inequality (23) together with the invariant expression (18) as the constitutive law for an isotropic material.

Introducing the angle ξ between the x axis and the principal direction of strain rate with the algebraically larger value gives

$$\left. \begin{array}{l} (V_{x,x}-V_{y,y}) = W\cos 2\xi \\ (V_{x,y}+V_{y,x}) = W\sin 2\xi \end{array} \right\} \qquad . \quad . \quad . \quad . \quad . \quad (24)$$

with $W = \sqrt{[(V_{x,x}-V_{y,y})^2 + (V_{x,y}+V_{y,x})^2]}$ where W is twice the strain rate deviator, which is always positive. Substituted into (23), these inequalities reduce to

$$-\cos 2(\xi-\psi)\sin\phi \leqslant \sin 2(\xi-\psi)\cos\phi \leqslant \cos 2(\xi-\psi)\sin\phi$$

which can be written as

$$-\tan\phi \leqslant \tan 2(\xi-\psi) \leqslant +\tan\phi \qquad . \quad . \quad . \quad . \quad (25)$$

by dividing the inequalities by $\cos 2(\xi-\psi)$, which is always positive because $W\cos 2(\xi-\psi) = (a+b)\cos\phi$ according to equations (20) and (24), whereas $(a+b)\cos\phi$ is positive because of inequalities (22).

Introducing the deviation angle i between the principal directions of the stress and strain rate tensors by the definition

$$i = \xi - \psi \qquad . \qquad . \qquad . \qquad . \qquad . \qquad . \qquad (26)$$

the inequality (25) shows that this deviation angle obeys

$$-\tfrac{1}{2}\phi \leqslant i \leqslant +\tfrac{1}{2}\phi \qquad . \qquad . \qquad . \qquad . \qquad . \qquad (27)$$

3+

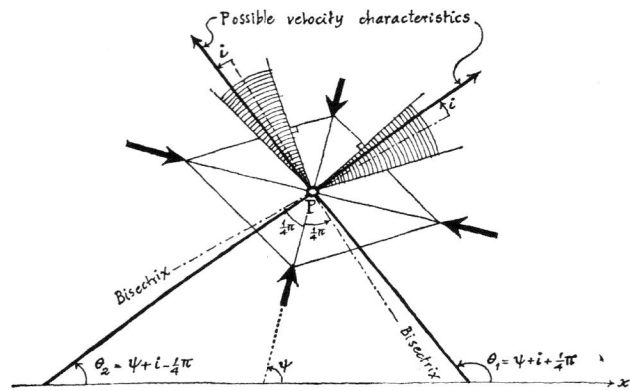

Fig. 3. Possible directions of velocity characteristics are limited to the shaded fans

HYPERBOLIC CHARACTER OF CONSTITUTIVE RELATIONS

The constitutive inequality (23) and equation (18) form a system of differential equations in V_x and V_y, by which the possible range of flow fields can be computed from boundary conditions.　The hyperbolic character of this system can be shown by the following analysis as suggested by Strack (1970).

Substituting equation (26) into (24) the fundamental inequality is given by

$$(V_{x,x}-V_{y,y}) \sin 2(\psi+i) = (V_{x,y}+V_{y,x}) \cos 2(\psi+i) \qquad . \quad . \quad . \quad (28)$$

This equation is an inequality because i obeys (27).　The equations (28) and (18) concern the derivatives of V_x and V_y and are of the form

$$\left. \begin{array}{l} A_1 V_{x,x}+B_1 V_{x,y}+C_1 V_{y,x}+D_1 V_{y,y} = 0 \\ A_2 V_{x,x}+B_2 V_{x,y}+C_2 V_{y,x}+D_2 V_{y,y} = 0 \end{array} \right\} \qquad . \quad . \quad . \quad . \quad (29)$$

where

$$\left. \begin{array}{l} A_1 = -D_1 = \sin 2(\psi+i) \\ B_1 = C_1 = -\cos 2(\psi+i) \\ A_2 = D_2 = +1 \\ B_2 = C_2 = 0 \end{array} \right\} \qquad . \quad . \quad . \quad . \quad (30)$$

From the theory of differential equations it is known that the system (29) is hyperbolic with characteristic directions θ_1 and θ_2 if

$$\tan \theta_{1,2} = [b \pm \sqrt{(b^2-4ac)}]/2a \qquad . \quad . \quad . \quad . \quad . \quad (31)$$

is real.　In this expression

$$a = A_1 C_2 - C_1 A_2 = \cos 2(\psi+i)$$

$$b = A_1 D_2 + B_1 C_2 - C_1 B_2 - D_1 A_2 = 2 \sin 2(\psi+i)$$

$$c = B_1 D_2 - D_1 B_2 = -\cos 2(\psi+i)$$

It follows that the root of equation (31) is real as

$$\sqrt{(b^2 - 4ac)} = \sqrt{[4 \sin^2 2(\psi + i) + 4 \cos^2 2(\psi + i)]} = 2$$

The system is therefore always hyperbolic regardless of the value of i.
The characteristic directions θ_1 and θ_2 follow from (31)

$$\tan \theta_{1,2} = [\sin 2 (\psi + i) \pm 1]/\cos 2(\psi + i)$$

giving for the angles

$$\left. \begin{array}{l} \theta_1 = \psi + i + \tfrac{1}{4}\pi \\ \theta_2 = \psi + i - \tfrac{1}{4}\pi \end{array} \right\} \qquad . \quad . \quad . \quad . \quad . \quad . \quad . \quad (32)$$

CONCLUSIONS

The velocity characteristics are everywhere perpendicular, because $\theta_1 - \theta_2 = \pi/2$, and this is in agreement with the volume conserving character of the model.

The characteristic directions are limited by fans whose boundaries deviate by $\phi/2$ from the bisectrices of the principal stress directions whose angles with the x axis are $\psi \pm \pi/4$ (see Fig. 3).

A boundary value problem which in the case $\phi = 0$ has a unique solution has a limited range of solutions if $\phi \neq 0$.

REFERENCES

DRESCHER, A. (1971). Private communication.
ERINGEN, A. C. (1962). *Non-linear theory of continuous media*, 477 pp. New York: McGraw-Hill.
GENIEV, G. A. (1958). Problems of the dynamics of a granular medium (in Russian). *Akad. Stroit. Archit.*, SSSR, Moscow.
DE JOSSELIN DE JONG, G. (1958). The undefiniteness in kinematics for friction materials. *Proc. Conf. Earth Pressure Probl., Brussels* **1**, 55–70.
DE JOSSELIN DE JONG, G. (1959). Statics and kinematics in the failable zone of a granular material. Thesis, University of Delft.
MANDEL, J. (1947). Sur les lignes de glissement et le calcul des déplacements dans la déformation plastique. *C.r.hebd. Séanc. Acad. Sci., Paris* **225**, 1272–1273.
MANDEL, J. (1966). Sur les equations d'écoulement des sols ideaux en deformation plane et le concept du double glissement. *J. Mech. Phys. Solids* **14**, 303–308.
MANDL, G. & FERNÁNDEZ LUQUE, R. (1970). Fully developed plastic shear flow of granular materials. *Géotechnique* **20**, No. 3, 277–307.
SPENCER, A. J. M. (1964). A theory of the kinematics of ideal soils under plane strain conditions. *J. Mech. Phys. Solids* **12**, 337–351.
STRACK, O. D. L. (1970). Private communication.
ZAGAINOV, L. S. (1967). On the equations of the plane stationary strain of a granular medium (in Russian). *Mech. Tverdovo Tela*, No. 2, 188–196.

J. Mech. Phys. Solids, 1972, Vol. 20, pp. 337 to 351. Pergamon Press. Printed in Great Britain.

PHOTOELASTIC VERIFICATION OF A MECHANICAL MODEL FOR THE FLOW OF A GRANULAR MATERIAL

By A. Drescher

Institute of Fundamental Technical Research, Warsaw, Poland

and

G. de Josselin de Jong

University of Technology, Delft, The Netherlands

(*Received* 13*th April* 1971)

SUMMARY

THIS PAPER describes experiments performed on an assembly of discs constituting a two-dimensional analogue of a granular material. The use of photo-elasticity techniques allows the determination of average stress and strain-rate tensors in the interior of the assembly. In this way, a comparison can be made with the behaviour predicted theoretically on the basis of a mechanical model. Test results indicate that the main features of the mechanical model, namely, the sub-division of the assembly into sliding elements, a possible non-coaxiality of stress and strain-rate tensors, and a free rotation of the elements are all indeed observed in practice.

1. INTRODUCTION

THE EXPERIMENTS described in this paper were undertaken in order to verify the flow rules developed for granular assemblies. The test set-up actually is a two-dimensional analogue of a granular medium because it consists of discs. The discs have different sizes and are stacked between two glass plates that prevent the stack from buckling side-ways. The assembly is loaded by bars to such an extent that the discs slide with respect to each other, thus causing deformations of the stack.

When viewed in circularly polarized light the discs, being photoelastic sensitive, show a pattern of isochromatics, from which can be deduced the forces that are transmitted through the contact points between the discs. By averaging these forces over a region in the interior of the assembly, it is possible to assign an average stress tensor for that region.

From the photographs of successive stages during a deformation cycle it is possible to determine the relative displacements of the individual discs. From these displacements an average velocity-gradient tensor (and its symmetric part, the average strain-rate tensor) can be deduced for the same region where the average stress tensor is determined.

Verification of the flow rules consists in relating the average velocity-gradient tensor to the average stress tensor that applies during the occurrence of the deformation. Because both tensors can be deduced from photographs the test results are not

24

disturbed by the measurements. Also, the disturbances existing at the boundaries of a test set-up can be eliminated in this case, because the region of consideration can be selected from a part of the interior where the average stress tensor is homogeneous.

The two-dimensional analogue of a granular material was introduced by SCHNEEBELI (1956) in the form of an assembly of metal rods. In such a test set-up only the displacements of the rods can be observed. Tests executed with these models have been reported by DE JOSSELIN DE JONG (1959), STUTZ (1963), DRESCHER, KWASZCYŃSKA and MRÓZ (1967). In general, however, the interpretation of test results is unsatisfactory because the force distribution in the interior has to be inferred from the boundary conditions, without the possibility of eliminating the disturbances at the boundaries.

DANTU (1957) and WAKABAYASHI (1957) suggested the use of optically-sensitive material for the rods or discs in order that the forces in the discs could also be determined. Analysis of the force distribution in such a test was described by DE JOSSELIN DE JONG and VERRUIJT (1969). Their procedure was adopted in the tests reported in this paper.

By using photoelasticity techniques the forces in the interior of the disc assembly can be measured without the introduction of disturbing foreign elements. A region in the interior of the assembly can be selected in which the stress state is homogeneous enough to serve as a test sample. Here, we shall call that region the *representative area*. Thus, it is possible to avoid the usual unsatisfactory procedure of determining the stress state from the boundaries and of inferring homogeneity, although that is doubtful because the boundaries always contain disturbances.

Because the polarizator was not large enough to cover the entire disc assembly, only a part of this was considered, from which only a circular area of 8 cm radius was finally selected as the representative area, because the stress in that region was sufficiently homogeneous. In that area the shear stress on a horizontal plane was not zero, although the horizontal loading plate was free to move in a horizontal direction and therefore the total horizontal force on the loading plate was zero. The reason for this discrepancy is the disturbance from homogeneity that exists at the corners of the total triangular disc assembly. These disturbances can be ignored by adopting the suggested procedure, and there will be no discussion in this paper of the relation between the forces to be measured on the three enclosing beams and the actual stress state in the representative circular area.

Since both stresses and velocities can be determined separately and without the interference of disturbing effects, an approximate check of the flow behaviour of the disc assembly can be made. By *flow* is to be understood here the deformation that takes place after the initial adjustment of stresses and strain rates has been developed and the assembly continues to move with large deformations under conditions of constant volume and constant stress.

It has been pointed out by P. W. Rowe (during the informal discussion after K. H. Roscoe's 1970 Rankine Lecture) that initial adjustment in the test reported here is obtained after deformations that are small with respect to the total deformation in every loading cycle, because the discs consist of rigid material. Since the test results elaborated in this paper are taken from the end of the loading cycles, only the behaviour of the material under conditions of flow is considered. So, the non-coaxiality being reported here later on is not in contradiction to the observations of the Cambridge group of soil mechanics workers who observed coaxiality in the initial stage of the deformation process.

Several theories have been proposed in the past to establish flow rules. Instead of surveying all the proposals that have been made, we shall pay attention here only to the mechanical model of double sliding, which is in accordance with Coulomb's initial ideas of internal friction. This model is based on the assumption that deformation of a granular assembly consists primarily of sliding movements along planes that coincide preferably with the stress characterstics.

According to that model the system of grains is sub-divided into elements consisting of many particles that remain as rigid bodies during deformation of the assembly because the particles in the elements preserve their respective orientation and contact points. The elements slide with respect to each other and are free to rotate as units. For a more detailed description of this model and its properties see DE JOSSELIN DE JONG (1959).

The test results reported here can be used to verify the equations developed for this double-sliding free-rotating model. In order to do so, the equations developed by DE JOSSELIN DE JONG (1971) are adequate because with these formulae the sliding rates and the rotation of the elements can be computed directly from the observed velocity-gradient tensor. The mode of deformation interpreted in terms of the model as obtained from test results turns out to be a sliding movement of elements that slide in one of the stress characteristic directions combined with a rotation of the elements.

A detailed investigation of the manner in which the individual discs behaved indeed shows that discs stayed together to form rigid elements that were elongated in the predicted direction and that executed the movements suggested by the theory.

The double-sliding mechanism has been considered separately by several investigators. MANDEL (1947), SPENCER (1964) and ZAGAINOV (1967) have developed a set of differential equations to describe the velocity field of that model. Their equations are all similar, but they differ in the physical interpretation of the rotation term. According to SPENCER (1964) this term is equal to the rotation of principal stresses, whereas MANDEL (1966) also includes with this term the rotations of the elements. Because there is no physical basis that substantiates Spencer's assumption, the present writers agree with the second interpretation which means according to MANDEL (1966) that the equations mentioned above cannot be used to solve boundary-value problems because the value of the rotation term is not known beforehand.

DE JOSSELIN DE JONG (1959, 1971) has developed an additional set of inequalities to describe the velocity field of the model based upon the thermodynamic requirement of energy dissipation. Since rotation is absent from these inequalities, the difficulty pointed out by MANDEL (1966) is circumvented, and the solution of boundary-value problems can be obtained with these formulae. Instead, however, of leading to a unique solution, the inequalities only provide a *range* of possible solutions. This is unattractive due to the lack of uniqueness, and it is of interest to know whether the difficulty mentioned by Mandel really exists. Crucial in this regard is the question of whether the elements can rotate separately from the principal stresses. The test results provide a means of verifying this particular point and show that rotations of principal stresses and elements can even be in *opposite senses*.

DE JOSSELIN DE JONG (1958, 1959) showed that a consequence of the double-sliding mechanism is the possibility that stress and strain-rate tensors are not coaxial. All the workers mentioned above agree upon this consequence, but in the past doubts have often been raised concerning this model, because non-coaxiality in isotropic materials is apparently prohibited. This result, however, is based upon the assumption that there exists a functional relationship between only *stress and strain-rate* tensors. Since, for the flow stage considered here, such a functional relationship does not exist, the result cited does not necessarily apply.

The test results presented here provide the possibility of verifying whether non-coaxiality exists in reality, and they show that deviation actually occurred in the experiments.

2. Experimental Procedure and Test Results

The detailed description of the technique of measurement, the test programmes, and the results obtained will be given elsewhere. In this paper, only the main results of the experiments will be briefly described.

Figure 1 presents schematically the system used for loading the disc assembly, this consisting of a fixed beam and a rotating beam hinged at the bottom. The

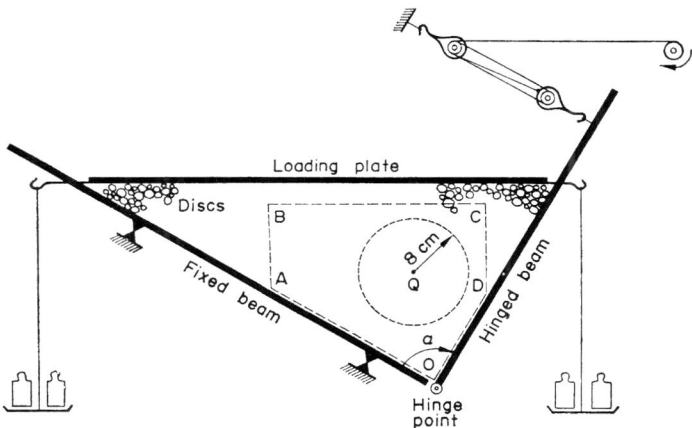

Fig. 1. Scheme of loading system.

wedge-shaped area of $1755\,\text{cm}^2$ was filled with approximately 1200 discs, made from 6 mm thick plate of CR-39 co-polymer, a relatively-sensitive photoelastic material. Six different diameters of discs were used, ranging from 8 to 20 mm. A movable loading plate was placed horizontally on the upper surface of the assembly.

The experiments consisted of rotating the hinged beam such that its angle α with the fixed beam changed slowly and gradually. The experiments began with a counter-clockwise rotation through 10° of the hinged beam, which initially made approximately a right angle with the fixed beam. Next, a clockwise rotation backwards through a similar angle was executed. During rotation of the hinged beam the loading plate could move freely and follow the deformation mode of the disc assembly.

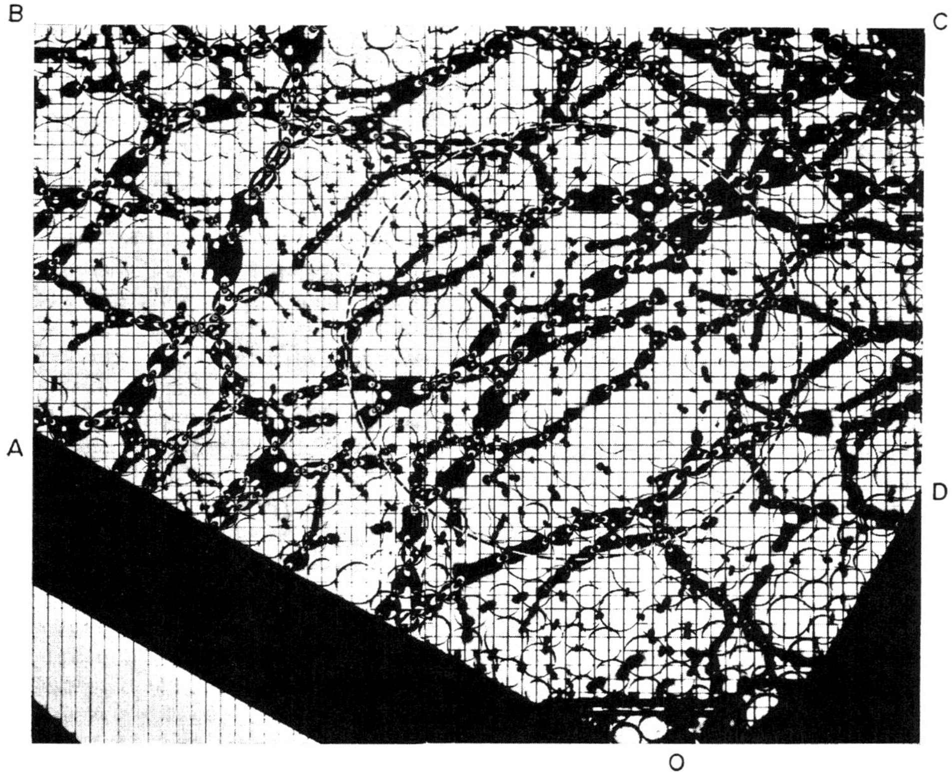

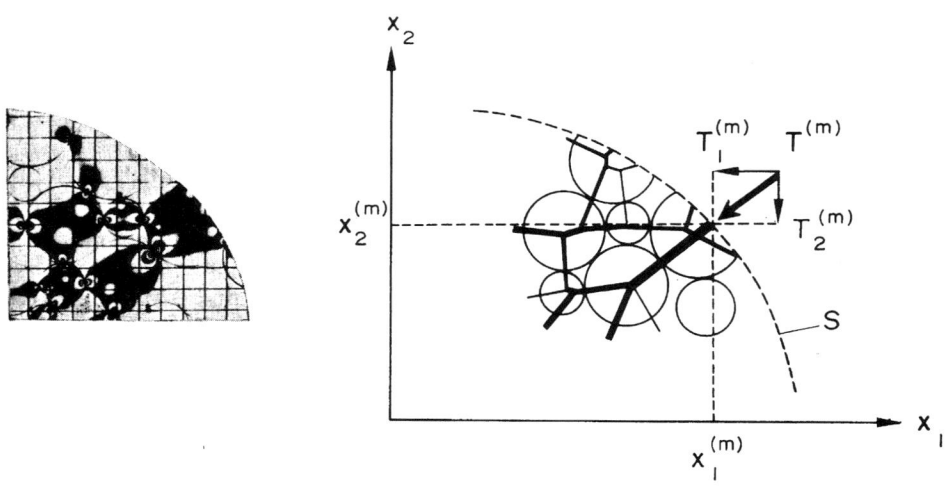

Fig. 2. Photoelasticity picture of the granular assembly.

Fig. 5. Components of surface traction vector in discrete assembly.

[facing page 340

During each test, several photographs of the model for different stages of deformation were taken.

Owing to the initial pre-stressing of the discs by the loading plate the interaction of the particles produced sufficiently high stresses within the discs during deformation to procure a well-developed isochromatic pattern observable in circularly polarized light. Figure 2 presents a portion of the tested model with the stressed discs at the moment of maximum counter-clockwise rotation.

From the discs only the rims are visible as circles. Within these circles isochromatics are seen as a pattern of black regions. Greater black-intensity means larger forces. From the pattern it can be deduced that forces are transmitted through the disc assembly along chains of discs. The oblique orientation of the chains of heavily-stressed discs, as shown in Fig. 2, remained virtually the same during the entire counter-clockwise rotation in every test, while a clockwise rotation produced chains positioned more or less vertically.

The magnitude and direction of the contact forces could be evaluated by measuring the geometry of the isochromatics in the vicinity of the contact points. In order to perform this evaluation, preliminary calibration tests were required. Using these results the calculation of forces was executed for the entire inner region OABCD (see Fig. 1) and for several stages of deformation. The correctness and accuracy of the determination was verified by constructing a Maxwell diagram of forces, and tracing the lines of action of the forces in the disc assembly. These two diagrams must consist of closed polygons, in order that equilibrium both in horizontal and vertical directions as well as equilibrium of moments is satisfied for each disc. All this was executed according to the procedure described by DE JOSSELIN DE JONG and VERRUIJT (1969).

Figure 3 presents the Maxwell diagram (consisting of about 600 individual forces) corresponding to Fig. 2. The fact that the Maxwell diagram obtained from the isochromatics consists of closed polygons is an indication that, if buckling of the disc assembly side-ways towards the glass plates created friction forces between glass plates and discs, these forces were so small that they are submerged within the overall accuracy.

Figure 4 shows the lines of action of forces throughout the assembly. The thickness of the lines is proportional to the magnitude of the transmitted forces.

The points OABCD in Fig. 3 refer to those in Figs. 1 and 4. Hence, the distance OA in the Maxwell diagram is equal to the resultant force acting on the sector OA of the fixed beam, etc. The forces indicated by heavy lines in Fig. 3 are the forces whose lines of action are intersected by the circle with radius 8 cm around Q (see Fig. 1). The more the whole heavy line approaches an ellipse, the more the force distribution is homogeneous.

3. TRANSITION TO TENSORS

It is customary to present flow rules in the form of a relation between two second-rank tensors, one describing the stress state and the other representing the velocity gradient or the strain rate. Such tensors are in fact second-order averages. The first is an average of the discrete forces acting on the discs, and the second is an average of the individual motions of the discs. If forces and motions are distributed more-or-less

Fig. 3. Maxwell diagram.

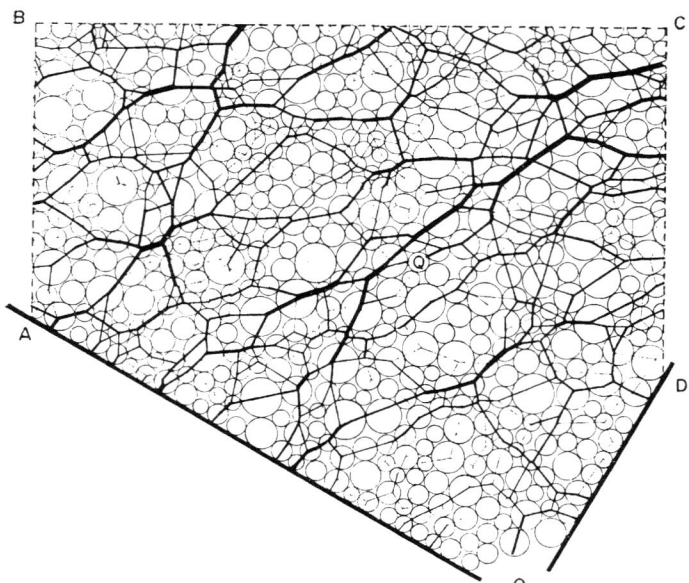

Fig. 4. A network of contact forces.

continuously an averaging procedure gives a fair description of the situation in the assembly. The photographs reveal, however, that reality cannot be compared to anything continuous at all.

The forces transmitted through the contact areas depend on the geometrical distribution of the contacts between adjacent particles. Chains of aligned contact points form rigid columns of particles which attract large forces because of their relative rigidity with respect to the surrounding particle groups. Since these columns are created by chance in a random pack, and the probability of obtaining a chain is small, the mutual distances between columns consist of several particles and as a consequence the magnitudes of the transmitted forces vary very much in adjacent contact points. This results in a discrete distribution of forces through the pack with high forces at great distances apart. Such an arrangement of discrete forces is completely different from a continuous distribution of force inherent in the concept of stress.

Similar discrete behaviour is observed in regard to the movements of the particles. If the direction of the force at a contact point deviates more from the direction of the normal to the contact area than friction allows, then a particular contact point will yield. The yielding of one contact is enough to create a movement of a great number of particles resulting in a possible re-orientation of many contact points and a redistribution of the forces. The movements of the particles have a discrete character, which differs basically from a continuous displacement field which is inherent in the concept of strain.

When soil engineering computations are to be executed for the prediction of the behaviour of a soil structure, it is impractical to attempt a calculation of the discrete forces and movements of individual particles and instead it is common practice to use stress and strain tensors. Stresses and strains, however, are misleading concepts if the basic properties determining the behaviour of a particle assembly are studied. They are averages that blur the real physical entities responsible for the mechanical actions working on the assembly.

If, however, stress and strain-rate tensors are ineffective averages of the discrete force and displacement distributions of the particles, a more serious drawback in dealing with a particle assembly is our incapacity to describe its geometry efficiently with continuous concepts. It is a well-known practice in soil mechanics to mention for a grain deposit only its density. Although a greater density entails the probability of a greater number of contact points between grains, it is insufficient for a complete description of contact-points distribution. This distribution of contact points is actually the essential property of a granular deposit that has to be known in order to be able to predict its deformation behaviour under action of forces. Presumably it will prove impossible to introduce contact-point geometry effectively with the use of tensors, which are averaging concepts only appropriate for continuous media.

Attempts have been made to improve the continuum concepts by introduction of the Cosserat continuum (NIKOLAEVSKII and AFANASIEV, 1969) or higher-order gradients of the velocity and multi-polar stress states. We shall not follow that approach here, because it is not our present purpose to propose better concepts for the description of the discrete distributions observed for forces, movements and contacts between particles. In order, however, to deduce a flow rule from the experimental evidence along the traditional lines, the discrete-force distribution and the movements of the discs have to be translated into a stress and a velocity-gradient tensor. It is irrelevant how these tensors were obtained, because they are only averages that ignore the real complexity of the aspects represented by them.

4. AVERAGE STRESS TENSOR

In a continuum, the stress is generally defined as the resultant force acting over a unit area. In the two-dimensional case, this reduces to the resultant force acting on a unit line. Since the forces acting in the tested disc assembly were not homogeneously distributed and were of varying magnitudes, every other unit line in the assembly would produce another value of stress, if defined on the basis of individual lines. This would be an inappropriate measure for the averages. In the present paper the averaging procedure over a representative elementary area is used, as proposed by HILL (1963) and WEBER (1966). Although the formulae introduced by these workers are different, it can be shown that they lead to equivalent results. Their proposals amount to the following.

If in a region V there is a stress state σ_{ij} which is in equilibrium, but otherwise may be arbitrarily distributed over V, then the average stress $\bar{\sigma}_{ij}$ is defined as

$$\bar{\sigma}_{ij} = \frac{1}{V} \int_V \sigma_{ij} \, dV. \tag{4.1}$$

Because $\sigma_{ij} = \delta_{ik}\sigma_{kj} = x_{i,k}\sigma_{kj}$ and σ_{ij} satisfies the equilibrium condition $\sigma_{kj,k} = 0$, (4.1) can be transformed by use of Gauss's divergence theorem into

$$\bar{\sigma}_{ij} = \frac{1}{V} \int_S x_i t_j \, dS \tag{4.2}$$

where S is the boundary of V, x_i is the i-coordinate of a point on S, and t_j is the j-component of the traction acting on S at that particular point. With (4.2) it is possible to determine the value of the average stress tensor $\bar{\sigma}_{ij}$ in the region V from the tractions t_j acting on the boundary of that region. Tensile normal stresses are taken as positive.

In our case the region V was taken to be the area enclosed by a circle with radius 8 cm and centre at the point Q (Figs. 1 and 4). Considering this circle as the boundary S of the region V, the tractions t reduce to the discrete forces $T^{(m)}$, whose lines of action are intersected by the circle (heavy lines in Fig. 3). The coordinates of the intersection points are $x_i^{(m)}$ (Fig. 5). The surface integral of (4.2) is then replaced by a sum over the μ forces intersected by the circle to give

$$\bar{\sigma}_{ij} = \frac{1}{V} \sum_{m=1}^{\mu} x_i^{(m)} T_j^{(m)}. \tag{4.3}$$

If $i \neq j$ the summation over $x_i^{(m)} T_j^{(m)}$ represents the moment couple exerted on the region V by the j-components of the forces acting on S. Since the determination of the force polygon and the lines of force network is such that equilibrium of moments is assured, this moment couple must be equal to $\sum x_j^{(m)} T_i^{(m)}$, the moment couple of the i-components of the forces. The consequence is that $\bar{\sigma}_{ij} = \bar{\sigma}_{ji}$ and the average stress tensor is symmetric. Because they must be equal, a computation of $\sum x_i^{(m)} T_j^{(m)}$ and $\sum x_j^{(m)} T_i^{(m)}$ separately gives a verification of the accuracy of the calculation.

Using (4.3), the average stresses $\bar{\sigma}_{11}$, $\bar{\sigma}_{12}$, $\bar{\sigma}_{22}$ were determined. From these stresses, the principal stresses $\bar{\sigma}_{\mathrm{I}}$ and $\bar{\sigma}_{\mathrm{II}}$, as well as the angle ψ between the major principal stress $\bar{\sigma}_{\mathrm{I}}$ (the smallest in compression) and the horizontal 1-axis, were computed. This was done for several steps of the experiment. Some results are given below. During a counter-clockwise rotation of the hinged beam over an angle $\Delta\alpha = 1.5°$,

$$\begin{aligned} \text{at} \quad \alpha = 90.5°&: \bar{\sigma}_{\mathrm{I}} = -7.15\,\mathrm{N\,cm}^{-1}, & \bar{\sigma}_{\mathrm{II}} &= -15.4\,\mathrm{N\,cm}^{-1}, & \psi &= 126.5°, \\ \text{at} \quad \alpha = 89°&: \bar{\sigma}_{\mathrm{I}} = -8.6\,\mathrm{N\,cm}^{-1}, & \bar{\sigma}_{\mathrm{II}} &= -20.7\,\mathrm{N\,cm}^{-1}, & \psi &= 119.5°. \end{aligned} \quad (4.4)$$

During a clockwise rotation of the hinged beam backwards over an angle $\Delta\alpha = 1.5°$,

$$\begin{aligned} \text{at} \quad \alpha = 90.5°&: \bar{\sigma}_{\mathrm{I}} = -6.4\,\mathrm{N\,cm}^{-1}, & \bar{\sigma}_{\mathrm{II}} &= -11.8\,\mathrm{N\,cm}^{-1}, & \psi &= 0°, \\ \text{at} \quad \alpha = 92°&: \bar{\sigma}_{\mathrm{I}} = -9.2\,\mathrm{N\,cm}^{-1}, & \bar{\sigma}_{\mathrm{II}} &= -17.4\,\mathrm{N\,cm}^{-1}, & \psi &= 6°. \end{aligned} \quad (4.5)$$

These average stresses refer to the circular region with radius 8 cm around Q (see Fig. 1). This region can be considered as a representative area with a homogeneous stress state because a similar analysis, for seven areas of smaller size and covering that region, produced practically the same value for the stresses, and the heavy line in Fig. 3 resembles an ellipse.

5. AVERAGE VELOCITY-GRADIENT TENSOR

From two successive photographs, displacement increments can be obtained. In the theory of flow of granular assemblies it is customary to use the term 'velocity' rather than 'displacement increment'. This infers that time may have an influence, but that is not the case here because movements were so slow that inertial effects can be disregarded and the particles were dry so that viscous effects were absent from the friction developed. The magnitude of the time lapse between photographs is therefore irrelevant, and time is only introduced to fix the direction of motion with respect to energy dissipation.

The particles moved individually and rotated. In order to obtain an average velocity gradient from two photographs it is necessary to substitute for the observed discrete displacement a continuous velocity field. This is done in the following arbitrary way.

The circular representative area, used above to determine the average stress tensor, is sub-divided into triangles by joining the disc centres by straight lines (Fig. 6). In each triangle the velocity is distributed linearly in such a manner that the velocity in the corner points is equal to the displacement of the disc centres divided by unit time. Such a velocity distribution is continuous at the sides of adjacent triangles.

In this procedure, both the rotations of the individual discs and the movements of the contact points between discs are disregarded. Since both of these quantities may be essential as regards the mechanical behaviour of the individual discs and the assembly, the procedure followed here is arguable. However, every averaging procedure annihilates the discrete character of the real particle assembly and has for the moment to be accepted.

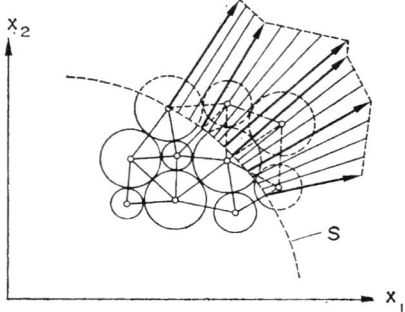

FIG. 6. Scheme of approximation of discontinuous velocity field.

Let u_i be the components of the substituted velocity field. Because the field is continuous the components have first-order derivatives $u_{i,j}$ at every point, which together form the components of the velocity-gradient tensor. Averaging over the representative area V gives the components of the average velocity-gradient tensor $\bar{u}_{i,j}$ according to

$$\bar{u}_{i,j} = \frac{1}{V} \int_V u_{i,j} \, dV, \tag{5.1}$$

which by use of Gauss's divergence theorem can be written as

$$\bar{u}_{i,j} = \frac{1}{V} \int_S u_i n_j \, dS. \tag{5.2}$$

Therefore, only the velocity distribution along the boundary as shown in Fig. 6 is required for the computation of $\bar{u}_{i,j}$.

The successive photographs, used for the computation of the average stress tensors, showed the following values for the average velocity-gradient tensor components. For the counter-clockwise rotation of the hinged beam over the interval $\alpha = 90.5°$ to $\alpha = 89°$,

$$\left.\begin{array}{ll} \bar{u}_{1,1} = -0.0083, & \bar{u}_{1,2} = -0.0178, \\ \bar{u}_{2,1} = +0.00425, & \bar{u}_{2,2} = +0.09067. \end{array}\right\} \tag{5.3}$$

For the clockwise rotation of the hinged beam backwards over the interval $\alpha = 90.5°$ to $\alpha = 92°$,

$$\left.\begin{array}{ll} \bar{u}_{1,1} = +0.00875, & \bar{u}_{1,2} = +0.01363, \\ \bar{u}_{2,1} = -0.0004, & \bar{u}_{2,2} = -0.00967. \end{array}\right\} \tag{5.4}$$

6. VERIFICATION OF DOUBLE-SLIDING FREE-ROTATING MODEL

In order to verify the double-sliding free-rotating model with the aid of the test results, relations will be used here that were developed for this model by DE JOSSELIN DE JONG (1971). These relations express the shear strain rates a, b and the rotation of the elements Ω as a function of the components $\bar{u}_{i,j}$ of the average velocity-gradient tensor, as follows:

$$a \sin(2\varphi) = -(\bar{u}_{1,1}+\bar{u}_{2,2})+(\bar{u}_{1,1}-\bar{u}_{2,2})\sin(2\psi+\varphi)-(\bar{u}_{1,2}+\bar{u}_{2,1})\cos(2\psi+\varphi), \quad (6.1)$$

$$b \sin(2\varphi) = -(\bar{u}_{1,1}+\bar{u}_{2,2})-(\bar{u}_{1,1}-\bar{u}_{2,2})\sin(2\psi-\varphi)+(\bar{u}_{1,2}+\bar{u}_{2,1})\cos(2\psi-\varphi), \quad (6.2)$$

$$2\Omega \sin\varphi = +(\bar{u}_{1,1}+\bar{u}_{2,2})\sin(2\psi)-(\bar{u}_{1,2}+\bar{u}_{2,1})\cos(2\psi)+(-\bar{u}_{1,2}+\bar{u}_{2,1})\sin\varphi. \quad (6.3)$$

Since the sliding movements in the model are such that volume remains constant during deformation, an additional requirement is that the volume increase is

$$\bar{u}_{1,1}+\bar{u}_{2,2} = 0. \quad (6.4)$$

The model requires that the shear strain rates a, b satisfy the thermodynamic requirements $a \geqslant 0$, $b \geqslant 0$, whereas, according to the present writers, Ω can have any value irrespective of the changes in the value of ψ.

In order to verify these relations, values have to be assigned to ψ and φ (φ is the angle of internal friction of the disc assembly). For φ, we shall take here the value of $32°$ because this value for φ (or even somewhat higher values) has been obtained by several methods for assemblies of the same discs.

Let us first consider the counter-clockwise rotation of the hinged beam over the interval $\alpha = 90·5°$ to $\alpha = 89°$. Photographs were only taken at the beginning and at the end of the interval, so we have no information of the stress history within the interval.

From motion pictures taken during many similar tests, it is known that the forces in the discs increase up to a maximum, which is achieved at the moment that the discs start to slide with respect to each other. Then, after the discs have re-adjusted, the forces fall off to build up again to a new maximum.

The average stress tensors determined from the photographs do not correspond to the maxima, because computation of angles of internal friction from the relation

$$\sin\varphi = (-\bar{\sigma}_{I}+\bar{\sigma}_{II})/(\bar{\sigma}_{I}+\bar{\sigma}_{II})$$

gives values $\varphi = 21·5°$ and $\varphi = 24·5°$, which are smaller than the known values for φ.

The motion pictures further reveal that the direction of the predominant forces remain essentially the same during these re-adjustments. Therefore, it may be assumed that the principal directions of the average stress tensor oscillate over only a few degrees.

The two photographs give two different values for ψ, from which the mean is

$$\bar{\psi} = \tfrac{1}{2}(126·5°+119·5°) = 123°. \quad (6.5)$$

This mean value for ψ will be used in verification by use of the formulae.

In the interval considered, ψ decreased, which means that the principal stresses rotated over an angle $\Delta\psi$ (positive for counter-clockwise rotation) given by

$$\Delta\psi = -126·5°+119·5° = -7°. \quad (6.6)$$

Using the values (5.3) for the components of the average velocity-gradient tensor and the values $\bar{\psi} = 123°$, $\varphi = 32°$ in (6.1) to (6.4), there results

$$a = +0·0204, \quad b = -0·0002, \quad \Omega = +0·0212, \quad \text{volume increase} = +0·0014 \quad (6.7)$$

These values are not in agreement with the theory, because b should be *positive* and the volume increase should be *zero*.

We remark, however, that the absolute values of these two quantities are respectively 1 and 7 per cent of a and Ω which predominate. Because they are only small fractions of the predominant quantities, a small correction in the value of φ and the introduction of a small uplift angle in the sliding mechanism are sufficient to reduce both b and the volume increase to zero. We shall not elaborate on that possibility but consider only a and Ω whose values change but a little by such corrections.

The interpretation of the result based upon the values of a and Ω from (6.7) is now the following. Sliding occurs predominantly along s_1-stress characteristics and superimposed on that sliding all elements rotate counter-clockwise.

This mode of deformation is shown schematically in Fig. 7. Figure 7(c) shows the original position with the s_1-lines, that divide the material into elements that are to

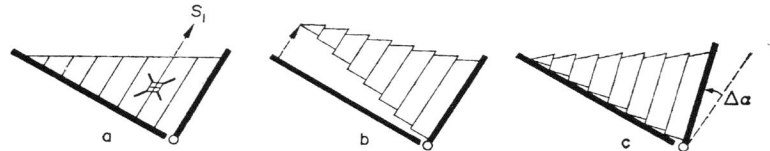

FIG. 7. Sequence showing how a counter-clockwise rotation of the hinged beam is responded to in the material by a shear strain in s_1-direction ($a \rightarrow b$) plus a counter-clockwise rotation ($b \rightarrow c$).

slide with respect to each other. Figure 7(b) shows the sliding of the elements, which creates a gap between the stationary beam at the left, because left-hand side elements slide upwards over the right-hand elements to satisfy the thermodynamic requirements of energy dissipation. This gap is closed by a counter-clockwise rotation of the material and the hinged beam at the right together, as shown in Fig. 7(c). In reality, the sliding and rotation occur simultaneously. The double-sliding free-rotating model admits many other combinations of slidings in two directions combined with rotation that satisfies the movements dictated by the boundaries. From all the possible combinations the one observed here requires the smallest force on the hinged beam. Since the disc assembly was forced to deform by that beam, the material seems to respond with a minimum resistance deformation mode.

A similar analysis applied to the photographs of the clockwise rotation of the hinged beam backwards over the interval $\alpha = 90\cdot5°$ to $\alpha = 92°$ gives the following results. Using a value of $\bar{\psi} = 3°$ for the mean value of ψ and $\varphi = 32°$ gives

$$a = +0\cdot0020, \quad b = +0\cdot0232, \quad \Omega = -0\cdot0187, \quad \text{volume increase} = -0\cdot0009, \quad (6.8)$$

In this case again, one of the two shear strain rates dominates, because a is 10 per cent of b. Being both positive, they satisfy the thermodynamic requirements of energy dissipation. Volume change again is not zero, but only 4 per cent of the predominant values of b and Ω. The mechanism now is a sliding movement along s_2-stress characteristics, in combination with a rotation of elements clockwise. A schematic representation of this deformation mode is shown in Fig. 8.

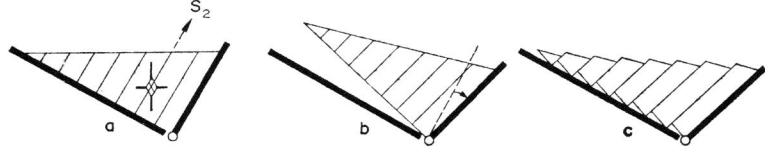

FIG. 8. Sequence showing how a clockwise rotation of the hinged beam is responded to in the material by a clockwise rotation of elements ($a \rightarrow b$) and a shear strain in s_2-direction ($b \rightarrow c$).

7. Comparison of Test Results and Theory

7.1 *Sub-division of the disc assembly into sliding elements*

The test results presented in Section 6, both for the counter-clockwise and for the clockwise rotation of the hinged beam, show a mode of sliding that has predominantly the direction of one family of stress characteristics. The disc assembly can only execute such a movement if the block-like elements, presumed in the sliding mechanism, are actually formed by the discs. In order to verify this assumption the displacements of the discs in the considered circular region were re-examined from the photographs.

By shifting and rotating the photographs for $\alpha = 90.5°$ and $\alpha = 89°$ with respect to each other, it is possible to match the discs and to visualize their relative movements. It turned out that sub-regions of many discs apparently stayed together as units and that these units moved as rigid blocks with respect to each other. The sub-regions are shown as shaded areas in Fig. 9 (which is a repetition of Fig. 4). It is seen that these

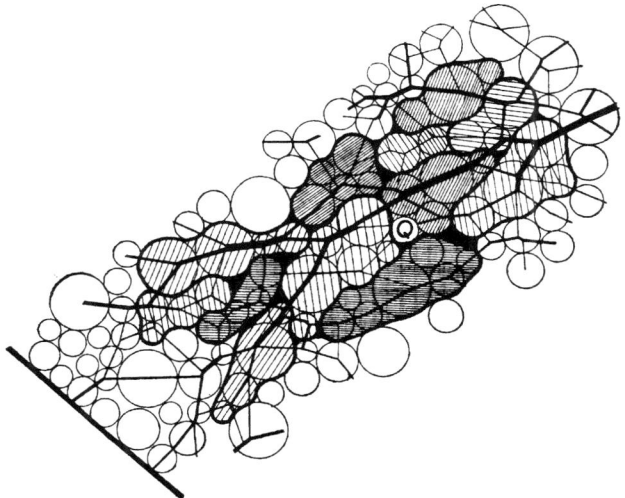

Fig. 9. Block-like sub-regions.

units show a predominant elongation parallel to the s_1-direction from Fig. 7. It seems that the units are created by the larger forces if they are normal to the contact surface. On the other hand, the impression exists that sliding of blocks along each other is possible because contact points are broken that transmitted forces whose directions deviate strongly from the normal to the contact area. As a consequence, the borders of the rigid disc units transmit a resultant force that makes a large angle with the border surfaces. This is in agreement with the observation that the division lines between units coincide with one family of stress characteristics.

7.2 *Non-coaxiality of stress and strain-rate tensors*

From the theory of the double-sliding free-rotating model there follows the possibility that stress and strain-rate tensors are non-coaxial. The strain-rate tensor is the symmetric part of the velocity gradient tensor and therefore its components are defined by

$$a_{ij} = \tfrac{1}{2}(\bar{u}_{i,j} + \bar{u}_{j,i}).$$

Being a symmetric tensor, its principal directions are orthogonal, and using the values mentioned in (5.3) and (5.4) we find for ξ, the direction of major principal strain-rate,

for $\alpha = 90.5°$ to $\alpha = 89°$: $\xi = 108.5°$,

for $\alpha = 90.5°$ to $\alpha = 92°$: $\xi = 17.5°$.

A comparison with (4.4) and (4.5) shows that there is a deviation angle i between the principal axes of stress and strain rate of a magnitude, respectively, as follows:

$$i = \xi - \bar{\psi} = 108.5° - 123° = -14.5°,$$
$$i = \xi - \bar{\psi} = 17.5° - 3° = +14.5°,$$

where for $\bar{\psi}$ the mean is taken of the values given by (4.4) and (4.5).

According to the theory, this deviation angle, i, is limited to the region

$$-\tfrac{1}{2}\varphi \leqslant i \leqslant +\tfrac{1}{2}\varphi,$$

and since φ is about $32°$, the values found are acceptable within the theory.

7.3 Rotation of elements

The quantity Ω defined by (6.3) is the rotation executed by the sliding elements. This can be seen by division throughout (6.3) by $2\sin\varphi$, because the formula then states that Ω is the difference between the asymmetric part of the velocity gradient tensor $\tfrac{1}{2}(\bar{u}_{2,1} - \bar{u}_{1,2})$ and a term of magnitude $a-b$. The term $a-b$ is due to the double-sliding mechanism, which according to de Josselin de Jong (1959) produces an asymmetric velocity-gradient tensor of that magnitude by sliding of the elements at different rates without rotation. If the asymmetric part of the velocity gradient is greater than the factor $a-b$, it means that the elements also execute a rotation here called Ω.

According to Spencer's (1964) interpretation, Ω must be equal to $D\psi/Dt$, the rotation of the principal stresses either in time at a point or by convection. In de Josselin de Jong's model, Ω is free to have any value independent of ψ.

In the experiments reported here, the change of ψ by convection can be considered to be zero because the stresses were homogeneous in the region considered, but there was a change of ψ in time. According to (6.3) this change was $\Delta\psi = -7°$ for the interval $\alpha = 90.5°$ to $\alpha = 89°$, indicating that the principal stresses rotated clockwise. According to (6.7) it was found for that interval that $\Omega = +0.0212$, indicating that the elements rotated counter-clockwise. This counter-clockwise rotation is also observed by super-imposing the respective photographs in such a way that the shaded-disc conglomerates shown in Fig. 9 match.

A similar analysis applied to the test interval $\alpha = 90.5°$ to $\alpha = 92°$ shows a counter-clockwise rotation of the principal stresses over $6°$ and a clockwise rotation of elements, because $\Omega = -0.0187$.

The conclusion drawn from these observations is that in both test intervals the rotation of the principal stresses and the rotation of the disc conglomerates were in opposite senses. So the tests indeed do indicate that Ω can be independent of $D\psi/Dt$.

REFERENCES

DANTU, P. 1957 *Proc. 4th Int. Conf. Soil Mech. Found. Engng.* **1**, 144.

DE JOSSELIN DE JONG, G. 1958 *Proc. Brussels Conf. Earth Press. Probl.* **1**, 55.

1959 *Statics and Kinematics in the Failable Zone of a Granular Material.* Waltman, Delft.

1971 *Géotechnique* **21**, 155.

DE JOSSELIN DE JONG, G. and VERRUIJT, A. 1969 *Cah. Gr. Franc. Rhéol.* **2**, 73.

DRESCHER, A., KWASZCZYŃSKA, K. and MRÓZ, Z. 1967 *Arch. Mech. Stos.* **19**, 99.

HILL, R. 1963 *J. Mech. Phys. Solids* **11**, 357.

MANDEL, J. 1947 *C.r. hebd. séanc. Acad. Sci.* **225**, 1272.

1966 *J. Mech. Phys. Solids* **14**, 303.

NIKOLAEVSKII, V. N. and AFANASIEV, E. F. 1969 *Int. J. Solids Struct.* **5**, 671.

SCHNEEBELI, G. 1956 *C.r. hebd. séanc. Acad. Sci.* **243**, 125.

SPENCER, A. J. M. 1964 *J. Mech. Phys. Solids* **12**, 337.

STUTZ, P. 1963 Thèse Doct. Spec., Université de Grenoble.

WAKABAYASHI, T. 1957 *Proc. 7th Jap. Nat. Congr. Appl. Mech.*, p. 153.

WEBER, J. 1966 *Cah. Gr. Franc. Rhéol.* **2**, 161.

ZAGAINOV, L. S. 1967 *Mech. Tverd. Tela* **2**, 188.

de Josselin de Jong, G. (1988). *Géotechnique* **38**, No.4, 1988, 533–555

Elasto–plastic version of the double sliding model in undrained simple shear tests

G. de JOSSELIN de JONG*

In this Paper it is shown how to use the double sliding, free rotating model for materials with internal friction to predict the stress history in undrained simple shear tests. In its original rigid plastic form this model could not be used, because there was no unique failure mode. By adding some elasticity to the prefailure stage (thus producing an elasto-plastic version of the model) this unique selection becomes possible. The extended model leads to explicit expressions for the stress history in a simple shear test. It is also shown how the failure mode taken by the model depends on the stress state at the start of the test. An active initial stress state leads to a 'toppling bookrow' mode of failure, while a passive initial stress state produces horizontal sliding planes. With the exception of elasticity, the other properties of the double sliding model, including dilatancy, are taken in their original form. The essential features of the stress history obtained from the analysis resemble those actually observed in tests.

KEYWORDS: constitutive relations; elasticity; plasticity; shear tests; strain rates; stress rates.

L'article montre comment utiliser le modèle à glissement double et rotation libre pour prédire l'histoire des contraintes dans des essais de cisaillement simple non-drainés. Dans sa forme plastique rigide originale ce modèle ne pouvait pas s'employer, car il n'y avait pas de mode unique de rupture. En ajoutant de l'élasticité à l'état précédant la rupture, produisant ainsi une version élastoplastique du modèle, cette sélection unique devient possible. Le modèle élargi conduit a des expressions explicites pour l'histoire des contraintes dans un essai de cisaillement simple. On démontre comment le mode de rupture choisi par le modèle dépend de l'état de contrainte au commencement de l'essai. Un état de contrainte initial actif conduit à un mode de rupture analogue à celui d'une rangée de livres qui s'écroulent, tandis qu'un état de contrainte initial passif produit des plans de glissement horizontaux. À l'exception de l'élasticité les autres propriétés du modèle à glissement double, y compris la dilatance, sont prises dans leur forme originale. Les caractéristiques essentielles de l'histoire des contraintes obtenues à partir de l'analyse ressemblent à celles observées au cours des essais.

NOTATION

a, b	conjugate shear strain rates (non-dilatancy)
a^*, b^*	conjugate shear strain rates (dilatancy)
s	$\frac{1}{2}(\sigma'_{xx} + \sigma'_{yy})$
t	$\frac{1}{2}[(\sigma'_{xx} - \sigma'_{yy})^2 + (\sigma_{xy} + \sigma_{yx})^2]^{1/2}$
i	angle of non-coaxiality
q_v, q_h	initial vertical and horizontal stresses in simple shear test
x, y	horizontal and vertical co-ordinates
$\dot{A}^e$	elastic energy dissipation rate
F	$(1 - \sin^2 \phi \sin^2 v^*)^{1/2}$
G	elastic shear modulus
M^*, N^*	simplifying strain rate combinations (*see* equations (79), (80))
T	time
V_x, V_y	velocity components
$V_{x,x}, V_{y,y}$	linear strain rates in x, y-directions
$V_{x,y} + V_{y,x}$	shear strain rate
$V_{y,x} - V_{x,y}$	material rotation
α	angle of potential sliding plane
θ	teeth uplift angle
λ	angle between force and normal to teeth
v	Poisson's ratio
v^*	angle of dilatancy
ξ, η	co-ordinates in principal stress directions
$\sigma'_{xx}, \sigma'_{yy}$	effective normal stresses, positive for compression
σ_{xy}, σ_{yx}	shear stresses
σ'_1, σ'_3	effective principal stresses, positive for compression
ϕ	angle of internal friction
ϕ_f	apparent angle of internal friction
$\tilde{\phi}$	auxiliary angle defined by equation (40)
ψ	angle between x-axis and plane of σ'_1

Discussion on this Paper closes on 1 April 1989. For further details see p. ii.
* Formerly with Delft University of Technology.

$\dot{\omega}$ $\frac{1}{2}(V_{y,x} - V_{x,y})$: material rotation
Ω structural rotation, rotation of the elements
*   distinguishes dilatancy from non-dilatancy
·   time derivative (d/dT)
v   objective co-rotational (Jaumann) stress rates

INTRODUCTION

In his article with Randolph (1981), his Rankine Lecture (1984) and again more recently (1987) Wroth considers the results of simple shear tests on clay as obtained by different investigators. As examples, the observation by Borin (1973) is reproduced in Fig. 1 and the test results of Ladd & Edgers (1972) in Fig. 20.

Wroth draws attention to the fact that the observed deformation at the onset of failure is apparently not the one with horizontal planes of maximum stress obliquity, as is often believed (*see* Fig. 1). Instead, failure appears to occur on vertical planes of maximum stress obliquity and, in addition to sliding on those vertical planes, a

rigid body rotation is executed to meet the boundary conditions. This failure mode is shown in Fig. 2. It can be visualized as a row of books, toppling over sideways when the left-hand support is removed.

The possible occurrence of a 'toppling bookrow' mode of failure was predicted (de Josselin de Jong, 1972) as a consequence of the double sliding, free rotating (DSFR) model for materials with internal friction. It was mentioned that the toppling bookrow is one case in a set of possible failure modes in a simple shear test. The usually accepted mode with horizontal sliding planes is another one of this set. The DSFR model admits both modes and all transition modes between these two extremes.

Considering simple shear tests with equal vertical normal effective stress σ' on horizontal planes, the two extreme modes possess different shear stresses at failure. In the toppling bookrow mode the shear stress is equal to $\sigma' \sin \phi \cos \phi/(1 + \sin^2 \phi)$; in the horizontal sliding mode it is $\sigma' \tan \phi$, with ϕ the angle of internal friction at failure. Combinations of the two modes give values between these two extremes. For $\phi = 23°$ the apparent angle of friction, the tangent of

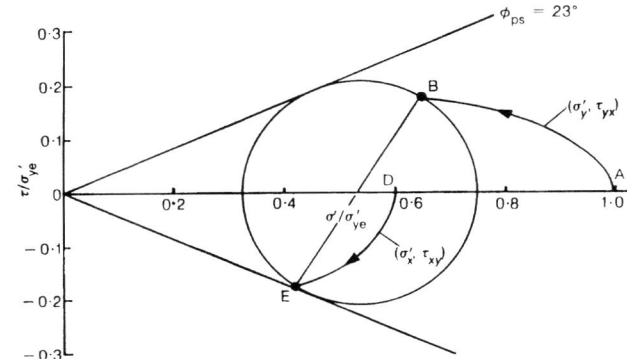

Fig. 1. Effective stress paths and the failure state from an undrained simple shear test on normally consolidated kaolin (data from Borin, 1973), test 10)

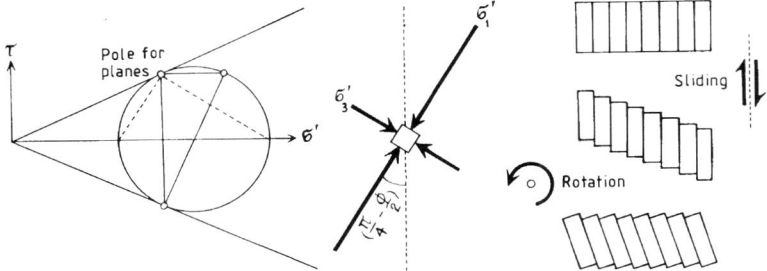

Fig. 2. Toppling bookrow mechanism

Selected Works of G. de Josselin de Jong   45

which is the shear stress divided by σ', is only 17·3° for the toppling bookrow mode, while it is 23° for the horizontal sliding mode. Ignoring the toppling bookrow mechanism leads to an underestimation of the angle of shearing resistance.

At that time it was not known how to select between the various modes of failure. The DSFR model cannot predict how a sample will behave in a test: a behaviour that is presumably unique. This created an uncertainty which was attributed to an incompleteness in the model thus prohibiting its practical use.

It was recognized by Vermeer (1980 and 1981) that the incompleteness of the model is due to its being rigid-plastic. By adding some elasticity in the pre-failure stage he created an elasto-plastic version of the DSFR model which produces unique solutions at failure. In this version the sample 'selects' between various modes according to its initial stress state. In this respect the DSFR model, which is based on internal friction, differs from perfect plasticity, where the unique response at failure is independent of the initial stress state.

It is the purpose of this Paper to demonstrate the use of the elasto-plastic DSFR model by examining the response of a sample in the undrained simple shear test and establishing the initial circumstances that lead either to the toppling bookrow mechanism or to horizontal sliding planes. The toppling bookrow was observed by Drescher (1976) in photo-elastic tests on crushed glass. This material can be considered as a model material with properties resembling sand. The undrained simple shear tests mentioned by Wroth were carried out on clays. In these tests the average normal effective stress reduces during the deformation. This indicates that the material is contractive, i.e. negative dilatant. The dilatant version of the elasto-plastic DSFR model has recently been implemented for computer use by Teunissen & Vermeer (1988). Their formulation is in terms of matrices and therefore differs in notation from the analysis given here. The relevant expressions are however, identical.

The DSFR model as developed by this Author in 1958, 1959 was specified mathematically in various later papers (de Josselin de Jong, 1971, 1977a and 1977b) and the principal features of this model will be recalled again here. In the first section of this Paper, the dilatant version of the DSFR model, which was described by this Author (1977a) using Rowe's (1962) stress dilatancy relation, is redeveloped using only the laws of friction. This leads to the plastic constitutive equations (14).

In the second section the elastic part of the constitutive equations is developed. The combined elasto-plastic constitutive relations are equation (26). How to divide the total strain rates into their plastic and elastic parts is presented visually in terms of vectors in the stress and strain rate spaces. The procedure using a minimum energy principle is set out mathematically in Appendix 2.

In the third section the response of an elasto-plastic DSFR sample in an undrained simple shear test is considered. The mathematical results obtained consist of explicit solutions for the stress paths in the Mohr diagram, the resemblance of which to the test results already mentioned is the reason for this publication.

SECTION 1

DOUBLE SLIDING MECHANISM WITH DILATANCY

The double sliding model describes the plane strain deformation of a material such as a soil, possessing internal friction, at the limit stress state. It is based on the simplification of subdividing the soil by parallel planes, on which sliding takes place, because on them the frictional shear resistance is exhausted. These are the potential sliding planes.

The sliding is called double, because there are two conjugate directions for the potential sliding planes, located symmetrically with respect to the principal stress directions. In Fig. 3 the two conjugate directions have both an angle α with the plane of the major principal compression stress, σ'_1. In order to distinguish between the two conjugate sliding possibilities, they are called a-sliding and b-sliding respectively as indicated in Figs 3(a) and 3(b).

The sliding planes divide a block of material into elements, that are liable to slide with respect to each other as shown in Fig. 3. The plastic deformation of the soil as observed from the

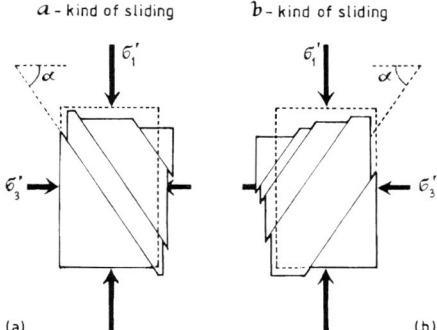

Fig. 3. **Sliding elements and potential sliding planes at angle** α

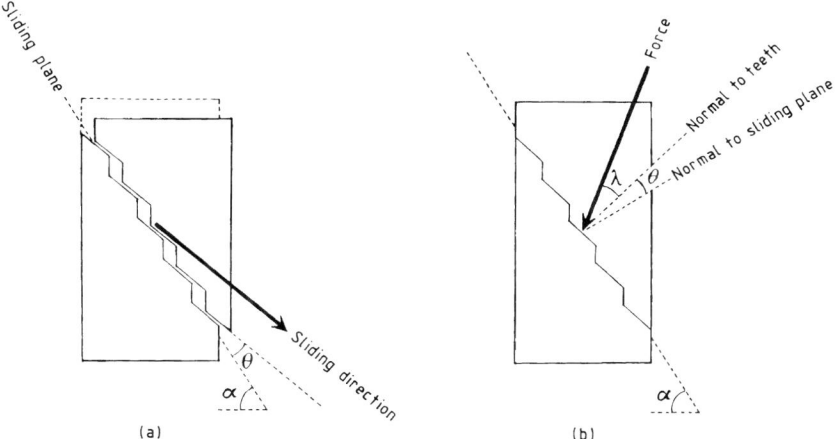

Fig. 4. Sliding planes with saw tooth protuberances: (a) relative displacement; (b) force on teeth

outside is due to this mutual sliding. In the original, rigid plastic version of the DSFR model the elements were supposed to be rigid. In the elasto-plastic version the elements are elastic and this looks after the elastic part of the strain rates that remain after the plastic part is accounted for by sliding of the elements.

It is impossible to visualize the two conjugate slidings occurring simultaneously without creating gaps or overlaps. Mathematically, however, double sliding is introduced in order to take account of conjugate slidings that occur successively, as observed in reality.

Discrete sliding planes as presented in Fig. 3 form elements of finite size. Mathematically it is complex to describe the discontinuous deformation of the soil created by their motions. By decreasing the thickness of the elements to infinitesimally thin slices, the sliding becomes shearing, and the deformation of a soil region can be described in terms of shear strain rates.

The magnitudes of the two conjugate shear strain rates are called a^* and b^* in this Paper.[1] By allowing these variables to become infinitely large, the discrete sliding of elements of finite size can be formulated mathematically. An essential feature of the DSFR model is that the values of a^* and b^* can be different. This is due to the principle that in the limit state of stress sliding displacements have arbitrary magnitudes.

By considering the sliding surfaces to be smooth the deformation is volume conserving. Dilatancy is obtained by assuming saw-teeth protuberances on the sliding planes, having uplift angles θ with the general direction α of the potential sliding planes (see Fig. 4(a)). The angles α and θ are assumed to be constants throughout a soil region.

The forces between the sliding elements are assumed to be transmitted exclusively by the faces along which the sliding occurs (see Fig. 4(b)). All forces on the teeth are taken to be parallel and inclined at an angle λ to the normal to the saw teeth.

Information on α, θ is obtained by considering the equilibrium of forces and the exhaustion of the shear resistance on the sliding planes. This is developed in the section 'force equilibrium' and Appendix 1.

In the section 'kinematics' the geometry of the sliding motion is treated. These motions represent the plastic part of the deformations. Expressions are developed for the velocity gradients in terms of the shear strain rates a^* and b^*, as required for establishing the plastic constitutive equations.

The relations obtained in these sections allow us to obtain expressions for the angles α, θ in terms of ϕ (the angle of internal friction) and v^* (the angle of dilatancy).

In this Paper only the homogeneous situation is considered, pertaining to a soil body in which the stresses and strains are constants over the entire region. Such a situation occurs in the simple shear test examined in this Paper. When

---

[1] In earlier papers the shear strain rates were denoted by a and b. Throughout this Paper asterisks are used, when dilatancy is involved. The notation a^*, b^* was introduced earlier (de Josselin de Jong, 1977a) and is used here to be consistent with that paper.

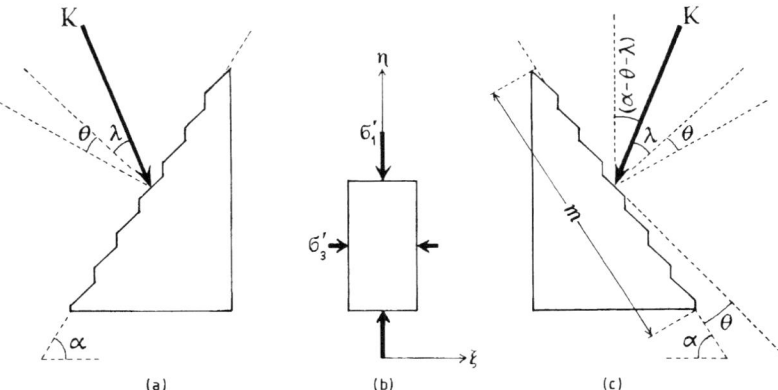

Fig. 5. Resultant forces on potential sliding planes

the stresses and strains are not homogeneous the expressions developed here are valid for an infinitesimally small region.

FORCE EQUILIBRIUM

A co-ordinate system (η, ξ) is taken as shown in Fig. 5(b) in the direction of the principal stresses σ'_1 and σ'_3, positive for compression with $\sigma'_1 > \sigma'_3$. The two conjugate potential sliding planes are located symmetrically with respect to the η-axis; both make an angle α with the ξ-axis.

Consider a portion of such a potential plane of length m (Fig. 5(c)). The resultant force K on that plane has vertical and horizontal components K_η, K_ξ created by the principal stresses on the horizontal and vertical projections of m. Their magnitudes are

$$\left. \begin{array}{c} K_\eta = \sigma'_1 m \cos \alpha \\ K_\xi = \sigma'_3 m \sin \alpha \end{array} \right\} \qquad (1)$$

The angle ϕ of maximum stress obliquity in the limit stress state is defined by[1]

$$\sin \phi = (\sigma'_1 - \sigma'_3)/(\sigma'_1 + \sigma'_3) \qquad (2)$$

Substituting K_η, K_ξ for σ'_1, σ'_3 gives

$$\sin \phi = \frac{(K_\eta \sin \alpha - K_\xi \cos \alpha)}{(K_\eta \sin \alpha + K_\xi \cos \alpha)} \qquad (3)$$

---

[1] For the case when cohesion is present, a term $2c \cot \phi$ has to be added to the denominator. The same results are found in principle by assuming a cohesion c' on the saw teeth of magnitude

$$c' = c[(1 - \sin \phi \sin v^*)/(1 + \sin \phi \sin v^*)]^{1/2}$$

The force K makes an angle $(\alpha - \theta - \lambda)$ with η; therefore

$$K_\eta = K \cos (\alpha - \theta - \lambda)$$
$$K_\xi = K \sin (\alpha - \theta - \lambda)$$

and

$$\sin \phi = \sin (\theta + \lambda)/\sin (2\alpha - \theta - \lambda) \qquad (4)$$

Because of the mirror symmetry with respect to the η-axis, the same force K acts at the same angle λ with respect to the normal on the saw teeth surfaces of the conjugate planes (see Fig. 5(a)).

Equation (4) gives, for a fixed value of ϕ, a set of possible value combinations for α and $(\theta + \lambda)$.

KINEMATICS

Let the regions represented in Figs 6(a) and 6(b) consist of many infinitesimally thin elements (like the pages of a book) that are parallel, all at an angle α. In Fig. 6(b) a constant shear strain rate a^* is assumed to occur, such that all elements slide with respect to each other, causing a homogeneous deformation. All have the same kind of protuberances making an angle θ with the potential sliding planes.

Considering only the plastic part of the deformation, the elements remain rigid during sliding. So the thin elements conserve their length and all points of the potential slip-line through Q have the same velocity V_Q relative to P. Thus V_Q makes an angle $(\alpha - \theta)$ with the ξ-axis, because of the teeth uplift angle θ. The shear strain rate a^* is defined such that the magnitude of V_Q is a^*n, where n is the perpendicular distance from P to the slip-line.

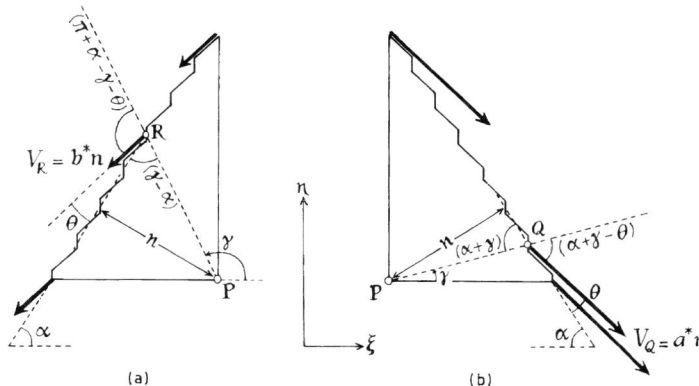

Fig. 6. Velocities of sliding planes

Consider the line PQ at an arbitrary angle γ with the ξ-axis. This line makes an angle $(\alpha + \gamma)$ with the slip-line and so its length is $n/\sin(\alpha + \gamma)$. The velocity V_Q makes an angle $(\alpha + \gamma - \theta)$ with PQ. The effect of V_Q on PQ is that the line is elongated with an extension rate $\dot{\varepsilon}(\gamma)$ and is rotated with a tilting rate $\dot{\rho}(\gamma)$ (positive counterclockwise) as given by the expressions

$$\left.\begin{aligned}
\dot{\varepsilon}(\gamma) &= + V_Q \cos(\alpha + \gamma - \theta)/\text{PQ} \\
&= + a^* \cos(\alpha + \gamma - \theta) \sin(\alpha + \gamma) \\
\dot{\rho}(\gamma) &= - V_Q \sin(\alpha + \gamma - \theta)/\text{PQ} \\
&= - a^* \sin(\alpha + \gamma - \theta) \sin(\alpha + \gamma)
\end{aligned}\right\} \quad (5)$$

where a dot represents the time derivative d/dT.

A similar analysis applied to PR with R on the conjugate slip-line in Fig. 6(a) gives

$$\left.\begin{aligned}
\dot{\varepsilon}(\gamma) &= + V_R \cos(\pi + \alpha - \gamma - \theta)/\text{PR} \\
&= - b^* \cos(\alpha - \gamma - \theta) \sin(\gamma - \alpha) \\
\dot{\rho}(\gamma) &= + V_R \sin(\pi + \alpha - \gamma - \theta)/\text{PR} \\
&= - b^* \sin(\alpha - \gamma - \theta) \sin(\gamma - \alpha)
\end{aligned}\right\} \quad (6)$$

Here b^* is the shear strain rate in the conjugate slip direction. This b^* may be different from a^* according to the double sliding model. When both shear strain rates a^* and b^* are active, the total elongation rate $\dot{\varepsilon}(\gamma)$ and the total tilt rate $\dot{\rho}(\gamma)$ are obtained by addition of the above expressions.

The rates $\dot{\varepsilon}(\gamma)$ and $\dot{\rho}(\gamma)$ are functions of γ. By giving γ the value zero, the above formulae give the elongation and tilt rates for lines in the ξ-direction. Let V_ξ, V_η be the velocities of material points in the ξ, η-directions. Then the elongation rate is $\partial V_\xi/\partial \xi$ and the tilt rate is $\partial V_\eta/\partial \xi$. These are written as $V_{\xi,\xi}$ and $V_{\eta,\xi}$ where a comma represents

differentiation with respect to the subscript variable following it. For $\gamma = 0$ the addition gives

$$\left.\begin{aligned}
V_{\xi,\xi} &= \dot{\varepsilon}(0) = +(a^* + b^*) \cos(\alpha - \theta) \sin \alpha \\
V_{\eta,\xi} &= \dot{\rho}(0) = -(a^* - b^*) \sin(\alpha - \theta) \sin \alpha
\end{aligned}\right\} \quad (7)$$

For $\gamma = \tfrac{1}{2}\pi$ the extension and tilt rates of lines in the η-direction are obtained. These are

$$\left.\begin{aligned}
V_{\xi,\eta} &= -\dot{\rho}(\tfrac{1}{2}\pi) = +(a^* - b^*) \cos(\alpha - \theta) \cos \alpha \\
V_{\eta,\eta} &= +\dot{\varepsilon}(\tfrac{1}{2}\pi) = -(a^* + b^*) \sin(\alpha - \theta) \cos \alpha
\end{aligned}\right\} \quad (8)$$

The derivatives $V_{\xi,\xi} \cdots$ etc. are the so-called velocity gradients.

In addition to the mutual sliding in two conjugate directions, the elements can also execute a rotation around P. This rotation is called the structural rotation Ω (positive counter-clockwise, see Fig. 7). The indication 'structural' was proposed to the Author by Drescher in order to distinguish Ω from the well-known material rotation $\dot{\omega}$, which is defined by $\dot{\omega} = \tfrac{1}{2}(V_{y,x} - V_{x,y})$ and represents the rotation of the total material as observed from the outside.

Ω creates additional velocity gradients of the form

$$V_{\xi,\xi} = 0; \quad V_{\eta,\xi} = \Omega; \quad V_{\xi,\eta} = -\Omega; \quad V_{\eta,\eta} = 0$$

Combining these results gives the strain rate components, which are

$$\left.\begin{aligned}
V_{\xi,\xi} + V_{\eta,\eta} &= (a^* + b^*) \sin \theta \\
V_{\xi,\xi} - V_{\eta,\eta} &= (a^* + b^*) \sin(2\alpha - \theta) \\
V_{\xi,\eta} + V_{\eta,\xi} &= (a^* - b^*) \cos(2\alpha - \theta)
\end{aligned}\right\} \quad (9)$$

and the rate of material rotation, which is

$$V_{\eta,\xi} - V_{\xi,\eta} = -(a^* - b^*) \cos \theta + 2\Omega = 2\dot{\omega} \quad (10)$$

Equation (10) shows the difference between material rotation $\dot\omega$ and the structural rotation Ω. When only sliding occurs and the elements themselves do not rotate, so that $\Omega = 0$, the total soil body nevertheless appears to rotate, in the case when a^* is not equal to b^*.

For example, when a^* is larger than b^*, the soil gives the impression of rotating clockwise, because $\dot\omega$ is negative. Hence the material rotation $\dot\omega$ may consist of two parts. One component is the effect of unequal shear strain rates, $a^* \neq b^*$; the other component Ω is due to rotation of the structure, i.e. of the elements between the sliding planes.

Dilatancy relation

The angle of dilatancy v^* is defined by

$$\sin v^* = (V_{\xi,\xi} + V_{\eta,\eta})/(V_{\xi,\xi} - V_{\eta,\eta}) \qquad (11)$$

where $V_{\xi,\xi}$ and $V_{\eta,\eta}$ are the linear strain rates of lines in the principal stress directions. Using equations (9) we find that

$$\sin v^* = \sin \theta/\sin (2\alpha - \theta) \qquad (12)$$

This relation is independent of the shear strain rates α^* and b^*. Equation (12) gives for a fixed value of v^* a set of possible combinations for the values of α and θ.

Thermodynamic requirement of energy dissipation

In the double sliding model it is assumed that a^* can differ from b^*. This is due to the principle in plasticity that, at failure, the strain rates are not bounded in magnitude—they can have any value. So also in the two conjugate directions they can have unequal values.

The shear strain rates a^* and b^* are only bounded by the requirement that the laws of friction are not violated. This would occur if the sliding directions were opposed to the shear stresses on the sliding planes. Considering Fig. 5 it is seen that a^* and b^*, as defined in Fig. 6, should be positive in order that the directions of the forces on the saw teeth correspond to the sliding directions. So the requirement on the shear strain rates is

$$\left.\begin{array}{c} a^* \geqslant 0 \\ b^* \geqslant 0 \end{array}\right\} \qquad (13)$$

and this is called the thermodynamic requirement of energy dissipation.

Mathematically it is possible to show that $(a^* + b^*)$ cannot be negative. The stronger requirements (13) are dictated by the above mentioned friction character of the system.

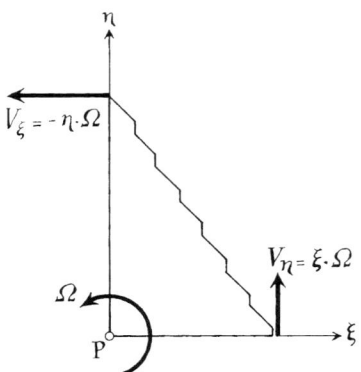

Fig. 7. **Structural rotation**

PLASTIC CONSTITUTIVE EQUATIONS

The expressions (9) for the strain rates can serve as constitutive equations for the plastic stage of deformation. However, before they can be used, the angles α, θ have to be replaced by the material properties ϕ, v^*. For this replacement, equations (4) and (12) are available. Of these, (4) contains the force inclination angle λ, which can be eliminated by using the rules of friction to guarantee that the shear resistance is exhausted on the surfaces of the teeth. This point is elaborated in Appendix 1, leading to equations (76). Using these in equations (9) results in the following plastic constitutive equations

$$\left.\begin{array}{l} (V_{\xi,\xi} + V_{\eta,\eta})^p = +(a^* + b^*)\cos\phi \sin v^*/F \\ (V_{\xi,\xi} - V_{\eta,\eta})^p = +(a^* + b^*)\cos\phi/F \\ (V_{\xi,\eta} + V_{\eta,\xi})^p = -(a^* - b^*)\sin\phi \cos v^*/F \end{array}\right\} \quad (14)$$

where $F^2 = 1 - \sin^2\phi \sin^2 v^*$ and the superscript p denotes plastic.

Although not belonging to the plastic constitutive equations proper, equation (10) for the rotation can be added to complete the system, giving

$$(V_{\eta,\xi} - V_{\xi,\eta}) = -(a^* - b^*)\cos v^*/F + 2\Omega = 2\dot\omega \qquad (15)$$

This is not a constitutive relation since the principle of objectivity prohibits a relation containing rotations to be constitutive. It is added here because it allows determination of the structural rotation at a later stage.

For solving a boundary value problem, equations (14) together with the limitations (13) on a^* and b^* are sufficient. They form a complete system of constitutive equations describing the plastic behaviour.

SECTION 2

ELASTIC PART OF THE ELASTO-PLASTIC DSFR MODEL

In the elastic-plastic DSFR model the strain rates that are imposed by the boundary conditions are divided into an elastic part and a plastic part. The plastic part consists of strain rates that obey relations (13) and (14) when the material is in the limit stress state. The remaining part is taken account of by the elastic stress rates. When the stresses are not at the limit state the plastic part vanishes, i.e. $a^* = 0$, $b^* = 0$, and the entire strain rates are accounted for by an elastic response to the stress rates. The elastic part is developed as follows.

STRESS STATE AND LIMIT STRESS STATE

Under plane strain conditions in the (x, y)-directions, the normal effective stress σ'_{zz} is a principal stress and intermediate between the two principal effective stresses in the (x, y)-plane. All derivatives in the z-direction are zero.

The normal effective stresses σ'_{xx}, σ'_{yy} are taken to be positive for compression.[1] In order to abbreviate the notation, the variables s, t, ψ are introduced, such that

$$\left.\begin{array}{l} \sigma'_{xx} = s - t \cos 2\psi \\ \sigma_{xy} = \sigma_{yx} = -t \sin 2\psi \\ \sigma'_{yy} = s + t \cos 2\psi \end{array}\right\} \quad (16)$$

Here, s is the effective stress level, t the radius of the Mohr circle and ψ the angle between the minor principal compression stress and the

---

[1] This sign convention is contrary to that generally adopted in mechanics, but is used here to suit soil mechanics readers.

x-axis, positive counter-clockwise. The variable t cannot be negative. These quantities are shown in Fig. 8, where ψ is shown with the negative value, occurring in the case of the simple shear test considered in this Paper.

The Coulomb-Mohr limit condition is adopted, with limit lines at an angle ϕ to the σ'-axis. Cohesion is assumed to be zero for simplicity; the limit lines pass through the origin of the σ, τ diagram. In terms of t the limit condition is expressed as follows.

(a) The material behaves *elastically*

$$\left.\begin{array}{l} \text{when } 0 \leqslant t < s \sin \phi, \text{ or} \\ \quad t = s \sin \phi \text{ and } \dot{t} < \dot{s} \sin \phi \\ \text{then } a^* = b^* = 0. \end{array}\right\} \quad (17)$$

(b) The material behaves *elasto-plastically*

$$\left.\begin{array}{l} \text{when } t = s \sin \phi \text{ and } \dot{t} = \dot{s} \sin \phi \\ \text{then } a^* \geqslant 0 \text{ and } b^* \geqslant 0. \end{array}\right\} \quad (18)$$

Stress rates

Considering s, t and ψ as time dependent variables, the expressions (16) become, when differentiated with respect to time

$$\left.\begin{array}{l} \dot{\sigma}'_{xx} = \dot{s} - \dot{t} \cos 2\psi + t \sin 2\psi\ (2\dot{\psi}) \\ \dot{\sigma}_{xy} = \dot{\sigma}_{yx} = -\dot{t} \sin 2\psi - t \cos 2\psi\ (2\dot{\psi}) \\ \dot{\sigma}'_{yy} = \dot{s} + \dot{t} \cos 2\psi - t \sin 2\psi\ (2\dot{\psi}) \end{array}\right\} \quad (19)$$

where a dot represents the time derivative d/dT.

When rotations of the material are involved, the objective co-rotational (Jaumann) stress rates $\overset{\triangledown}{\sigma}_{xx} \ldots$ etc. are required. These are related to the time derivatives $\dot{\sigma}'_{xx} \ldots$ etc. by replacing $2\dot{\psi}$ by $(2\dot{\psi} - 2\dot{\omega})$ where $\dot{\omega}$ is the material rotation.

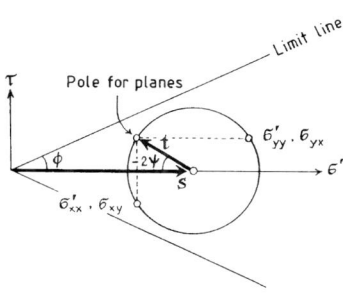
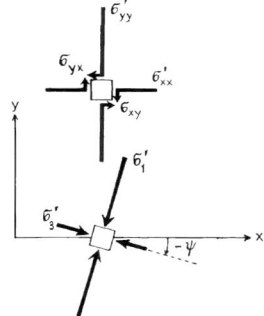

Fig. 8. Stresses in elastic stage

They are given by

$$
\left.
\begin{aligned}
\breve{\sigma}'_{xx} &= \dot{s} - \dot{t} \cos 2\psi + t \sin 2\psi\,(2\dot{\psi} - 2\dot{\omega}) \\
\breve{\sigma}'_{xy} &= \breve{\sigma}'_{yx} = -\dot{t} \sin 2\psi - t \cos 2\psi\,(2\dot{\psi} - 2\dot{\omega}) \\
\breve{\sigma}'_{yy} &= \dot{s} + \dot{t} \cos 2\psi - t \sin 2\psi\,(2\dot{\psi} - 2\dot{\omega})
\end{aligned}
\right\} (20)
$$

Elasticity

Although perfect elasticity is not a good approximation for soils, it is adopted here for convenience because the formulation of elastic stress-strain relations is well known and undisputed. It reduces the validity of the model considered here to situations where the elastic strain rates are small compared with the plastic strain rates.

In plane strain the constitutive equations are

$$
\left.
\begin{aligned}
-V_{x,x}{}^{e} &= [(1 - v)\breve{\sigma}'_{xx} - v\breve{\sigma}'_{yy}]/2G \\
-(V_{x,y} + V_{y,x})^{e} &= \breve{\sigma}'_{xy}/G = \breve{\sigma}'_{yx}/G \\
-V_{y,y}{}^{e} &= [-v\breve{\sigma}'_{xx} + (1 - v)\breve{\sigma}'_{yy}]/2G
\end{aligned}
\right\} (21)
$$

where the superscript e denotes elastic. The terms $V_{x,x}{}^{e}$, $V_{y,y}{}^{e}$ are the extension rates $\dot{\varepsilon}_{xx}{}^{e}$, $\dot{\varepsilon}_{yy}{}^{e}$; $(V_{x,y} + V_{y,x})^{e}$ is the shear strain rate $\dot{\gamma}_{xy}{}^{e}$. The shear modulus G is related to the modulus of elasticity E and Poisson's ratio v in the usual way by $G = E/2(1 + v)$. The minus signs arise from the soil mechanics sign convention.

From the velocity gradients $V_{x,x}{}^{e}$ ... etc. the strain rates in the principal stress directions ξ, η (which are rotated ψ with respect to x, y) are obtained from the following relations

$$
\left.
\begin{aligned}
(V_{\xi,\xi} + V_{\eta,\eta})^{e} &= +(V_{x,x} + V_{y,y})^{e} \\
(V_{\xi,\xi} - V_{\eta,\eta})^{e} &= +(V_{x,x} - V_{y,y})^{e} \cos 2\psi \\
&\quad + (V_{x,y} + V_{y,x})^{e} \sin 2\psi \\
(V_{\xi,\eta} + V_{\eta,\xi})^{e} &= -(V_{x,x} - V_{y,y})^{e} \sin 2\psi \\
&\quad + (V_{x,y} + V_{y,x})^{e} \cos 2\psi
\end{aligned}
\right\} (22)
$$

For the material rotation the transformation is

$$
(V_{\eta,\xi} - V_{\xi,\eta}) = (V_{y,x} - V_{x,y}) = 2\dot{\omega} \tag{23}
$$

Elastic constitutive equations

Combining equations (20), (21), (22) gives the elastic constitutive equations

$$
\left.
\begin{aligned}
(V_{\xi,\xi} + V_{\eta,\eta})^{e} &= -\dot{s}(1 - 2v)/G \\
(V_{\xi,\xi} - V_{\eta,\eta})^{e} &= +\dot{t}/G \\
(V_{\xi,\eta} + V_{\eta,\xi})^{e} &= +t(2\dot{\psi} - 2\dot{\omega})/G
\end{aligned}
\right\} (24)
$$

In the limit relations (18) apply and the elastic constitutive equations become

$$
\left.
\begin{aligned}
(V_{\xi,\xi} + V_{\eta,\eta})^{e} &= -\dot{s}(1 - 2v)/G \\
(V_{\xi,\xi} - V_{\eta,\eta})^{e} &= +\dot{s} \sin \phi/G \\
(V_{\xi,\eta} + V_{\eta,\xi})^{e} &= +s \sin \phi(2\dot{\psi} - 2\Omega)/G
\end{aligned}
\right\} (25)
$$

In the last expression $\dot{\omega}$ from equations (20) is replaced by the structural rotation Ω defined by equation (15). The reason is that the rotation which has to be introduced in the co-rotational stress rate formulation must reflect the rotation of the material that deforms elastically. In the elasto-plastic version of the DSFR model it is the elements between the sliding planes that deform elastically. Thus, only their rotation, i.e. the structural rotation Ω, has to enter in the corotational formulation and not the material rotation $\dot{\omega}$, which contains an apparent rotational part, when a^* is unequal to b^*.

ELASTO-PLASTIC DSFR MODEL

The elasto-plastic concept is that in the limit stress state, as defined by relations (18), the total strain rates, as determined from boundary conditions, are divided into elastic and plastic parts such that the plastic constitutive equations (13), (14), (15) are satisfied in the first place. The remaining part of the strain rates is elastic and obeys equations (25). The total strain rates are the sum of elastic and plastic parts as follows

$$
\left.
\begin{aligned}
V_{\xi,\xi} + V_{\eta,\eta} &= +(a^* + b^*) \cos \phi \sin v^*/F \\
&\quad - \dot{s}(1 - 2v)/G \\
V_{\xi,\xi} - V_{\eta,\eta} &= +(a^* + b^*) \cos \phi/F \\
&\quad + \dot{s} \sin \phi/G \\
V_{\xi,\eta} + V_{\eta,\xi} &= -(a^* - b^*) \sin \phi \cos v^*/F \\
&\quad + s \sin \phi(2\dot{\psi} - 2\Omega)/G \\
V_{\eta,\xi} - V_{\xi,\eta} &= -(a^* - b^*) \cos v^*/F + 2\Omega \\
&= 2\dot{\omega}
\end{aligned}
\right\} (26)
$$

Equations (26) apply for a point in (x, y) where the stress state (in the limit) is known. This means that s and 2ψ are given quantities. The question is then to establish how the total strain rates will change the stress state in the course of time, i.e. how large $\dot{s}$ and $2\dot{\psi}$ are, when $V_{\xi,\xi}$ ... etc. are given.

The first two relations of (26) give $\dot{s}$ explicitly and 2Ω can be eliminated from the last two. So there remain two equations for the three unknowns $2\dot{\psi}$, a^*, b^*. Additional information is given by the requirements (13) which state that the shear strain rates a^* and b^* cannot become

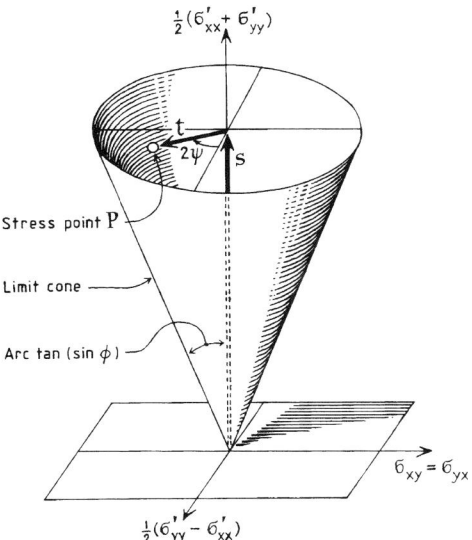

$$\tfrac{1}{2}(\sigma'_{xx} + \sigma'_{yy})$$

Stress point P

Limit cone

Arc tan (sin ϕ)

$\sigma_{xy} = \sigma_{yx}$

$$\tfrac{1}{2}(\sigma'_{yy} - \sigma'_{xx})$$

Fig. 9. Stress space with limit cone

negative. So there are two equations and two inequalities for determining three unknowns.

This is an unusual system. Vermeer (1980) proposed to solve for the unknowns by requiring the rate of elastic energy dissipation to be a minimum. The corresponding procedure is examined mathematically in Appendix 2. Since that analysis is rather complex, it is perhaps helpful to present the procedure visually by means of vectors in the stress and strain spaces in the following sections.

STRESS SPACE

The stress space is defined by three perpendicular axes, which have as co-ordinates: $\tfrac{1}{2}(\sigma'_{xx} + \sigma'_{yy})$ vertical; $\tfrac{1}{2}(\sigma'_{yy} - \sigma'_{xx})$ and $\sigma_{xy} = \sigma_{yx}$ horizontal (see Fig. 9). A point P in this space represents a stress state. It is located in a horizontal plane at a height s, at the end of a radius vector of length t rotated through an angle 2ψ. This corresponds to equations (16). In this stress space the variables s, t, 2ψ are in fact cylindrical co-ordinates.

In time, the stress point P moves through the stress space and its path is called the stress path. The material reaches the limit stress state when the stress point reaches the limit cone. This cone cuts the horizontal $(t, 2\psi)$ co-ordinate plane in a circle with radius $t_{\text{limit}} = s \sin \phi$. Therefore, the semi-vertex angle of the limit cone has a magnitude of arctan (sin ϕ). When cohesion is zero, the

vertex V of the cone lies in the horizontal plane for $\tfrac{1}{2}(\sigma'_{xx} + \sigma'_{yy}) = 0$ (see Fig. 9).

STRAIN RATE SPACE

In a similar manner it is possible to define a strain rate space with orthogonal co-ordinates: $(V_{x,x} + V_{y,y})$ vertical, $(V_{x,x} - V_{y,y})$ and $(V_{x,y} + V_{y,x})$ horizontal. Vectors in this space represent strain rates. Fig. 10 shows such an example.

The origin of the strain rate space is usually placed at the stress point P of the stress space, wherever P may be located temporarily. Corresponding stress and strain rate axes are oriented parallel but opposite in direction because of the sign convention adopted for the stresses.

Figure 10 demonstrates the combination of stress and strain rate spaces. Point V is the origin of the stress space and the vertex of the limit cone. Point P is the current stress point. Since it lies on the limit cone, this figure represents a limit stress state in which the strain rate consists of a combination of elastic and plastic parts. This strain rate is plotted as the vector PQ in the strain rate space with P as origin. It is shown how this PQ is decomposed into its plastic and elastic parts, respectively PR and RQ. It is assumed here that PQ is a vector of known magnitude and direction.

In the horizontal plane through P the axes $(V_{\xi,\xi} - V_{\eta,\eta})$, $(V_{\xi,\eta} + V_{\eta,\xi})$ are shown. According to equations (22) these are rotated through 2ψ with respect to $(V_{x,x} - V_{y,y})$, $(V_{x,y} + V_{y,x})$. They are therefore respectively collinear with, and perpendicular to, the t co-ordinate of the stress space.

In Fig. 10 the t co-ordinate is divided by the shear modulus G in order that the stress rates correspond to the elastic strain rates (see equations (24)). In the vertical direction the s coordinate has to be adjusted by a factor $(1 - 2v)/G$. By this adaptation of the stress space, the elastic part of the strain rate vector coincides with the tangent to the stress path.

The adapted limit stress cone is more obtuse; its vertex angle is equal to arctan (sin $\phi/(1 - 2v)$). The stress path through P cannot trespass outside the limit cone. When the material deforms plastically it will remain in its tangent plane. So the elastic part of the strain rate is a vector parallel to the plane, that is tangential to the adapted cone in P.

A vector in this tangent plane consists of two components: one is in the direction of the line connecting P to the cone vertex, the other is tangential to the circular cross-section of the cone. The first may be called the $\dot{s}$-component, because only s changes when the stress path has that direction; the second is the $\dot{\psi}$-component, because ψ changes along it.

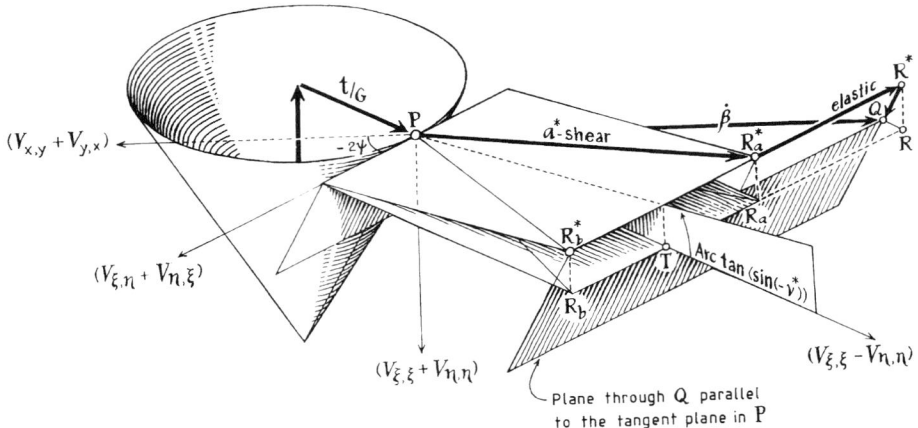

Fig. 12. Strain rate space for dilatant double sliding (vector PQ of length $\dot\beta$ is strain rate in undrained simple shear test)

to the left of R_a. Decomposing PR vectorially in directions at angle ϕ would give a component $R_m R$ in the negative b-direction producing a negative value for b. This is not allowed by the thermodynamic requirement of energy dissipation (13). Instead PR is the vectorial addition of the vectors PR_a and $R_a R$, where $R_a R$ is a $\dot\psi$-component.

Before showing this it is necessary to return to Fig. 10. A case (ii) is shown, where PQ represents the known total strain rate. It is composed of a plastic part PR in the horizontal plane and an elastic part RQ in the plane through Q parallel to the tangent plane through P. The point R lies on the intersection line of these two planes. The requirement of minimal elastic work is satisfied when QR is the smallest distance between Q and the intersection line. So the construction of R is to project Q perpendicularly on to that line.

In Fig. 10 the projection R of Q lies between R_a and R_b. So a case (ii) is involved and PR can be decomposed into two positive shear strain rates a and b. In that case the elastic stress path of P, being parallel to RQ, consists only of an $\dot s$-component. Pointing downwards it means that s decreases in the course of time.

Dilatant or contractive behaviour

In Fig. 12 the material is contractive, i.e. with a negative angle of dilatancy v^*. According to equations (14) the plastic shear strain rates a^*, b^* produce vectors that are located in the plane $R_a^* P R_b^*$ at an angle arctan (sin v^*) with the horizontal plane. This plane may be called the plane of the a^*, b^* shearings. Since v^* is negative

the plane tilts upwards, i.e. in the negative $(V_{\xi,\xi} + V_{\eta,\eta})$-direction.

The shear strain rate a^* produces a vector along PR_a^* and b^* along PR_b^*, but unfortunately these vectors do not have the lengths of a^* and b^*. It is the projections PR_a and PR_b of these lines, in the horizontal plane, that have lengths a^* and b^*, respectively. This follows from the last two lines of equations (14) because F defined there satisfies $F^2 = \cos^2 \phi + \sin^2 \phi \cos^2 v^*$. The apex angle of triangle $R_a P R_b$ has a magnitude of 2 arctan (tan ϕ cos v^*).

In Fig. 12 the strain rate, imposed on the material in the undrained simple shear test, is shown. It is, as demonstrated in the section below, the vector PQ of length $\dot\beta$ in the negative $(V_{x,y} + V_{y,x})$-direction. In order to decompose PQ into plastic and elastic parts, the plane is drawn through Q, parallel to the tangent plane in P, and the intersection line $R_b^* R_a^* R^*$ of this plane with the tilted plane for a^*, b^* shearings is constructed. To determine a^*, b^* it is then necessary to project R_a^* and R_b^* on the horizontal plane. This leads to complicated mathematical expressions, since planes at different angles are involved in this procedure. The formulae are somewhat simplified in the sections below and Appendix 2 by calling M^*, N^* the co-ordinates of point R in the horizontal plane, such that $M^* = $ PT and $N^* = $ TR.

In Fig. 12 a situation is shown where the point R^* lies outside the allowable range $R_a^* R_b^*$ and its projection R outside the range $R_a R_b$. This situation resembles Fig. 11(b) and is accordingly called a case (iii).

Minimal elastic work is produced when the elastic strain rate vector is the shortest distance from Q to the allowable range $R_a^* R_b^*$. In this case the vector QR_a^* is the shortest distance and so R_a^*Q is the elastic part of the strain rate. The plastic part is the vector PR_a^* and that means that only the shear strain mode a^* is active and $b^* = 0$. How this decomposition may be derived mathematically is shown in Appendix 2.

The elastic vector R_a^*Q can be decomposed into a $\dot{\psi}$-component $R_a^*R^*$ and an $\dot{s}$-component R^*Q. The $\dot{\psi}$-component is in the negative ($V_{\xi,\eta}$ + $V_{\eta,\xi}$)-direction, which according to the last line of equations (25) produces a negative ($2\dot{\psi} - 2\Omega$). The $\dot{s}$-component is downward, which means a decrease of s in the course of time. The vector R_a^*Q is parallel to the tangent of the stress path in P. That path is consequently a downward turning helical line on the cone surface.

SECTION 3

EXAMPLE OF UNDRAINED SIMPLE SHEAR TEST

The undrained simple shear test is used here as an example to show how the elasto-plastic DSFR model predicts stress paths. The deformations in a simple shear test can be considered to be homogeneously distributed over the sample, when the deformations are not too large. This leads to simple equations for the imposed strain rates and closed analytic solutions for the stress paths.

Figure 13 shows a typical result of a stress path plotted from such a solution. The heavy line is the path of the stress point, representing the effective stresses σ'_{yy}, σ_{yx} on horizontal planes, as traced in the Mohr diagram. Initially the principal stresses are vertical and horizontal, equal respectively to q_v and q_h. These form the initial conditions. The stress point is in P_0.

The dotted initial stress circle is smaller than the limit circle. So the test starts with an elastic stage. In this prefailure stage the Mohr circles remain concentric, but expand, and the stress point moves vertically upwards.

The end of the elastic and beginning of the plastic stage is at P_1 when the corresponding stress circle touches the limit lines. After P_1 the sample behaves plastically; all stress circles remain tangential to the limit lines.

During the plastic stage the total strain rates consist of a plastic part and an elastic part. The magnitude of the plastic shear strain rates a^* and b^* is determined in such a way that the remaining elastic strain rates dissipate a minimum amount of energy. Applying the procedure developed in Appendix 2 leads us to identify three cases, called respectively case (ii), case (iii) and case (iv). The location of P_1 on the limit stress circle determines which of those cases is occurring. The relevant regions are indicated in Fig. 13.

In all three cases that part of the total strain rates which produces volume changes, is taken account of by an elastic reduction of the effective stress level. The model is assumed to be contractive (negative dilatant). Since the test is undrained, the volume remains constant and the porewater pressure increases to prevent the sample from contracting. Accordingly, the effective stress level reduces in the course of time, causing the stress path to turn left.

Case (ii) is obtained when P_1 lies on the middle arc of aperture $2\bar{\phi}$, where $\bar{\phi}$ is an auxiliary angle characteristic for the simple shear test, defined by equation (40). In case (ii) both a^* and b^* will develop. The non-coaxiality occurring in this case is uniquely determined by the initial stresses q_v, q_h. The stress paths during the plastic stage are straight lines (see Fig. 17).

Case (iii) is involved when P_1 lies on the right-hand arc of aperture $(\frac{1}{2}\pi - \bar{\phi})$. Then only a^*-type sliding occurs and $b^* = 0$. The sliding planes are vertical. This leads to a toppling bookrow mechanism. The solution for s as a function of 2ψ leads to curved stress paths from which one example is shown in Fig. 13. More curves of this family are given in Fig. 18. In case (iii) q_v is larger than q_h: an initial situation which may be called active. The apparent angle of internal friction is smaller than ϕ.

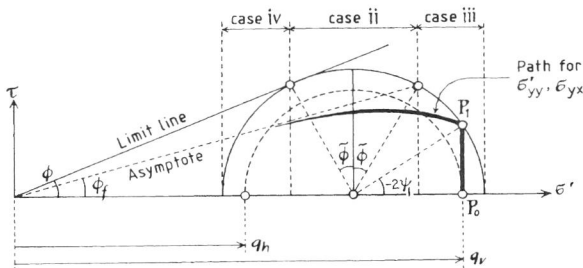

Fig. 13. Example of stress path in undrained simple shear test

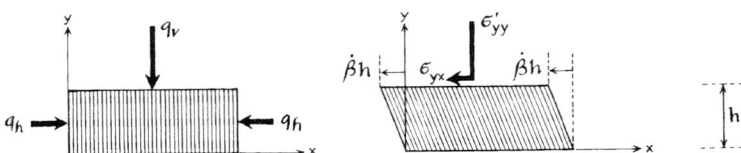

Fig. 14. Geometry of undrained simple shear test

Case (iv) arises when P_1 lies on the left-hand arc of aperture $(\frac{1}{2}\pi - \bar{\phi})$. Then $a^* = 0$ and only b^*-type sliding occurs, with horizontal sliding planes. The apparent angle of internal friction is almost equal to ϕ. In this case $q_v < q_h$ initially, a situation that may be termed passive.

IMPOSED STRAIN RATES IN UNDRAINED SIMPLE SHEAR TEST

When the sample is completely saturated the pore water prevents volume changes. In the simple shear apparatus a plane strain condition in the z-direction is maintained. Then the sample cross-section in the (x, y)-plane conserves its area, while deforming from a rectangle into a parallelogram (see Fig. 14).

The deformation is assumed to be homogeneously distributed over the sample. Then the upper plane remains horizontal and keeps its height, h. Let the velocity of the upper plane be β towards the left. The velocity components of points in the sample are then given by

$$V_x = -\beta y; \quad V_y = 0 \quad \text{with } \beta > 0 \quad (28)$$

In this analysis β is taken to be positive and furthermore it is independent of x and y because of the assumed homogeneity of the deformation. The minus sign allows us to compare the computed stress paths in the Mohr diagram with the test results in Figs 1 and 20 (taken from Wroth et al.) and this is an advantage. A disadvantage is that negative values of 2ψ occur in the analysis below.

The terms that represent the velocity gradients $V_{x,x} = \partial V_x/\partial x \ldots$ etc. are obtained by partial differentiation of V_x and V_y with respect to x and y. It follows that

$$V_{x,x} = V_{y,x} = V_{y,y} = 0; \quad V_{x,y} = -\beta \quad (29)$$

The major principal direction of the strain rates accordingly makes an angle θ with the x-axis, where

$$\tan 2\theta = \frac{V_{x,y} + V_{y,x}}{V_{x,x} - V_{y,y}} = -\infty;$$

$$\text{i.e.} \quad \theta = -\frac{1}{4}\pi \quad (30)$$

The combinations (22) and (23) of the strain rates in the principal stress directions (ξ, η) are given by

$$\left.\begin{aligned} V_{\xi,\xi} + V_{\eta,\eta} &= 0 \\ V_{\xi,\xi} - V_{\eta,\eta} &= -\beta \sin 2\psi \\ V_{\xi,\eta} + V_{\eta,\xi} &= -\beta \cos 2\psi \\ V_{\eta,\xi} - V_{\xi,\eta} &= \beta = 2\dot{\omega} \end{aligned}\right\} \quad (31)$$

Comparing the above results with the coordinates in Fig. 12, it can be seen that the strain rate vector imposed on the material in the undrained simple shear test is the vector PQ of length β.

Initial conditions

At time $T = T_0$ the test starts. At that moment the principal effective stresses are vertical and horizontal. They are positive and called respectively q_v and q_h (see Fig. 14). So the initial stress state is given by

$$\sigma'_{xx}(0) = q_h; \quad \sigma'_{yy}(0) = q_v; \quad \sigma_{xy} = \sigma_{yx}(0) = 0 \quad (32)$$

ELASTIC STAGE

In the beginning the material is elastic and the velocity gradients from equations (29) are

$$V_{x,x}{}^e = V_{y,x}{}^e = V_{y,y}{}^e = 0; \quad V_{x,y}{}^e = -\beta \quad (33)$$

Then equations (21) give

$$\overset{\triangledown}{\sigma}'_{xx} = \overset{\triangledown}{\sigma}'_{yy} = 0; \quad \overset{\triangledown}{\sigma}_{xy} = \overset{\triangledown}{\sigma}_{yx} = \beta G$$

In order to simplify the presentation, the co-rotational stress rates $\overset{\triangledown}{\sigma}'_{xx} \ldots$ etc. are replaced here by the time derivatives $\dot{\sigma}'_{xx} \ldots$ etc. The more complicated, co-rotational solution is described elsewhere. The difference is in the term $(2\dot{\psi} - 2\dot{\omega})$ and $2\dot{\psi}$ in equations (19) and (20). Introducing the strain rates (equations (31)) in the last line of equations (24) gives during the elastic stage: $(2\dot{\psi} - 2\dot{\omega}) = -\beta(G/t) \cos 2\psi$ with $2\dot{\omega} = \beta$. Since (from relations (17)) $t \leqslant s \sin \phi$ this indicates that

$2\dot{\psi}$ exceeds $2\dot{\omega}$ by a factor of the order of (G/s). Disregarding $2\dot{\omega}$ compared with $2\dot{\psi}$ is reasonable when G is large compared with the effective stress level s.

The simplified equations are

$$\dot{\sigma}'_{xx} = \dot{\sigma}'_{yy} = 0; \quad \dot{\sigma}_{xy} = \dot{\sigma}_{yx} = \beta G$$

These can be integrated directly and with the initial conditions (equation (32)) give

$$\left. \begin{array}{l} \sigma'_{xx} = \text{constant} = \sigma'_{xx}(0) = q_h \\[4pt] \sigma'_{yy} = \text{constant} = \sigma'_{yy}(0) = q_v \\[4pt] \sigma_{xy} = \sigma_{yx} = \displaystyle\int_{T_0}^{T} \beta G \, dT = \beta G(T - T_0) \end{array} \right\} \quad (34)$$

It follows from the constancy of σ'_{yy} that the stress path in Fig. 13 is a vertical straight line. In the co-rotational solution it is slightly curved to the right. From equations (34) and (16) it follows that

$$\left. \begin{array}{l} s = s_0 = \tfrac{1}{2}(q_v + q_h) \\[4pt] t \cos 2\psi = \tfrac{1}{2}(q_v - q_h) \\[4pt] t \sin 2\psi = -\beta G(T - T_0) \end{array} \right\} \quad (35)$$

Figures 15(a) and 15(b) show the behaviour of the stresses in the elastic stage. In Fig. 15(a) an active initial stress state is shown, where $q_v > q_h$. The principal stresses turn clockwise, i.e. $\psi < 0$. The angle ψ is initially zero and enters the region between 0 and $-\tfrac{1}{4}\pi$.

Figure 15(b) shows a passive initial stress state, where $q_v < q_h$. The principal stresses turn counter-clockwise, i.e. $\psi > 0$. The angle ψ starts from $-\tfrac{1}{2}\pi$ and enters the region between $-\tfrac{1}{2}\pi$ and $-\tfrac{1}{4}\pi$. In the Mohr diagram the pole for planes is indicated, and the heavy line is the stress path for $\sigma'_{yy}, \sigma_{yx}$.

Considering the second of equations (35), the values of ψ mentioned above are accounted for by the expression

$$2\psi = -\tfrac{1}{2}\pi + \arcsin \left[\tfrac{1}{2}(q_v - q_h)/t \right] \quad (36)$$

where t is limited by relations (17) to the values $0 \leqslant t < s \sin \phi$.

End of elastic and beginning of plastic stage

At time T_1 the elastic stage ends. The point for $\sigma'_{yy}, \sigma_{yx}$ in Fig. 13 has reached P_1, the stress point for which the accompanying stress circle touches the limit line. Then from relations (18) t_1 equals $s_1 \sin \phi$ and the stress state characterized by s_1, ψ_1 is from equations (35)

$$\left. \begin{array}{l} s_1 = s_0 = \tfrac{1}{2}(q_v + q_h) \\[4pt] \cos 2\psi_1 = \tfrac{1}{2}(q_v - q_h)/s_1 \sin \phi \\[4pt] \sin 2\psi_1 = -\beta G(T_1 - T_0)/s_1 \sin \phi \end{array} \right\} \quad (37)$$

Equation (36) then becomes

$$2\psi_1 = -\tfrac{1}{2}\pi + \arcsin \left(\frac{q_v - q_h}{(q_v + q_h)\sin \phi} \right) \quad (38)$$

SECTION 4

PLASTIC STAGE

After T_1 the material behaves elasto-plastically and the equations developed in Appendix 2 are then applicable. These are expressed in terms of the quantities M^* and N^* defined by equations (79) and (80). Using equation (31) with $\psi = \psi_1$ at

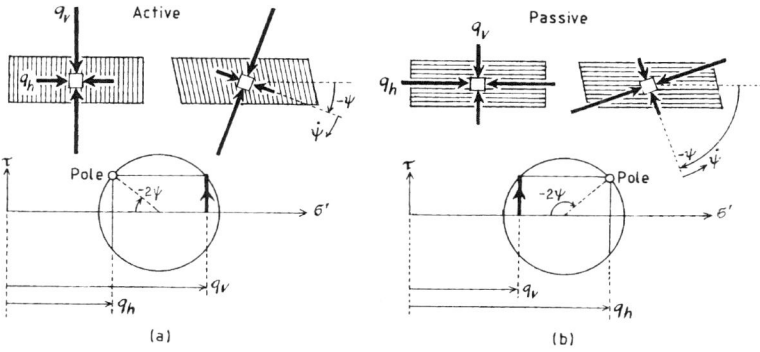

Fig. 15. Stresses in elastic stage: (a) starting from active initial stress state, $q_v > q_h$; (b) starting from passive initial stress state, $q_v < q_h$

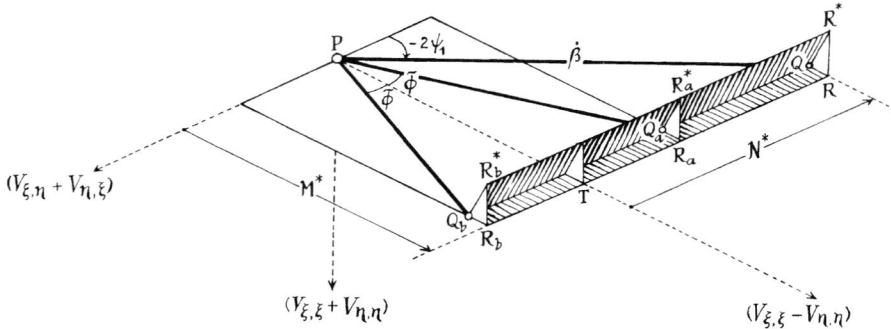

Fig. 16. Detail of Fig. 12 showing character of auxiliary angle $\tilde{\phi}$

time T_1 these quantities are in the beginning of the plastic stage

$$M^* = -\beta \sin 2\psi_1 \frac{(1-2v)}{(1-2v+\sin\phi\sin v^*)} \left.\right\}$$
$$N^* = +\beta \cos 2\psi_1 \qquad\qquad\qquad (39)$$

At this point it is convenient to introduce an auxiliary angle $\tilde{\phi}$ defined by

$$\tan \tilde{\phi} = \frac{\tan\phi \cos v^*(1-2v)}{(1-2v+\sin\phi\sin v^*)} \qquad (40)$$

The auxiliary angle $\tilde{\phi}$ has no special meaning, but is used here because it simplifies the mathematical results. In Fig. 16, which illustrates a part of Fig. 12, it is the angle between PT and the lines PQ_a and PQ_b. Further, it reduces to ϕ in the non-dilatant case when $v^* = 0$.

Using this angle $\tilde{\phi}$ the variables M^* and N^* from equations (39) are

$$M^* = -\beta \sin 2\psi_1 \tan\tilde{\phi}/\tan\phi\cos v^* \left.\right\}$$
$$N^* = +\beta \cos 2\psi_1 \qquad\qquad\qquad (41)$$

and the combinations $(M^* \tan\phi \cos v^* \pm N^*)$ which are of particular significance in Appendix 2 are

$$M^* \tan\phi \cos v^* + N^* =$$
$$+ \beta \cos (2\psi_1 + \tilde{\phi})/\cos\tilde{\phi}$$
$$M^* \tan\phi \cos v^* - N^* = \left.\right\} \quad (42)$$
$$- \beta \cos (2\psi_1 - \tilde{\phi})/\cos\tilde{\phi})$$

As $\sin 2\psi_1$ is negative (equations (37)) M^* (equations (39)) is always positive here. This means (see Appendix 2) that case (i) does not occur. After T_1 the material behaves differently in

the three remaining cases (ii), (iii) and (iv) distinguished in Appendix 2. Only $\dot{s}$, the rate of the effective stress level s, is the same for all three.

Effective stress level

According to equation (77) of Appendix 2, it is found from equations (31) and (40) that for the undrained simple shear test

$$\dot{s} = -\beta G \frac{\sin 2\psi \tan\tilde{\phi}\tan v^*}{(1-2v)\tan\phi} \qquad (43)$$

Since $\sin 2\psi$ is negative at time $T_1 > T_0$ (equations 37)), $\dot{s}$ has the sign of v^*, the angle of dilatancy. When the material is contractive, v^* is negative and so the effective stress level reduces in the course of time.

The relevant parameters a^*, b^*, $2\dot{\psi}$ differ for the three different cases.

CASE (ii)

When q_v, q_h satisfy the inequalities

$$-\sin\tilde{\phi} \leqslant (q_v - q_h)/(q_v + q_h)\sin\phi \leqslant \sin\tilde{\phi} \quad (44)$$

the value of $2\psi_1$ is (according to equation (38)) limited by

$$-\tfrac{1}{2}\pi - \tilde{\phi} \leqslant 2\psi_1 \leqslant -\tfrac{1}{2}\pi + \tilde{\phi} \qquad (45)$$

and this implies that $-1 \leqslant \sin 2\psi_1 \leqslant -\cos\tilde{\phi}$ and $-\sin\tilde{\phi} \leqslant \cos 2\psi_1 \leqslant \sin\tilde{\phi}$.

Considering equations (41) it is found that M^*, N^* satisfy the inequalities

$$M^* \tan\phi \cos v^* \geqslant \beta \sin\tilde{\phi} \left.\right\}$$
$$-\beta\sin\tilde{\phi} \leqslant N^* \leqslant \beta\sin\tilde{\phi} \qquad (46)$$

So the condition of equation (85) is satisfied and a case (ii) occurs.

Values of a^*, b^*, $2\dot\psi$ in case (ii)

When case (ii) occurs the plastic behaviour is described by equation (89). Using equations (42) the strain rates are

$$\left.\begin{aligned}
a^* &= +\beta \cos (2\psi_1 + \tilde\phi) \\
&\quad \times F/2 \cos \tilde\phi \sin \phi \cos v^* \\
b^* &= -\beta \cos (2\psi_1 - \tilde\phi) \\
&\quad \times F/2 \cos \tilde\phi \sin \phi \cos v^*
\end{aligned}\right\} \quad (47)$$

Considering relations (45) it is verified that both a^*, b^* are non-negative and that therefore the thermodynamic requirements (13) are satisfied.

The last line of equations (89) requires $2\dot\psi$ to be equal to 2Ω and this latter can be determined by the use of the last lines of equations (26), (31) and (47), giving

$$2\Omega = 2\dot\psi = \beta(1 + \cos 2\psi_1/\sin \phi) \quad (48)$$

This shows that $\dot\psi$ is of the order of β. Equation (43) shows that $\dot s$ is of the order βG. Since $\dot s$ is negative and s reduces in time, its total change will be smaller than s. So $\beta G(T - T_1) < s$ and the total change of 2ψ will be of the order s/G which is assumed to be small.

In order to simplify the analysis, $2\dot\psi$ will be taken to be zero. This means that 2ψ remains constant and equal to $2\psi_1$ in the plastic stage. As a consequence the stress paths are straight lines directed towards the origin in the Mohr diagram. A family of these case (ii) curves is shown in Fig. 17.

Non-coaxiality in case (ii)

From equation (30) the angle θ between the x-axis and the major principal direction of the strain rate tensor equals $-\frac14\pi$. The angle between the x-axis and the minor principal compression stress is ψ. The angle of non-coaxiality (which was called i in previous work) is therefore given by $i = \theta - \psi = -\frac14\pi - \psi$. Using equation (38) it is deduced that

$$\sin 2i = -\cos 2\psi = -\cos 2\psi_1$$
$$= -(q_v - q_h)/(q_v + q_h) \sin \phi \quad (49)$$

This shows that the non-coaxiality (in case (ii) plastic behaviour) is uniquely determined by the initial principal stresses q_v, q_h.

Coaxiality (i.e. $i = 0$) is to be expected only in the special case when $q_v = q_h$, i.e. that the initial principal stresses are equal (indicated in Fig. 17 by an arrow). An expression for i in terms of the shear strain rates a^*, b^* is found by the use of equations (78), (81) and (41). Since $(2\psi - 2\Omega)$ is zero in case (ii), it is found that

$$\frac{b^* - a^*}{b^* + a^*} = \frac{-N^*}{M^* \tan \phi \cos v^*} = \cot 2\psi_1 \cot \tilde\phi$$

As $2i = -\frac12\pi - 2\psi_1$ it follows that

$$\tan 2i = \tan \tilde\phi(b^* - a^*)/(a^* + b^*)$$

This relation reduces for the case of volume conserving non-dilatancy to the well-known

$$\tan 2i = \tan \phi(b - a)/(a + b)$$

Apparent angle of internal friction in case (ii)

From Fig. 17 the stress paths for σ'_{yy}, σ_{yx} are straight lines towards the origin. These suggest apparent angles of internal friction ϕ_f defined by

$$\tan \phi_f = \frac{\sigma_{yx}}{\sigma'_{yy}} = \frac{-t \sin 2\psi}{s + t \cos 2\psi}$$
$$= -\sin \phi \sin 2\psi/1 + \sin \phi \cos 2\psi$$

In this case 2ψ is constant and equal to $2\psi_1$. Furthermore, $2\psi_1$ can (depending on q_v, q_h) have any value between $(-\frac12\pi - \phi)$ and $(-\frac12\pi + \phi)$

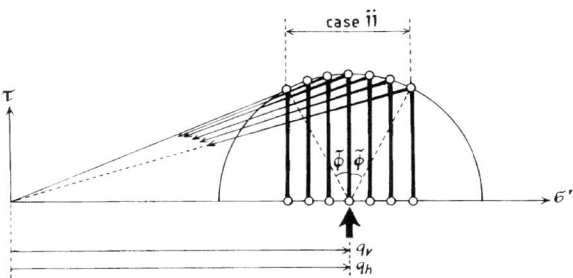

Fig. 17. Family of case (ii) stress paths (arrow indicates coaxiality)

(*see* relations (45)). Therefore $\tan \phi_f$ can have values between

$$\left.\begin{array}{l} \tan \phi_f = \sin \phi \cos \tilde{\phi}/(1 + \sin \phi \sin \tilde{\phi}) \\ \text{and} \\ \tan \phi_f = \sin \phi \cos \tilde{\phi}/(1 - \sin \phi \sin \tilde{\phi}) \end{array}\right\} \quad (50)$$

For $v^* = 0$, $\tilde{\phi} = \phi$, and these values for ϕ_f correspond to the prediction mentioned earlier (de Josselin de Jong, 1972).

By using equation (38) it is possible to express the apparent angle of internal friction in terms of q_v, q_h only, showing the important influence of the initial horizontal stress on the test results.

CASE (iii), ACTIVE CASE

When q_v, q_h satisfy the inequalities

$$\sin \tilde{\phi} < (q_v - q_h)/(q_v + q_h) \sin \phi < 1 \quad (51)$$

the value of $2\psi_1$ is, according to equation (38), limited by

$$(-\tfrac{1}{2}\pi + \tilde{\phi}) < 2\psi_1 < 0 \quad (52)$$

and this implies that $-\cos \tilde{\phi} < \sin 2\psi_1 < 0$ and $\sin \tilde{\phi} < \cos 2\psi_1 < 1$.

Considering equations (41) it is found that M^*, N^* then satisfy the inequalities

$$M^* \tan \phi \cos v^* < \beta \sin \tilde{\phi}$$

$$N^* > \beta \sin \tilde{\phi}$$

Therefore the condition (86) is satisfied and case (iii) occurs.

Values of a*, b*, $2\dot{\psi}$ in case (iii)

When case (iii) occurs, the plastic behaviour is described by equations (90). With M^*, N^* given by equations (41) the strain rates are

$$\left.\begin{array}{l} a^* = -\beta \sin 2\psi \tan \tilde{\phi} F/\sin \phi \cos v^* \\ b^* = 0 \end{array}\right\} \quad (53)$$

This is single sliding behaviour.

Using these values of a^*, b^* in the last of equations (26), with $2\dot{\omega} = \beta$ from (31), gives

$$2\Omega = \beta(1 - \sin 2\psi \tan \tilde{\phi}/\sin \phi) > 0 \quad (54)$$

The last of equations (90) with M^*, N^* from equations (41) gives

$$2\dot{\psi} = -\beta(G/s) \cos (2\psi - \tilde{\phi})/\sin \phi \cos \tilde{\phi} \quad (55)$$

In this expression 2Ω is disregarded compared with the other terms, because 2Ω is of the order (s/G) smaller.

Solution for the stress paths in case (iii)

In order to establish 2ψ as a function of time, equation (55) has to be integrated. This is,

however, not possible directly, because this expression contains also the variable s. Elimination of β with equation (43) produces the following differential equation

$$\frac{\sin 2\psi}{\cos (2\psi - \tilde{\phi})} \frac{d(2\psi)}{dT} = \frac{(1 - 2v) \cot v^*}{\cos \phi \sin \tilde{\phi}} \frac{ds}{s \, dT} \quad (56)$$

Since the variables s and 2ψ are separated, each side can be integrated directly, giving

$$-\cos \tilde{\phi} \log [\cos (2\psi - \tilde{\phi})] + 2\psi \sin \tilde{\phi}$$

$$= (1 - 2v) \frac{\cot v^*}{\cos \phi \sin \tilde{\phi}} \log s + \text{const.} \quad (57)$$

Determining the integration constant by using the boundary conditions at T_1 gives the solution

$$\log \left[\frac{\cos (2\psi_1 - \tilde{\phi})}{\cos (2\psi - \tilde{\phi})} \right] + 2(\psi - \psi_1) \tan \tilde{\phi}$$

$$= (1 - 2v) \frac{\cot v^*}{\cos \phi \sin \tilde{\phi} \cos \tilde{\phi}} \log \left[\frac{2s}{q_v + q_h} \right] \quad (58)$$

where ψ_1 given by equation (38) is a function of q_v, q_h alone.

The stress paths for σ'_{yy}, σ_{yx} of Fig. 18 are plotted from this solution, using equation (16) with $t = s \sin \phi$. The material constants are taken to be $\phi = 23°$; $v^* = -3°$; $(1 - 2v) = 0·1$. These values give $\tilde{\phi} = 28·05°$ (equation (40)). Curves have been drawn for $(q_v/q_h) = 1·45$; $1·7$; $1·96$; $2·28$, which correspond to $2\psi_1 = -61·95°$; $-48·5°$; $-33·5°$; $0°$ respectively.

Apparent angle of internal friction

As shown by equation (55) $2\dot{\psi}$ vanishes for $2\psi = -\tfrac{1}{2}\pi + \tilde{\phi}$. So the curves of Fig. 18 approach this value asymptotically and the curve for $2\psi_1 = -\tfrac{1}{2}\pi + \tilde{\phi}$, which is a straight line, is their asymptote. The apparent angle ϕ_f then has the value

$$\tan \phi_f = -\sin \phi \sin 2\psi_1/(1 + \sin \phi \cos 2\psi_1)$$

$$= \sin \phi \cos \tilde{\phi}/(1 + \sin \phi \sin \tilde{\phi}) \quad (59)$$

This is one of the limits for ϕ_f in case (ii) (see equation (50)).

Toppling bookrow mechanism

The plastic behaviour in all case (iii) situations is the toppling bookrow mechanism. This can be seen as follows. According to relation (51) the initial stress states in case (iii) are active, $q_v > q_h$ and so Fig. 15(a) applies to them. Since $b^* = 0$, only a^* sliding occurs (*see* equations (53)). Comparison with Fig 3(a) shows that then only vertical sliding planes develop. This corresponds to Fig. 2.

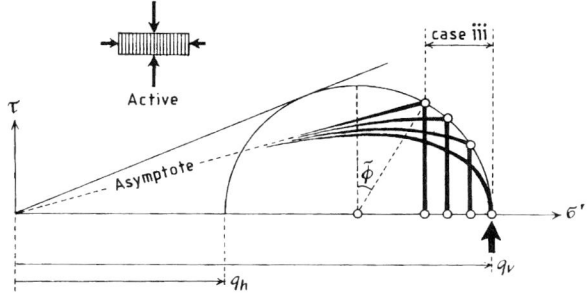

Fig. 18. Family of case (iii) stress paths

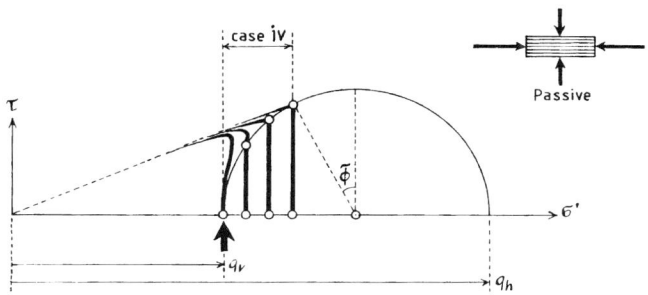

Fig. 19. Family of case (iv) stress paths

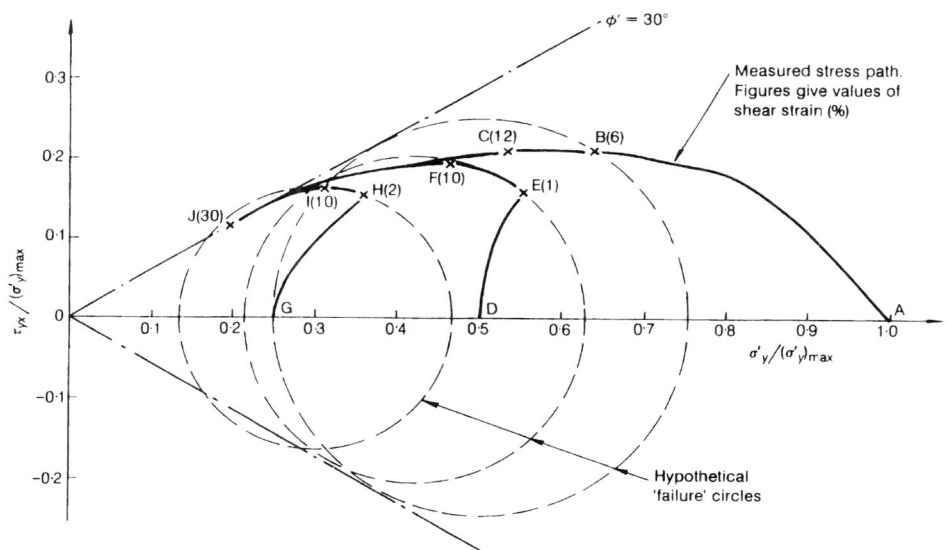

Fig. 20. Stress paths from simple shear tests on Boston blue clay (from Ladd & Edgers, 1972)

Furthermore, the structural rotation Ω given by equation (54) is positive, because $\sin 2\psi$ is initially negative according to relation (52) and does not change sign afterwards. This means a counter-clockwise rotation of the sliding elements as indicated in Fig. 2.

It might be noted that the structural rotation Ω is opposed to the principal stress rotation ψ. In equation (55) $\cos (2\psi - \tilde{\phi})$ is positive for $2\psi_1$ limited by equation (52) causing 2ψ to be negative to start with. The principal stresses start to rotate clockwise and continue to do so thereafter. This falsifies the postulate that structural rotation and principal stress rotation are to be identified.

CASE (iv), PASSIVE CASE

When q_v, q_h satisfy the inequalities

$$-1 < (q_v - q_h)/(q_v + q_h) \sin \phi < -\sin \tilde{\phi} \quad (60)$$

a case (iv) situation occurs. An analysis similar to that for case (iii) gives the results

$$\left.\begin{array}{l} a^* = 0 \\[2mm] b^* = -\beta \sin 2\psi \tan \tilde{\phi} F/\sin \phi \cos v^* \\[2mm] 2\dot{\psi} = -\beta(G/s) \cos (2\psi + \tilde{\phi})/\sin \phi \cos \tilde{\phi} \end{array}\right\} \quad (61)$$

The solution for s as a function of ψ, giving the stress paths, is

$$\log \left[\frac{\cos (2\psi_1 + \tilde{\phi})}{\cos (2\psi + \tilde{\phi})} \right] + 2(\psi_1 - \psi) \tan \tilde{\phi}$$

$$= (1 - 2v) \frac{\cot v^*}{\cos \phi \sin \tilde{\phi} \cos \tilde{\phi}} \log \left[\frac{2s}{q_v + q_h} \right] \quad (62)$$

Now $2\dot{\psi}$ vanishes for $2\psi = -\frac{1}{2}\pi - \tilde{\phi}$, so the stress paths approach the line asymptotically at an angle ϕ_f defined by

$$\tan \phi_f = \sin \phi \cos \tilde{\phi}/(1 - \sin \phi \sin \tilde{\phi}) \quad (63)$$

Since $\tilde{\phi}$ differs only slightly from ϕ, the apparent angle of internal friction ϕ_f is almost equal to ϕ (see Fig. 19).

Horizontal sliding planes

According to relation (60) case (iv) occurs when the initial stress state is passive, i.e. $q_v < q_h$. So Fig 15(b) applies to them. Since $a^* = 0$, only b^*-sliding occurs. Comparison with Fig. 3(b) shows that then horizontal sliding planes develop.

This is the mechanism of sliding that is generally believed to be the only possible failure mode to arise in simple shear tests.

CONCLUSION

Figures 17, 18 and 19 show the families of stress paths in undrained simple shear tests for hypothetical samples that obey the laws of the contractive elasto-plastic DSFR model. The material properties of the samples are the same and they have the same angle of internal friction ϕ. Their behaviour differs in the tests and this is due only to the differences in their initial stresses, i.e. the vertical and horizontal normal stresses q_v and q_h.

Consider for simplicity the stress paths marked by the arrows in Figs 17, 18 and 19. The marked curve in Fig. 17 starts with $q_v = q_h$. In Fig. 18 where $q_v > q_h$, the extreme active case is marked. In Fig. 19 the extreme passive case with $q_v < q_h$ is marked. There is a remarkable resemblance between these three curves and the test results from Ladd & Edgers (1972) reproduced in Fig. 20. In this figure the initial stresses are not indicated, but the overconsolidation ratios of 1, 2, 4 suggest that the three curves correspond to q_h values that are, respectively, smaller than, equal to and larger than q_v.

The theoretical stress paths start with vertical straight lines and have sharp bends at the transition from elastic to plastic. This is due to the oversimplification of perfect elasticity in the pre-failure stage. Introducing dilatancy and a gradual decrease of the shear modulus G may soften the sharp bends. It was, however, not the objective of this study to try to match test results by matching material properties.

The purpose of this study was only to show that the original, unaltered DSFR model, if extended to its elasto-plastic version, predicts the unique failure behaviour observed and that the chosen failure mode depends on the initial stress state. Because of the preponderant influence of the ratio q_v/q_h, it seems evident that in the execution of simple shear tests the horizontal stresses should no longer be disregarded.

APPENDIX 1: DETERMINATION OF α, θ, λ FOR GIVEN VALUES OF ϕ, v^*

The three parameters α, θ, λ describe the geometry of the sliding mechanism and the forces acting upon it. The equations developed in section 4, relating these parameters to the properties ϕ, v^* of the particle assembly as a whole, are equations (4) and (12) which repeated here are

$$\sin \phi = \sin (\theta + \lambda)/\sin (2\alpha - \theta - \lambda) \quad (64)$$

$$\sin v^* = \sin \theta/\sin (2\alpha - \theta) \quad (65)$$

These are only two equations for solving for the three unknown parameters α, θ, λ as functions of ϕ, v^*, which are material properties observable from the outside. The objective of the analysis below is to show how α, θ, λ can be determined by use of the additional information that the sliding mechanism is bound to obey the rules of friction. In terms of the interparticle friction angle ϕ_μ these rules are

(a) sliding will *not* occur when λ is smaller than ϕ_μ
(b) sliding *can* occur when λ equals ϕ_μ
(c) it is impossible for λ to exceed ϕ_μ.

From all sliding geometries having value combinations of α, θ that produce v^* according to (65), only a particular set of α, θ combinations can be expected to be active. That particular set is the set that introduced into equation (64) gives the observed value ϕ for a value of λ which equals ϕ_μ. However, since λ cannot exceed ϕ_μ, this set of α, θ combinations is not allowed to contain values of λ exceeding ϕ_μ. This means that λ considered as a function of α, θ must have an extreme value λ_{extr} which is a maximum, such that $\lambda_{\text{extr}} = \phi_\mu$.

Elimination of θ from equations (64) and (65) gives

$$\tan \lambda = (\sin \phi - \sin v^*) \sin 2\alpha/m \qquad (66)$$

with

$$m = (1 + \sin \phi \sin v^*) + (\sin \phi + \sin v^*) \cos 2\alpha$$

This gives λ as a function of α. In establishing λ_{extr} by differentiation of equation (66) with respect to α, only α is a variable; the values of ϕ and v^* are fixed quantities. Taking the derivative then gives

$$(d\lambda/d\alpha)/\cos^2 \lambda = 2(\sin \phi - \sin v^*)n/m^2 \qquad (67)$$

with

$$n = (\sin \phi + \sin v^*) + (1 + \sin \phi \sin v^*) \cos 2\alpha \quad (68)$$

A maximum occurs, when $d\lambda/d\alpha = 0$ or $n = 0$. This gives

$$\cos 2\alpha = -\frac{\sin \phi + \sin v^*}{1 + \sin \phi \sin v^*} \qquad (69)$$

Replacement of $\sin \phi$, $\sin v^*$ by use of equations (64) and (65) gives

$$\cos 2\alpha = \frac{\cos 2\alpha \cos \lambda - \cos (2\alpha - 2\theta - \lambda)}{\cos \lambda - \cos 2\alpha \cos (2\alpha - 2\theta - \lambda)}$$

or

$$\sin^2 2\alpha \cos (2\alpha - 2\theta - \lambda) = 0 \qquad (70)$$

Since α cannot be zero (*see* Figs 3 and 4) the solution is

$$\lambda_{\text{extr}} = 2\alpha - 2\theta - \tfrac{1}{2}\pi \qquad (71)$$

This result is equivalent to Rowe's stress dilatancy relation (1962).

In order to verify whether λ_{extr} indeed produces a maximum, the second derivative of λ with respect to α is required. The derivative $d\lambda/d\alpha$ as expressed by equation (67) can be written in the form

$$d\lambda/d\alpha = nf \qquad (72)$$

with

$$f = 2 (\sin \phi - \sin v^*) \cos^2 \lambda/m^2 \qquad (73)$$

As ϕ is always larger than v^*, f is positive. The second derivative of λ is

$$\frac{d^2\lambda}{d\alpha^2} = \frac{dn}{d\alpha} f + n \frac{df}{d\alpha} \qquad (74)$$

and since $n = 0$ for λ_{extr}, this gives, with n given by equation (68)

$$\left(\frac{d^2\lambda}{d\alpha^2}\right)_{\lambda = \lambda_{\text{extr}}} = -2f(1 + \sin \phi \sin v^*) \sin 2\alpha$$

As α is smaller than $\tfrac{1}{2}\pi$, this result shows that the second derivative is negative. So the solution $\lambda_{\text{extr}} = 2\alpha - 2\theta - \tfrac{1}{2}\pi$ represents a maximum and this satisfies the rules of friction. Introducing λ_{extr} in equation (64) gives

$$\sin \phi = -\cos (2\alpha - \theta)/\cos \theta \qquad (75)$$

Inversion from this relation and equation (65) gives the following expressions for θ and $(2\alpha - \theta)$

$$\left.\begin{array}{l} \sin \theta = \cos \phi \sin v^*/F \\[4pt] \cos \theta = \cos v^*/F \\[4pt] \sin (2\alpha - \theta) = \cos \phi/F \\[4pt] \cos(2\alpha - \theta) = -\sin \phi \cos v^*/F \end{array}\right\} \qquad (76)$$

with

$$F^2 = 1 - \sin^2 \phi \sin^2 v^*$$

APPENDIX 2: DECOMPOSITION OF TOTAL STRAIN RATES INTO PLASTIC AND ELASTIC PARTS

Equations (26) express the components of the total strain rates in terms of a^*, b^*, $\dot{s}$ and $(\dot{\psi} - \Omega)$. The unknown $\dot{s}$ can be found directly from the first two by elimination of $(a^* + b^*)$. This gives

$$\dot{s}(1 - 2v + \sin \phi \sin v^*)/G =$$
$$-(V_{\xi,\xi} + V_{\eta,\eta}) + (V_{\xi,\xi} - V_{\eta,\eta}) \sin v^* \quad (77)$$

The connection of this relation with vectors in Fig. 12 is that the right-hand side divided by $(1 + \sin \phi \sin v^*/(1 - 2v))$ equals RR^*, the vertical component of QR^*, which itself is parallel to VP.

Solving for a^*, b^*
From the first two of equations (26) it follows that

$$(a^* + b^*) \cos \phi/F = M^* \qquad (78)$$

with

$$M^* = \frac{(V_{\xi,\xi} + V_{\eta,\eta}) \sin \phi + (V_{\xi,\xi} - V_{\eta,\eta})(1 - 2v)}{(1 - 2v + \sin \phi \sin v^*)} \quad (79)$$

Let a quantity N^* be defined by

$$N^* = -(V_{\xi,\eta} + V_{\eta,\xi}) \qquad (80)$$

The third of equations (26) then gives

$$(a^* - b^*) \sin \phi \cos v^*/F$$
$$= N^* + \sin \phi(s/G)(2\dot{\psi} - 2\Omega) \quad (81)$$

Using equations (78) and (81) the shear strain rates a^* and b^* can be expressed in terms of M^* and N^*. There results

$$2a^* \sin \phi \cos v^*/F - \sin \phi(s/G)(2\dot{\psi} - 2\Omega)$$
$$= M^* \tan \phi \cos v^* + N^* \quad (82)$$

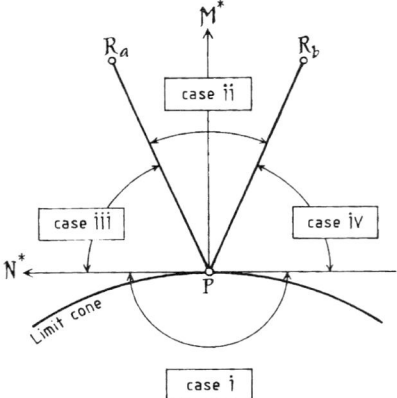

Fig. 21. Horizontal plane of Fig. 12, showing regions of R in the cases (i), (ii), (iii) and (iv)

$2b^*\sin\phi\cos v^*/F + \sin\phi(s/G)(2\dot\psi - 2\Omega)$

$$= M^*\tan\phi\cos v^* - N^* \quad (83)$$

Eliminating $(a^* - b^*)$ between equation (81) and the fourth of equations (26) gives

$$\sin\phi\,(s/G)(2\dot\psi - 2\Omega) = (2\Omega - 2\dot\omega)\sin\phi - N^* \quad (84)$$

The quantities M^* and N^*, as defined by equations (79) and (80), are the co-ordinates of R in the horizontal plane of Fig. 12. It may be verified that $M^* = PT$ and $N^* = TR$. Fig. 21 shows this horizontal plane as viewed from above. There are four regions in which the point R can be located. These regions correspond to the cases (i), (ii), (iii) and (iv) mentioned in this Paper.

Case (i)
When the components of the imposed strain rate are such that $(V_{\xi,\xi} - V_{\eta,\eta})(1 - 2v)$ is smaller than $-(V_{\xi,\xi} + V_{\eta,\eta})\sin\phi$, the value of M^* is negative according to equation (79) and $\dot t$ is smaller than $\dot s\sin\phi$ (according to the first two of equations (24)). When the stresses are in the limit state, then a situation occurs as indicated by the second of equations (17): the stress point P re-enters the elastic region and deformation is purely elastic. The corresponding strain rate vector points downwards in Fig. 21 into the region indicated as case (i). In that case the stress history being purely elastic is described by equations (24). Equations (26) are not valid and $(a^* + b^*)$ solved in the manner of equation (78) does not apply. Actually (78) with M^* negative is unacceptable, because that violates the thermodynamic requirement (13).

In the undrained simple shear test as described by equations (31) M^* defined by (79) is always positive due to the third line of (37). So case (i) does not occur and there remain three cases—(ii), (iii) and (iv)—which are relevant to this study.

Cases (ii), (iii) and (iv)
M^* and N^* are the co-ordinates of R in the horizontal plane in Fig. 12. As the apex angle of triangle $R_a\,PR_b$

is 2 arctan (tan ϕ cos v^*) the magnitude of N^* with respect to $M^*\tan\phi\cos v^*$ determines how R is located with respect to that triangle. The three different possibilities are indicated as cases (ii), (iii) and (iv) in Fig. 21. They are as follows.

(ii)   When R lies on the section $R_a\,R_b$ then

$$-M^*\tan\phi\cos v^* \leqslant N^* \leqslant M^*\tan\phi\cos v^* \quad (85)$$

(iii)   When R lies to the left of R_a then

$$N^* > +M^*\tan\phi\cos v^* \quad (86)$$

(iv)   When R lies to the right of R_b then

$$N^* < -M^*\tan\phi\cos v^* \quad (87)$$

Determination of $(2\dot\psi - 2\Omega)$
When the total strain rate is given, M^* and N^* are known from equations (79) and (80). Then only two relations—(82) and (83)—are available to determine the three unknowns a^*, b^*, $(2\dot\psi - 2\Omega)$, since (84) does not contribute any useful information, since Ω being also unknown. Vermeer's proposal for solving this system is to require that the rate of elastic energy dissipation $\dot A^e$ is a minimum.

Using this principle, together with the thermodynamic requirement of energy dissipation, it is argued below on logical grounds that one of the three unknowns is zero in each of the three cases (ii), (iii) and (iv) mentioned above.

The quantity $\dot A^e$ is defined by

$$\dot A^e = -V_{x,x}\,{}^e\bar\sigma'_{xx} - V_{y,y}\,{}^e\bar\sigma'_{yy} - (V_{x,y} + V_{y,x})\,{}^e\bar\sigma_{xy}$$

and this can, by using equations (20) and (21) be transformed into

$$\dot A^e = [(1 - 2v)\dot s^2 + \dot t^2 + t^2(2\dot\psi - 2\dot\omega)^2]/G$$

Comparing equations (24) with the limit values of equations (25), $\dot A^e$ becomes for the limit stress state

$$\dot A^e = [(1 - 2v + \sin^2\phi)\dot s^2 + s^2\sin^2\phi(2\dot\psi - 2\Omega)^2]/G \quad (88)$$

Since relation (77) establishes $\dot s$ independently, a minimum of $\dot A^e$ is obtained by choosing the lowest possible value of $(2\dot\psi - 2\Omega)^2$. This is tantamount to requiring that the absolute value $|2\dot\psi - 2\Omega|$ is a minimum. Together with the thermodynamic requirement (13) the conclusions which follow can be drawn.

Resulting expressions
Case (ii). The minimum value for $(2\dot\psi - 2\Omega)^2$ is obtained when $(2\dot\psi - 2\Omega) = 0$. This value is acceptable, when relations (85) hold, because then (82) and (83) show that neither a^* nor b^* are negative. In this case the solution is

$$\left.\begin{array}{l} a^* = (M^*\tan\phi\cos v^* + N^*)F/2\sin\phi\cos v^* \\ b^* = (M^*\tan\phi\cos v^* - N^*)F/2\sin\phi\cos v^* \\ 2\dot\psi - 2\Omega = 0 \end{array}\right\} \quad (89)$$

Case (iii). In this case $(2\dot\psi - 2\Omega)$ cannot be zero, because (86) introduced into (83) would then produce a negative b^*, which violates requirements (13). A minimum for $|2\dot\psi - 2\Omega|$ is obtained by choosing b^* to be zero. So the solution is

$$\left. \begin{array}{l} a^* = M^*F/\cos \phi > 0 \\ b^* = 0 \\ (2\dot{\psi} - 2\Omega) \sin \phi(s/G) \\ \quad = (+M^* \tan \phi \cos v^* - N^*) < 0 \end{array} \right\} \quad (90)$$

Case (iv). In a similar manner (87) and (81) require in this case that a^* be zero. So the solution is

$$\left. \begin{array}{l} a^* = 0 \\ b^* = M^*F/\cos \phi > 0 \\ (2\dot{\psi} - 2\Omega) \sin \phi(s/G) \\ \quad = (-M^* \tan \phi \cos v^* - N^*) > 0 \end{array} \right\} \quad (91)$$

REFERENCES

Borin, D. L. (1973). *The behaviour of saturated kaolin in the simple shear apparatus.* PhD thesis, University of Cambridge.

de Josselin de Jong, G. (1958). The undefiniteness in kinematics for friction materials. *Proc. Conf. Earth Pressure Probl., Brussels* **1**, 55–70.

de Josselin de Jong, G. (1959). *Statics and kinematics in the failure zone of a granular material.* Thesis, University of Delft.

de Josselin de Jong, G. (1971). The double sliding free rotating model for granular assemblies. *Géotechnique* **21**, No 2, 155–163.

de Josselin de Jong, G. (1972). Discussion, Session II, Roscoe Memorial Symposium. *Stress-strain behaviour of soils* (ed. Parry, R. H. G.) pp. 258–261. Cambridge: Foulis.

de Josselin de Jong, G. (1977a). Mathematical elaboration of the double sliding free rotating model. *Archives of Mechanics* **29**, No. 4, 561–591.

de Josselin de Jong, G. (1977b). Constitutive relations for the flow of a granular assembly in the limit state of stress. *Proc. 9th Int. Conf. Soil Mech. Fdn Engng, Tokyo*, pp. 87–95.

Drescher, A. (1976). An experimental investigation of flow rules for granular materials using optically sensitive glass particles. *Géotechnique* **26**, No. 4, 591–601.

Ladd, C. C. & Edgers, L. (1972). *Consolidated undrained direct simple shear tests on saturated clays.* Research report T 72–82, Massachusetts Institute of Technology.

Randolph, M. F. and Wroth C. P. (1981). Application of the failure state in undrained simple shear to the shaft capacity of driven piles. *Géotechnique* **31**, No. 1, 143–157.

Rowe, P. W. (1962). The stress dilatancy relation for static equilibrium of an assembly of particles in contact. *Proc. Roy. Soc.* **A269**, 500–527.

Teunissen, J. A. M. & Vermeer, P. A. (1988). Analysis of double shearing in frictional materials. *Int. J. Numer. & Analy. Meth. Geomech.* **12**, 323–340.

Vermeer, P. A. (1980). Double sliding within an elasto-plastic framework. *L.G.M.mededelingen* **21**, 199–207.

Vermeer, P. A. (1981). A formulation and analysis of granular flow. *Proc. Int. Conf. Mech. Behaviour Struct. Media.* **B**, 325–339.

Wroth, C. P. (1984). The interpretation of in situ soil tests. *Géotechnique* **34**, No. 4, 449–489.

Wroth, C. P. (1987). The behaviour of normally consolidated clay as observed in undrained direct shear tests. *Géotechnique* **37**, No. 1, 37–43.

de Josselin de Jong, G. (1989). *Géotechnique* **39**, No. 3, 1989, 565–566

DISCUSSION

Elasto-plastic version of the double sliding model in undrained simple shear tests

G. de JOSSELIN de JONG (1988). *Géotechnique* **38**, No. 4, 1988, 533–555

R. E. Gibson, *Golder Associates (UK) Ltd*

On p. 544 of this Paper the Author states that the vector PQ (in Fig. 12) represents '. . . the strain rate imposed on the material in the undrained simple shear test'. It is, however, not clear to me which of the three cases (ii), (iii) and (iv), is referred to here.

Comparing Figs. 12 and 21 it would appear that case (iii) is shown, since R in Fig. 12 lies in the region indicated as case (iii) in Fig. 21. If this is correct it implies that a stress path of the type shown in Fig. 13 is followed, so that $2\psi_1$ decreases gradually from 0 to $[-(\pi/2) + \tilde{\phi}]$. This gives the impression that the stress point P moves along the cone in Fig. 12 (as viewed from above) in a counterclockwise direction. The following

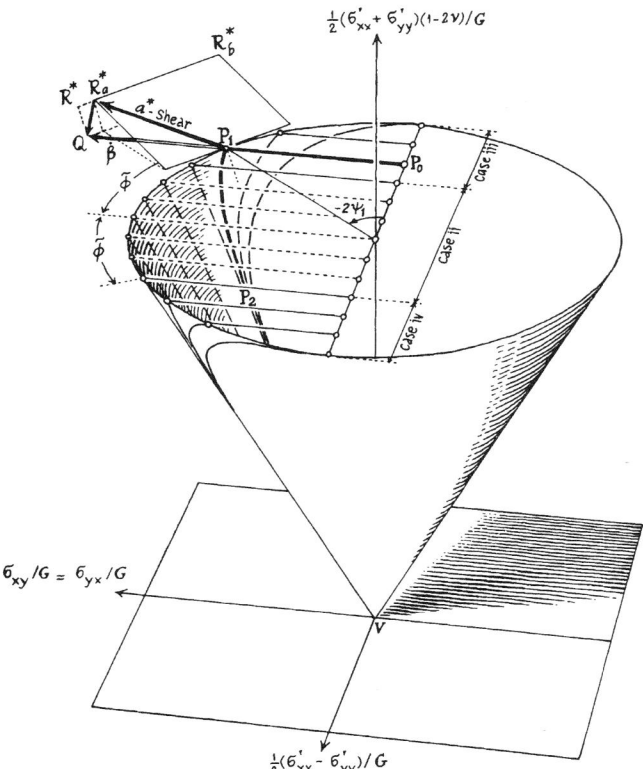

Fig. 22. Adapted stress space with limit cone, similar to Fig. 12, but shown from other side

565

question then arises: what stress path will P follow in the remaining cases (ii) and (iv)? Perhaps the Author will favour us with one of his elegant drawings to elucidate this point.

Authors' reply

The case represented in Fig. 12 on p. 544 is indeed a case iii. In order to demonstrate this explicitly the situation of Fig. 12 is redrawn here in Fig. 22. In this Fig. the adapted stress space with the limit cone is shown from the other side by rotating the space over 180 degrees to permit a better view of the stress paths. The strain rate vector, as imposed by the undrained simple shear test, is a vector of length β, parallel to the axis of $\sigma_{xy}/G = \sigma_{yx}/G$ and in its direction. So it points towards the left, here. It is shown only once as the vector $P_1 Q$ for the case that the stress point is located in P_1.

Let the stress path $P_0 P_1 P_2$ be considered in more detail, first. Point P_0 represents the initial stress state, in which according to the stresses, indicated on the co-ordinate axes, $\sigma_{xy}(0) = \sigma_{yx}(0) = 0$ and $\sigma'_{xx}(0) < \sigma'_{yy}(0)$. According to equations (32) this implies $q_v > q_h$ and so an active, initial situation denoted as case iii in the paper, is involved.

The part $P_0 P_1$ is the stress path during the elastic stage. This part has the direction of the imposed strain rate vector of length β. The elastic stage ends when the stress point has reached the limit cone in the point P_1. This path corresponds to the point P of Fig. 12 in the paper. The tilted plane $R_a^* P_1 R_b^*$ is the plane of the a^*, b^* shearings. The imposed strain rate vector $P_1 Q$ of

length β is decomposed into a plastic component $P_1 R_a^*$, consisting of only a^* shear, and an elastic component $R_a^* Q$, parallel to the tangent plane of the cone in P_1.

The elastic component $R_a^* Q$ can be decomposed into a ψ-component $R_a^* R^*$ parallel to the circular cross section of the cone and an $\dot{s}$-component $R^* Q$ parallel to the line connecting P to the cone vertex V. The vector $R_a^* R^*$ points towards the left and that gives the stress point P a motion in a counterclockwise direction as viewed from above. This agrees with the correct impression of Professor Gibson.

The component $R^* Q$ points downward and that produces a downward movement, in the form of a helical line on the cone surface, that spirals downwards. At point P_1 the tangent to this line has the direction of the vector $R_a^* Q$. Two other of those helical lines are shown on the far side of the cone surface. Both belong to case iii initial stress states.

Case iv initial stress states produce similar stress paths, that are shown on this side of the cone surface. The helical lines on both sides of the cone are similar, but they are each other mirror image. When the centre angle $2\psi_1$ of the radius to P_1 has values between $(-\frac{1}{2}\pi + \tilde{\phi})$ and $(-\frac{1}{2}\pi - \tilde{\phi})$, then case ii initial stress states are involved, originally. For such a case ii the corresponding dotted stress paths, after reaching the cone rim, turn downward and follow a straight line path towards the cone vertex. The curved helical stress paths of the cases iii or iv initial stress states have the border lines between case ii and cases iii or iv as asymptotes.

Reprinted from *Géotechnique*, June 1978

TECHNICAL NOTES

Improvement of the lowerbound solution for the vertical cut off in a cohesive, frictionless soil

G. DE JOSSELIN DE JONG*

INTRODUCTION

One of the classical problems in soil mechanics is the determination of the depth h, to which a soil can be excavated by a vertical cut off before collapse occurs. When the soil is rigid plastic with cohesion c and no internal friction, $\phi = 0$, and its flow properties obey normality, the upper and lowerbound theorems of plasticity are applicable to h. Let a parameter α be defined by

$$h = \alpha c / \gamma$$

where γ is the proper weight of the soil, and let α_{coll} be the value of α corresponding to the collapse height h_{coll}. Then according to the upperbound theorem a value of α larger than α_{coll} is obtained by computing h with a kinematically admissible velocity field, and according to the lowerbound theorem a value of α smaller than α_{coll} is obtained by using a statically admissible stress field.

The solutions known from literature (Heyman, 1973) give $\alpha = 2 \cdot 83$ for the lowerbound and $\alpha = 3 \cdot 83$ for the upperbound. The purpose of this Technical Note is to present a statically admissible stress field corresponding to $\alpha = 3 \cdot 39$, which improves the existing lowerbound value.

A stress distribution is statically admissible if it obeys the following requirements:

(a) The stresses are everywhere in equilibrium with the soil weight. With tension stresses positive and γ in negative y-direction, the equilibrium equations are

$$(\partial \sigma_{xx} / \partial x) + (\partial \sigma_{yx} / \partial y) = 0$$
$$(\partial \sigma_{xy} / \partial x) + (\partial \sigma_{yy} / \partial y) = \gamma \qquad \qquad \text{. . . . . . . (1)}$$
$$\sigma_{xy} = \sigma_{yx}$$

(b) Stress discontinuities are allowed, provided that equilibrium is not violated. With local coordinates n, t normal and tangential to a discontinuity line separating regions (a) and (b) this is expressed by

$$\sigma_{nn}{}^{(a)} = \sigma_{nn}{}^{(b)}; \quad \sigma_{nt}{}^{(a)} = \sigma_{nt}{}^{(b)} \qquad \qquad \text{. . . . . . . (2)}$$

(c) The stress state in every point of the interior is within or on the limits imposed by the yield condition

$$(\sigma_{xx} - \sigma_{yy})^2 + (\sigma_{xy} + \sigma_{yx})^2 \leq 4c^2 \qquad \qquad \text{. . . . . . . (3)}$$

(d) The stresses comply with loads applied to the boundaries. In the case of the vertical cut off the stresses on all boundaries are zero.

LIMIT STATE OF STRESS EVERYWHERE

The idea was to use a computer for the generation of a statically admissible stress field in

Discussion on this Technical Note closes 1 September, 1978. For further details see inside back cover.
* Department of Civil Engineering, University of Technology, Delft, The Netherlands.

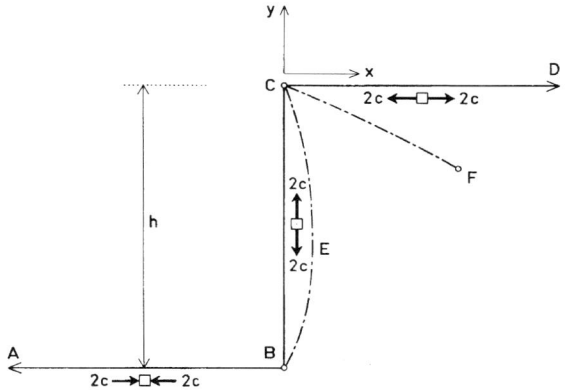

Fig. 1.   Discontinuity lines, when the entire region adjacent to the boundary is in the limit state of stress

order to avoid the usual procedure of predetermining intuitively stress distributions, which restrict the analysis to a limited class of solutions.

The stress field was generated by using Kötter's equations, which are based on equilibrium and the limit state of stress everywhere.   The equations are integrated numerically along stress characteristics s_1 and s_2 giving the mean normal stress and the principal stress directions in successive nodal points of the network of characteristics.   The shape of the network is obtained by computation of the locations of the nodal points assuming arcs of circles between them.   Being a solution of Kötter's equations the stress distribution satisfies requirements (a) and (c).   Where necessary stress discontinuity lines were introduced and the stresses on them were made to satisfy (b).

The construction of the stress characteristic network starts from the boundaries, where the stresses are known.   On the boundary the solution has to be chosen to be either strong or weak.   Since shear stresses on the boundaries are zero in this case, the principal stresses are parallel and perpendicular to the boundaries.   The principal stress perpendicular to the boundary is zero, but the principal stress parallel to the boundary can be either a tension or a compression.   When the stresses along the boundary are in the limit state, the tension or compression have both the same absolute value, $2c$.

The choice was made to start along AB with a compression of $2c$ for the principal stress parallel to AB and along CD with a tension of $2c$ (Fig. 1).   So below CD the stress state is given by

$$\sigma_{xx} = \gamma y + 2c; \quad \sigma_{xy} = \sigma_{yx} = 0; \quad \sigma_{yy} = \gamma y \qquad . \quad . \quad . \quad . \quad (4)$$

Satisfying (1) and (3) with the equality sign.

A first trial was made by assuming maximum tension $2c$ for the principal stress all along BC, also.   Constructing the field of characteristics leads in point C to overlaps, which resolve by introduction of two stress discontinuities CE and CF, making angles $\pi/8$ with CB and CD respectively in the corner point, C.   The discontinuity line CE has a bended form and cuts the vertical in a point B located at a depth $3,24\ c/\gamma$.   In this solution the entire region CEB is in the limit state of stress.

In order to satisfy the requirements (a) and (c) in every point of the interior, it is necessary to extend the stress field over the entire soil body.   It proved impossible, however, to con-

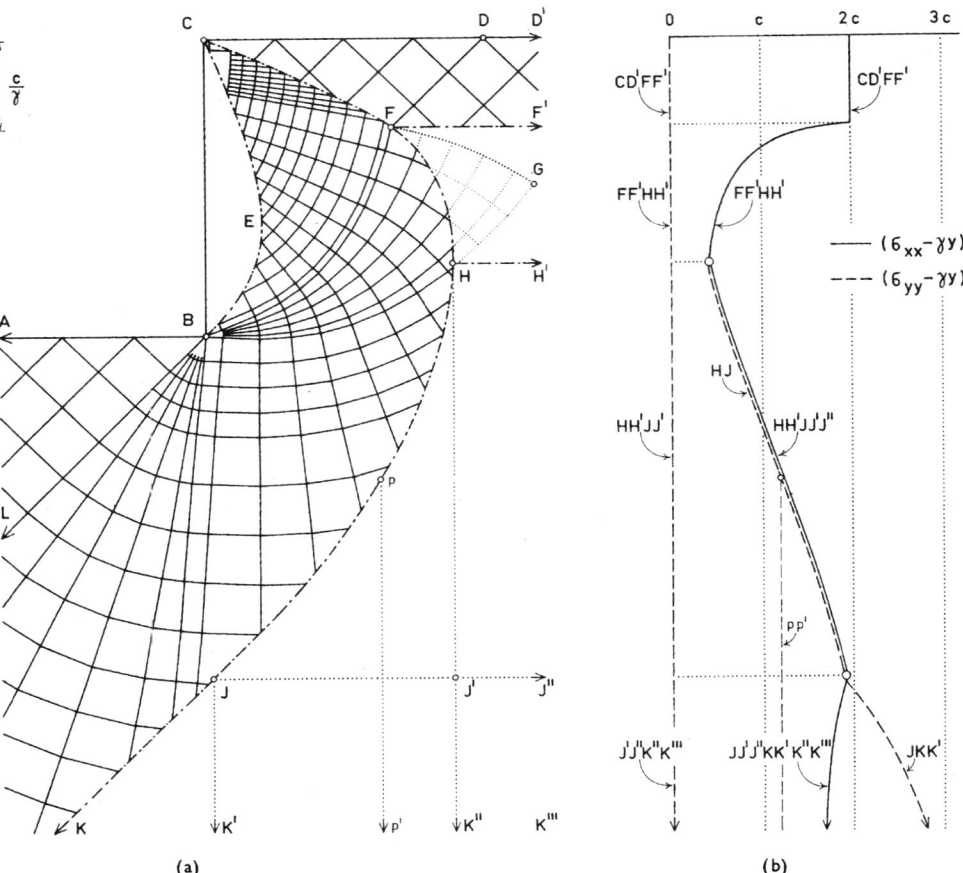

(a)   (b)

Fig. 2 (a). Field of stress characteristics, when the region adjacent to BC is not in the limit state of stress; (b) horizontal and vertical principal stresses up to infinity in the region to the right of FHJK

tinue the solution in the region around B and downwards, because the sheet of stress characteristics folded backwards. Therefore the solution mentioned above was not acceptable.

STRESSES IN CEB NOT IN LIMIT STATE

In order to render the situation in B better suited for adjustment to the passive Rankine State in the region below AB, it seemed appropriate to consider a non limit stress state in the region CEB expressed by

$$\sigma_{xx} = 0; \quad \sigma_{xy} = \sigma_{yx} = 0; \quad \sigma_{yy} = \gamma y + 2c \quad . \quad . \quad . \quad . \quad (5)$$

This stress state is statically admissible because it satisfies (1) everywhere, and for $(-4c/\gamma) <$ $y < 0$ the condition given by (3). Also (d) is satisfied because $\sigma_{xx} = \sigma_{xy} = 0$. Since for (3) the inequality sign holds for most of the region CEB, this region is not in the limit state of stress.

The construction of the field of stress characteristics starts again in C with the same two discontinuity lines at $\pi/8$ mentioned above. The discontinuity line CEB is constructed by extending the stress characteristics s_2 from the line CF up to the region CEB, where the stress state given by (5) exists. A next nodal point on CEB is found from a previous one by intersection with a subsequent s_2 stress characteristic and bending both in such a manner, that the stress state required by Kötter's equation along the s_2 line is in equilibrium across the discontinuity line with the known stress state (5) according to relation (2). The line CEB curves and crosses the vertical through C in a point B located a a depth $3.39\ c/\gamma$. The s_2 line through B intersects the other discontinuity line emanating from C in the point F (see Fig. 2a).

For this solution it proved possible to continue below B with a field in the limit state of stress by introducing two Prandtl wedges with their centres in B. The pattern of stress characteristics obtained by starting from the line AB, the two wedges in B and the line BF, can be extended downwards towards infinity. The pattern is bounded above by a s_1 line shown in Fig. 2(a) as the dotted line, FG.

STRESSES BELOW FF'

A conflict arises in extending the solution in the region above FG, because the pattern of stress characteristics will fold and solving Kötter's equation beyond such a fold requires complicated arrangements. The analysis was simplified by assuming a stress field with principal stresses vertical and horizontal, in the region below the horizontal line FF' and to the right of a discontinuity line FHJK, with K at infinity. Since shear stresses are zero on vertical and horizontal planes, integration of the equilibrium equations (1) then give

$$\sigma_{xx} = \gamma y + f(y); \quad \sigma_{xy} = \sigma_{yx} = 0; \quad \sigma_{yy} = \gamma y + g(x) \quad . \quad . \quad . \quad (6)$$

with $f(y)$ and $g(x)$ arbitrary, adjustable functions.

In order to have equilibrium on the line FF', between the stress distributions (4) and (6), the value of $g(x)$ has to be zero along FF'. As a consequence $g(x)$ is zero in the entire region to the right of FHJ'K'.

From F to H the value of $f(y)$ is adjusted in such a manner, that equilibrium exists through the discontinuity line between the stresses in the established field of characteristics BFG and the stresses (6). In every point of BFG there is one possible direction for such a discontinuity line, and FH follows as a trajectory by plotting successive tangents. The line FH is unique.

At point H the tangent is vertical and the discontinuity line is also principal stress trajectory. From H to J the discontinuity line is arbitrarily chosen to continue as principal stress trajectory. Then shear stress on HJ is zero and equilibrium across HJ requires that $f = g$ on HJ, because only then shear stress to the right of HJ is zero, the stress state being isotropic.

In order that (6) satisfies requirement (c) it is necessary, that

$$(f(y) - g(x)) \leqq 2c \quad . \quad . \quad . \quad . \quad . \quad . \quad . \quad . \quad (7)$$

in the entire region FHJKK'J''H'F'. In Fig. 2(b) a plot of $f(y)$ full line and $g(x)$ dashed line is given. A vertical line pp' in the region HKK' has a constant value for $g(x)$ and this gives a vertical dashed line in Fig. 2(b).

The point J is at such a height, that the line J'J'' is critical having $(f - g) = 2c$. Below J the discontinuity line JK is taken at $\pi/4$ with the horizontal giving values of $f(y)$ and $g(x)$ as shown in Fig. 2(b). From this figure it follows that (7) is satisfied everywhere.

CONCLUSION

Constructing a stress field by characteristics with stresses in the limit state everywhere did not lead to an acceptable solution. It was necessary to assume in the region CEB stresses that were not in the limit state. The stress field obtained then, satisfies the requirements (a) to (d) in the entire soil body and therefore it is statically admissible stress field. The value $\alpha = 3.39$ is therefore a lowerbound to α_{coll}.

That the solution gives a safe height follows from the consideration of the kinematics. A potential slipline would be the s_2 line BFD. Slip cannot occur along this line, though, because of the slope discontinuity in F.

The solution was verified in 1976 by C. H. Engels with a different computer program, based on the relation for stress discontinuities developed in her graduation study.

REFERENCE

Heyman, J. (1973). Simple Plastic Theory applied to Soil Mechanics. *Proc. Symp. on the role of plasticity in soil mechanics*, 161–172. Cambridge.

De Josselin De Jong, G. (1980). *Géotechnique* **30**, No. 1, 1–16

Application of the calculus of variations to the vertical cut off in cohesive frictionless soil

G. DE JOSSELIN DE JONG

The collapse height of a vertical cut off is computed by use of the variational calculus assuming the existence of a real slip line at collapse. A class of lines containing the real slip line is defined by total and local equilibrium conditions of the limit stress state. The extremal of the class is found to be an involute. Verification of the solution shows that the extremal gives either an unsafe estimate of the collapse height or corresponds to no extremum at all. These disappointing results are a consequence of the inadequate formulation of slope stability problems, when slip lines are computed by the calculus of variations.

La hauteur, correspondant à la rupture d'un talus vertical, est déterminée à l'aide du calcul de variations en présupposant l'existence d'une ligne de glissement unique en cas de rupture. La classe de lignes, contenant cette ligne de glissement, est definie par l'équilibre total et local sous condition d'état de contraintes limites. L'extrémale de la class possède la forme d'une involute. En vérifiant la solution, il est démontré que l'extrémale produit une hauteur de talus plus élevée que la hauteur de rupture, où une hauteur qui ne correspond pas du tout à un extremum. Ces résultats décevants sont engendrés par la formulation inepte des problèmes de stabilité, quand des lignes de glissement sont déterminées à l'aide du calcul de variations.

INTRODUCTION

The calculus of variations provides mathematical procedures to find the shape of an extremal, the curve that maximizes or minimizes the value of an integral along that line. For the reader who is not familiar with the subject the books by Bolza (1973) and Petrov (1968) are recommended here. The first because it treats all aspects of the parametric solution employed in this article, the second because it gives a comprehensive and convenient description of the subject matter and its practical use.

Because it can be used for establishing the shape of a line with particular properties, the calculus of variations seems of interest for soil mechanics, especially for solving slope stability problems. Since Coulomb, it has become a standard procedure to solve plane strain stability problems by using minimalization procedures in search for the line that represents a failure plane. The variational calculus could be applied to determine the shape of such a line, if the mechanical requirements of failure along the line can be formulated in the form of integrals, whose extreme values are related to the stability. This idea has occurred to several investigators, e.g. Revilla and Castillo (1977), Ramamurthy *et al.* (1977), Baker and Garber (1977, 1979). The approaches proposed by these different writers differ and demonstrate that there is not one unique manner to formulate the minimalization problem.

Also the Author was intrigued a few years ago by the possibility of using the variational calculus for establishing slip lines, but withheld the results from publication when it became clear that the solutions were of a disappointing, unacceptable character. The reason for presenting this work now however is that an analysis as published by Baker and Garber (1978) is incomplete and produces results that prove to be not meaningful when the analysis is properly pursued. It is the objective of this Paper to demonstrate an analysis that is concluded by investigating the

---

Discussion on this paper closes 1 June, 1980. For further details see inside back cover.
* Department of Civil Engineering, University of Technology, Delft.

conditions for an extremum and to show the kind of disappointments that are encountered when the calculus of variations is applied to determine slip lines.

In order to be specific and to deal with explicit results, the relatively simple case is treated here of a vertical cut off in cohesive, frictionless, non-dilatant soil. The collapse behaviour of the vertical cut off in such a soil has been studied by reliable procedures based on the rigorous proofs of the theory of plasticity. The exact solution is not yet available, but it is known that the collapse height h_{coll} is unique and expressed in terms of the cohesion c and the specific weight γ, is between the following limits

$$3{\cdot}64c/\gamma \leqslant h_{coll} < 3{\cdot}83c/\gamma$$

The upper limit corresponds to a Fellenius solution with a circular slip line, satisfying a kinematically admissible velocity field. The lower limit corresponds to a statically admissible stress distribution determined numerically by Pastor (1978). Because of this information on the collapse height, the vertical cut off is an appropriate example for testing the determination of slip lines by variational methods.

CONCEPTS BASIC TO THE ANALYSIS

In its simplest form the calculus of variations is used to determine the shape of one particular line, which is called extremal, because it can produce an extreme value of a definite integral. The analysis described in this Paper is an application of this simplest form of variational calculus to the slope stability problem of establishing a safe estimate for the height h of a vertical cut off.

The one line analysis is based on a few concepts that are decisive for the formulation of the problem and are determinative for the resulting solution. The first concept is the assumption that at collapse, there exists one particular slip line, which is called here the real slip line. The second concept is a class of potential slip lines, defined in such a manner that it includes the real slip line. The third concept is the presumption that a safe estimate of h can be found by determining the extremal.

Real slip line

In the plane strain case of an embankment in z direction, the slip line in a vertical x, y plane represents a failure plane perpendicular to x, y. Failure planes exist at collapse. They are required to separate soil masses that slide with respect to each other and with respect to the stationary soil mass below. When sliding occurs over the failure planes, the stresses along it are in the limit state of stress. Since the soil is taken to be frictionless and non-dilatant, the lines representing the failure planes in the two-dimensional x, y-plane are stress characteristics. There are two families of conjugate stress characteristics. For slip to develop, it is enough that sliding takes place along one of the two conjugate lines. This line should intersect the boundaries in at least two points in order to separate a soil mass that is free to move. For sliding it is necessary that the slip line has a continuous slope.

The assumption is that at collapse there is at least one continuous smooth slip line which corresponds to a stress characteristic and which intersects the boundaries in two points. This line is called here the real slip line. The shear stress along the real slip line has the maximum available value which is equal to the cohesion c for a frictionless soil.

Class of potential slip lines

For the analysis a class of lines is required that contains the real slip line. This class is obtained by selecting lines in such a manner that their stress distribution satisfies the same equilibrium conditions as the real slip line. At incipient collapse, the separated soil mass that is about to slide is still in equilibrium. Its total weight is balanced by the stresses along the real slip line. This is called the requirement of total equilibrium in the following sections. This requirement can be formulated in the form of definite integrals along the line. Furthermore, the real slip line is a stress characteristic. Therefore the stresses along it satisfy Kötter's equation, which represents local equilibrium for the limit stress state, in a direction tangential to the stress characteristic. The requirement of local equilibrium is satisfied by introducing Kötter's equation in the integral expressions.

The requirements of total and local equilibrium produce three definite integrals as shown in the section headed ' Determination of the integrals'. The value of one of these integrals corresponds to the height of the cut off. The other two are zero. All lines that give these values to the integrals are called potential slip lines, because each of them might be the real slip line.

Extremal

The calculus of variations provides methods to find the extremal. That is a particular line of the class of potential slip lines, computed in such a manner, that it can give an extreme value to h, the height of the cut off. In the section headed ' Determination of the extremal ', the solution for this line is produced by standard variational procedures. The extremal of the potential slip lines is not necessarily the real slip line. It can be any line satisfying the equilibrium conditions imposed.

The value of h, computed for the extremal is called h_{em1} here. This value is an extremum for h, if it is a maximum or a minimum. If it is a minimum, this indicates that h_{em1} is smaller than the values of h corresponding to all other lines of the class investigated. If this class is large enough to contain the real slip line (the line that is assumed to correspond to h_{coll}), it may be concluded that h_{em1} is smaller than or equal to h_{coll}. The analysis then procures a safe estimate for the building height of the embankment. The analysis presented here was initiated originally in the supposition that a minimum would result.

Verification

The basic assumptions mentioned in this section may occur to be plausible for soil mechanical investigators because they are commonly postulated in slip line analysis. There are two improvements with respect to common procedures: the shape of the line is a result of the analysis and the stress distribution along the slip line satisfies limit stress state conditions by applying Kötter's equation.

The assumptions are, however, not self-evident and are even debatable. This is revealed by a verification of the solution using procedures that are standard in the calculus of variations. The character of the extremum is established in the section headed ' Investigation of the solution ' by use of these verification methods, showing that either a weak maximum or no extremum at all is involved. This is disappointing because the conclusion can only be that the analysis was not meaningful.

Unfortunately, it is not possible to establish, *a priori*, whether the basic assumptions underlying the analysis will lead to a meaningful end result. It is only after the solution has been obtained that its character can be verified. Therefore, first the analysis, based on the three plausible assumptions mentioned in this section, is carried out in the following text. In the final section the result is discussed and the basic assumptions are reconsidered.

DETERMINATION OF THE INTEGRALS

The vertical cut off is shown in Fig. 1. The soil is limited by the following stress free boundaries:

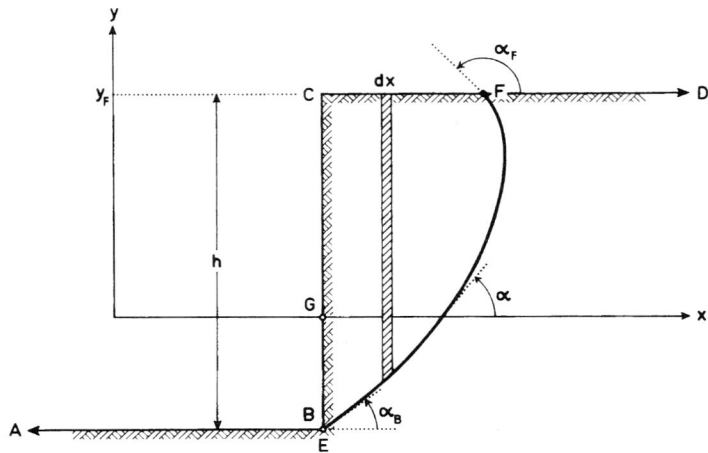

Fig. 1. Vertical cut off with potential slip line

the vertical plane BC and the horizontal planes AB and CD. The problem is to determine an acceptable height h of the cut off. The soil properties involved are the specific weight γ, the cohesion c and a vanishing angle of internal friction $\phi = 0$. No external loads are acting on the boundary ABCD. Failure is due only to the weight of the sliding soil mass.

Class of potential slip lines

If the class of lines considered is to consist of potential slip lines, the lines have to intersect the free surface in two points, E and F. The lower point E will coincide with the lower corner point B, and F is on the upper surface CD. So the lines in the analysis are lines BF.

The shape of the line BF will be treated in parametric form, such that the horizontal x coordinate and the vertical y coordinate are both functions of a parameter α that increases from α_B in B to α_F in F. This can be written as

$$x = x(\alpha), \quad y = y(\alpha) \quad \text{for } \alpha_B \leqslant \alpha \leqslant \alpha_F \quad . \quad . \quad . \quad . \quad . \quad (1)$$

In the analysis the parameter α will be taken to be the local angle between the line and the x direction (see Fig. 1). Differentiation with respect to α is indicated by a prime such that

$$x' = dx/d\alpha, \quad y' = dy/d\alpha \quad . \quad . \quad . \quad . \quad . \quad . \quad (2a)$$

$$x'' = d^2x/d\alpha^2, \quad y'' = d^2y/d\alpha^2 \quad . \quad . \quad . \quad . \quad . \quad (2b)$$

A small element of BF with length ds subtends in horizontal and vertical directions distances dx, dy (see Fig. 2) given by

$$dx = \cos \alpha \, ds, \quad dy = \sin \alpha \, ds \quad . \quad . \quad . \quad . \quad . \quad (3)$$

In order to be a potential slip line, it is necessary that the shear stress acting on the line BF has the maximum available value, which is equal to the cohesion c. The normal stress on BF will be denoted as p and is taken positive for compression. These quantities will be used to evaluate the integrals that are developed in the following.

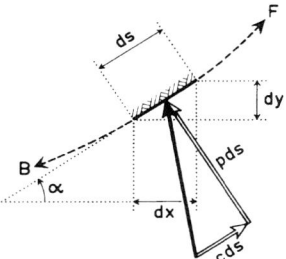

Fig. 2. Force $d\mathbf{R}$ (black arrow) acting on element ds is composed of a shear component $c\,ds$ and a normal component $p\,ds$ (white arrows)

Total equilibrium

In the state of incipient failure, the soil and all its subdivisions satisfy equilibrium. So the forces acting on the part BCF, separated from the main soil body by the line BF in Fig. 1, form an equilibrium system. This means that the force $\mathbf{Q}$ created by the weight of the part BCF is in equilibrium with the resistive force $\mathbf{R}$ due to the stresses acting along BF. This equilibrium will be called total equilibrium because it refers to the equilibrium of the total mass BCF, in contrast to local equilibrium which is considered later and refers to the equilibrium at all points of the line BF.

Total equilibrium is satisfied when the forces $\mathbf{Q}$ and $\mathbf{R}$ annihilate each other. Expressed in their x, y components: X_Q, Y_Q; X_R, Y_R and their moments M_Q, M_R around the origin of co-ordinates, this requires that

$$X_Q + X_R = 0, \quad Y_Q + Y_R = 0, \quad M_Q + M_R = 0 \quad . \quad . \quad . \quad . \quad . \quad (4)$$

Since the soil weight acts only in vertical direction downwards, the horizontal component X_Q of $\mathbf{Q}$ is zero and the vertical component can be obtained by integrating over slices of width dx as shown in Fig. 1, this gives

$$X_Q = 0, \quad Y_Q = -\int_B^F \gamma(y_F - y)\,dx \quad . \quad . \quad . \quad . \quad . \quad . \quad (5)$$

The moment of $\mathbf{Q}$ around the origin, positive for counter-clockwise rotation, is

$$M_Q = -\int_B^F \gamma(y_F - y)x\,dx \quad . \quad . \quad . \quad . \quad . \quad . \quad (6)$$

In these expressions x, y are the coordinates of the line BF and integration is along BF.

The components of the resistive force $\mathbf{R}$ can also be expressed in the form of integrals along BF. Let dX_R, dY_R be the x, y components of the force $d\mathbf{R}$ (black arrow in Fig. 2) produced by the stresses on the strip of unit width corresponding to the line element ds. This elementary force $d\mathbf{R}$ consists of a component $c\,ds$ tangential to ds and a normal component $p\,ds$ (white arrows in Fig. 2). Decomposed into x, y directions these give according to Fig. 2 and using equation (3)

$$dX_R = c \cos \alpha\,ds - p \sin \alpha\,ds = c\,dx - p\,dy \quad . \quad . \quad . \quad . \quad (7a)$$

$$dY_R = c \sin \alpha\,ds + p \cos \alpha\,ds = c\,dy + p\,dx \quad . \quad . \quad . \quad . \quad (7b)$$

From these the components X_R, Y_R of $\mathbf{R}$ are obtained by integration along BF giving

$$X_R = \int_B^F dX_R = \int_B^F (c\,dx - p\,dy) \quad . \quad . \quad . \quad . \quad . \quad (8a)$$

$$Y_R = \int_B^F dY_R = \int_B^F (c\, dy + p\, dx) \quad . \quad . \quad . \quad . \quad . \quad . \quad (8b)$$

The moment M_R of **R** around the origin is

$$M_R = \int_B^F (x\, dY_R - y\, dX_R) = \int_B^F [c(x\, dy - y\, dx) + p(x\, dx + y\, dy)] \quad . \quad . \quad . \quad (9)$$

The three equations (4), that represent total equilibrium, can now be expressed in the form of integrals. The first two are transformed into the relations (10a) and (10b) by use of the equations (5) and (8). The last becomes relation (10c) by using equations (6) and (9). This gives

$$\int_B^F (c\, dx - p\, dy) = 0 \quad . \quad . \quad . \quad (10a)$$

$$\int_B^F [c\, dy + p\, dx - \gamma(y_F - y)\, dx] = 0 \quad . \quad . \quad . \quad (10b)$$

$$\int_B^F [c(x\, dy - y\, dx) + p(x\, dx + y\, dy) - \gamma(y_F - y)x\, dx] = 0 \quad . \quad . \quad . \quad (10c)$$

These relations represent total equilibrium and agree with the equation (5) in the paper by Baker and Garber (1979). The analysis continues in the following sections by introducing Kötter's equation along the line BF. This is an essential difference between that paper and the analysis followed here.

Local equilibrium

In a frictionless material (for which $\phi = 0$) a potential slip line coincides with a stress characteristic. Local equilibrium along stress characteristics is expressed by two Kötter's equations, one for each of the two conjugate stress characteristics. The equation corresponding to the line BF is

$$(dp/ds) - 2c(d\alpha/ds) + \gamma \sin \alpha = 0 \quad . \quad . \quad . \quad . \quad . \quad . \quad . \quad (11)$$

This is a differential equation that can be integrated along s. Adjusting the integration constant to fit the situation in point F gives after integration

$$(p - p_F) - 2c(\alpha - \alpha_F) + \gamma(y - y_F) = 0 \quad . \quad . \quad . \quad . \quad . \quad . \quad (12)$$

in which p_F, α_F and y_F stand for values of the quantities p, α and y respectively in the point F. Let β_F be defined by

$$\beta_F = (p_F - 2c\alpha_F)/2c \quad . \quad . \quad . \quad . \quad . \quad . \quad . \quad . \quad (13)$$

then p can be solved from equation (12) to give

$$p = 2c(\beta_F + \alpha) + \gamma(y_F - y) \quad . \quad . \quad . \quad . \quad . \quad . \quad . \quad (14)$$

It may be remarked here that it is impossible to solve for p in a similar manner from Kötter's equation, when the soil has internal friction, such that ϕ is unequal to zero. Therefore, the analysis developed below cannot be applied directly for soils with internal friction.

Equation (14) is used to eliminate p from the integral expressions (10). This gives then

$$\int_B^F [c\, dx - 2c(\beta_F + \alpha)\, dy - \gamma(y_F - y)\, dy] = 0 \quad . \quad . \quad . \quad (15a)$$

$$\int_B^F [c\, dy + 2c(\beta_F + \alpha)\, dx] = 0 \quad . \quad . \quad . \quad (15b)$$

$$\int_B^F [c(x\,dy - y\,dx) + 2c(\beta_F + \alpha)\,(x\,dx + y\,dy) + \gamma(y_F - y)y\,dy] = 0 \quad . \quad . \quad . \quad (15c)$$

Since $(y_F - y_B)$ is equal to h, the height of the vertical cut off, the parts containing γ become

$$\int_B^F \gamma(y_F - y)\,dy = \tfrac{1}{2}\gamma h^2 \qquad . \quad . \quad . \quad . \quad . \quad . \quad . \quad (16a)$$

$$\int_B^F \gamma(y_F - y)y\,dy = \tfrac{1}{6}\gamma h^2(y_F + 2y_B) \qquad . \quad . \quad . \quad . \quad . \quad (16b)$$

By taking the origin of coordinates in point G on one-third of the height of the cut off (Fig. 1) such that

$$y_F = \tfrac{2}{3}h, \quad y_B = -\tfrac{1}{3}h, \quad x_B = 0 \quad . \quad . \quad . \quad . \quad . \quad (17)$$

the integral (16b) vanishes and the integrals (15) reduce to

$$\Gamma_0 = \int_B^F [dx - 2(\beta_F + \alpha)\,dy] = \tfrac{1}{2}\gamma h^2/c \qquad . \quad . \quad . \quad . \quad . \quad (18a)$$

$$\Gamma_1 = \int_B^F [dy + 2(\beta_F + \alpha)\,dx] = 0 \qquad . \quad . \quad . \quad . \quad . \quad (18b)$$

$$\Gamma_2 = \int_B^F [(x\,dy - y\,dx) + 2(\beta_F + \alpha)\,(x\,dx + y\,dy)] = 0 \quad . \quad . \quad . \quad (18c)$$

At this stage it is convenient to introduce the parametric manner of describing these integrals. From the relations (2a) it follows that dx, dy can be written as functions of the parameter α in the following manner

$$dx = x'\,d\alpha, \quad dy = y'\,d\alpha \quad . \quad . \quad . \quad . \quad . \quad . \quad (19)$$

Further, relations (3) indicate that $\tan \alpha = (dy/dx)$ and using relations (19) this can be written as $\tan \alpha = (y'/x')$ or

$$\alpha = \text{arc tan}\,(y'/x') \quad . \quad . \quad . \quad . \quad . \quad . \quad . \quad (20)$$

So the integral expressions (18) can be written as

$$\Gamma_0 = \int_B^F G_0(\alpha)\,d\alpha = \int_B^F [x' - 2(\beta_F + \text{arc tan}\,y'/x')y']\,d\alpha = \tfrac{1}{2}\gamma h^2/c \quad . \quad (21a)$$

$$\Gamma_1 = \int_B^F G_1(\alpha)\,d\alpha = \int_B^F [y' + 2(\beta_F + \text{arc tan}\,y'/x')x']\,d\alpha = 0 \quad . \quad . \quad (21b)$$

$$\Gamma_2 = \int_B^F G_2(\alpha)\,d\alpha = \int_B^F [(xy' - x'y) + 2(\beta_F + \text{arc tan}\,y'/x')\,(xx' + yy')]\,d\alpha = 0 \quad . \quad . \quad (21c)$$

The functions $G_0(\alpha)$, $G_1(\alpha)$, $G_2(\alpha)$ introduced on the left-hand sides of equations (21) are functions of x, y, x', y', defined by the integrals between brackets. Because of equations (1) and (2) they can be considered to be functions of α only.

These three integrals represent the three equilibrium conditions for balancing the total driving force $\mathbf{Q}$ and the total resistive force $\mathbf{R}$. In addition, local equilibrium is guaranteed in all points of the potential slip line BF in a direction parallel to the line, the line being a stress characteristic.

DETERMINATION OF THE EXTREMAL

The integral expressions (21) are suitable for application of the variational calculus in order to find an extreme value of h, the height of the vertical cut off for given values of c and γ. The analysis consists of determining the functions $x(\alpha)$ and $y(\alpha)$ that represent an extremal in parameter form. Such an extremal is the curve that produces an extreme value for Γ_0, the integral defined by (21a), under the additional conditions that the integrals Γ_1 and Γ_2 defined by (21b) and (21c) vanish. In the calculus of variations this is a so-called isoperimetric problem that can be solved by the use of Lagrange multipliers. The following analysis is based on the Weierstrass theory for parameter problems as described by Bolza (1904).

General expression for the extremal

According to the theory, the curve $x(\alpha)$, $y(\alpha)$ is found by solving Euler's differential equation, which in Weierstrass's form is

$$H_{xy'} - H_{yx'} + H_{x'x'}(x'y'' - x''y')/(y')^2 = 0 \ . \quad . \quad . \quad . \quad . \quad . \quad . \quad (22)$$

In this equation subscripts stand for partial differentiation with respect to the subscript variable, $H_x = \partial H/\partial x$, etc. The quantity H is an auxiliary function given by

$$H = G_0 + g_1 G_1 + g_2 G_2$$

where G_0, G_1, G_2 are the integrands of the equations (21) and g_1, g_2 are Lagrange multipliers that are constants in this case. Written out, the function H has the form

$$H = [(g_1 + g_2 x)y' + (1 - g_2 y)x'] + 2(\beta_F + \text{arc} \tan y'/x') [(g_1 + g_2 x)x' - (1 - g_2 y)y'] \quad (23)$$

Performing the partial differentiations on function H required in equation (22), this equation becomes

$$(x'y'' - x''y') [(g_1 + g_2 x)y' + (1 - g_2 y)x'] = g_2[(x')^2 + (y')^2)]^2 \ . \quad . \quad . \quad . \quad (24)$$

The solution of this differential equation satisfying the parametric requirement $y'/x' = \tan \alpha$ (according to (20)) represents the extremal. It is not so simple to solve equation (24) in a straightforward manner. The solution, however, is as follows

$$x = -(g_1/g_2) - a \cos \alpha + (b - \alpha a) \sin \alpha \quad . \quad . \quad . \quad . \quad (25a)$$

$$y = +(1/g_2) - a \sin \alpha - (b - \alpha a) \cos \alpha \quad . \quad . \quad . \quad . \quad (25b)$$

This can be verified by substitution, using equation (2) which gives for, instance,

$$x' = dx/d\alpha = (b - \alpha a) \cos \alpha, \quad y' = dy/d\alpha = (b - \alpha a) \sin \alpha \quad . \quad . \quad (26a)$$

$$x'y'' - x''y' = (b - \alpha a)^2, \quad (x')^2 + (y')^2 = (b - \alpha a)^2 \quad . \quad . \quad . \quad (26b)$$

Since equation (24) is a second-order differential equation, solution (25) possesses two integration constants. These are the quantities a and b that have to be determined from boundary conditions.

Shape of the extremal

Equation (25) represents a curve that is called an involute, with a circle as evolute. In Fig. 3 the line $P_0BP_\alpha QFP_1 \ldots$ is such a curve. It is the path followed by the end point P of a string of length $N_0P_0 = b$, attached in N_0 and wound around the circle with radius $N_0M = a$. For an arbitrary value of α the part N_0N_α of the string with length αa is contiguous with the circle and the remaining part of length $(b - \alpha a)$ is the straight line $N_\alpha P_\alpha$. The centre of the circle is the point M, with coordinates

$$x_M = -(g_1/g_2), \quad y_M = +(1/g_2) \ . \quad . \quad . \quad . \quad . \quad . \quad (27)$$

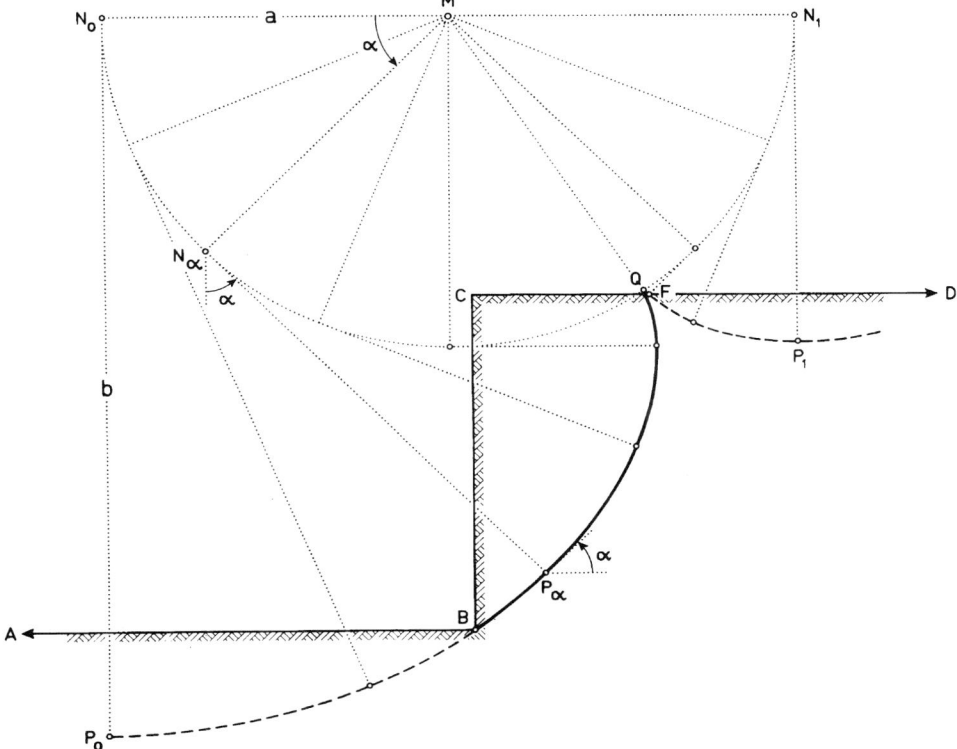

Fig. 3. Extremal BP_aQF, satisfying boundary conditions and constraints. The line $P_0P_aQP_1$ is an involute with circle $N_0N_aQN_1$ as evolute

The involute reaches the circle in the point Q, when the value of α is equal to

$$\alpha_Q = b/a \quad . \quad . \quad . \quad . \quad . \quad . \quad . \quad . \quad . \quad . \quad (28)$$

Beyond that point, the line continues towards $FP_1 \dots$ etc. as the path of the end point of a string that unwinds from the circle on the other side. At point Q the curve has a cusp.

Determination of the constants of the extremal

Of all curves that obey the relations (25), the relevant extremal is obtained by computing the values of the integration constants a, b and the Lagrange multipliers g_1, g_2. In addition, the two end points B, F are to be localized on the curve by establishing the value of the parameter α in those points. These are α_B and α_F. In total there are six constants a, b, g_1, g_2, α_B, α_F to be determined and this requires six relations between these six variables.

The first two relations are the constraints which are expressed by the requirement that both Γ_1, Γ_2 given by the integrals (21b and c) are zero. Since x, y, x', y' are all known goniometric functions of the parameter α by relations (25) and (26), it is possible to integrate the integrals (21b and c) with respect to α from α_B to α_F. The resulting formulae are fairly long and therefore, not written out in full length here. They are known functions of a, b, g_1, g_2, α_B, α_F, though which are called Γ_1^* and Γ_2^* for brevity. So the first two relations are

$$\Gamma_1^* \, (a, b, g_1, g_2, \alpha_B, \alpha_F) = 0 \qquad \qquad \text{(29a)}$$

$$\Gamma_2^* \, (a, b, g_1, g_2, \alpha_B, \alpha_F) = 0 \qquad \qquad \text{(29b)}$$

The following two relations are obtained by requiring that the extremal passes through the point B. The coordinates of B have the values (17) and combined with equations (25) there results

$$x_B = -(g_1/g_2) - a \cos \alpha_B + (b - \alpha_B a) \sin \alpha_B = 0 \qquad \qquad \text{(30a)}$$

$$y_B = +(1/g_2) - a \sin \alpha_B - (b - \alpha_B a) \cos \alpha_B = -\tfrac{1}{3}h \qquad \qquad \text{(30b)}$$

The last two relations are obtained from the boundary conditions at the point F. This point is located on the free upper surface and therefore the magnitude of y_F is known to be $\tfrac{2}{3}h$ (see equation (17)). Introduced into equation (25b) this gives

$$y_F = +(1/g_2) - a \sin \alpha_F - (b - \alpha_F a) \cos \alpha_F = +\tfrac{2}{3}h \qquad \qquad \text{(31)}$$

The location of the point F on the upper surface is not known, however, so x_F is an unknown. The second relation to be obtained from the boundary condition in F is due to the stress state in F, which produces a value for the inclination of the potential slip line, by adapting limit stress state to the requirement, that the horizontal upper surface is free of stresses.

There are two possible cases represented by the Figs 4(a) and 4(b), which correspond, respectively, to maximum compression and maximum tension on vertical planes. In the compression case (Fig. 4(a)) the value of p, the normal compressive stress on the potential slip line, is equal to $p_F = +c$ and the direction of the potential slip line is

$$\alpha_F = 3\pi/4 \qquad \qquad \text{(32)}$$

In the case of maximum tension (Fig. 4(b)), assuming a soil which can support a tension of $2c$, the normal compressive stress on the slip line is $p_F = -c$ and the corresponding direction of the potential slip line is $x_F = \pi/4$. It can be shown that a solution for this second value of a_F, satisfying all boundary conditions and the constraints (equation (29)), does not exist.

For the compression case in F we have $p_F = +c$ and $\alpha_F = (3\pi/4)$, which introduced in β_F defined by (13) gives

$$\beta_F = \tfrac{1}{2}(1 - 3\pi/2) \qquad \qquad \text{(33)}$$

This value is needed in the expressions (29) for Γ_1^* and Γ_2^*. Finally, there are six equations for the determination of the constants a, b, g_1, g_2, α_B, α_F. These are the relations (29a and b), (30a and b), (31 and 32). Since these relations are implicit goniometric relations, it is not possible to give explicit equations for the constants. A solution satisfying all six relations was obtained numerically. The pertinent values of the constants are

$$\begin{aligned} a &= 0{\cdot}988h, \quad g_1 = 0{\cdot}0424, \quad \alpha_B = 0{\cdot}604 \\ b &= 2{\cdot}140h, \quad g_2 = 0{\cdot}6674/h, \quad \alpha_F = 2{\cdot}356 \end{aligned} \qquad \text{(34)}$$

Equations (25) for the coordinates, with the values of the constants mentioned above, form the solution for the extremal satisfying the Weierstrass–Euler condition (22), the constraints (21b) and (21c) and the boundary conditions. This solution is unique.

The value for h_{eml}, corresponding to this solution of the extremal, is obtained by introducing

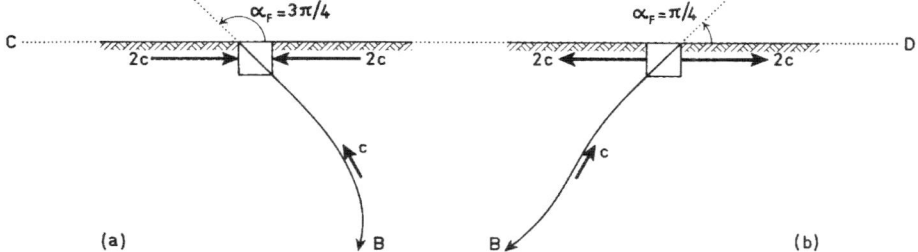

Fig. 4. Boundary conditions for stress characteristic at the point F on the upper surface CD: (a) maximum compression; (b) maximum tension

expressions (25) and (26) into the integral (21a) and integrating. The result is too long to reproduce here. Insertion of the values (34) finally gives

$$h_{em1} = 3 \cdot 783 c/\gamma \quad . \quad . \quad . \quad . \quad . \quad . \quad . \quad . \quad (35)$$

Description of the solution for the extremal

The shape of the extremal computed is shown in Fig. 3 as the line $BP_\alpha QF$. This line contains the point Q, which was mentioned above to be a cusp. That Q lies indeed within the interval BF follows from a verification of the magnitude of α_Q. Using the relation (28) and the values (34) it is found that

$$\alpha_Q = h/a = 2 \cdot 166 \quad . \quad . \quad . \quad . \quad . \quad . \quad . \quad (36)$$

and this is a value between α_B and α_F given in (34). Further it can be computed that Q lies outside the soil boundaries. Using the value (36) in the expression (25b) for the y coordinate gives, with the values (34) for the constants, $y_Q = 0 \cdot 6801 h$. This shows that Q lies a distance $0 \cdot 0134 h$ above the free surface which is located at $y = 2h/3$.

The fact that the extremal BF contains the point Q, where the curve has a cusp and which lies outside the soil mass, indicates that the solution does not represent the real slip line. For a real slip line it is physically illegitimate to form a cusp and to exceed the soil body. Therefore the solution (35) is not the collapse height. It is either too high or too low, that depends on the character of the extremum, whether it is a maximum or a minimum.

INVESTIGATION OF THE SOLUTION

The solution of the Weierstrass–Euler equation (22) is a unique curve given in parameter form by the expressions (25) with the values (34) for the relevant constants. This curve is called an extremal because it satisfies condition (22) and that is necessary for the curve to produce an extremum for h. However, satisfying condition (22) is not yet sufficient. Before it can be concluded that the value (35) computed for h_{em1} with the extremal, indeed represents an acceptable extreme value, there are three additional investigations to be made. These verifications are required for establishing the character of the solution and completing the analysis.

In the first place the Legendre condition has to be verified because that condition indicates whether the solution represents a maximum or a minimum of the height h. In the second place the Jacobi condition has to be verified because that condition indicates whether there exists an extremum at all. In the last place the Weierstrass E function has to be investigated because that function shows whether the extremum (if it exists at all) is strong or weak.

Legendre condition

The first point to investigate is the Legendre condition. This condition is satisfied if the sign of a quantity H_1, defined below, is the same for every point of the extremal, $x(\alpha)$, $y(\alpha)$, between B and F. For a maximum to be involved it is necessary that H_1 is negative, for a minimum that H_1 is positive. The definition of the function H_1 can be written in three different ways which because of homogeneity are equivalent, i.e.

$$H_1 = H_{x'x'}/(y')^2 = -H_{x'y'}/x'y' = H_{y'y'}/(x')^2 \quad . \quad . \quad . \quad . \quad ., \quad (37)$$

where H is the function described by (23). Elaboration gives

$$H_1 = -4[(g_1+g_2x)y'+(1-g_2y)x']/[(x')^2+(y')^2)]^2$$

and using relations (24) and (26b) this can be written as

$$H_1 = -4g_2/(x'y''-x''y') = -4g_2/(b-\alpha a)^2$$

According to the solution (34), the value of g_2 is positive and this indicates that H_1 is negative for every point of the extremal. So the result is

$$H_1 < 0, \quad \text{for every } \alpha \quad . \quad . \quad . \quad . \quad . \quad . \quad . \quad . \quad . \quad (38)$$

Therefore, the Legendre condition is satisfied and the extremum involved is presumably a maximum. However, compliance with the Legendre condition is not yet sufficient for the solution to be a maximum. It is also necessary that the Jacobi condition is satisfied.

Jacobi condition

The second point of investigation is to verify whether the Jacobi condition is satisfied. According to the theory no extremum, maximum or minimum, exists when a function u, defined below, vanishes in a point of the extremal between B and F. The function u is defined by

$$u = C_1u_1 + C_2u_2 \quad . \quad . \quad . \quad . \quad . \quad . \quad . \quad (39a)$$

with

$$u_1 = y'x_a - x'y_a \quad . \quad . \quad . \quad . \quad . \quad . \quad . \quad (39b)$$

$$u_2 = y'x_b - x'y_b \quad . \quad . \quad . \quad . \quad . \quad . \quad . \quad (39c)$$

where x_a, etc. are partial derivatives of the functions (25) with respect to the integration constants a and b. The constants C_1 and C_2 are to be chosen in such a manner that for the point B, common to all extremals, the values of u and u' are as follows:

$$u_B = 0, \quad u_B' = 1 \quad . \quad . \quad . \quad . \quad . \quad . \quad . \quad (40)$$

Elaboration of u_1 and u_2 with equation (25) gives

$$u = (-\alpha C_1 + C_2)(b-\alpha a)$$

in which the constants C_1, C_2 can be solved by use of relations (40). It is then found that

$$u = (\alpha - \alpha_B)(b - \alpha a)/(b-\alpha_B a) \quad . \quad . \quad . \quad . \quad . \quad . \quad (41)$$

The function u expressed by equation (41) vanishes in the cusp point Q, because there α equals $\alpha_Q = b/a$) (see equation (28)). If, therefore, the extremal BF contains the cusp Q, the Jacobi condition is violated and neither a maximum nor a minimum exists at all.

Weierstrass condition

The third point of investigation is the Weierstrass E function, which indicates whether an extremum is weak or strong. For the curve determined here, this investigation could be omitted because the violation of the Jacobi condition already showed that no extremum for h is involved at all. However, the analysis of the E-function is mentioned here for completeness.

The Weierstrass's E function is defined by

$$E(x, y; x', y'; \tilde{x}', \tilde{y}') = \tilde{x}'[H_{x'}(x, y; \tilde{x}', \tilde{y}') - H_{x'}(x, y; x', y')]$$
$$+ \tilde{y}'[H_{y'}(x, y; \tilde{x}', \tilde{y}') - H_{x'}(x, y; x', y')]$$

where x', y' are the derivatives of the computed extremal and $\tilde{x}'$, $\tilde{y}'$ are the derivatives of another comparison curve. The extremum is strong when the sign of E is independent of the magnitude of $(\tilde{x}' - x')$ and $(\tilde{y}' - y')$. It is weak when the sign of E is constant only for small values of $(\tilde{x}' - x')$ and $(\tilde{y}' - y')$, but changes if the values are large. A negative value of E corresponds to a maximum, a positive value to a minimum.

Elaboration of the E function for H defined in equation (23) results in

$$E = [(g_1 + g_2 x)x' - (1 - g_2 y)y'] [\arctan(\tilde{y}'/\tilde{x}') - \arctan(y'/x') + (\tilde{x}'\tilde{y}' - x'y')(x'^2 + y'^2)]$$
$$+ [(g_1 + g_2 x)(\tilde{x}' - x') - (1 - g_2 y)(\tilde{y}' - y')] [\arctan(\tilde{y}'/\tilde{x}') - \arctan(y'/x')]$$

For small values of $(\tilde{x}' - x')$ and $(\tilde{y}' - y')$ it is found that E is negative. This indicates that a maximum is involved. However, the presence of the term $\arctan(\tilde{y}'/\tilde{x}')$ creates the possibility that the sign of E changes for a comparison curve with arbitrary $\tilde{x}'$, $\tilde{y}'$ values. Therefore, the computed curve would produce a weak maximum for h, if the Jacobi condition had not been violated.

DISCUSSION OF THE RESULT

The result of the variational analysis is an extremal in the form of an involute. Its mathematical expression is given in parameter form by equations (25). The constants a, b, g_1, g_2 in these expressions have the values (34). The shape of the line is shown in Fig. 3 as the line BP_aQF.

The involute is an extremal because it satisfies in every point the Euler condition (equation 22). It is an extremal of the class of potential slip lines that contains the real slip line. This is achieved by defining integrals that impose similar conditions of total and local equilibrium in the limit stress state on the lines of that class as are satisfied by the real slip line.

The extremal, represented by the line BP_aQF, is unique. It is the only involute that satisfies the constraints (21b and c) and the boundary conditions in B and F. The corresponding height of the vertical cut off has the value $h_{em1} = 3 \cdot 783 c/\gamma$, which is high, but could be a valid bound for the collapse height, h_{coll}. All these points are in favour of the analysis and suggest that it is an attractive method of slip line determination. There are, however, some disappointments.

Verification of the conditions for the extremum in the previous section (Investigation of the solution) shows that the analysis produces a weak maximum in general, whereas in the special case that the extremal contains the cusp point Q, there is no extremum at all. These verification results indicate that the analysis is apparently not meaningful. But before this conclusion can be drawn, it is necessary to consider these two points in more detail.

Weak maximum

When a maximum is involved, this means that the value of h_{em1} is higher than every value of h corresponding to any line of the class investigated. When this class contains the real slip line, then h_{em1} is larger than the height corresponding to the real slip line, which presumably is the collapse height, h_{coll}. So being a maximum, the computed height h_{em1} is an unsafe estimate for h_{coll}.

It is essential for this conclusion that the class contains the line that produces h_{coll}. If, for instance, a class is considered that gives for all lines values lower than h_{coll}, it would be interesting to establish the maximum of that class. This would give the best, safe value for h. In the case considered here, the real slip line, if such a line exists, is still in the class because the stress distributions of the lines in the class satisfy the same equilibrium conditions as the stress distribution along the real slip line. Therefore a maximum corresponds here to an unsafe estimate.

The maximum is called weak in variational terms. This means that it is a maximum with regard to comparison lines that deviate from the extremal only in such a manner that their inclination is almost the same. As a consequence another maximum or even a minimum could exist for completely different lines. In textbooks examples are given of unexpected shapes in this respect, consisting of discontinuous solutions.

The examination of discontinuous solutions has not beer ~~cted exhaustively for the problem of the slip line analysis. Only the case was consider﹃ ˙two involutes with a discontinuous second derivative, forming a smooth **S**-shaped curve. Such a curve has piecewise different values for the circle centres x_M, y_M and so according to (27) the Lagrange multipliers g_1, g_2 are not constant along the line. Since these multipliers have to be constants, such an **S** shaped curve cannot be an extremal. There are other possibilities for discontinuous solutions. It could be envisaged to study a combination of slip lines consisting piecewise of conjugate stress characteristics. However, such explorations were not pursued.

No extremum at all

In the previous section (Investigation of the solution) it is established that an extremal consisting of an involute violates the Jacobi condition if the line contains the cusp point Q. It is possible that for embankments or slopes with other boundary conditions the cusp is avoided. For the particular case of the vertical cut off, the cusp is not avoided and is located within the integration interval of the extremal.

This indicates that in this particular case no extremum is associated with the extremal. Being an extremal means that the line satisfies in every point the Euler condition (22). This is necessary for the line to give an extremum, but not yet sufficient. It is also necessary that the line satisfies the Jacobi condition and if that second condition is violated, then according to the variational calculus no extremum is involved at all.

Since this conclusion seems curious, it is revealing to mention some additional computations that can be reproduced readily to verify the lack of extremum. To that end it is convenient to consider smooth **S** shaped lines, consisting of two circles, that have the same tangent in their meeting point. The integrals (21a–c) can be evaluated in closed form for such curves. For every height of the meeting point there is one unique combination of circles that satisfies the constraints (21b and c) and the boundary conditions in B and F. The values of h corresponding to these **S** shaped lines are above or below h_{em1} for lines with a meeting point respectively above or below $y \approx 0.33h$.

By this verification it is established that the class of potential slip lines is not limited in a meaningful manner. Apparently the conditions of total and local equilibrium of the limit stress state along a line are not restrictive enough. This is further substantiated by an examination of the stress state in the vicinity of the point B. The stress state in the line should be compatible with the stress-free vertical boundary BC. Compatibility can be verified by introducing a fan of stress characteristics centred in B. The result is that the **S** shaped lines with a meeting point above $y = 0.17h$ are unacceptable because the stresses near B surpass the limit stress state, and this limits the height for **S** shaped slip lines to $h \leqslant 3\cdot69c/\gamma$. We will not elaborate this point here further because **S** shaped curves have no special status. They were arbitrarily chosen for convenience to verify the absence of an extremum.

Another example of a line that might be called an extremal because an Euler equation is satisfied, but apparently produces no valid extreme value for h, is encountered in the analysis of Baker and Garber (1978). In that analysis the procedure is to determine the critical slip line by considering an extremal, with the location $y(x)$ and the normal stress distribution $\sigma'(x)$ as the two relevant variables. The stationary value of the integral is determined with respect to these two variables, by two Euler equations. The first involves σ and σ', the second involves y and y'. The authors surmise that satisfaction of the first Euler equation only is enough for a line to form a useful bound for the critical slip line. Their solution reduces to a circle for the case of a frictionless soil.

In fact, this solution is only a regular extremal if also the second Euler equation is satisfied. Introducing the circle in this relation gives a differential equation for the normal stress σ which determines the distribution of σ. Now the curious situation occurs that for the vertical cut off it is impossible to find a circle that satisfies the boundary conditions and the constraints, possessing in addition a normal stress distribution that corresponds to the second Euler equation. So there exists no circle that can be claimed to be a regular extremal.

It is possible to indicate circles that satisfy moment equilibrium only and to adjust the normal stress distribution in such a manner that the other two constraints of vertical and horizontal force equilibrium are satisfied. But then the second Euler equation is violated. This class of circles has a minimum height corresponding to $h_{\min} = 3\cdot8c/\gamma$. From the theory of plasticity it is known that this height is an upper bound based on a kinematically admissible velocity field. So all circles correspond to heights that are above h_{coll}. This result indicates that the class of circles, which are lines that satisfy the first Euler equation only, does not contain the critical slip line.

It is difficult to infer directly from the variational calculus that the real slip line is outside the class of extremals determined by the first Euler equation only. An indication could be that there are no terms with σ' in the integrands. Therefore, all derivatives with respect to σ' vanish, also the second derivatives required in the Legendre condition. This indicates that a degenerated extremum for h is involved.

Real slip line

Up to this point it has been assumed that there exists a real slip line and that the collapse state is sufficiently characterized by considering the stresses along that line only. For the variational analysis it is a convenient assumption because the analysis is then restricted to the determination of the shape of one line and the elementary variational calculus provides all tools for such a determination. The one-line assumption is, however, questionable because satisfying equilibrium and limit stress state along one line does not guarantee that the stress states in all other points of the soil body are within acceptable limits. An example of considering regions outside the line was mentioned above with regard to **S** shaped curves and the region in the vicinity of B. But one small region is not enough. For a line to be considered a potential slip line it is necessary to verify the entire soil body.

Possibly the disappointments revealed by evaluating the results of the variational analysis in this study are due to an inadequate formulation of the problem. The collapse state may be associated with a distribution of stress characteristics and regions of slip lines of a more complex character than assumed so far. Characterizing the stress state and the sliding mechanism may require the formulation of a more sophisticated system of integrals before the calculus of variations can be used meaningfully.

CONCLUSION

The variational calculus seems an intriguing tool for investigating slope stability problems. The formulation of the problem by defining a class of lines satisfying total and local equilibrium and assuming that the extremal of that class is the real slip line is unsatisfactory. The class is not specified in such a manner that a bound is found for the real slip line, and it is even questionable whether or not there exists one single real slip line at collapse. The analysis based on total and local equilibrium of a single line leads to a weak maximum (or no extremum at all) and therefore it produces unsafe predictions for slope stability problems.

REFERENCES

Baker, R. & Garber, M. (1977). Variational approach to slope stability. *Proc. Ninth Int. Conf. Soil Mech. Fdn Engng, Tokyo* 3/2, 9–12.
Baker, R. & Garber, M. (1978). Theoretical analysis of the stability of slopes. *Géotechnique* 28, No. 4, 395–411.
Bolza, O. (1973). *Lectures on the calculus of variations.* New York: Chelsea Publishing Company.
Pastor, J. (1978). Analyse limite: Détermination de solutions statiques complètes. Applications au talus vertical. *Journal de Méchanique Appliquée* 2, No. 2, 167–196.
Petrov, I. P. (1968). *Variational methods in optimum control theory.* New York and London: Academic Press.
Ramamurthy, T., Narayan, C. G. P. & Bhatkar, V. P. (1977). Variational method for slope stability analysis. *Proc. Ninth Int. Conf. Soil Mech. Fdn Engng, Tokyo* 3/26, 139–142.
Revilla, J. & Castillo, E. (1977). The calculus of variations applied to stability of slopes. *Géotechnique* 27, No. 1, 1–11.

A variational fallacy

G. DE JOSSELIN DE JONG*

It is unfortunate that valuable information concerning fallacies in soil mechanics can get lost in the course of time. This has happened with the use of variational calculus in slope stability problems.

The variational method, for determining the critical slip surface as the surface that minimizes the load at rupture, was presented in soil mechanics by Kopácsy (1957, 1961). He published in the 1957 London conference a three-dimensional solution, and in the 1961 Paris conference the two-dimensional version of it. The shape of the surface is suitably established in the 1957 paper by vector analysis and described by equation (16) of that paper which, with $\bar{\omega}$ defined by equation (20), can be expressed in words as follows.

The normal to the critical slip surface shown in Fig. 1 at some generic point P, has a constant angle ϕ with a plane through the line PQ, which connects P to its projection Q on an axis along a vector $\bar{\mu}$ fixed in space. This plane is further tilted with respect to $\bar{\mu}$ by an angle, which is equal to the angle between the lines PQ and PR, with R located on the axis of $\bar{\mu}$ at a constant distance of magnitude $(\bar{\lambda} . \bar{\mu})/(\bar{\mu} . \bar{\mu})$ from Q.

The location in space of the surface and of the axis of $\bar{\mu}$, and the coefficients of the vectors $\bar{\lambda}$ and $\bar{\mu}$ are obtained from equilibrium and boundary conditions. In the conclusion of the 1957 paper it is remarked that equation (16) defines a screw surface and, in the 1961 paper, it is shown that this solution reduces to a logarithmic spiral in the two-dimensional case.

This elegant result shows that the variational method is worthy of attention. But, although available for many years and apparently useful for solving a basic problem in soil mechanics, the method was not pursued by leading scientists. The reason for this, which may be unknown to the general soil mechanics public, is that the minimization aspect of Kopácsy's analysis was shown to be false in 1961.

The falsification concerns not Kopácsy's papers

directly, but an independent study by Gibson & Morgenstern, which was rather similar in character and also resulted in logarithmic spirals. Their analysis was examined by a mathematician who drew attention to a subtle fallacy in their reasoning. It was shown that such curves are not limitative, but only have the property that the moment is independent of the normal stress distribution. It was pointed out that the method breaks down, but how this is related to a defect in the variational sense was not explained. Neither their work nor the refutation of it was ever published, but reference to it was given by Morgenstern (1977) in his general report at the Tokyo Conference.

The essence of the defect is that a degenerate functional is involved. This can be readily verified since Petrov's (1968) book appeared. In this book, Sections 29 and 30 are devoted to degenerate cases (which are hardly considered in other textbooks) and their properties are extensively treated. The degenerations are classified according to the theory of Krotov, who distinguishes five kinds. Petrov's description is followed here. For the details of the slipline analysis it is convenient to refer to Baker & Garber's (1978) paper.

Their treatment is a modification of Kopácsy's 1961 analysis, improved by the introduction of a safety factor. The essential defect is not removed and the result is still a logarithmic spiral.

The intermediate function g, defined by equation (9.1) in the paper by Baker & Garber, has two variables: the normal stress distribution $\sigma(x)$ and the height of the slip surface $y(x)$. The functional is degenerate, because g does not contain σ' (a prime indicates a derivative with respect to x) and is linear in y'. In the Krotov sense this creates a degeneration of the third kind with respect to both σ and y.

In Petrov's book the simultaneous occurrence of two degenerations is not treated, but it can be expected that such a combination reinforces the degenerate character, rather than ameliorating the situation. Consider therefore only σ. Because of the absence of σ' the functional G, defined by equation (9.1), has its extreme value only for particular

---

* University of Technology, Delft.

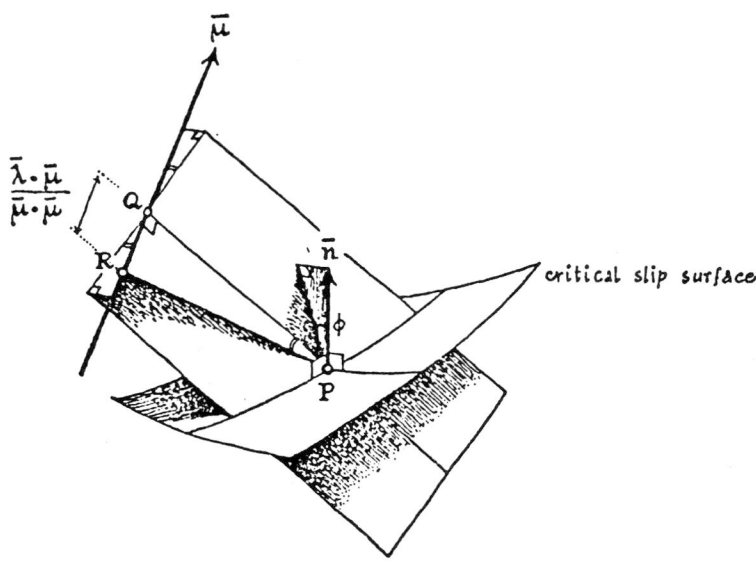

Fig. 1. Kopácsy's solution

distributions of σ, and the possibility exists that discrete jumps between these distributions are involved. In order to establish these distributions it is necessary to investigate $g_{\sigma\sigma}$ (the second partial derivative with respect to σ). But g is only linear in σ, so $g_{\sigma\sigma}$ vanishes. In that case there is only an extremum of G (maximum or minimum) if σ is bounded, and the extreme value is obtained by giving σ one of the bounding values. In the mechanical sense σ is not bounded, because stress states are limited in stress space by a cone, which widens with increasing values of σ. There is no maximum value for σ and there is no upper bound. Therefore, no extremum exists and the method breaks down. There is a lower bound for σ, which is zero for cohesionless soils or tensile if there is cohesion. This lower bound is irrelevant here, because it produces no valid solution for the entire slip surface.

At this point it can be concluded that Kopácsy's analysis is not meaningful, because a functional is considered which has a degenerate character and possesses no minimum. This is due to his particular formulation of the slope stability problem in terms of the variational calculus. It is unfortunate that the falsification remained unknown. It would have prevented the recent revival of this variational fallacy.

REFERENCES

Baker, R. & Garber, M. (1978). Theoretical analysis of the stability of slopes. *Géotechnique* **28**, No. 4, 395–411.

Kopácsy, J. (1957). Three-dimensional stress distribution and slip surfaces in earth works at rupture. *Proc. 4th Int. Conf. Soil. Mech., London* **3a**, 17.

Kopácsy, J. (1961). Distribution des contraintes à la rupture, forme de la surface de glissement et hauteur théorique des talus. *Proc. 5th Int. Conf. Soil Mech., Paris* **6**, 23.

Morgenstern, N. R. (1977). *Proc. 9th Int. Conf. Soil. Mech., Tokyo* **3**, 319.

Petrov, Iu. P. (1968). *Variational methods in optimum control theory.* New York: Academic Press.

Reprinted from JOURNAL OF APPLIED PHYSICS, Vol. 24, No. 7, 922–928, July, 1953

Consolidation Around Pore Pressure Meters

G. DE JOSSELIN DE JONG

Laboratorium voor Grondmechanica, Oostplantsoen 25, Delft, The Netherlands

(Received March 26, 1953)

Response of pore pressure meter on variations in loading conditions of surrounding soil is retarded by the necessity that pore water has to enter the instrument. This property is introduced as an instruments coefficient influencing the boundary conditions. With regard to the surface of the instrument, two types are considered: a rigid type and a cavernous type. At $t=0$ for unit step loading, consolidation has not yet started and the response of the instrument depends on shear modulus of soil only. Calculation of response as a function of time involves three-dimensional consolidation theory and is established with the aid of spherical solutions for simple harmonic and unit step loading conditions.

1. INTRODUCTION

THE newest type of pore pressure meter in use with the Delft Soil Mechanics Laboratory* consists of a cylindrical instrument of about 20 cm in length and 3.6 cm in diameter embedded in the soil mass. The body of the instrument contains a diaphragm-type pressure transducer connected by a coaxial flexible cable with the measuring apparatus above soil surface.

The response time of previous nonelectronic types amounted to several days, because of the compliance of the diaphragm and the very great flow resistance encountered in the soil by the water that has to actuate the manometer. By constructing the diaphragm as rigid as practically feasible the response time was reduced to 15 minutes.

A further reduction of the response time to a fraction of a second was obtained by a device resulting from an analysis of pore-water movement and soil deformation. This study forms the subject of this paper.

We shall treat the influence of simple harmonic loading conditions of the surrounding soil mass and by integration over the whole range of frequencies, using Fourier's integral, obtain the influence of unit step loading.

---

* Capacitive pressure-transducer and electronic circuitry developed by S. L. Boersma, Consulting Engineer, Delft, The Netherlands.

To simplify the mathematical treatment the instrument is idealized and instead of cylindrical symmetry a spherical symmetry is assumed. If idealized in this way, the pore pressure meter consists of a sphere with radius r_0 (cm). The following two types will be considered.

1.1 The Rigid Type

Here the diaphragm is in contact with a water volume enclosed in a rigid chamber. Holes in the wall form the connection between this water volume and the water in the pores of the soil.

We shall represent this schematically as a rigid, pervious sphere (Fig. 1a).

1.2 The Cavernous Type

Here the diaphragm is in contact with a water volume that fills a cavity in the soil. The connection between this water volume and the pore water is direct.

We shall represent this schematically as a spherical cavity (Fig. 1b).

Indicating by ΔV (cm³) the quantity of water necessary to account for the deformation of the diaphragm under water pressure, we introduce analogous to the compressibility coefficient a of the soil (see Sec. 5.1) a coefficient b (cm²/kg) for the instrument in such a way that

$$\Delta V/(\tfrac{4}{3})\pi r_0^3 = bw_0, \qquad (1.1)$$

in which w_0 (kg/cm²) indicates the pressure in the water enclosed in the sphere.

When the surrounding soil mass is loaded, water pressures are generated in the pore water and in order that the pore pressure meter may register this, a volume of water ΔV must enter the chamber.

For the rigid type the volume of water enters by percolation through the holes in the wall. This water can only be supplied by the pore water squeezed out of the surrounding soil mass. For the cavernous type the volume ΔV arises principally from the deformation of the cavity, while an additional part is furnished by pore water coming out of the surrounding soil.

We shall proceed to formulate this action mathematically by first establishing the boundary conditions.

2. BOUNDARY CONDITIONS AT $r = r_0$

2.1 Water pressure   at $r = r_0$

The *rigid type*. If we may apply Darcy's law of percolation. The water volume Q_1 cm³ entering through the pervious wall with a surface area $4\pi r_0^2$ is related to the pressure gradient in the pore water $\partial w/\partial r$ in the soil at $r = r_0$, by the expression:

$$\gamma_w(\partial Q_1/\partial t) = k4\pi r_0^2(\partial w/\partial r), \qquad (2.1.1)$$

in which γ_w (kg/cm³) denotes the density of water and k (cm/sec) the permeability of the soil (the permeability of the wall of the pore pressure meter is considered large in comparison with k of the soil).

Since for the rigid type $\Delta V = Q_1$, we obtain by using (1.1)

$$\tfrac{1}{3}br_0(\partial w/\partial t) = (k/\gamma_w)(\partial w/\partial r) \quad \text{at} \quad r = r_0. \quad (2.1.2)$$

It is assumed here that w_0, the pressure in the meter, is equal to w, the pore pressure in the soil directly surrounding the meter, $w_0(r = r_0 - 0) = w(r = r_0 + 0)$.

The *cavernous type*. Here the water amount ΔV is the sum of Q_1 and the volume change produced by the deformation of the cavity Q_2. If we introduce the displacement u_r (cm) of the surface of the cavity in a radial direction, this volume is given by $Q_2 = -4\pi r_0^2 u_r$. Thus we obtain

$$\tfrac{1}{3}br_0(\partial w/\partial t) = (k/\gamma_w)(\partial w/\partial r) - (\partial u_r/\partial t)$$
$$\text{at} \quad r = r_0. \quad (2.1.3)$$

These expressions relate to the spherically symmetrical case.

In the case when the variables w and u_r are axially symmetrical and are functions of ψ (Fig. 2) the second member of Eqs. (2.1.2) and (2.1.3) must be integrated over the range $0 < \psi < \pi$. (See Sec. 4.)

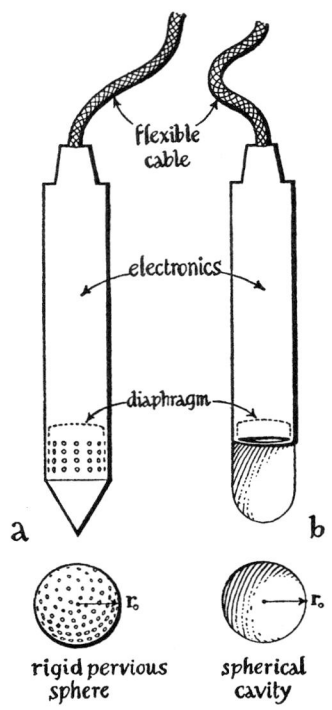

FIG. 1. Pore pressure meters of two different types.

2.2 Soil skeleton   at $r = r_0$

In the *rigid type* the boundary condition at $r = r_0$ is imposed on the soil skeleton by the rigidity of the sphere:

$$u_r = 0 \quad \text{at} \quad r = r_0. \qquad (2.2.1)$$

In the *cavernous type* the wall of the cavity can displace itself freely and the stresses are determined by the condition that there is equilibrium between the water pressure w_0 in the cavity and the radial stress in the soil mass σ_r^* (negative sign for compression).

$$\left.\begin{array}{r} w_0 = -\sigma_r^* \\ 0 = \tau_{r\psi}^* \end{array}\right\} \quad \text{at} \quad r = r_0. \qquad (2.2.2)$$

Here $-\sigma_r^*$ indicates the total radial stress acting on the soil skeleton and water combined, so we can divide $-\sigma_r^*$ into $-\sigma_r$, the radial stress on the grains only, and w the water pressure in the pores[†] by putting

$$-\sigma_r^* = -\sigma_r + w. \qquad (2.2.3)$$

These stresses act in the soil, that is to say for $r > r_0$, while w_0 is the water pressure in the cavity $r < r_0$.

[†] We shall use an asterisk to denote stresses and elastic constants associated with the combined system, water and soil skeleton, and omit it in the case of the grain skeleton alone.

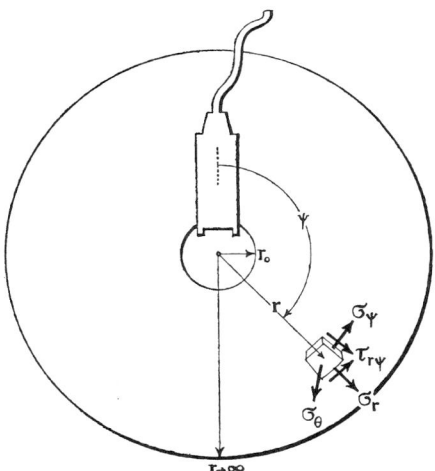

Fig. 2. Definitions of symbols used.

It follows from (2.2.2) and (2.2.3) that

$$w_0(r_0-0)=-\sigma_r+w(r_0+0). \qquad (2.2.4)$$

Normally there is no difference between $w(r_0-0)$ and $w(r_0+0)$, except when the loading conditions are discontinuous with regard to time, as for instance at the moment $t=0$ for unit step loading. We shall treat this special situation in Sec. 4 and use condition (2.2.2) then. For the case of continuous loading conditions $w(r_0-0)$ equals $w(r_0+0)$ and the boundary stress condition becomes

$$\sigma_r=\tau_{r\psi}=0 \quad \text{at} \quad r=r_0. \qquad (2.2.5)$$

3. BOUNDARY CONDITIONS AT $r \to \infty$

We shall assume that the dimensions of the pore pressure meter are so small in comparison with the distance to the boundary of the soil mass wherein it is embedded that we may consider the stress conditions of the soil as homogeneous. To simplify the formulas we shall also suppose that this homogeneous stress condition holds up to $r \to \infty$. As the disturbing effect diminishes proportional to r^{-3} the errors introduced by these assumptions are negligible.

In addition we shall suppose that in the surrounding soil mass the rate of squeezing out of pore water is slow in comparison with the response studied here. We may therefore simplify the condition for $r \to \infty$ to an impermeable coat, upon which the exterior loads act. Thus, we may write

$$\partial w/\partial r=0 \quad \text{at} \quad r \to \infty. \qquad (3.1)$$

Regarding the stresses on the boundary at $r \to \infty$ we must consider the general case where there are three

principal stresses of unequal magnitude: S_1, S_2, S_3. If we know the influence of one principal stress we may obtain by superposition their combined influence. Taking the principal stress S in the direction $\psi=0$, π we obtain as boundary condition:

$$\sigma_r{}^*=-S\cos^2\psi \qquad \sigma_\psi{}^*=-S\sin^2\psi \qquad \sigma_\theta{}^*=0$$
$$\tau_{r\psi}{}^*=S\sin\psi\cos\psi \qquad \tau_{r\theta}{}^*=\tau_{\theta\psi}{}^*=0.$$
$$\text{at} \quad r \to \infty. \qquad (3.2)$$

Because of the simpler treatment involved we shall finally derive a solution to the case where a uniform pressure P acts in all directions. In such a case we have

$$\sigma_r{}^*=\sigma_\theta{}^*=\sigma_\psi{}^*=-P, \quad \tau_{r\psi}{}^*=\tau_{r\theta}{}^*=\tau_{\theta\psi}{}^*=0,$$
$$\text{at} \quad r \to \infty. \qquad (3.3)$$

The conditions (2.1.2), (2.1.3), (2.2.1), (2.2.2), (2.2.5), (3.1), (3.2), and (3.3) are necessary and sufficient to determine the solution to the problem explicitly in the different cases indicated.

4. SITUATION AT $t=+0$ FOR UNIT STEP LOADING

Because of its elucidatory character we shall first determine the response of the pore pressure meter to unit step loading at the moment of loading $(t=+0)$. This is very simply feasible by virtue of the following considerations.

By neglecting mass acceleration and viscous retardation, we shall suppose, as is usual in the literature on consolidation, that any loading increases instantaneously the stresses in the soil mass, generating elastically the deformations that are therefore instantaneous too.

Assuming that in comparison with the compressibility of the grain skeleton of the soil the pore water is incompressible, at the instant of loading no volume change of the soil will take place. This may be expressed by putting

$$\nu^*=\tfrac{1}{2} \quad \text{for} \quad t=+0, \qquad (4.1)$$

where ν^* is the Poisson's ratio of the combined system, water and soil skeleton. As the water is gradually expelled from an element at any point in the loaded soil, the Poisson's ratio ν^* at this point decreases and tends to the value of ν, where $\nu(<\tfrac{1}{2})$ is the value of the Poisson's ratio of the grain skeleton alone.

A result of some importance concerning the value of E^* may be derived in the following way. The water enclosed in the pores can prevent volume changes from occurring, but not shearing strains. This implies that shear stress in the combined system is carried by the grain skeleton alone. It follows directly that $\tau^*=\tau$, and hence that the shear moduli G^* and G are equal. Then by a well-known relation in elasticity theory,

$$G^*=E^*/2(1+\nu^*)=E/2(1+\nu)=G \quad \text{at all times}, \qquad (4.2)$$

and in particular

$$\tfrac{2}{3}E^*=E/(1+\nu) \quad \text{at} \quad t=+0. \qquad (4.3)$$

The relations (4.1) and (4.3) enable us to describe the condition at $t=+0$ as a function of the elasticity constants of the grain skeleton E and ν.

The relations between the stresses in the combined system and grain skeleton only, which have already been stated, may be condensed to

$$-\sigma^* = -\sigma + w \quad \text{and} \quad \tau^* = \tau. \qquad (4.4)$$

By adding the normal stresses in three perpendicular directions we obtain

$$-(\sigma_1^* + \sigma_2^* + \sigma_3^*) = -(\sigma_1 + \sigma_2 + \sigma_3) + 3w.$$

As the volume dilatation ϵ, which is known to be equal to

$$\epsilon = (1 - 2\nu)(\sigma_1 + \sigma_2 + \sigma_3)/E, \qquad (4.5)$$

is zero at $t=+0$, we obtain, by virtue of the fact that $\nu \neq \frac{1}{2}$,

$$-(\sigma_1^* + \sigma_2^* + \sigma_3^*) = 3w = -(\sigma_I^* + \sigma_{II}^* + \sigma_{III}^*)$$
$$\text{at} \quad t = +0. \qquad (4.6)$$

In this expression w_0 is the pressure in the water contained in the cavity and caused by the volume change ΔV, which is, in turn, caused by the displacement of the wall of the cavity. We may obtain ΔV by integration over the surface of the sphere

$$-\Delta V = \int_\pi^0 2\pi r_0 \sin\psi \cdot u_r r_0 d\psi. \qquad (4.9)$$

$$-\Delta V = \{[(1+\nu^*)/2E^*](-\tfrac{1}{3}S + w_0)$$
$$-[(1-2\nu^*)/3E^*]S\}4\pi r_0^3. \qquad (4.10)$$

By the use of (4.1) and (4.2) it follows that

$$-\Delta V = 4\pi r_0^3[-\tfrac{1}{3}S + w_0]/4G^*$$
$$= 4\pi r_0^3[-\tfrac{1}{3}S + w_0]/4G. \qquad (4.11)$$

However, we have a relation (1.1) between ΔV and w_0, by which we may eliminate ΔV obtaining:

$$w_0 = \tfrac{1}{3}S/\{1 + \tfrac{4}{3}bG\}. \qquad (4.12)$$

This result indicates that the cavernous pore pressure meter registers a pressure of nearly $\frac{1}{3}S$ so long as b is small in comparison with $1/G$. As in the newest type b is of the order of $1/100G$, where G is the shear modulus of the stiffest clay we have to deal with, the error involved is less than 1 or 2 percent.

We have shown in (4.6) that the water pressure in the pores of the soil is one-third of the sum of the principal stresses. Therefore, neglecting the influence of the denominator of (4.12), it follows that the water pressure, as measured by the cavernous type meter, is

in which σ_1^* etc., indicate the *principal* stresses in the combined system.

With the aid of these general considerations we shall proceed to determine the conditions at $t=+0$ in our special case.

Under the influence of the stress system (3.2) and the boundary condition for the cavity (2.2.5), the stress distribution in the soil can be determined by the aid of stress functions.[1] We shall not enlarge on these computations because they are easy to carry out and by applying the following expression for the radial displacement as derived from a stress function, Φ:

$$u_r = \frac{1+\nu^*}{E^*}\left\{ 2(1-\nu^*)\cos\psi \nabla^2 \Phi \right.$$

$$\left. - \frac{\partial}{\partial r}\left[\cos\psi \frac{\partial \Phi}{\partial r} - \sin\psi \frac{1}{r}\frac{\partial \Phi}{\partial r} \right] \right\}, \quad (4.7)$$

the reader may verify that u_r is given by

$$u_r = \frac{1+\nu^*}{E^*}r_0\left\{ \left[\frac{\nu^*}{1+\nu^*} - \cos\psi \right]S + \tfrac{1}{2}w_0\left(\frac{r_0}{r}\right)^2 + \frac{[(6-5\nu^*)+5(4\nu^*-5)\cos^2\psi](r_0/r)^2 - [3-9\cos^2\psi](r_0/r)^4}{2(7-5\nu^*)}S \right\}. \quad (4.8)$$

effectively the actual water pressure *in* the pores of the soil.

In the case of spherically symmetrical stress P at infinity all expressions have a simpler form, and it is easy to verify that the boundary conditions (2.2.2) and (3.3) are satisfied, together with those of equilibrium and compatibility, by taking

$$u_r = -[(1+\nu^*)/2E^*](P-w_0)(r_0^3/r^2)$$
$$-[(1-2\nu^*)/E^*]Pr = -(P-w_0)r_0^3/4Gr^2, \quad (4.13)$$

$$\sigma_r^* = -P + (P-w_0)(r_0/r)^3, \qquad (4.14)$$

$$\sigma_\theta^* = -P - \tfrac{1}{2}(P-w_0)(r_0/r)^3. \qquad (4.15)$$

By virtue of the identity of (4.10) and (4.13) the resulting water pressure w_0 in the instrument is the same as given by (4.12) with $P=\frac{1}{3}S$.

The water pressure in the soil mass itself is $w = -\frac{1}{3}(\sigma_r^* + 2\sigma_\theta^*) = P$. There is therefore a jump in the water pressure at $r=r_0$ with a magnitude of $\frac{4}{3}PbG$. It follows therefore that, the water gradient $\partial w/\partial r$ is infinite at that surface and percolation starts with infinite velocity giving a vertical slope of the water pressure registered by the meter as a function of time. The same effect occurs in the rigid type, but in this case the starting point for the water pressure at $t=+0$ is zero, because no increase of water pressure is possible before water entered the holes by percolation. In Fig. 3 an account is given of the stresses around the instruments in the different cases.

[1] S. Timoshenko, *Theory of Elasticity* (McGraw-Hill Book Company, Inc., New York, 1934), p. 326.

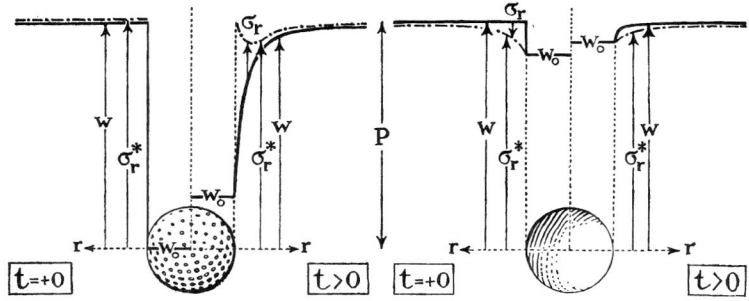

FIG. 3. Stresses and water pressure around pervious sphere and spherical cavity
(arrow pointing upwards indicates compression).

5. DETERMINATION OF THE CONSOLIDATION PROCESS AROUND THE PORE PRESSURE METER

To determine the water pressure registered by the meter as a function of time we need a consolidation theory in 3 dimensions. Such a theory has been developed by Biot.[2] For the case of saturated soil the basic equation is derived in the form

$$(k/a\gamma_w)\nabla^2\epsilon = \partial\epsilon/\partial t, \qquad (5.1)$$

where

$a=$ compressibility coefficient $=(1-2\nu)(1+\nu)/(1-\nu)E$ (cm²/kg,)

$k=$ permeability (cm/sec), $\gamma_w=$ density of water (kg/cm³),

$\epsilon=$ volume dilatation of grain skeleton, and $t=$ time.

The physical meaning of this result may be appreciated if we consider the well-known relation from elasticity[3] in which equilibrium and compatibility conditions are taken into account,

$$(1/a)\nabla^2\epsilon + (\partial X/\partial x) + (\partial Y/\partial y) + (\partial Z/\partial z) = 0. \quad (5.2)$$

The body forces X, Y, Z in our case are furnished by the water gradients $X = -\partial w/\partial x$ etc., so that

$$(1/a)\nabla^2\epsilon = \nabla^2 w. \qquad (5.3)$$

Now, by extending Darcy's law of percolation to the case of 3 dimensions it is easy to deduce that the rate of increase of volume $\partial\epsilon/\partial t$ necessary to store the excess of water at a certain point is given by

$$\partial\epsilon/\partial t = (k/\gamma_w)\nabla^2 w. \qquad (5.4)$$

A combination of (5.3) and (5.4) gives the basic equation (5.1).

In order to obtain an agreeable treatment of the problem we shall consider simple harmonic loading conditions for $r\to\infty$.

$$P = \tilde{P}\exp(i\omega t) \text{ in cond. (3.3)}.$$

All variables influenced by this harmonic effect as for instance ϵ, will be written $\epsilon = \bar\epsilon\exp(i\omega t)$. So we can divide all expressions throughout by $\exp(i\omega t)$ and retain the overdashed characters. We get for (5.1) in the case of axially symmetrical stress distribution

$$\nabla^2\bar\epsilon = \frac{1}{r^2}\left[\frac{\partial}{\partial r}\left(r^2\frac{\partial}{\partial r}\right) + \frac{1}{\sin\psi}\frac{\partial}{\partial\psi}\left(\sin\psi\frac{\partial}{\partial\psi}\right)\right]\bar\epsilon = q^2\bar\epsilon,$$

where   (5.5)

$$q^2 = i\omega\left(\frac{a\gamma_w}{k}\right),$$

which has the solutions

$$\bar\epsilon = r^{-\frac{1}{2}}[A_1 J_{n+\frac{1}{2}}(iqr) + A_2 N_{n+\frac{1}{2}}(iqr)]\Psi_n, \qquad (5.6)$$

where J and N denote, respectively, Bessel and Neumanns functions,[4] and Ψ Legendres polynomials.[5] To fit this solution to the purpose of our problem we need such a combination of J and N that for $r\to\infty$ the expression vanishes. From the theory of Bessel functions it is known that this is accomplished if $A_2 = iA_1$. We then obtain from (5.6)

$$\bar\epsilon = A_1 i^{(3-n)}[\tfrac{1}{2}\pi i q]^{-\frac{1}{2}} r^{-1}\exp(-qr)S_{n+\frac{1}{2}}(2qr)\Psi_n, \quad (5.7)$$

in which S are polynomials in $(1/qr)$.[6]

This equation for $\bar\epsilon$ permits us to compute the solution to the axial symmetrical problem that arises when introducing the true form of boundary conditions. For instance location of the holes on rigid type, upper part of the cavity formed by the instrument, etc. Computation would then require a treatment with many stress functions Φ derived from (5.7) by

$$\bar\epsilon = [(1-2\nu)(1+\nu)/E]\{\cos\psi\,\partial/\partial r$$
$$-\sin\psi r^{-1}\partial/\partial\psi\}\nabla^2\Phi. \quad (5.8)$$

[2] M. Biot, J. Appl. Phys. 12, 155 (1941).
[3] See reference 1, p. 198.
[4] Jahnke and Emde, Tables of Functions (Dover Publications, New York, 1945), pp. 144, 146.
[5] See reference 4, p. 107.
[6] See reference 4, p. 136.

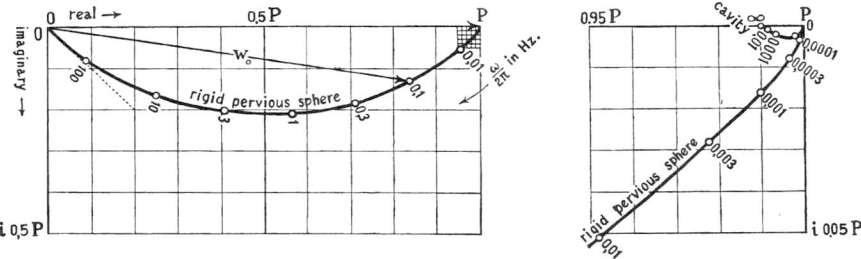

FIG. 4. Graphical representation of formula (6.6): response of pore pressure meter on simple harmonic loading. Assumptions: $b/3a = 10^{-2}$, $r_0 = 1$ cm, $a\gamma_w/k = 10^{-3}$ sec/cm^2, $\nu = \frac{1}{3}$.

In general: in all cases, where the solution of stress distribution under static loading is known as derived from stress functions, the consolidation problem is solved by replacing the dilatation-stress functions which satisfy $\nabla^2\nabla^2\Phi = 0$ by corresponding functions of the form (5.7). The functions $\nabla^2\Phi = 0$ need no alterations, because $\bar{\epsilon} = 0$ also satisfies (5.5), neither the rotation-stress-functions.

We shall limit ourselves, however, to the solution of complete spherical symmetry, because this leads to a simplification in the treatment. The same solution for w_0 as a function of time is then obtained for 3 equal stresses ($P = \frac{1}{3}S$) at $r \to \infty$ as well as for 1 principal stress (S) there. This conclusion has already been proved for the situation at $t = +0$ for unit loading (see the identity of (4.10) and (4.13)) and that it holds here may be accepted without proof.

6. SPHERICAL SYMMETRY

The solution needed is that of order $n = 0$

$$\bar{\epsilon} = \tilde{A} r^{-1} \exp(-qr). \qquad (6.1)$$

We shall not use a stress function now, but u_r as basic variable as all stresses and boundary conditions can be written as functions of u_r by virtue of the symmetry. For instance $\bar{\epsilon} = r^{-2}\partial/\partial r(\tilde{u}_r r^2)$, which gives after integration:

$$\tilde{u}_r = -\tilde{A}[(qr)^{-1} + (qr)^{-2}]\exp(-qr) - \tilde{B}r^{-2}. \qquad (6.2)$$

In order to describe the boundary conditions with $\tilde{u}_r$ we must firstly dispose of $\tilde{w}$. This is simple by virtue of (5.3) which can be written $(1/a)\partial^2/\partial r^2(\bar{\epsilon}r) = \partial^2/\partial r^2(\tilde{w}r)$, and which after integration becomes $(1/a)\bar{\epsilon} = \tilde{w} - C_1 - C_2 r^{-1}$. We may omit the term $C_2 r^{-1}$, which contributes a water-flow radiating from a point source at the origin and is irrelevant here. Therefore, it follows that

$$(1/a)\bar{\epsilon} = \tilde{w} - \tilde{C} \quad \text{or} \quad \tilde{w} = \tilde{C} + (1/a)r^{-2}\partial/\partial r(\tilde{u}_r r^2). \qquad (6.3)$$

Also $\bar{\sigma}_r$ and $\bar{\sigma}_r{}^*$ may be written in terms of $\tilde{u}_r$ in the following way:

$$\bar{\sigma}_r{}^* = \bar{\sigma}_r - \tilde{w} = E/(1+\nu)[\partial\tilde{u}_r/\partial r + \bar{\epsilon}\nu/(1-2\nu)] - \tilde{w}. \qquad (6.4)$$

We can now express the boundary conditions in terms of $\tilde{u}_r$. For the rigid type we need: (2.1.2), (2.2.1), (3.1), and (3.3). For the cavernous type: (2.1.3), (2.2.5), (3.1), and (3.3). The condition (3.1) is already satisfied by making $A_2 = iA_1$, in (5.6). There then remain three conditions to be satisfied for each type. The integration constants to adjust are $\tilde{A}$ and $\tilde{B}$ in (6.2), and $\tilde{C}$ in (6.3). It can be verified that their values are

$$\tilde{A} = -\bar{P}\tfrac{1}{3}bq^2 r_0{}^3 \exp(qr_0)/N,$$
$$\tilde{B} = \bar{P}\tfrac{1}{3}br_0{}^3[(\mu_2-\mu_1)(qr_0)^2 + (qr_0) + 1]/N, \qquad (6.5)$$
$$\tilde{C} = \bar{P} \quad \text{with} \quad N = [\mu(qr_0)^2 + (qr_0) + 1],$$

where

$\mu = \mu_1$ for rigid type $= b/3a$,
$\mu = \mu_2$ for cavernous type $= b/3a + [(1-\nu)/2(1-2\nu)]$.

With these values we obtain finally for the water pressure $\tilde{w}_0$ at $r = r_0$ (that is the pressure registered by the meter)

$$\tilde{w}_0 = \bar{P}\{1 - (b/3a)(qr_0)^2/N\}. \qquad (6.6)$$

The complex value of q [see (5.5)] leads to a complex value of $\tilde{w}_0$, $\tilde{w}_0$ being out of phase with $\bar{P}$. In Fig. 4 is shown the relation between $\bar{P}$ and $\tilde{w}_0$ in the complex plane as a function of the frequency $\omega/2\pi$. The range of frequencies where the rigid type can be used is from 10^{-2} Herz down to zero, while the cavernous type can be used for all frequencies.

The response of the pore pressure meter on unit step loading is obtained by using the Fourier integral

$$w_0 = \frac{1}{2\pi i}\int_{-\infty}^{+\infty} \tilde{w}_0 \frac{\exp(i\omega t)}{\omega} d\omega. \qquad (6.7)$$

Putting $\tilde{w}_0$ in a more manageable form

$$\tilde{w}_0 = \bar{P}\left\{1 - \frac{b}{3a}\frac{q^2}{\mu}\frac{1}{(q+\alpha)(q+\beta)}\right\}$$
$$= \bar{P}\left\{1 - \frac{b}{3a}\frac{q^2}{\mu}\frac{1}{(\alpha-\beta)}\left[\frac{1}{(q+\beta)} - \frac{1}{(q+\alpha)}\right]\right\}, \qquad (6.8)$$

Soil Mechanics and Transport in Porous Media

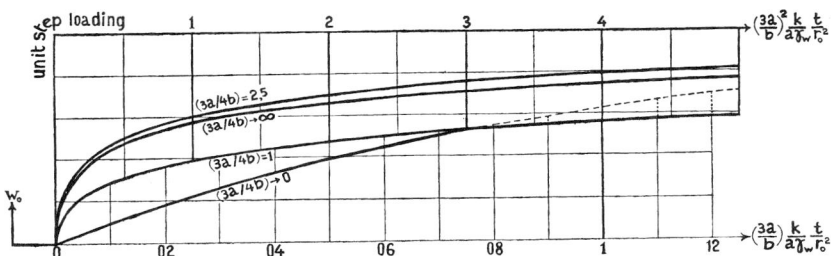

FIG. 5. Graphical representation of formula (6.9). Response of pore pressure meter on unit step loading. Rigid pervious type with some typical values of (3a/4b).

the integral may be expressed in terms of tabulated functions in the form (see Fig. 5):

$$w_0 = P\{1 - (b/3a)r_0(1-4\mu)^{-\frac{1}{2}}[\alpha \exp(\alpha^2\kappa t) \operatorname{erfc}(\alpha^2\kappa t)^{\frac{1}{2}} \\ - \beta \exp(\beta^2\kappa t) \operatorname{erfc}(\beta^2\kappa t)^{\frac{1}{2}}]\}, \quad (6.9)$$

which is valid for all $t > 0$,

where P = value of step loading at $r \to \infty$

$$\alpha = [1 + (1-4\mu)^{\frac{1}{2}}]/2\mu r_0,$$

$$\beta = [1 - (1-4\mu)^{\frac{1}{2}}]/2\mu r_0,$$

$$\operatorname{erfc}\xi = 2\pi^{-\frac{1}{2}} \int_{\xi}^{\infty} \exp(-\lambda^2)d\lambda,$$

$$\kappa = k/a\gamma_w \text{ (see (5.1))}.$$

For the definition of μ see (6.5) and of $b/3a$ see (1.1) and (5.1).

Putting $\mu = \mu_2$ and $t = 0$ the expression (6.9) reduces to (4.12). Since b is small in comparison with $3a$ the expression (6.9) for the rigid type becomes approximately

$$w_0 = P\{1 - \exp(\gamma^2\kappa t) \operatorname{erfc}(\gamma^2\kappa t)^{\frac{1}{2}}\}$$

with

$$\gamma = (3a/br_0). \quad (6.10)$$

Application of Stress Functions to Consolidation Problems

Application des Fonctions d'Airy aux Problèmes de Consolidation

by G. DE JOSSELIN DE JONG, Ir., Sub-director of the Soil Mechanics Laboratory, Delft, Netherlands

Summary

The number of stress functions necessary and sufficient for the solution of consolidation problems in axial symmetry is three: one for compression, one for rotation and a function satisfying $\nabla^2 F = 0$. The boundary conditions referring to the grain skeleton, as well as to the pore water, are then accounted for.

The use is shown for the case of a rigid sphere embedded in an infinite soil mass and loaded with a vertical force, and for the case of a semi-infinite solid loaded uniformly over a circular area of its surface, considering both an impervious and a pervious boundary.

Sommaire

Le nombre de fonctions de tension necessaire et suffisant pour resoudre les problèmes de consolidation, en symétrie axiale, est de trois. La première se rapportant à la compression, la seconde à la rotation et la troisième répondant à la condition $\nabla^2 F = 0$. Les conditions aux limites sont alors satisfaites, aussi bien pour la phase solide que pour l'eau interstitielle.

Une application est faite au cas d'une sphère rigide, placée au sein d'un massif de terre indéfini, et chargée verticalement, d'une part, et au cas d'un massif semi-infini supportant une charge circulaire uniforme, à sa surface, d'autre part. Ces cas sont traités dans les deux hypothèses de surfaces de contact perméables ou imperméables.

A useful method in elasticity for determining stress-distributions is the application of stress functions. In order to satisfy the conditions of equilibrium and compatibility these functions, F, must be solutions of the equations

$$\nabla^2 \nabla^2 F = 0 \quad \text{or} \quad \nabla^2 F = 0 \qquad \dots (1)$$

From these functions the stresses and strains are obtained by differential operations (LOVE, 1892).

When body forces act inside an elastic medium, the basic equations for the stress functions change, and it has been shown (by RAYLEIGH, 1894: LAMB, 1895) in the case of vibrations that two types of stress function are needed, one for compression and the other for rotation.

When pore water is present in a soil, this water causes hydrodynamic pressures which, by their gradients, act as body forces on the soil.

It has been shown (DE JOSSELIN DE JONG, 1953) how this conception leads to BIOT's (1941) basic equation for consolidation in three dimensions

$$c \nabla^2 \epsilon = \partial \epsilon / \partial t \qquad \dots (2)$$

where $c = k / a \gamma_w =$ consolidation coefficient; $k =$ permeability; $\gamma_w =$ density of water; $a =$ compressibility of grain skeleton; $\epsilon =$ volume dilatation of grain skeleton; and $t =$ time.

For the water pressure w we then find

$$(1/a) \nabla^2 \epsilon = \nabla^2 w \qquad \dots (3)$$

These formulae refer to the compression and are based on the supposition that the pore water is incompressible compared with the soil skeleton. The pore water therefore takes part of the normal stresses, and because of its viscosity the dissipation of superfluous water is slow and the water pressure decreases gradually.

Similarly the viscosity of the water may influence the deformations of the skeleton itself, especially when the stresses between the soil particles are supported by bounded water of greater viscosity (as in clay). We will, however, in this study disregard the time dependence of the deformation moduli and suppose that all strains of the grain skeleton take place simultaneously with the stresses. Because by these simplifying suppositions the shear strains are not retarded, the stress functions for the rotation are equal to those in elasticity and we will, in

our results, obtain instantaneous deformations due to shear strains.

By separating the effects of shear and compression we can appreciate their importance and decide which of the two must be studied the more according to the soil properties.

TAN (1954) has shown how the time dependence of the deformation moduli can be introduced from the beginning. However, the computations become too intricate even with over-simplified flow properties of the soil.

Finally we will suppose the grain skeleton to be an elastic medium with shear modulus G and Poisson's ratio μ.

Thus the previous introduced compressibility coefficient, a, is equal to

$$a = (1 - 2\mu)/2(1 - \mu)G$$

In reality there is no resilience and the relationship between stresses and strains is not linear.

Axial Symmetry

For brevity we will treat here the case of axial symmetry. In terms of displacements u_r and u_z ($u_\theta = 0$ because of symmetry) we can express the volume dilation ϵ and the rotation ω_{rz} as

$$\epsilon = \frac{1}{r} \frac{\partial}{\partial r}(r u_r) + (\partial u_z/\partial z) \qquad \dots (4)$$

$$\omega_{rz} = (\partial u_r/\partial r) - (\partial u_r/\partial z)$$

We now introduce the stress functions E and Ω in the following way:

$$2G u_r = (1 - 2\mu)(\partial^2 E/\partial r \partial z) - 2(1 - \mu)(\partial^2 \Omega/\partial r \partial z)$$

$$2G u_z = (1 - 2\mu)(\partial^2 E/\partial z^2) + 2(1 - \mu)\left(\frac{1}{r}\frac{\partial}{\partial r}\left[r \frac{\partial \Omega}{\partial r}\right]\right) \qquad \dots (5)$$

By inserting these expressions for the displacements in expression 4 we obtain:

$$2G\epsilon = (1 - 2\mu)\frac{\partial}{\partial z} \nabla^2 E$$
$$2G\omega_{rz} = 2(1 - \mu)\frac{\partial}{\partial r} \nabla^2 \Omega \qquad \dots (6)$$

which indicate how the stress functions are separated according to compression ϵ and rotation ω_{rz}.

320

The operator ∇^2 in the case of axial symmetry is equal to

$$\left[\frac{1}{r}\frac{\partial}{\partial r}\left(r\frac{\partial}{\partial r}\right) + \frac{\partial^2}{\partial z^2}\right]$$

The determination of the stresses from the stress-functions is obtained by application of the usual expressions for the stresses in terms of displacements, as given by the theory of elasticity:

$$\sigma_z = 2G[(\mu/(1 - 2\mu))\epsilon + \partial u_z/\partial z], \text{ etc.}$$

Elaboration of these relations then gives

$$\sigma_z = \frac{\partial}{\partial z}\nabla^2[\mu E + 2(1 - \mu)\Omega] + \frac{\partial^3}{\partial z^3}[(1 - 2\mu)E - 2(1 - \mu)\Omega]$$

$$\sigma_r = \frac{\partial}{\partial z}\nabla^2[\mu E] + \frac{\partial}{\partial z}\frac{\partial^2}{\partial r^2}[(1 - 2\mu)E - 2(1 - \mu)\Omega]$$

$$\quad \dots (7)$$

$$\sigma_\theta = \frac{\partial}{\partial z}\nabla^2[\mu E] + \frac{\partial}{\partial z}\frac{1}{r}\frac{\partial}{\partial r}[(1 - 2\mu)E - 2(1 - \mu)\Omega]$$

$$\tau_{rz} = \frac{\partial}{\partial r}\nabla^2[(1 - \mu)\Omega] + \frac{\partial}{\partial r}\frac{\partial^2}{\partial z^2}[(1 - 2\mu)E - 2(1 - \mu)\Omega]$$

So far compatibility and elasticity have been accounted for. The further consideration of equilibrium necessitates that

$$(\partial\sigma_r/\partial r) + [(\sigma_r - \sigma_\theta)/r] + (\partial\tau_{rz}/\partial z) = R \quad \dots (8)$$
$$(\partial\sigma_z/\partial z) + (\partial\tau_{rz}/\partial r) + (\tau_{rz}/r) = Z$$

where R and Z are the body forces exerted by the water gradients in radial and axial directions respectively, so that $R = \partial w/\partial r$ and $Z = \partial w/\partial z$.
A combination of expressions 7 and 8 gives

$$(1 - \mu)\left[\frac{\partial^2}{\partial r\partial z}\nabla^2 E - \frac{\partial^2}{\partial r\partial z}\nabla^2\Omega\right] = \partial w/\partial r \quad \dots (9)$$

$$(1 - \mu)\left[\frac{\partial^2}{\partial z^2}\nabla^2 E + \frac{1}{r}\frac{\partial}{\partial r}\left(r\frac{\partial}{\partial r}\nabla^2\Omega\right)\right] = \partial w/\partial z$$

If $\nabla^2\nabla^2\Omega = 0$, it is easily verified that equations 9, 3 and the first of 6 are satisfied by

$$w = (1 - \mu)[(\partial/\partial z)\nabla^2 E - (\partial/\partial z)\nabla^2\Omega] \quad \dots (10)$$

In equations 5, 7 and 10 all displacements and stresses involved are expressed in terms of the stress functions E and Ω. In addition, we need the basic equations which have to be satisfied by E and Ω in order to know which forms of stress function are applicable. From equations 2 and 6 it can be seen that E has to satisfy

$$\nabla^2 E_1 = c(\partial E_1/\partial t) \quad \text{or} \quad \nabla^2 E_2 = 0 \quad \dots (11)$$

A~ ~~th respect to the rotation we use an expression for ω_{rz}
deri   from 8 by elimination of R and Z, and by introduction
of tl   splacements we obtain:

$$r\frac{\partial}{\partial r}\left(\frac{1}{r}\frac{\partial\omega_{rz}}{\partial r}\right) + \frac{\partial^2\omega_{rz}}{\partial z^2} = 0$$

Introduction of equation 6 shows that this condition may be satisfied if Ω is a solution of one of the two following equations:

$$\nabla^2\nabla^2\Omega_1 = 0 \quad \text{or} \quad \nabla^2\Omega_2 = 0 \quad \dots (12)$$

From these basic equations 11 and 12 we obtain 4 types of stress function: $E_1, E_2, \Omega_1, \Omega_2$. The two functions E_2 and Ω_2 are not different from the function F_2 satisfying $\nabla^2 F_2 = 0$, and if we take $E_2 = \Omega_2$, the strains and stresses derived from them are identical to those derived from F_2, as is easily verified. So there are essentially 3 stress functions to satisfy the boundary conditions.

In elasticity there are always, for axial symmetry, two stress functions and two boundary conditions: 2 stresses are given, or 1 stress and 1 displacement, or 2 displacements. In the consolidation problem there is one additional boundary condition which refers to the pore-water; so the third stress function is not superfluous but is necessary for the analysis to be uniquely determined.
To show the use of the stress functions we will first give the solution for a rigid sphere embedded in an infinite soil mass and loaded with a vertical force, and then the solution for a semi-infinite body loaded over a circular area of its surface. In both problems we will consider two different boundary conditions with respect to the pore-water: (A) the boundary is pervious and the pore-water pressure there is zero; (B) the boundary is impervious so the gradient of water pressure normal to the surface is zero.

Rigid Sphere Embedded in Soil

For the case of a spherical body embedded in a soil mass we need solutions of equations 11 and 12 in the form of spherical harmonics, which are of the type:

$$E_1 = AR^{-1}e^{-qR}S_{n+\frac{1}{2}}(2qR)\Psi_n$$
$$\Omega_1 = BR^{(-n+1)}\Psi_n \quad \dots (13)$$
$$E_2 = \Omega_2 = CR^{(-n-1)}\Psi_n$$

We adopted the following notations $R = (r^2 + z^2)^{\frac{1}{2}} =$ spherical coordinate; $S_{n+\frac{1}{2}} =$ polynomial in qR as defined by TAN (1954, p. 136); $\Psi_n =$ Legendre polynomial of first kind, TAN (1954, p. 107); $A, B, C =$ arbitrary constants, adjustable according to boundary conditions; $q = (s/c)^{\frac{1}{2}}$ where s is the variable introduced in Heaviside or operational calculus to replace $\partial/\partial t$.
The solution for E_1 has been introduced by Rayleigh in dealing with vibration problems. Below we will indicate the Laplace transformations of the displacements and stresses by a bar over the symbol. Loads are supposed to be applied step-wise at $t = 0$, so the load P when transformed equals P/s.
Considering a rigid sphere with radius R_0, loaded at time $t = 0$ with a force P in the Z-direction, we have to satisfy the following boundary conditions at $R = R_0$. The rigidity of the sphere gives

$$\bar{u}_r = 0 \qquad \bar{u}_z = \text{constant} \qquad \dots (14)$$

The force P is supported by the sum of all the stresses from water and soil acting together on the surface. This gives

$$(P/s) = 2\int_0^{\pi/2} 2\pi R^2 \sin\Psi \, d\Psi [\sin\Psi \cdot \bar{\tau}_{rz} + \cos\Psi (\bar{\sigma}_z - \bar{w})]$$
$$\dots (15)$$

These two conditions refer to the grain skeleton and are sufficient in the case of an elastic medium to determine the stresses uniquely. The additional condition referring to the pore-water is: (A) the sphere is pervious; (B) the sphere is impervious giving respectively

$$\text{(A)} \quad \bar{w} = 0 \quad \text{(B)} \quad \partial\bar{w}/\partial R = 0 \quad \dots (16)$$

These three boundary conditions are satisfied by taking the solutions 13 with $n = 0$, giving

$$E_1 = AR^{-1}e^{-qR}, \quad \Omega_1 = BR, \quad E_2 = \Omega_2 = CR^{-1}$$
$$\dots (17)$$

Computation of the displacements and stresses may be performed by the introduction of these stress functions in the expressions 5, 7 and 10. By introduction of these stresses and displacements further in the conditions 14, 15 and 16, it is easy

321

to verify that the boundary conditions are satisfied by the following values for the constants:

$$B = P/8\pi(1 - \mu)s$$

$$A = 2Be^{qR_0}/q^2N$$

$$C = \tfrac{2}{3}B[(1 -\mu)R_0^2 + (1 - 2\mu)(3 + 3qR_0 + (qR_0)^2)]/q^2N$$

where $N = 1 + qR_0$ for the case (A); $N = 1 + qR_0 + \tfrac{1}{2}(qR_0)^2$ for the case (B). Inserting these in the expression for the vertical displacement gives finally

$$2G\bar{u}_z = (P/3\pi R_0 s)[1 + (1 - 2\mu)/4(1 - \mu)N]$$

From this result the reaction of the sphere to a stepwise loading of P is obtained by the inverse transformation giving for case (A)

$$2Gu_z = (P/3\pi R_0)\{1 + ((1 - 2\mu)/4(1 - \mu))[1 - e^{\tau}erfc(\tau)^{\frac{1}{2}}]\} \tag{18a}$$

and for case (B)

$$2Gu_z = (P/3\pi R_0)\{1 + ((1 - 2\mu)/4(1 - \mu))$$
$$[1 - \cos (2\tau)(1 - S(2\tau)) + \sin (2\tau)(1 - C(2\tau))]\} \tag{18b}$$

where S and C indicate Fresnel's integrals (MANDEL, 1953, p. 35) and $\tau = tc/R_0^2$.

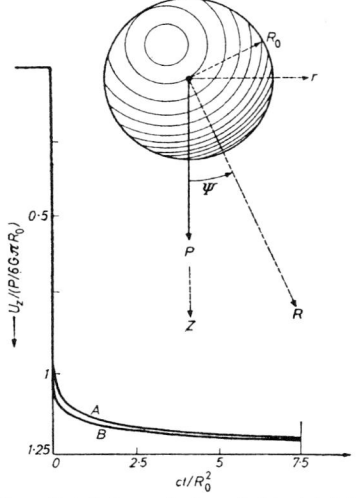

Fig. 1   The vertical displacement of a vertical loaded rigid sphere which is, (A) pervious, (B) impervious
Déplacement vertical d'une sphère rigide sous charge verticale (A) perméable, (B) imperméable

Fig. 1 shows these displacements as a function of time, for $\mu = 0$. The term between the square brackets in equation 18 in both cases (A) and (B) equals 0 for $t = 0$, increases rapidly at first and then gradually attains a value of 1 at $t = \infty$ (without oscillating). So the displacements become: in cases (A) and (B)

$$2Gu_z = (P/3\pi R_0) \qquad\qquad t = 0$$
$$2Gu_z = (P/3\pi R_0)[(5 - 6\mu)/4(1 - \mu)] \quad t = \infty$$

We notice that there is an instantaneous displacement and an after-effect, which is zero if $\mu = \tfrac{1}{2}$ (i.e. the medium is incompressible), and which attains a maximum value of $\tfrac{1}{4}$ of the

322

instantaneous settlement for $\mu = 0$ (i.e. the extreme case of compressibility for a material). The instantaneous displacement is due to the shear deformation which, by our initial supposition, can take place unhindered by the enclosed water.

We see, therefore, that for the embedded sphere the shear properties of the soil dominate the course of the displacement and the consolidation effect is very small.

Surface Loading of a Semi-infinite Body

For the case of a circular loaded area on the surface of a semi-infinite body, the boundary conditions at $z = 0$ are as follows: A total load P uniformly distributed over a circle of radius r_0 gives a distribution of normal stress σ_z as a function of r equal to

$$\bar{\sigma}_z = - (P/2\pi r_0)\int_0^\infty J_0(\lambda r)J_1(\lambda r_0)d\lambda \quad \dots (19)$$

This expression implies that the stress is zero outside the loaded area. J_n is Bessel's function of first kind, order n. The condition may be that there are no shear stresses over the whole surface, giving

$$\bar{\tau}_{rz} = 0 \qquad\qquad \dots (20)$$

The third boundary condition refers to the pore-water and we will consider two different cases: (A) the surface is pervious and the drainage of squeezed out water is so effective that the pressure is zero. This gives

$$\bar{w} = 0 \qquad\qquad \dots (21)$$

(B) the surface is covered with an impervious layer, so that

$$\partial\bar{w}/\partial z = 0 \qquad\qquad \dots (22)$$

The stress functions that may be used are now (see also MANDEL, 1953)

$$E_1 = Ae^{-\alpha z}J_0(\lambda r) \qquad (\partial/\partial z)\Omega_1 = Be^{-\lambda z}J_0(\lambda r)$$
$$E_2 = \Omega_2 = Ce^{-\lambda z}J_0(\lambda r) \qquad \dots (23)$$

where

$$\alpha^2 = \lambda^2 + q^2 = \lambda^2 + s/c$$

Again displacements and stresses can be computed using these stress functions, and the boundary conditions are applied in order to determine the values of the constants A, B and C. We then obtain

$$A = (P/s\pi r_0)\int_0^\infty J_1(\lambda r_0)d\lambda/\alpha(\alpha - \lambda)M$$

$$B = A\alpha^2(\alpha^2 - \lambda^2)/2\lambda^2$$

$$C = A(1 - 2\mu)\alpha^2/2(1 - \mu)\lambda^2$$

with $M = [(1 - \mu)(\alpha + \lambda) - (1 - 2\mu)\lambda]$ (for the case A), and $M = [(1 - \mu)(\alpha + \lambda)(\alpha/\lambda) - (1 - 2\mu)\lambda]$ (for the case B). So the displacement of the surface becomes

$$2G\bar{u}_z = (P/s\pi r_0)\int_0^\infty \{(1/\lambda) + (1 - 2\mu)/M\}J_0(\lambda r)J_1(\lambda r_0)d\lambda \tag{24}$$

The inverse transformation leads finally to the reaction of the soil to stepwise loading and gives with the aid of solutions from the operational calculus

$$2Gu_z = (P/\pi r_0)\{(1 - \mu)[2L(0,0) - L(0,1)] + \mu L\left(\frac{1 - 2\mu}{(1-\mu)^2}, \frac{\mu}{1-\mu}\right)\} \tag{25a}$$

$$2Gu_z = (P/\pi r_0)\{(1-\mu)[2L(0,0) - L(0,1)]$$ $\quad\quad \dots (25b)$

$$+ \frac{\mu(n+\tfrac{1}{2}) - (1-\mu)}{2n}\Big[2L\Big(\frac{\mu}{1-\mu} + n - \tfrac{1}{2}, 0\Big)$$

$$- L\Big(\frac{\mu}{1-\mu} + n - \tfrac{1}{2}, n - \tfrac{1}{2}\Big)\Big]$$

$$+ \frac{\mu(n-\tfrac{1}{2}) + (1-\mu)}{2n} L\Big(\frac{\mu}{1-\mu} - n - \tfrac{1}{2}, n + \tfrac{1}{2}\Big)\Big\}$$

where $n = \tfrac{1}{2}\{(5 - 9\mu)/(1-\mu)\}^{\tfrac{1}{2}}$ and

$$L(p,q) = \int_0^\infty e^{-pu^2ct/r_0^2} erfc\{qu(ct/r_0^2)^{\tfrac{1}{2}}\} J_0\Big(u\frac{r}{r_0}\Big)\frac{J_1(u)}{u}du$$

Because for $r = 0$, $L(p,q)$ equals 1 for $t = 0$ and zero for $t = \infty$ (if $-p < q^2$) and $L(0,0) = 1$ independently of time, both

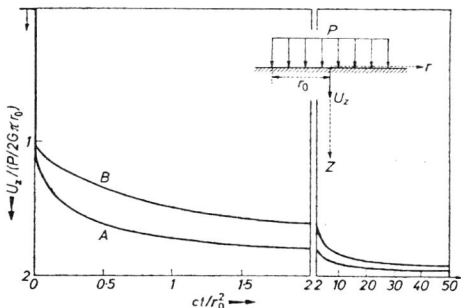

Fig. 2   The vertical displacement of the centre of circular area loaded by a uniformly distributed pressure, (A) pervious surface, (B) impervious surface

Déplacement vertical du centre d'une aire circulaire, chargée uniformément, (A) surface perméable, (B) surface imperméable

expressions 25a and 25b give the same settlement for the centre of the loaded area at $t = 0$ and $t = \infty$, being respectively

$$2Gu_z = P/\pi r_0 \quad \text{and} \quad 2Gu_z = 2(1-\mu)P/\pi r_0$$

Here again an instantaneous settlement is obtained in the solution corresponding to the deformation of the soil under shear. The after-effect is again zero for an incompressible soil

($\mu = \tfrac{1}{2}$) and has a maximum value for $\mu = 0$ which, at $t = \infty$, becomes equal to the immediate displacement.

Fig. 2 shows how the settlement of the centre of the loaded area progresses with time, for a soil having $\mu = 0$, in both cases A and B, indicating the retardation due to the impermeability of the surface.

Conclusion

In the cases considered the solutions for the variation of settlement with time show an instantaneous displacement at the moment of load application ($t = 0$) and an after-effect.

In the extreme case of compressibility ($\mu = 0$), this after-effect is equal to $\tfrac{1}{4}$ of the initial movement for the rigid sphere, and is equal to the initial movement for the loaded surface. The after-effect consists of a compressive strain which is retarded by the enclosed water but shows no great influence on the way in which superfluous water can escape.

The instantaneous settlement is due to shear deformation caused by shear stresses which, according to the previous supposition of elasticity of the soil, appear immediately. In reality the shear deformation has a character resembling viscous flow, and it is therefore retarded by a time effect.

As a combined computation of consolidation and viscous flow properties necessitates a still more complex mathematical treatment, it seems preferable to separate both effects and to determine first whether the consolidation has an appreciable effect, and if not (as in the case of the rigid sphere) to concentrate on the viscous flow properties only, which are independent of the geometry of the loaded soil mass.

The author wishes to acknowledge the kindness of Dr R. E. Gibson for his helpful discussion of this study.

References

BIOT, M. A. (1941).   Theory of three-dimensional consolidation. *J. Appl. Phys.*, **12**

JAHNKE and EMDE (1945).   *Tables of functions.*   New York; Dover

DE JOSSELIN DE JONG, G. (1953).   Consolidation around pore pressure meters.   *J. Appl. Phys.*, **24**, 7

LAMB, H. (1895).   *Hydrodynamics.*   Cambridge University Press

LOVE, A. E. H. (1892).   *The Mathematical Theory of Elasticity.*   Cambridge University Press

MANDEL, J. (1953).   Tassement d'un couche d'argile saturée d'eau sous l'effet d'une force concentrée à la surface du sol.   *Proc. 3rd International Conference on Soil Mechanics and Foundation Engineering*, Vol. 1, p. 413

RAYLEIGH, J. W. (1894).   *Theory of Sound*, Section 341

TAN, T. K. (1954).   *Onderzoekingen over de rheologische eigenschappen van klei.*   Doctorsthesis, Delft

323

DE JOSSELIN DE JONG, PROF. DR. IR G., 1968. *Géotechnique*: 18, 195–228.

CONSOLIDATION MODELS CONSISTING OF AN ASSEMBLY OF VISCOUS ELEMENTS OR A CAVITY CHANNEL NETWORK

PROFESSEUR DR IR G. DE JOSSELIN DE JONG*

SYNOPSIS

A cavity channel network consisting of many cavities with different compressibility interconnected by channels with different conductivity can serve as a model for a consolidating soil in both the primary and the secondary periods of consolidation.

The abundance of the constituting elements is introduced as a continuous frequency function using the spring dashpot assembly as a model because it produces similar effects. It is shown how this frequency function can be determined from test results.

Un réseau de cavités et de canaux comprenant de nombreuses cavités à compressibilités différentes reliées entre elles par des canaux à conductivités différentes peut servir de modèle pour un sol de consolidation aux deux périodes, primaire et secondaire de consolidation.

L'abondance des éléments prenant part à la constitution est introduite comme fonction de fréquence continue en utilisant l'ensemble à amortisseur comme modèle parce qu'il produit les mêmes résultats. On montre comment cette fonction de fréquence peut être établie à partir des résultats d'essais.

INTRODUCTION

In his book *Grondmechanica*, Keverling Buisman (1940) describes his observation that soils show time settlement effects which differ from the original Terzaghi theory of consolidation and which are especially pronounced in the secondary period. This discovery was mentioned in his paper to the 1936 Conference, but in his book and other publications (1938) he develops in more detail the possible causes and the consequence of this phenomenon. Since, however, all this was written in Dutch and never translated, some of his ideas which formed the basis of the present considerations are reproduced here.

Buisman called the secondary settlement the 'secular effect', from the Latin 'seculum' which means century. He coined the word 'secular' to point out to engineers that settlement could continue to develop after excess pore pressure had disappeared, and could continue to do so for a long time, perhaps centuries.

Because his work is only thirty years old a centennial verification is not available. However, for practical purposes his suggestion to extrapolate the settlement time curve as a straight line, when settlement is plotted on a linear scale and time on a logarithmic scale, gives an estimate which was verified to be reasonable in many instances. He called the corresponding settlement time relation the logarithmic time law and gave it the mathematical form

$$\zeta = ph(\alpha_p + \alpha_s \log t) \qquad \qquad . \quad . \quad . \quad . \quad . \quad . \quad (1)$$

where ζ is the settlement, p the effective stress increase by loading, h the layer thickness, α_p and α_s the settlement parameters, and t is the time in days after loading.

It was his opinion that the soil skeleton follows such a time settlement law, starting from the moment when an effective stress increase is created. Therefore the secular settlement also operates during the primary period of consolidation when excess pore pressures still prevent the effective stress from carrying the total load. Keverling Buisman suggested as a first approximation to use Terzaghi's theory for the determination of excess pore pressure and the effective stresses as a function of time. From these he calculated the settlement as produced by the gradually increasing effective stresses by adopting the validity of a superposition principle. He assumed that the response to unit step loading always conforms to equation (1).

---

* Professor of Soil Mechanics, Civil Engineering Department, Technological University, Delft.

195

His analysis to include this secular settlement within the framework of the hydrodynamic theory of consolidation was only a crude approximation, because at that time the powerful method of integral transforms was not available for the solution of consolidation problems. The importance of his analysis, however, is in the introduction of a time retarded soil skeleton in both primary and secondary periods of consolidation.

To give credit to Keverling Buisman, the word 'secular' could be reserved for those effects and mechanisms which delay settlement both in the primary and secondary periods of consolidation, and are responsible for deviations in time-settlement behaviour from Terzaghi's linear theory. In this Paper 'secular' is used in this sense.

Another important aspect of consolidation which he examines in his book is the explicit description of two mechanisms that could be responsible for the time-delayed reaction of the soil skeleton. One secular mechanism could be the viscous character of the pore water bound by the particles, the other mentioned is the size difference of the pores which have to transmit the expelled water in order to allow settlement to occur.

Both these mechanisms will be shown in this Paper to lead to the deviations from Terzaghi's theory which are frequently observed in tests after the excess pore pressure has reduced to zero. It will be seen that the pore system with different channel widths is to be preferred as a model to the viscous soil skeleton, for the former will allow the data to be fitted to the theory over a wider time range than is possible with the latter.

Taylor and Merchant (1940) introduced secular effects in their consolidation theory. As Barden (1965) has pointed out they obtained results which are representative for spring dashpot systems, but they did not specify the mechanism in terms of a rheological model.

Tan (1957) substituted a specific spring dashpot system for the grain contacts in order to introduce a secular effect according to Keverling Buisman's viscous secular mechanism. Tan (1957) and Gibson and Lo (1961) give solutions for discrete values of the relaxation properties of the spring dashpot combination. By taking three kinds of elements with enough difference in the relaxation time, Schiffman, Ladd and Chen (1964) obtained for the secondary period a time settlement curve, which on a semilogarithmic scale has an undulating shape with three successive waves. Barden (1965) introduced the non-linear behaviour of the viscous soil skeleton and using a power law for the relation between shear stress and shear strain rate he obtained a secondary time settlement curve. By plotting the difference between the final value of settlement and the settlement at an intermediate time, a power time law is obtained, which differs from the power time law used in this Paper.

These workers have therefore developed mathematically the first mechanism proposed by Keverling Buisman. The viscous character of the soil skeleton has been incorporated in the consolidation process and exact solutions have been obtained. It may be worth mentioning that the cases considered by Schiffman obey differential equations of so complicated a character that only electronic computers can solve them.

In this Paper the notion of a viscous grain skeleton represented by elements containing springs and dashpots is reconsidered. Instead of introducing a limited number of discrete parameters for the elements as was done in the previous work, the treatment will be concerned with a distribution of element parameters which is continuous and stochastic. This means that the properties of the individual elements vary randomly and that by their abundance it is possible to represent the frequency of occurrence of their parameters by a continuous function.

Besides the introduction of a continuous frequency function, the object of this study is to determine the form of this frequency function for a particular soil by analysing its settlement time response to loading. Actually this is the inverse analysis of the previous investigators, who determined settlement time response for a given element distribution. The motive for such an inverse procedure is the opportunity it offers us to study soil structure on a microscale.

Detailed information is not to be expected from a settlement observation which is the combined reaction of all elements working simultaneously. Besides this general reason for inaccuracy, the inverse analysis provides limited information, because the settlement time curve is only known by a finite number of observations in a limited time period and the viscous retardation mechanism acting at the grain contact point is only surmised.

In order to execute the inverse analysis, it is necessary to choose a functional relationship between settlement and time which is mathematically convenient and describes test results closely enough. In this case the choice was to approximate the settlement time curves by straight lines if plotted in a double logarithmic diagram. Since a straight line at a slope α in such a diagram gives a power law for the settlement time relationship, such that ζ is proportional to t^α, the corresponding law can be called a power time law. In order to include also settlement time curves that are not exactly straight lines in a double logarithmic diagram and settlements that reach an end value within a reasonable time, the more general case is considered that the observed curve can be approximated by a succession of straight line portions. Depending on the mathematical forms adopted below, these settlement time laws are called the sectional power time law and the product of power time law.

Although mathematical convenience was the main reason for adopting power time laws, the choice was inspired by finding that several settlement time observations gave better straight lines when plotted in a double logarithmic diagram than in the single logarithmic diagram used by Keverling Buisman. For practical use the power time law has the additional advantage of giving a safer prediction of settlement than the logarithmic time law because straight line extrapolation on a double logarithmic diagram gives greater values for the settlement than on a single logarithmic diagram.

The determination of the frequency function pertaining to the occurrence of spring dashpot elements proved possible by an inverse analysis of power time settlement curves in the secondary period of consolidation, i.e. after excess pore pressure had vanished. The corresponding theoretical considerations are developed in the following two sections.

In the subsequent section the consolidation process is considered which occurs when the spring dashpot assembly is considered to be immersed in pore water. The stochastic distribution of elements can be introduced in the consolidation theory by use of the frequency function. The result is a differential equation which resembles the original Terzaghi equation sufficiently to use the well known solutions as is shown by an example.

In the final section the cavity channel network suggested by Keverling Buisman as a second possibility to account for secular effects is considered in detail. This second secular mechanism seems never to have been developed theoretically. While studying settlement curves from peat in 1941 the Author observed in the secondary period an undulating time settlement behaviour on double logarithmic plots, similar to the curves obtained mathematically by Schiffman, Ladd and Chen (1964). For different load increments three successive waves were found, and for each increment these were repeated at the same times after loading. It was conjectured that interconnected channel systems of different size could be responsible for this behaviour (de Josselin de Jong, 1942). It is reasonable to suppose that the stems and veins of foliage have left larger channels embedded in more degraded material containing smaller pores. At the time a crude graphical procedure was used to separate the action of several communicating channel systems. By means of this analysis the observed settlement behaviour could be described by three different interconnecting systems.

For clay the undulating character of the settlement curve in the secondary period was not so pronounced. It was, therefore, not possible to separate in clays the action of discrete channel systems. However, it is reasonable to imagine that in natural clays also, channels of different size and conductivity exist. The concept of unequal pore size was used by Olsen (1960) to account for discrepancies observed in the permeability of clays. His work does not cover, however, the consolidation process.

In the subsequent analysis the possibility of describing secular effects by a model of cavities with various compressibility interconnected with channels of various conductivity will be examined. Again, the properties of the cavities and the channels will be considered to be stochastic and their occurrence will be described by continuous frequency functions.

The curious result obtained will be that mathematically the cavity channel elements behave very similarly to the spring dashpot elements. Their time-delaying effect on compressibility is in fact identical. The consequence is that, in order to match test results for large t by secular mechanisms, the same frequency functions are required for the two different systems.

The essential difference between the two systems is the interpretation of excess pore pressure and permeability. In the spring dashpot model the concept of excess pore pressure is the same as in the original consolidation theory of Terzaghi. Excess pore pressure has a well defined value at every moment and at every point of the body. This value is computed in the analysis. Discharge of pore water obeys Darcy's law with a well defined permeability coefficient.

In the cavity channel model excess pore pressure is a stochastic variable. It is not possible to compute the excess pore pressure for the individual cavities since the equations do not describe these various individual pore pressures explicitly. Instead an *average* excess pore pressure $\hat{u}$ appears in the analysis, which is a weighted average of the pore pressures in the surrounding cavities. The weighting function turns out to be proportional to the conductivity of the connecting channels. This quantity $\hat{u}$ will be called the average ambient excess pore pressure. A pore pressure weighted in this manner may be closely related to the pore pressure measured by a piezometric device, although it is probably not the same.

The differential equation governing the average ambient excess pore pressure has the same structure as the original Terzaghi consolidation equation and resembles the consolidation equation obtained for the spring dashpot assembly. This permits use of all the results obtained for spring dashpots for the cavity channel network as well. The difference is in the permeability, which in the consolidation equation appears now as a time dependent entity.

The effect of a difference in permeability is apparent only at the beginning of the settlement process. The normal theory of consolidation gives for small values of t the familiar root time law as a first approximation. According to the theory with secular effects developed here, the approximation is of the form $t^{(1-\epsilon)/2}$, where ϵ includes time dependence of both compressibility and permeability.

This approximate solution was verified with test results obtained from spherical samples loaded by an all round pressure (de Josselin de Jong and Verruijt, 1965). It turned out that a positive value for ϵ of 0·1 gave a close fit of test results with an accuracy of the third order of refinement in the approximation. The coefficients in the terms of the approximate expression were determined by use of Biot's theory of three dimensional consolidation. The fact that ϵ is positive according to observations rules out the model of a spring dashpot assembly submerged in pore water, and therefore favours the cavity channel model.

The tests on spherical samples gave for the settlement, in the secondary period, power time laws of the form t^α with α equal to about 0·1. The values for α and ϵ mentioned above as obtained from the spherical compression tests are used as examples. The spherical compression test is specifically adapted to the study of settlement increase during long time periods because the test results are not obscured by side friction as may occur in oedometer tests.

STOCHASTIC ASSEMBLY OF SPRING DASHPOT ELEMENTS

The analysis developed in this Paper departs from previous theories primarily by the introduction of stochastic elements, whose abundance permits us to consider them as being distributed according to a continuous frequency function. In order to explain the theory this

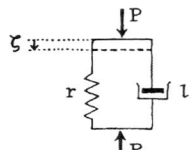

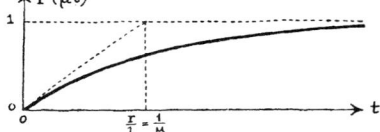

Fig. 1 (left). Spring dashpot element

Fig. 2 (right). Retardation function for linear spring dashpot (4)

special aspect is treated first by demonstrating how a continuous distribution of spring dashpot elements can produce a time dependent settlement.

Physical mechanism of spring dashpot element

The spring dashpot analogy has been introduced to simulate the response of the solid phase of a soil to an increase of the effective stresses. The dashpot in parallel with the spring (Fig. 1) retards its compression and can therefore be considered to produce a retarding effect similar to that which the physico-chemical bonds between soil particles exercise on the deformation of the grain skeleton.

For mathematical convenience the mechanical properties of spring and dashpot are assumed to be linear. This may be a crude approximation to reality, and better mechanisms could be substituted without difficulty when the real phenomena are better understood; a revised formula for the retardation function would then result, but the considerations developed in this Paper would not be altered substantially.

Let r be the compressibility of the spring and l the fluidity of the dashpot. A reduction in height ζ of the element requires a force ζ/r in the spring and a force $(1/l)(d\zeta/dt)$ in the dashpot. The total force P acting in the element is then

$$P = \frac{\zeta}{r} + \frac{1}{l}\frac{d\zeta}{dt} \qquad \ldots \ldots \ldots \quad (2)$$

which can be solved directly. If the initial condition is $\zeta=0$ for $t=0$, then

$$\zeta = Pr[1-\exp(-\mu t)] \qquad \ldots \ldots \ldots \quad (3)$$

where $\mu = l/r$.

The bracketed expression, the retardation function, will be called $F(\mu t)$, because it is this function which represents the retarding effect of the spring dashpot element (*see* Fig. 2). Therefore

$$F(\mu t) = 1 - \exp(-\mu t) \qquad \ldots \ldots \ldots \quad (4)$$

By taking the values of (μt) to be zero and infinite in this expression it is seen that

$$\left.\begin{array}{ll} F(\mu t) = 0 & \text{for } \mu t = 0 \\ F(\mu t) = 1 & \text{for } \mu t = \infty \end{array}\right\} \quad \ldots \ldots \ldots \quad (5)$$

In general $F(\mu t)$ can be any non-decreasing function which exhibits retarding properties. But whatever the mechanism, the possible retardation function will have to obey equations (5) in order to ensure that the settlement is zero at the start and reaches a terminal value at infinity. In order to include also other possible secular mechanisms, we will retain the generality of $F(\mu t)$ in the analysis as long as the treatment allows. In general, therefore, the settlement of the element will be

$$\zeta = PrF(\mu t) \qquad \ldots \ldots \ldots \quad (6)$$

The variable of the function F is the product (μt), which is dimensionless, so μ always has the dimension of reciprocal time. Actually μ is the retardation parameter whose reciprocal is a measure of the timescale involved in the retardation process. In equation (4) a small value of μ indicates that the element needs a long time before its compressibility is completely mobilized, because $F(\mu t) \to 1$ only if exp $(-\mu t)$ is reduced to a small number, and this is the case if (μt) has become large.

The grain skeleton can be considered to consist of rigid grains connected by many retarding elements at the contact points.   The elements are orientated randomly in all directions, each having its own r or l value.   In order to give expression to this randomness, the elements will be considered as stochastic entities.   Since the analysis presented in this Paper deviates from previous theories primarily by the introduction of randomness, this aspect will be examined first.

Stochastic distribution of elements

The constituent elements are called stochastic because their occurrence is arbitrary.   The probability of finding an element with special properties at a certain location is proportional to the number of elements possessing that property, expressed as a fraction of the total number of elements present.   The frequency of occurrence of elements possessing a certain material property expressed as a fraction of the total is by definition the frequency function of that property.

In this analysis it will be assumed that the number of elements is so great that the frequency distribution can be considered to be a continuous function.   This assumption is justified if it is possible to subdivide the soil mass into volumes in which the stress condition is nearly homogeneous but the number of elements large.   Such a volume will be called a representative elementary volume and its size will be L^3.   This volume will be considered to contain N^3 elements with N a large number.

The block in Fig. 3 is loaded by a vertical pressure p.   This load is transmitted primarily by the elements located at the vertical contact points between the grains.   For simplicity the elements are thought of as being arranged in columns, which transmit the force in a vertical direction.   This is an over-simplification because the cross links between the columns are ignored.   The effect of the cross links would be to redistribute the forces between the columns, but incorporating them in the analysis leads only to a second order refinement.   The forces in the columns are therefore considered to be equal throughout their height.

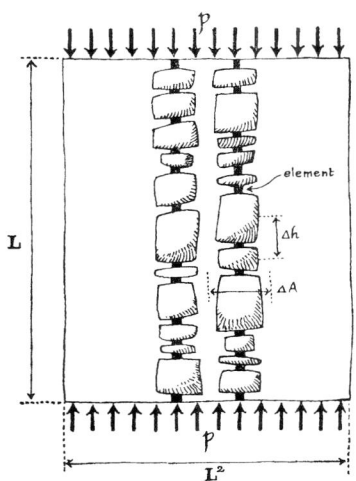

Fig. 3.   The representative elementary volume L^3 is subdivided in columns of grains.   At the contact points between the grains spring dashpot elements are distributed stochastically

On the average there are N^2 columns in a cross section of area L^2, so the columns occupy on the average an area

$$\Delta A = L^2/N^2 \qquad \ldots \ldots \ldots \ldots \quad (7)$$

and carry a force

$$P = p(L^2/N^2) \qquad \ldots \ldots \ldots \ldots \quad (8)$$

Deviations from the average are also ignored in this study because they also require a second order refinement.

Settlement of a stochastic assembly of retarded elements

Let the ith element of an assembly have a spring compressibility r_i and a dashpot fluidity l_i, while its retardation parameter is $\mu_i = l_i/r_i$. The height reduction of the ith element is then, using equations (6) and (8)

$$\zeta_i = p(L^2/N^2)r_i F(\mu_i t) \qquad \ldots \ldots \ldots \quad (9)$$

Introducing the volume compressibility ρ_i of the element instead of the spring compressibility r_i, the settlement of a soil block of height h can be written as

$$\zeta = ph(N^2/L^3) \sum_{i=1}^{N} \rho_i F(\mu_i t) \qquad \ldots \ldots \quad (10)$$

The justification of this expression is explained in the marked[1] paragraphs.

The volume compressibility ρ_i is defined as the volume reduction ΔV_i of the ith element divided by the vertical pressure p acting on it. If the element is prohibited from expanding laterally, the entire volume reduction is created by the height reduction ζ_i only, and obtained by multiplying ζ_i by the cross section area occupied by the columns ΔA. This gives

$$\Delta V_i = \zeta_i \Delta A = \zeta_i(L^2/N^2)$$

By definition $\rho_i = \Delta V_i/p$ and $r_i = \zeta_i/P$, therefore it follows from equation (8) that

$$\rho_i = \frac{\Delta V_i}{p} = \frac{\zeta_i L^2/N^2}{PN^2/L^2} = r_i \frac{L^4}{N^4}$$

Changing equation (9) to include the volume compressibility ρ_i, the height reduction of the ith element is

$$\zeta_i = p(N^2/L^2)\rho_i F(\mu_i t)$$

The height reduction of the total column is the summation of all the height reductions of the elements in the column. For a soil block of height h, the columns contain as an average Nh/L elements. The summation contains therefore Nh/L terms and the height reduction of the block is

$$\zeta = \sum_{i=1}^{Nh/L} \zeta_i = p(N^2/L^2) \sum_{i=1}^{Nh/L} \rho_i F(\mu_i t) \qquad \ldots \ldots \quad (11)$$

It is assumed that the number of elements in a column, of any length considered, is so large that although the elements are distributed at random their combined response is a fair representation of the average. This assures that the Nh/L elements contained in h produce a response which is h/L times the response of N elements. So the summation in equation (11) over Nh/L terms is equal to h/L times the summation over N terms, and this changes equation (11) into equation (10).

Final settlement for a stochastic assembly

In this section an expression for the final settlement of the soil block is developed as equation (15), which contains $\hat{a}$, the overall compressibility of the soil. This entity $\hat{a}$ is

---

[1] The reader can at first omit the marked paragraphs without impeding comprehension of the subject matter.

4

expressed in relation to the volume compressibility of the individual elements ρ_i by equations (13) and (14).

Because the dashpot continues to retard the process for an infinitely long time, the final settlement is only reached for $t=\infty$ when $\exp(-\mu_i t)$ is zero and $F(\mu_i t)=1$. Therefore the end value of the settlement is

$$\zeta(\infty) = ph(N^2/L^3) \sum_{i=1}^{N} \rho_i \qquad . \quad . \quad . \quad . \quad . \quad . \quad (12)$$

The same formula is obtained for any other retardation function provided it obeys equation (5).

At this stage $\bar{\rho}$, the mean[1] compressibility, is introduced, defined by

$$\bar{\rho} = (1/N) \sum_{i=1}^{N} \rho_i \qquad . \quad . \quad . \quad . \quad . \quad . \quad (13)$$

The only requirement on N is that it is sufficiently large to ensure that the sample contains a representative distribution of ρ_i. Inserted in equation (12) this becomes

$$\zeta(\infty) = ph(N^3/L^3)\bar{\rho}$$

The compressibility $\hat{a}$ of the entire block is now introduced as the average value of the compressibility $\bar{\rho}$ of its component elements multiplied by the number of elements per unit volume, N^3/L^3, namely

$$\hat{a} = (N^3/L^3)\bar{\rho} \qquad . \quad . \quad . \quad . \quad . \quad . \quad (14)$$

which then gives for the final settlement

$$\zeta(\infty) = p\hat{a}h \qquad . \quad . \quad . \quad . \quad . \quad . \quad (15)$$

Frequency function of the element distribution

The settlement of an assembly of elements was obtained in equation (10) as a summation over many elements with different values of ρ_i and μ_i. Because of their abundance, the elements can be considered to be distributed according to a continuous frequency function. This allows the settlement equation (10) to be converted from a summation into an integral.

How the transition from summation to integral can be developed is explained in the subsequent marked paragraphs, with equation (20) as a result. Fig. 4 helps to visualize the concept of the frequency function in this case.

In order to construct a frequency function the large number of N^3 elements are subdivided into J portions, labelled by j, a number running from 0 to J. Each portion contains the elements possessing a retardation parameter with a value between μ_j and $\mu_j+\Delta\mu_j$. The number J of the portions depends on the manner of subdivision. It is irrelevant here how many there are and also how the total interval between $\mu_0=0$ and $\mu_J=\infty$ is subdivided. The size of $\Delta\mu_j$ can, for instance, be different. The number of elements in the jth portion is called ΔM_j and so

$$\sum_{j=0}^{J} \Delta M_j = N^3 \qquad . \quad . \quad . \quad . \quad . \quad . \quad (16)$$

This summation includes all the values of μ that are to be found in the N^3 elements from μ_0 to μ_J. The terms of the summation in equation (10) will now be rearranged in ascending order of μ. This means that the serial number i is re-allocated in a manner such that $\mu_i < \mu_{i+1}$. Also let M_j be the number of elements with μ values smaller than μ_j. The interval $\Delta\mu_j$ is then covered when the serial number i runs from M_j to $M_{j+1}=M_j+\Delta M_j$ according to the definition of ΔM_j above.

---

[1] The circumflex (^) on a quantity is used to denote some average value of it, defined on each occasion.

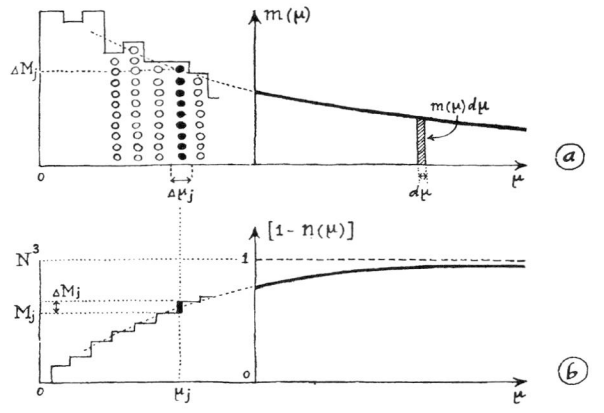

Fig. 4. (a) Black dots are elements in interval $\Delta\mu_j$. Numbers on left side divided by N^3 give continuous curve of frequency function $m(\mu)$. $N^3m(\mu)d\mu$ is number of elements in $d\mu$ (b) $n(\mu)$, cumulative distribution function is integral of $m(\mu)$ according to equation (36)

In equation (10) the summation contains N terms, and is after that multiplied by N^2. So it contains N^3 terms, from which ΔM_j terms have the required values for μ_j in the interval $\Delta\mu_j$. Equation (10) can therefore be rewritten as

$$\zeta = (ph/L^3) \sum_{j=0}^{J} \left[\sum_{i=M_j}^{M_j+\Delta M_j} \rho_i F(\mu_i t) \right] \quad \cdot \quad \cdot \quad \cdot \quad \cdot \quad \cdot \quad (17)$$

The interval $\Delta\mu_j$ will be taken small enough to allow the approximation of considering $F(\mu_i t)$ a constant in that interval and equal to $F(\mu_j t)$. Then

$$\sum_{i=M_j}^{M_j+\Delta M_j} \rho_i F(\mu_i t) = F(\mu_j t) \sum_{i=M_j}^{M_j+\Delta M_j} \rho_i$$

As a consequence the number ΔM_j is only a small fraction of N^3. It is assumed, however, that N^3 is so large that although ΔM_j is only a small fraction of N^3, it is still a large number. Further it is assumed that there is no correlation between ρ_i and μ_i. Then the ΔM_j elements contain by their abundance a fair sample of the ρ_i distribution in the entire system. One is then justified in using the average compressibility $\hat\rho$ according to equation (13) with ΔM_j substituted for N, and writing

$$\sum_{i=M_j}^{M_j+\Delta M_j} \rho_i = \hat\rho \Delta M_j$$

If there is correlation between ρ_i and μ_i this correlation can be taken care of in the frequency function defined below. It is assumed here that there is no correlation. These considerations applied to equation (17) yield

$$\zeta = (ph/L^3) \sum_{j=0}^{J} F(\mu_j t) \sum_{i=M_j}^{M_j+\Delta M_j} \rho_i = (ph/L^3)\hat\rho \sum_{j=0}^{J} F(\mu_j t)\Delta M_j$$

or using equation (14)

$$\zeta = p\hat{a}h \sum_{j=0}^{J} F(\mu_j t)(1/N^3)\Delta M_j \quad \cdot \quad \cdot \quad \cdot \quad \cdot \quad \cdot \quad (18)$$

This expression suggests an integral over μ_j, because all values of μ_j are incorporated, while j ranges from 0 to J. In order to obtain an integral a frequency function $m(\mu)$ is introduced in such a manner that $m(\mu_j)\Delta\mu_j = \Delta M_j/N^3$. Then equations (16) and (18) become

$$\sum_{j=0}^{J} m(\mu_j)\Delta\mu_j = 1 \text{ and } \zeta = p\hat{a}h \sum_{j=0}^{J} m(\mu_j) F(\mu_j t)\Delta\mu_j$$

If now $\Delta\mu$ is diminished to an infinitesimal quantity, $d\mu$, the summation becomes an integral which extends over all values of μ from 0 to ∞.

The last two expressions then become

$$\int_0^\infty m(\mu)d\mu = 1 \qquad \ldots \ldots \ldots \ldots (19)$$

and

$$\zeta = p\hat{a}h \int_0^\infty m(\mu)F(\mu t)d\mu \qquad \ldots \ldots \ldots (20)$$

The frequency function $m(\mu)$, as defined above to be related to ΔM_j can be considered as the fraction of elements whose values of μ cover the interval between μ and $\mu+d\mu$, divided by the size of the interval $d\mu$. It indicates how the elements are distributed with respect to μ (*see* Fig. 4(a)). Equation (20) describes how the settlement is created by an assembly of elements that are distributed according to $m(\mu)$, all having the same retardation function $F(\mu t)$.

TIME DEPENDENT SETTLEMENT PRODUCED BY SPRING DASHPOT ASSEMBLY
DURING THE SECONDARY PERIOD OF CONSOLIDATION

Equation (20) gives the relation between settlement $\zeta(t)$ and $m(\mu)$ the frequency function of the spring dashpot assembly as an integral expression with $m(\mu)$ in the integrand. If one wants to know $m(\mu)$ explicitly to study the soil structure, it is necessary to solve equation (20) for $m(\mu)$.

In the subsequent sub-sections two different procedures will be treated which permit determination of $m(\mu)$. The first procedure gives a crude approximation but simple expressions. The second is more refined and gives a better approximation, but more complicated expressions.

Both procedures require a particular mathematical structure for the settlement-time relationship observed in the secondary period of the oedometer test after excess pore pressures have vanished. Both these will be treated in the next sub-section and in the subsequent sub-sections they will be used for the determination of $m(\mu)$.

Settlement time law in the form of a combination of power terms

The two mathematical procedures required in the subsequent analysis for the determination of $m(\mu)$ demand a curve fitting which in both cases will consist of approximating the settlement time curve plotted in a double logarithmic diagram by portions which are straight lines.

These straight line portions give power terms of the form of equation (22).

Let $\theta_1, \theta_2, \ldots$ be the times corresponding to the points where the straight lines intersect and let $\alpha, \beta, \gamma \ldots$ be the slopes of the sections (*see* Fig. 5(a)). Then, if $x = \log t$ and $y = \log \zeta$ are the co-ordinates in the double logarithmic $t-\zeta$ diagram, these straight lines are

$$\left.\begin{array}{ll} y = \alpha x + \log A & 0 < t < \theta_1 \\ y = \beta x + \log B & \theta_1 < t < \theta_2 \\ y = \gamma x + \log C & \theta_2 < t < \theta_3 \ldots \text{etc.} \end{array}\right\} \quad \ldots \ldots (21)$$

The first two lines intersect at $t = \theta_1$, if y in the first two expressions is equal for $x = \log \theta_1$. This gives $\alpha \log \theta_1 + \log A = \beta \log \theta_1 + \log B$ or $B = A\theta_1{}^{\alpha-\beta}$. So all successive coefficients, B, C, and so on, can be expressed in terms of A. Since A has an inconvenient dimension, introduce θ_0, such that $A = h\theta_0{}^{-\alpha}$, with h being the thickness of the sample. That the forms in (21) are power time laws can be seen by reintroducing t and ζ, instead of x and y. For instance, the first then gives $\zeta = At^\alpha = ht^\alpha\theta_0{}^{-\alpha}$. Written in terms of ζ and t

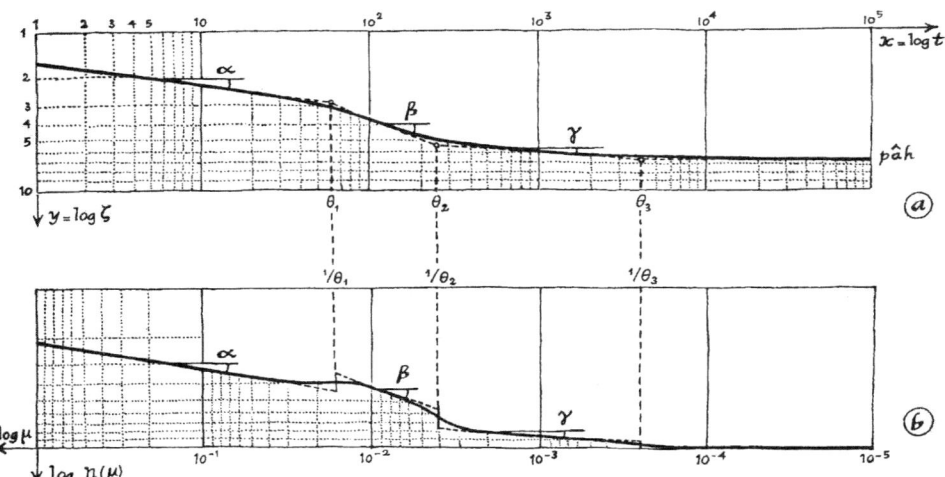

Fig. 5. (a) Settlement time curve in double logarithmic diagram. Dashed line is sectional power time law, solid line is product of power terms time law
(b) The cumulative distribution function $n(\mu)$ has sectionally the same slope in the double logarithmic diagram as the corresponding settlement

the equations (21) give a combination of power laws valid in successive time intervals, as follows:

$$
\left.
\begin{aligned}
\zeta(t) &= ht^{\alpha}\theta_0^{-\alpha} & 0 < t < \theta_1 \\
\zeta(t) &= ht^{\beta}\theta_0^{-\alpha}\theta_1^{\alpha-\beta} & \theta_1 < t < \theta_2 \\
\zeta(t) &= ht^{\gamma}\theta_0^{-\alpha}\theta_1^{\alpha-\beta}\theta_2^{\beta-\gamma} & \theta_2 < t < \theta_3 \ldots \text{etc.}
\end{aligned}
\right\} \quad \ldots \ldots \quad (22)
$$

The way in which this expression for $\zeta(t)$ was obtained shows that an arbitrary settlement observation, plotted against time on a double logarithmic diagram, can provide the values of all the parameters by approximating the actual curve by a few straight lines. The slopes of the lines give the parameters $\alpha, \beta, \gamma, \ldots$ and so on, the intersection points the parameters $\theta_1, \theta_2, \ldots$ and so on. The parameter θ_0 is equal to the time corresponding to a settlement of magnitude h on the extrapolated first straight line.

The form (22) may be called the *sectional power time law* because it represents a curve consisting of sections, from which each obeys a power law. This representation by power laws will allow a crude approximate determination of $m(\mu)$ valid for any retardation function.

A more refined theory based on integral transforms is possible only if the retardation function is known. The special function pertaining to the linear spring dashpot element will be considered as an example. The analysis then requires for $\zeta(t)$ a function that can be continued analytically in the complex t-plane enclosed by $-\pi < \arg t < +\pi$.

A function that satisfies this condition and resembles (22) satisfactorily is the following equation which will be called the *product of power terms time law*

$$
\zeta(t) = h\left(\frac{t}{\theta_0}\right)^{\alpha}\left(\frac{t+\theta_1}{\theta_1}\right)^{-\alpha+\beta}\left(\frac{t+\theta_2}{\theta_2}\right)^{-\beta+\gamma} \ldots \text{etc.} \quad \ldots \ldots \quad (23)
$$

This product power law is approximated by (22) if the intervals between $\theta_1, \theta_2, \theta_3 \ldots$ and so forth on the logarithmic diagram are large enough, such that

$$
\theta_1 \ll \theta_2 \ll \theta_3 \ll \ldots \text{etc.} \quad \ldots \ldots \ldots \quad (24)
$$

This can be seen by considering first a time t which is small with respect to θ_1. Then the term $(t+\theta_1)/\theta_1$ is almost unity, and so also are the subsequent terms. Thus the settlement curve is practically

$$\zeta(t) = ht^\alpha\theta_0^{-\alpha} \qquad\qquad 0 < t \ll \theta_1 \quad . \quad . \quad . \quad . \quad (25)$$

If t is large compared with θ_1, but small compared with θ_2, then the second term in the product becomes $(t/\theta_1)^{-\alpha+\beta}$, whereas the rest remains unity. So the product is approximately

$$\zeta(t) = ht^\beta\theta_0^{-\alpha}\theta_1^{\alpha-\beta} \qquad\qquad \theta_1 \ll t \ll \theta_2 \quad . \quad . \quad . \quad (26)$$

These results show that equation (23) coincides with (22) for the greater part of each of the intervals. Again, the parameters $\alpha, \beta, \gamma \ldots, \theta_1, \theta_2$, and so on can be obtained directly from the double logarithmic $t - \zeta$ plot. In the vicinity of $t = \theta_1$, $t = \theta_2$, and so on functions (22) and (23) differ. The product power law is a smooth curve, which blends the sharp intersections of straight lines appearing in a double logarithmic diagram, if (22) is plotted (*see* Fig. 5(a)).

The reasons for introducing the sectional power time law or the product power time law are mathematical convenience and convertibility of test results into formulae. The adoption of the equations is not dictated by the physics of the problem.

In the subsequent sub-sections the Laplace transform of the settlement will be needed. This will be denoted by a bar over the particular variable and be defined as

$$\bar{\zeta}(s) = \int_0^\infty \zeta(t) \exp(-st)dt \quad . \quad . \quad . \quad . \quad . \quad . \quad (27)$$

It is not possible to give an explicit expression for the transform of the product of power terms settlement time law. For the sectional power time settlement law, however, the transform can be obtained directly.

This gives by integrating sectionally

$$\bar{\zeta}(s) = hs^{-1-\alpha}\theta_0^{-\alpha}[\Gamma(1+\alpha) - \Gamma\{(1+\alpha; s\theta_1)\}]$$
$$+ hs^{-1-\beta}\theta_0^{-\alpha}\theta_1^{\alpha-\beta}[\Gamma\{(1+\beta; s\theta_1) - \Gamma\{(1+\beta; s\theta_2)\}] + \ldots \quad . \quad . \quad . \quad (28)$$

where $\Gamma(x)$ is the complete gamma function and $\Gamma\{x; y\}$ is the incomplete gamma function, both defined in the appendix. Using the approximations (125) and (126) this can be simplified to

$$\bar{\zeta}(s) = hs^{-1-\alpha}\theta_0^{-\alpha}\Gamma(1+\alpha) \qquad\qquad 0 < s^{-1} \ll \theta_1 \,$$
$$\bar{\zeta}(s) = hs^{-1-\beta}\theta_0^{-\alpha}\theta_1^{\alpha-\beta}\Gamma(1+\beta) \qquad \theta_1 \ll s^{-1} \ll \theta_2 \Big\} \quad . \quad . \quad (29)$$
$$\ldots \text{ etc.}$$

and if a final settlement $\zeta(t) = h\hat{a}p$ is reached for $t \to \infty$ the last term of this row will be

$$\bar{\zeta}(s) = h\hat{a}p \, s^{-1} \qquad\qquad s^{-1} \to \infty \quad . \quad . \quad . \quad . \quad (30)$$

The approximations used to obtain this result are valid over the entire region of the complex s-plane required in the elaboration of equation (76).

Approximate determination of frequency function and cumulative distribution function for general retardation mechanism

It is stated above that, if the settlement time relation is represented by a power law, $m(\mu)$ can be determined for any $F(\mu t)$. The result is a power law for $m(\mu)$ given by equation (42).

However, in order for the analysis to be executed the retardation function cannot be completely arbitrary. Physically a retardation function should be a non-decreasing function that is zero at $t = 0$, and unity at $t = \infty$ as indicated by equation (5). Moreover, the analysis also imposes requirements on E, the first derivative of F. The following, not so severe, conditions have to be satisfied:

$$0 < F(\lambda) < a_1\lambda \qquad E(\lambda) < a_1 \qquad \lambda \ll 1 \left.\right\}$$
$$1 > F(\lambda) > 1 - a_2\lambda^{-1} \qquad E(\lambda) < a_2\lambda^{-2} \qquad 1 \ll \lambda \left.\right\} \quad . \quad . \quad (31)$$

where a_1 and a_2 are positive numbers and

$$E(\lambda) = dF(\lambda)/d\lambda \quad . \quad . \quad . \quad . \quad . \quad . \quad (32)$$

From (31) it can be deduced that

$$F(\lambda) = 0 \qquad E(\lambda) < a_1 \qquad \lambda = 0 \left.\right\}$$
$$F(\lambda) = 1 \qquad E(\lambda) = 0 \qquad \lambda = \infty \left.\right\} \quad . \quad . \quad . \quad (33)$$

The retardation functions of linear spring dashpot elements and cavity channel elements obey these requirements.

The analysis is based on the fact that the time derivative of the settlement contains the settlement itself, when a power time settlement law is valid. From (22) it is seen, for instance, for the second interval, that

$$d\zeta/dt = h\beta t^{\beta - 1}\theta_0^{-\alpha}\theta_1^{\alpha - \beta} = \beta(\zeta/t) \qquad \theta_1 \ll t \ll \theta_2 \quad . \quad . \quad (34)$$

In order to bring in $m(\mu)$, the variable ζ in equation (34) is replaced by the expression (20). Since in this expression $F(\mu t)$ is the only function containing t, differentiation of ζ with respect to t requires only the time derivative of $F(\mu t)$. Using equation (32) this derivative is

$$dF(\mu t)/dt = \mu[dF(\mu t)/d(\mu t)] = \mu E(\mu t)$$

and so expression (20) gives for the left side of equation (34)

$$d\zeta/dt = \hat{p}\hat{a}h \int_0^\infty \mu m(\mu)E(\mu t)d\mu \quad . \quad . \quad . \quad . \quad . \quad (35)$$

On the other hand it is possible to obtain from equation (20) by integration by parts an expression for (ζ/t), which resembles equation (35). In order to obtain this expression it is convenient to introduce the cumulative distribution function $n(\mu)$ (see Fig. 4(b)), defined by

$$n(\mu) = \int_\mu^\infty m(\mu')d\mu' \quad . \quad . \quad . \quad . \quad . \quad . \quad (36)$$

By this definition $n(\mu)$ is the probability that an element possesses a relaxation parameter larger than μ. From equation (19) it is seen that

$$n(\mu) = 1 \qquad \text{for} \quad \mu = 0 \left.\right\}$$
$$n(\mu) = 0 \qquad \text{for} \quad \mu = \infty \left.\right\} \quad . \quad . \quad . \quad . \quad (37)$$

Differentiating equation (36) with respect to μ gives

$$dn(\mu)/d\mu = -m(\mu) \quad . \quad . \quad . \quad . \quad . \quad . \quad (38)$$

So integration by parts of equation (20) using equation (32) gives

$$\zeta = \hat{p}\hat{a}h\left[-n(\mu)F(\mu t)\Big|_0^\infty + t\int_0^\infty n(\mu)E(\mu t)d\mu\right]$$

The first term between the brackets is zero, because (33) and (37) indicate that the product is zero on both limits. So it follows that

$$(\zeta/t) = \hat{p}\hat{a}h \int_0^\infty n(\mu)E(\mu t)d\mu \quad . \quad . \quad . \quad . \quad (39)$$

A combination of equations (34), (35) and (39) gives

$$\int_0^\infty \mu m(\mu)E(\mu t)d\mu = \beta \int_0^\infty n(\mu)E(\mu t)d\mu \quad . \quad . \quad . \quad . \quad (40)$$

and this relation is only valid in the time interval $\theta_1 \ll t \ll \theta_2$.

Although the integrals extend from zero to infinity, only the range $1/\theta_2 \ll \mu \ll 1/\theta_1$ of the integral contributes to its value. It would take too much space to enter here into all the details of the proof, but with (24), (31) and (43) it is possible to obtain the following conclusions.

The part of the integral from $1/\theta_1$ to infinity gives a small contribution, because $E(\mu t)$ is small there. The part of the integral from zero to $1/\theta_2$ gives a small contribution, because the interval is small. This means that, using the values of $m(\mu)$ and $n(\mu)$ valid for $1/\theta_2 \ll \mu \ll 1/\theta_1$, but extending the integrals over $0 < \mu < \infty$, an error is introduced which can be disregarded if $\theta_1 \ll t \ll \theta_2$.

So equation (40) is satisfied for every $E(\mu t)$, and $F(\mu t)$, if

$$\mu m(\mu) = \beta n(\mu) = \beta \int_\mu^\infty m(\mu')d\mu'$$

Differentiating with respect to μ gives

$$\mu[dm(\mu)/d\mu] + m(\mu) = -\beta m(\mu)$$

$$\frac{dm(\mu)}{m(\mu)} = (-1-\beta)\frac{d\mu}{\mu}$$

$$m(\mu) = C_2\mu^{-1-\beta} \qquad \ldots \ldots \ldots \quad (41)$$

This argument can be repeated for every interval of (21) giving for the frequency function

$$\left.\begin{array}{lll} m(\mu) = C_1\mu^{-1-\alpha} & 1/\theta_1 \ll \mu < \infty \\ \quad\quad C_2\mu^{-1-\beta} & 1/\theta_2 \ll \mu \ll 1/\theta_1 \\ \quad\quad C_3\mu^{-1-\gamma} & 1/\theta_3 \ll \mu \ll 1/\theta_2 \ldots \text{etc.} \end{array}\right\} \quad \ldots \ldots \quad (42)$$

From these the cumulative distribution function defined by (36) is obtained as

$$\left.\begin{array}{lll} n(\mu) = (C_1/\alpha)\mu^{-\alpha} & 1/\theta_1 \ll \mu < \infty \\ \quad\quad (C_2/\beta)\mu^{-\beta} & 1/\theta_2 \ll \mu \ll 1/\theta_1 \\ \quad\quad (C_3/\gamma)\mu^{-\gamma} & 1/\theta_3 \ll \mu \ll 1/\theta_2 \ldots \text{etc.} \end{array}\right\} \quad \ldots \ldots \quad (43)$$

The constants $C_1, C_2, C_3 \ldots$ cannot be determined unless the function $F(\mu t)$ is specified, as shown in the next sub-section. There is, however, the general restriction on these constants that equation (19) has to be satisfied.

In Fig. 5(b) $n(\mu)$ is plotted in a double logarithmic diagram such that the μ-axis runs in the opposite direction to the t-axis. Then the line sections have the same slope as the corresponding intervals of the settlement curve. This remarkable result is not unexpected. The reason is that the retardation mechanisms are considered to be the same for all elements, so that the value of (μt), for which the retardation function reaches a certain large part of its end value, is the same for all. Let this be for $\mu t = a$. The elements that have completed their task at time t have then retardation parameters which are larger than a/t. Since settlement is the cumulative reaction of all these elements, it is proportional to the cumulative distribution function of these elements whose μ is inversely proportional to t.

More refined determination of frequency function for retardation mechanism consisting of linear spring dashpots

If the retardation function is the special function (4) valid for linear spring dashpot elements, a more refined analysis is possible, using a Laplace transform. The result is equation (46), a complicated formula that is approximated by equation (47), an expression resembling (42). This result is obtained as follows.

With equation (4) the settlement expression (20) can be written as

$$\zeta(t) = p\hat{a}h \int_0^\infty m(\mu)[1-\exp(-\mu t)]d\mu$$

For t infinite, exp $(-\mu t)$ reduces to zero and $\zeta(\infty)=p\hat{a}h\int_0^\infty m(\mu)d\mu$, which by equation (19) coincides with equation (15).

The difference between the terminal value $\zeta(\infty)$ of the settlement and an intermediate settlement $\zeta(t)$ is therefore

$$\zeta(\infty)-\zeta(t) = p\hat{a}h\int_0^\infty m(\mu)\exp(-\mu t)d\mu \qquad \ldots \ldots \quad (44)$$

This shows that $[\zeta(\infty)-\zeta(t)]$ is the Laplace transform of $m(\mu)$. Then the theory of integral transforms gives as a solution for $m(\mu)$

$$m(\mu) = \frac{1}{2\pi i}\int_{c-i\infty}^{c+i\infty}\frac{\zeta(\infty)-\zeta(t)}{p\hat{a}h}\exp(+\mu t)\,dt \qquad \ldots \ldots \quad (45)$$

The fact that $[\zeta(\infty)-\zeta(t)]$ is a Laplace transform limits the class of functions that can be used as representations of $\zeta(t)$. In particular, $\zeta(t)$ must be a function of t that is analytic for all complex values of t whose real part is larger than c, i.e. to the right of line AB in Fig. 6. In this case c can be taken as zero.

In practice the information about $\zeta(t)$ will consist of ζ values for a limited number of t-values. These values of t are all real and positive. From this information an analytical function has to be constructed which obeys the condition of regularity to the right of AB.

The sectional power time law, which by its simplicity was convenient in the previous analysis, is not suitable here because the discontinuities in $t=\theta_1$, θ_2, and so on introduce singularities to the right of AB.

The more complicated power product law equation (23) is preferred here. This function has its singularities in $t=-\theta_1$, $-\theta_2$ and so on, which are, actually, branch points.

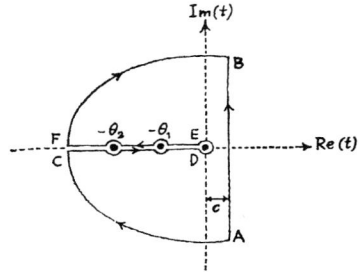

Fig. 6. **Integrative paths in complex** t **plane**

Evaluation of the integral is obtained by shifting the integral path from the vertical line AB towards the line CDEF which runs as a hairpin on both sides of the negative t-axis. The integral paths AC and FB give no contribution, because $[\zeta(\infty)-\zeta(t)]$ tends towards zero for large values of t.

By this shift the integral is brought into a form which leads to tabulated hypergeometric functions, if the power product consists of two terms. If the product (23) contains more terms, two adjacent terms are separated out in the way (26) was obtained from (23). There, one term, t^β, remained by considering $\theta_1 \ll t \ll \theta_2$. Here, two terms, $(t+\theta_1)^\beta$, $(t+\theta_2)^{-\beta+\gamma}$, can be dealt with simultaneously by considering $0 < t \ll \theta_2$. In virtue of (24) all other terms of equation (23) become constants. The consequence is that the values of $\zeta(t)$ in the vicinity of $t = \theta_1$, are represented more accurately, and therefore also the values of $m(\mu)$ in the vicinity of $\mu = 1/\theta_1$. The result is

$$m(\mu) = \frac{1}{p\hat{a}}\sum \frac{\alpha\Gamma(1-\alpha+\beta)}{\Gamma(1-\alpha)\Gamma(2+\beta)}\left(\frac{\theta_1}{\theta_0}\right)^\alpha \theta_1 \, _1F_1(1+\alpha;2+\beta;-\mu\theta_1)$$

$$+\frac{\beta\Gamma(1-\beta+\gamma)}{\Gamma(1-\beta)\Gamma(2+\gamma)}\left(\frac{\theta_1}{\theta_0}\right)^\alpha\left(\frac{\theta_2}{\theta_1}\right)^\beta \theta_2\, e^{-\mu\theta_1}\, _1F_1(1+\beta;2+\gamma;-\mu\theta_2)$$

$$+\frac{\gamma\Gamma(1-\gamma+\delta)}{\Gamma(1-\gamma)\Gamma(2+\delta)}\left(\frac{\theta_1}{\theta_0}\right)^\alpha\left(\frac{\theta_2}{\theta_1}\right)^\beta\left(\frac{\theta_3}{\theta_2}\right)^\gamma \theta_3\, e^{-\mu\theta_2}\, _1F_1(1+\gamma;2+\delta;-\mu\theta_3)$$

$$+\ldots\text{etc.} \qquad \ldots \ldots \quad (46)$$

where $_1F_1$ is the hypergeometric function (*see* equation (127)).

This summation of terms reduces to a simple form because of (24). By use of the approximation formulae for $_1F_1$ (equations (128) and (129)) it can be shown that for values of μ between $(1/\theta_1)$ and ∞ the first term exceeds all other terms and reduces to $\mu^{-1-\alpha}$ multiplied by a constant. In the interval between $(1/\theta_1)$ and $(1/\theta_2)$ the second term is the largest, giving $\mu^{-1-\beta}$ and so on. The frequency function (46) can therefore be approximated by

$$
\begin{aligned}
m(\mu) = {} & \frac{1}{\hat{p}\hat{a}}\frac{\alpha}{\Gamma(1-\alpha)}\left(\frac{1}{\theta_0}\right)^{\alpha}\mu^{-1-\alpha} && 1/\theta_1 \ll \mu < \infty \\[2mm]
& \frac{1}{\hat{p}\hat{a}}\frac{\beta}{\Gamma(1-\beta)}\left(\frac{\theta_1}{\theta_0}\right)^{\alpha}\left(\frac{1}{\theta_1}\right)^{\beta}\mu^{-1-\beta} && 1/\theta_2 \ll \mu \ll 1/\theta_1 \\[2mm]
& \frac{1}{\hat{p}\hat{a}}\frac{\gamma}{\Gamma(1-\gamma)}\left(\frac{\theta_1}{\theta_0}\right)^{\alpha}\left(\frac{\theta_2}{\theta_1}\right)^{\beta}\left(\frac{1}{\theta_2}\right)^{\gamma}\mu^{-1-\gamma} && 1/\theta_3 \ll \mu \ll 1/\theta_2 \dots \text{etc.}
\end{aligned}
$$

$$\cdots\quad(47)$$

A comparison with (42) shows that this is the same function of μ, a result that was to be expected, because $F(\lambda) = 1-\exp(-\lambda)$ satisfies the conditions (31) required in the analysis there. The difference is that the coefficients $C_1, C_2 \dots$ are obtained explicitly now.

Power time law with an asymptotic value

In order to demonstrate the use of these results, they can be applied to the example of a settlement curve which follows a power law up to the time θ_1 and then levels off towards a final value. This is a type of settlement observed in tests. In a double logarithmic time settlement diagram the curve can be shown diagrammatically by two straight lines, the first at a slope, say α, the second horizontal, at a slope zero (see Fig. 7(a)).

The product power law in that case would be

$$\zeta(t) = h\left(\frac{t}{\theta_0}\right)^{\alpha}\left(\frac{t+\theta_1}{\theta_1}\right)^{-\alpha}\qquad\cdots\qquad(48)$$

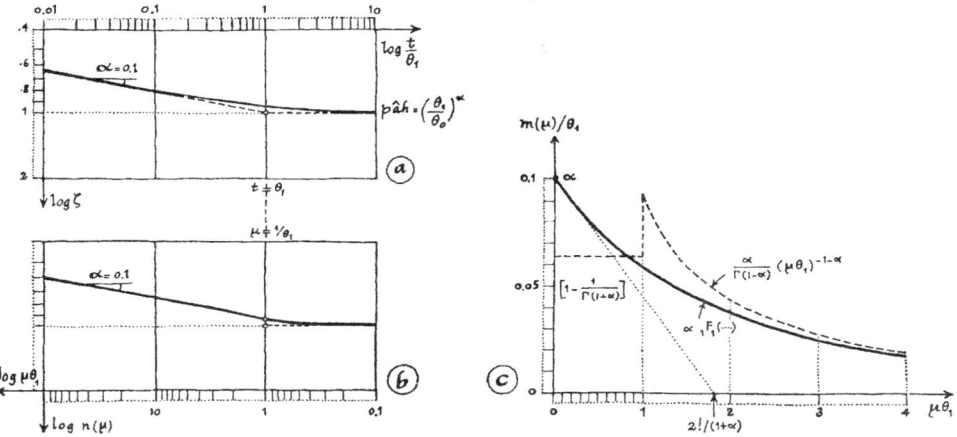

Fig. 7.   (a) Settlement
(b) Cumulative distribution function $n(\mu)$
(c) Frequency function $m(\mu)$
All solid lines are from refined theory, dashed lines from crude analysis

because β, γ and so on are all zero. The end value for the settlement is obtained for $t \gg \theta_1$ giving $\zeta(\infty) = h(\theta_1/\theta_0)^\alpha$, and so according to equation (15)

$$(\theta_1/\theta_0)^\alpha = p\hat{a} \qquad \ldots \ldots \ldots \quad (49)$$

From (46) the frequency function is, since $\beta = \gamma = \cdots = 0$ and $\Gamma(2) = 1$

$$m(\mu) = \alpha\theta_1 \, {}_1F_1(1+\alpha; 2; -\mu\theta_1) \qquad \ldots \ldots \ldots \quad (50)$$

This is a tabulated function plotted in Fig. 7(c). The following approximations for $m(\mu)$ for large and small values of μ are obtained using equations (128) and (129):

$$\left.\begin{aligned} m(\mu) &= [\alpha/\Gamma(1-\alpha)]\theta_1{}^{-\alpha}\mu^{-1-\alpha} & 1/\theta_1 \ll \mu \\ m(\mu) &= \alpha\theta_1\{1 - [(1+\alpha)/1!\,2!](\mu\theta_1) \\ &\quad + [(1+\alpha)(2+\alpha)/2!\,3!](\mu\theta_1)^2 - \ldots\} & \mu \ll 1/\theta_1 \end{aligned}\right\} \quad . \quad (51)$$

and for the cumulative distribution function

$$\left.\begin{aligned} n(\mu) &= [1/\Gamma(1-\alpha)](\mu\theta_1)^{-\alpha} & 1/\theta_1 \ll \mu \\ n(\mu) &= \int_\mu^\infty m(\mu')d\mu' = 1 - \int_0^\mu m(\mu')d\mu' \\ &= \{1 - \alpha(\mu\theta_1) + [\alpha(1+\alpha)/2!\,2!](\mu\theta_1)^2 - \ldots\} & \mu \ll 1/\theta_1 \end{aligned}\right\} \quad . \quad (52)$$

The values for large μ correspond well with (42), (43) and (47). The values for small μ are irregular in so far as they do not fit in the pattern of (42), (43) and (47). It could be surmised that the slope being zero, $m(\mu)$ would be proportional to μ^{-1}. Instead $m(\mu)$ tends to $a\theta_1$. The reason for this irregularity is that the analysis based upon equation (40) is not valid for zero slope. The analysis of the past sub-section, on the contrary, remains valid whatever the slope is.

From this result it can be seen that if an end value of the settlement is reached beyond a time θ_n the frequency function and the accumulative distribution function are defined as

$$\left.\begin{aligned} m(\mu) &= C_n \\ n(\mu) &= 1 - C_n\mu \end{aligned}\right\} \text{for } \mu \ll 1/\theta_n \qquad . \quad \ldots \ldots \quad (53)$$

with C_n a constant, which can be evaluated if the retardation function $F(\mu t)$ is known.

It is instructive to use the example of this sub-section to show how the constants C_1, C_2 and so on of (42) can be evaluated if $F(\mu t)$ is known. From the analysis there, together with the information on how to treat an end value with zero slope, with (42) and (53) it is possible to write immediately that

$$m(\mu) = C_1\mu^{-1-\alpha} \qquad n(\mu) = (C_1/\alpha)\mu^{-\alpha} \qquad 1/\theta_1 \ll \mu < \infty . \quad (54)$$

$$m(\mu) = C_2 \qquad\qquad n(\mu) = 1 - C_2\mu \qquad\quad 0 < \mu \ll 1/\theta_1 . \quad (55)$$

We will now make the crude but simple approximation that (54) and (55) are valid up to $\mu = 1/\theta_1$. It is then possible to determine the settlement, because the retardation function is known and given by equation (4), being $F(\mu t) = 1 - \exp(-\mu t)$, $E(\mu t) = \exp(-\mu t)$. Using equation (39) to avoid unsatisfactory integrals gives

$$\zeta(t) = h\left(\frac{\theta_1}{\theta_0}\right)^\alpha\left[1 + \frac{C_2}{t} - \left(1 + \frac{C_2}{t} + \frac{C_2}{\theta_1}\right)\exp\left(-\frac{t}{\theta_1}\right) + (C_1/\alpha)t^\alpha\Gamma\{(1-\alpha); (t/\theta_1)\}\right] \quad . \quad (56)$$

Using the approximation (125) for small values of t gives

$$\zeta(t) = h(\theta_1/\theta_0)^\alpha t^\alpha \alpha C_1/\Gamma(1-\alpha) \qquad t \ll \theta_1 \quad \ldots \quad (57)$$

From equation (48) it follows that this should equal $h(t/\theta_0)^\alpha$ so that C_1 should have the value

$$C_1 = [\alpha/\Gamma(1-\alpha)]\theta_1^{-\alpha} \qquad . \quad \ldots \ldots \quad (58)$$

This is exactly the same as equation (51).

The next step is to evaluate C_2. This can be done by use of the requirement (19) on $m(\mu)$. By integrating $m(\mu)$ of equation (55) from zero to $(1/\theta_1)$ and of (54) from $(1/\theta_1)$ to infinity the result must be unity. This gives

$$(C_2/\theta_1) + (C_1/\alpha)\theta_1^\alpha = 1 \qquad \ldots \ldots \ldots \quad (59)$$

which together with equation (58) requires that

$$C_2 = \{1 - [1/\Gamma(1-\alpha)]\}\theta_1 \qquad \ldots \ldots \ldots \quad (60)$$

According to equation (51) this should be $\alpha\theta_1$. In a typical test result α is $0\cdot1$, then $\Gamma(1-\alpha) = 1\cdot0682$, and so C_2 is $0\cdot0638\,\theta_1$ instead of $0\cdot1\,\theta_1$. In order to show the difference between this crude analysis and the more refined theory, Fig. 7 contains the settlement, the frequency function and the cumulative distribution functions as determined in this sub-section. The difference is small enough to permit the use of the simpler analysis.

Conclusion

The above analysis allows us to determine from the secondary period of an oedometer test how the relaxation parameters of an assembly of spring dashpots have to be distributed in order to show a settlement time behaviour similar to that of the soil under the influence of effective stress. Consolidation effects created by excess pore pressures have not been introduced here.

The approximate result is that the frequency function (42) and the cumulative function (43) are power laws of μ, if the settlement obeys a power law of t, (22). These power laws have the same coefficients but a reciprocal character. The more refined theory gives as a result (46), a complicated formula for the frequency function, which reduces to the same power law by approximation. It may be emphasized here that the information about the frequency function, to be extracted from test results, is limited to values of μ that are the reciprocal of the observation times.

The example of the past sub-section shows that the difference between the crude analysis and the more refined is small. This suggests that the simpler analysis may be preferred in a study of soil structure where uncertainties about the retarding mechanism obviate the necessity for mathematical refinements.

CONSOLIDATION OF SPRING DASHPOT ASSEMBLY

So far the settlement considered was caused by an increase of the effective stress with a magnitude p for all elements starting at $t = 0$. If the spring dashpot elements are immersed in water, the effective stress gradually increases as the excess pore pressure decreases, and during the consolidation process the effective stresses differ at different elevations.

In this section the basic equation of consolidation is developed for the one dimensional case, taking account of excess pore pressure and the retarding response to effective stresses caused by the secular mechanism of spring dashpot elements.

The effect of secular mechanisms on consolidation is discussed by use of an example.

Basic equation of consolidation for retarded elements immersed in pore water

To develop the basic equation for one dimensional consolidation a layer of thickness H is considered, which is compressed in the z direction. The differential equation will be developed for a slice of thickness Δz which, although of infinitesimal thickness, will contain enough elements to permit the use of the integral expression (20) for evaluating its settlement.

Assuming that Darcy's law applies to the excess pore pressure, u (where k is the permeability and γ_w is the specific weight of water), the amount of water ΔV expelled from the layer over an area, A, during a time, dt, is equal to

$$d(\Delta V) = -(k/\gamma_w)(\partial^2 u/\partial z^2)\Delta z\, dt\, A \quad \cdots \cdots (61)$$

If the soil is completely saturated this amount of water is equal to the reduction in volume of the grain system. In one dimensional consolidation the elements are prohibited from expanding laterally so the volume reduction ΔV divided by the area A is equal to the height reduction $d(\Delta\zeta)$ of the slice Δz, so equation (61) becomes

$$d(\Delta\zeta) = -(k/\gamma_w)(\partial^2 u/\partial z^2)\Delta z\, dt$$

Introducing the time rate of settlement $\partial(\Delta\zeta)/\partial t$, which is equal to the height reduction $d(\Delta\zeta)$ divided by the time interval dt necessary to produce that settlement, gives

$$\partial(\Delta\zeta)/\partial t = -(k/\gamma_w)(\partial^2 u/\partial z^2)\Delta z \quad \cdots \cdots (62)$$

In establishing Terzaghi's equation for consolidation the next step is to relate the rate of settlement $\partial(\Delta\zeta)/\partial t$ to the increase of effective stress by introducing the compressibility as a coefficient which is time independent. In the analysis here, however, the compressibility is time-dependent. From the moment when the effective stress increases a mechanism begins to operate which produces settlement in a retarded manner.

By using the Laplace transform the analysis can be simplified, because the rôle of time in the problem is then incorporated in the transformation parameter, s, which can be considered as a constant parameter in the following procedure.

A bar over a variable denotes its Laplace transform, which is a variable in s instead of the original variable in t (see equation (27)). The transform of equation (62) then becomes

$$s\Delta\bar{\zeta}(s) = -(k/\gamma_w)(d^2\bar{u}(s)/dz^2)\Delta z \quad \cdots \cdots (63)$$

For the transformation of $\partial(\Delta\zeta)/\partial t$ use is made of the fact that the settlement is zero at time $t=0$.

The next step is to produce the relation between the transform of the settlement and the transform of the effective stress. This can be obtained by starting from equation (2), the differential equation for one element. Instead of a time-independent force P, the load on one element will now be a time variable effective stress σ' which acts over an area ΔA given by equation (7). The Laplace transform of equation (2) then becomes for the ith element,

$$\bar{\sigma}'(s)\frac{L^2}{N^2} = \frac{\bar{\zeta}_i(s)}{r_i} + s\frac{\bar{\zeta}_i(s)}{l_i}$$

Since s is an inert parameter, this equation, with $\mu=l/r$, becomes

$$\bar{\zeta}_i(s) = \bar{\sigma}'(s)\frac{L^2}{N^2}\frac{r_i\mu_i}{s+\mu_i} \quad \cdots \cdots \cdots (64)$$

That this expression is the transform of equation (9) can be verified by substituting, in place of $\sigma'(t)$ the time independent pressure, p. The transform of such a time invariant quantity is p/s, so equation (64) for this case becomes

$$\bar{\zeta}_i(s) = p\frac{L^2}{N^2}\frac{r_i\mu_i}{s(s+\mu_i)} = p\frac{L^2}{N^2}r_i\left[\frac{1}{s}-\frac{1}{s+\mu_i}\right]$$

The inverse of this is

$$\zeta_i(t) = p(L^2/N^2)r_i[1-\exp(-\mu_i t)]$$

and this is the same as equation (9) with $F(\mu_i t)$ given by equation (4).

In a similar way the transform of equation (10) gives

$$\bar{\zeta}(s) = ph(N^2/L^3)\sum_{i=1}^{N}[\rho_i\mu_i/s(s+\mu_i)] \quad \cdots \cdots (65)$$

By introducing the frequency function $m(\mu)$, expression (20) was finally obtained. In that analysis neither p nor $F(\mu t)$ changed, so all the manipulations involved can be applied to the transformed settlement $\bar{\zeta}_i(s)$ from equation (64) as well, without changing $\bar{\sigma}'(s)$, which stands for p/s, or $\mu_i/(s+\mu_i)$ which stands for s times the transform of $F(\mu_i t)$. The transformed settlement of the layer of thickness Δz can then be written

$$\Delta\bar{\zeta}(s) = \bar{\sigma}'(s)\hat{a}\Delta z \int_0^\infty [\mu m(\mu)/(s+\mu)]d\mu \quad . \quad . \quad . \quad . \quad (66)$$

The time dependence of the compressibility caused by the stochastically distributed elements is incorporated in the integral over μ. This will be denoted by $g(s)$, so

$$g(s) = \int_0^\infty [\mu m(\mu)/(s+\mu)]d\mu \quad . \quad . \quad . \quad . \quad . \quad (67)$$

and equation (66) becomes

$$\Delta\bar{\zeta}(s) = \bar{\sigma}'(s)\hat{a}g(s)\Delta z \quad . \quad . \quad . \quad . \quad . \quad . \quad (68)$$

This expression describes the transformed relation between settlement and effective stress and involves the use of a compressibility $\hat{a}g(s)$ which incorporates the time dependence by the function $g(s)$.

Although $g(s)$ is defined here by equation (67) as a function of the frequency function $m(\mu)$, it is not necessary to know $m(\mu)$ if $g(s)$ is required. The direct way to obtain $\hat{a}g(s)$ is by use of settlement observations during a period when the effective stress does not vary any more, i.e. after excess pore pressure has disappeared. Expressed in a sectional power law this settlement has a Laplace transform given approximately by equations (29) and (30).

Use of equation (68) with $\bar{\sigma}'(s)=p/s$ and $\Delta z=h$, then gives

$$\left.\begin{aligned}
g(s) &= \frac{1}{\hat{p}\hat{a}}\left(\frac{1}{\theta_0}\right)^\alpha s^{-\alpha}\Gamma(1+\alpha) && 1/\theta_1 \ll s \ll \infty \\
g(s) &= \frac{1}{\hat{p}\hat{a}}\left(\frac{\theta_1}{\theta_0}\right)^\alpha\left(\frac{1}{\theta_1}\right)^\beta s^{-\beta}\,\Gamma(1+\beta) && 1/\theta_2 \ll s \ll 1/\theta_1 \\
&\ldots \text{etc.} && \ldots \\
g(s) &= 1 && s \to 0
\end{aligned}\right\} \quad . \quad . \quad (69)$$

It is now possible to continue the development of the differential equation for the consolidation process, by substituting equation (68) into equation (63), which gives

$$s\bar{\sigma}'(s)\hat{a}g(s) = -(k/\gamma_w)(d^2\bar{u}(s)/dz^2)$$

The effective stress can be expressed as the difference of the load p and excess pore pressure u, which becomes after Laplace transformation

$$\bar{\sigma}'(s) = (p/s) - \bar{u}(s) \quad . \quad . \quad . \quad . \quad . \quad (70)$$

So, finally, the transformed consolidation equation can be written

$$(k/\gamma_w)(d^2\bar{u}/dz^2) = \hat{a}g(s)[s\bar{u}-p] \quad . \quad . \quad . \quad . \quad (71)$$

and using $(k/\gamma_w\hat{a})=\hat{c}_v$ this becomes

$$[\hat{c}_v/g(s)](d^2\bar{u}/dz^2) = s\bar{u}-p \quad . \quad . \quad . \quad . \quad . \quad (72)$$

Discussion of the consolidation equation by use of an example

The expression (72) is a differential equation for $\bar{u}(z, s)$, which can be solved for particular boundary conditions. While $\bar{u}(z, s)$ is solved as a function of z, the other quantities including s remain constant. Therefore the factor $[k/\gamma_w\hat{a}g(s)]$ in equation (71) is a constant coefficient at this stage, and actually it plays the rôle of the consolidation coefficient c_v of the normal consolidation theory. Therefore $(k/\gamma_w\hat{a})$ was indicated by $\hat{c}_v$ in equation (72), and so the consolidation factor there is $[\hat{c}_v/g(s)]$. This factor excepted, equation (72) is identical to the Laplace transform of the Terzaghi consolidation equation. The complete secular effect is incorporated in the coefficient of consolidation, because this coefficient contains the function $g(s)$, which introduces the retardation mechanism as is seen from its definition (67). The value of $g(s)$ is obtained directly from test results in the form (69).

Solution of the geometric part of the problem requires no specification of $g(s)$ because s is inert at that stage. Only at the end, when the solution of $\bar{u}(z, s)$ is inverted to obtain the excess pore pressure $u(z, t)$ as a function of t, is the form $g(s)$ required. This means that all known transformed solutions of boundary value problems in normal consolidation can be used. Only in the last operation of inversion will the result deviate from the known results.

How this works out will be shown by a one dimensional example of a layer of thickness H, resting on an impermeable base at $z=0$ and loaded at $t=0$ with a pressure p.

The transformed boundary conditions are

$$\bar{u} = 0, \qquad z = H$$
$$d\bar{u}/dz = 0, \qquad z = 0$$

The solution of equation (72) gives

$$\bar{u}(z, s) = \frac{p}{s}\left[1 - \frac{\cosh(\hat{q}z)}{\sinh(\hat{q}z)}\right] \qquad \cdots \qquad (73)$$

with

$$\hat{q} = [sg(s)/\hat{c}_v]^{1/2} \qquad \cdots \qquad (74)$$

The settlement is $\zeta = \int_0^H \bar{\sigma}'dz$ which, using equations (70) and (73) becomes

$$\zeta(s) = (p\hat{a}/s\hat{q})g(s)\tanh(\hat{q}H) \qquad \cdots \qquad (75)$$

This solution of the transformed settlement is identical to the solution obtained in the normal case. The final step is to construct from $\zeta(s)$ the settlement $\zeta(t)$ by use of the inversion integral

$$\zeta(t) = (p\hat{a}/2\pi i)\int_{c-i\infty}^{c+i\infty}[g(s)/s\hat{q}]\tanh(\hat{q}H)\exp(st)ds \qquad \cdots \qquad (76)$$

Evaluation is effected by determining the location of the poles of the integrand. Besides the branch point at $s_0=0$ these are given by

$$\hat{q}_m = \pm\tfrac{1}{2}i\pi(2m-1), \qquad m = 1 \to \infty \qquad \cdots \qquad (77)$$

so that the poles all lie on the imaginary $\hat{q}$ axis.

Up to this point the analysis is identical to the procedure followed in a normal consolidation problem, and solutions that are worked out along these lines can be used without any revisions. The difference comes out in the transition from $\hat{q}$ to s. In the consolidation theory without secular effects $g(s)$ is unity and according to equation (74) q is $(sa\gamma_w/k)^{1/2}$, a function which contains s only as a square root. Therefore the imaginary q-axis is mapped into the negative s-axis and all poles lie on the negative s-axis.

In the case of a secular effect, $g(s)$ is some complicated function approximated by the simpler expressions (69). Therefore the relation (74) between $\hat{q}$ and s is an intricate function which maps the imaginary $\hat{q}$-axis on to two undulating lines symmetric to the negative s-axis. These lines, indicated as dotted lines in Fig. 8, carry the poles.

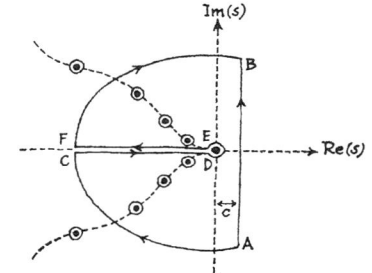

Fig. 8.   Integration paths in complex s plane

Instead of integrating from $c-i\infty$ to $c+i\infty$ along the straight line AB, it is more convenient to shift the integration path to ACDEFB. The parts AC and FB give no contribution, but there remains the hairpin around the negative s-axis and the residues of the poles on the undulating dotted lines. There are no other singularities in the region swept by the shifting of the integration path because it can be shown that the function $g(s)$, as obtained from the time settlement observations, behaves regularly up to the line CDEF. The solution then becomes

$$\zeta(t) = (p\hat{a}/2\pi i) \int_{CDEF} [g(s)/s\hat{q}] \tanh(\hat{q}H) \exp(st)ds$$

$$-p\hat{a}H \frac{8}{\pi^2} \sum_{m=1}^{\infty} \frac{1}{(2m-1)^2} \left\{ \frac{g(s^+) \exp(s^+t)}{1+s^+[g'(s^+)/g(s^+)]} + \frac{g(s^-) \exp(s^-t)}{1+s^-[g'(s^-)/g(s^-)]} \right\} \quad . \quad (78)$$

with

$$s^+g(s^+) = \hat{c}_v[+\tfrac{1}{2}i\pi(2m-1)]^2/H^2$$
$$s^-g(s^-) = \hat{c}_v[-\tfrac{1}{2}i\pi(2m-1)]^2/H^2$$
$$g'(s) = dg(s)/ds$$

In the normal solution $g(s)=1$, then from the integral over CDEF in equation (78) only the circle around the branch point in $s=0$ survives to give $2\pi iH$, because the integrand becomes equal on both sides of the negative s-axis, CD and EF. The summation over the residues of the poles finally gives the well-known solution

$$\zeta(t) = p\hat{a}H \left\{ 1 - \frac{8}{\pi^2} \sum_{m=1}^{\infty} \frac{1}{(2m-1)^2} \exp\left[-\left(\frac{2m-1}{2}\frac{\pi}{H}\right)^2 c_v t \right] \right\} \quad . \quad (79)$$

The solution (78) can be evaluated numerically after separating the real and imaginary parts of s^+ and s^-; the result is not shown here.

An impression of the structure of $\zeta(t)$ can be obtained conveniently by considering the solutions for large and small t. These solutions follow directly from approximations of equation (75) for small and large s respectively and taking the inverse of these approximations. This yields:

Approximation for large t is obtained by developing $\tanh(\hat{q}H)$ from equation (75) in a series expression for small values of $\hat{q}H$, which, because of equation (74) corresponds to small values of s. The transformed settlement then becomes

$$\bar{\zeta}(s) \approx (p\hat{a}/s\hat{q})g(s)[(\hat{q}H) - \tfrac{1}{3}(\hat{q}H)^3 + \ldots] \quad . \quad . \quad . \quad . \quad (80)$$

The first term between brackets gives $p\hat{a}Hs^{-1}g(s)$. This has the form of equation (68) and represents the transformed settlement as created by an effective stress of magnitude p, because the transform of p is p/s. Its inversion gives the settlement time relation, when excess pore pressure has vanished, as explained in the past sub-section.

From the second term can be deduced when excess pore pressure has diminished enough to permit disregarding its influence, by considering the value of s that makes the second term small with respect to the first.

Approximation for small t is obtained by developing $\tanh(\hat{q}H)$ from equation (75) in an expression valid for large values of $\hat{q}H$. This expression is $[1-2\exp(-2\hat{q}H)]$. So retaining only the first term, which is equal to unity, the transformed settlement in the first approximation becomes

$$\bar{\zeta}(s) \approx (p\hat{a}/s\hat{q})g(s) = p\hat{a}[\hat{c}_v g(s)]^{1/2}s^{-3/2} \quad . \quad . \quad . \quad . \quad (81)$$

This can be evaluated by taking $g(s)$ from equation (69). Considering the region $1/\theta_1 \ll s < \infty$

$$\bar{\zeta}(s) \approx [\hat{c}_v\Gamma(1+\alpha)/p\hat{a}\theta_0^\alpha]^{1/2}s^{-(3+\alpha)/2}$$

The inverse of this is

$$\zeta(t) \approx [\hat{c}_v\Gamma(1+\alpha)/p\hat{a}\theta_0^\alpha]^{1/2}t^{(1+\alpha)/2}/\Gamma(\tfrac{3}{2}+\tfrac{1}{2}\alpha) \quad . \quad . \quad . \quad (82)$$

According to this solution the settlement for small t follows a power time law with t to the power $(\frac{1}{2}+\frac{1}{2}\alpha)$. In the spherical compression test it was observed that the power of t was less than a half. So matching of test results requires a negative value of α, i.e. a positive power of s in $g(s)$.

The mathematical implications of this requirement cannot be explained here in detail, but from the theory of Stieltjes transforms it can be deduced that no physically acceptable frequency function $m(\mu)$ exists which introduced in equation (67) produces a positive power of s in $g(s)$. This incompatibility of theory and test results indicates that the present model of spring dashpot elements immersed in pore water is incapable of reproducing reality.

Conclusion

The basic equation of consolidation in its Laplace transform for a spring dashpot assembly is identical to the transform of the original Terzaghi equation, except the coefficient of consolidation, which contains the compressibility $\hat{a}g(s)$ as a function of s. Solution of boundary value problems with secular effects therefore follows the same lines as for consolidation without secular effect as long as only geometry is involved. Only the final inversion of the result is more elaborate because of $\hat{a}g(s)$.

For large values of t the solution reduces to the case when the load is carried entirely by effective stresses, as treated in the third section. For small values of t the solution reduces to a form which contradicts test results, and, therefore, indicates that the present model of spring dashpot elements immersed in pore water is not entirely acceptable.

In the following section a model will be treated that consists of cavities interconnected by channels. For this model the transformed basic equation of consolidation is shown to be slightly different, in so far as the permeability also is a function of s. This second function of s opens the possibility to choose frequency functions for the physical entities involved in a manner such that for small t the settlement also will conform to the test results.

CONSOLIDATION OF STOCHASTIC ASSEMBLY OF CAVITY CHANNEL ELEMENTS

The second secular mechanism proposed by Keverling Buisman (1938) consists of a network of compressible pores interconnected by channels. In the following analysis the stochastic character of such a system will again be introduced by considering the cavities to possess different compressibilities and the channels' different conductivities. The difficulty now arises that the pore pressure itself becomes a stochastic variable, because every pore has its own pore pressure. This difficulty is overcome by introducing an average pore pressure, which is not a physical quantity that can be determined in a test, but a convenient mathematical device that permits the continued use of the classical theory of consolidation.

For further treatment of consolidation problems the time dependence of compressibility and conductivity are involved.

Physical mechanism of cavity channel element

In a system of cavities interconnected by channels, every cavity is connected by a number of channels to cavities surrounding it (Fig. 9). Consider a sample cavity with a number k. When finally a great number, N, of cavities is considered, a summation is effected where k runs from 1 to N. The compressibility of the kth cavity is called ρ_k; this means that the cavity expels a volume ρ_k of fluid when there is a unit pressure decrease of fluid pressure u_k in this cavity.

From the kth cavity ν channels emanate and these channels connect the kth cavity to surrounding cavities numbered m. In a summation m will run from 1 to ν, ν having an order of magnitude of 10 in a three dimensional channel network.

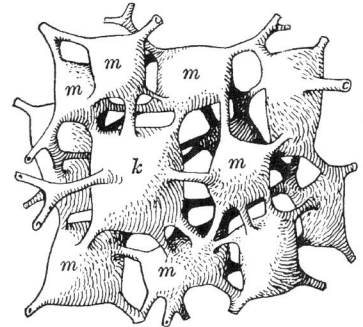

Fig. 9. Cavity channel network consisting of compressible pores interconnected randomly

The conductivity of a channel connecting the kth cavity to an mth cavity will be denoted by λ_{km}. This means that a volume of fluid λ_{km} will flow per unit time through this channel if the pressure difference of the fluids in these cavities ($u_k - u_m$) is unity.

If the fluid pressure in the kth cavity changes by an amount $du_k = (\partial u_k / \partial t) dt$ during dt, the volume of fluid expelled from the cavity during that time is

$$dV_k = -\rho_k du_k = -\rho_k(\partial u_k/\partial t)dt \qquad \cdots \cdots (83)$$

The volume of fluid that can be discharged through the surrounding channels during dt is,

$$dV_m = \sum_{m=1}^{v} [\lambda_{km}(u_k - u_m)]dt \qquad \cdots \cdots (84)$$

If there is no gas in the fluid, continuity requires that $dV_k = dV_m$. This gives

$$\rho_k(\partial u_k/\partial t) = \sum_{m=1}^{v} \lambda_{km}(u_m - u_k) \qquad \cdots \cdots (85)$$

In this expression u_k and u_m can be considered to represent the excess pore pressure. Let the initial excess pore pressure be the same for the different pores, and equal to p. Then the Laplace transform of equation (85) is

$$\rho_k(s\bar{u}_k - p) = \sum_{m=1}^{v} \lambda_{km}(\bar{u}_m - \bar{u}_k) \qquad \cdots \cdots (86)$$

Solving for $\bar{u}_k$, the transformed excess pore pressure, yields

$$\bar{u}_k[s + (1/\rho_k)\sum_{m=1}^{v} \lambda_{km}] = p + (1/\rho_k)\sum_{m=1}^{v} (\lambda_{km}\bar{u}_m) \qquad \cdots \cdots (87)$$

In this equation two summations of different character appear. In order to simplify the notation the following abbreviations are introduced

$$\mu_k = (1/\rho_k)\sum_{m=1}^{v} \lambda_{km} \qquad \cdots \cdots \cdots (88)$$

$$\hat{u}_k = \left[\sum_{m=1}^{v} \lambda_{km}\bar{u}_m\right] \Big/ \left[\sum_{m}^{v} \lambda_{km}\right] \qquad \cdots \cdots (89)$$

These two quantities have a physical meaning.

The character of μ_k, as defined by equation (88), is understood by considering the case when the kth cavity has an excess pore pressure p at time $t = 0$, and all the surrounding cavities are maintained at zero excess pore pressure. Then equation (87) reduces to

$$\bar{u}_k = p/(s + \mu_k)$$

with the inverse

$$u_k = p \exp(-\mu_k t) \qquad \cdots \cdots \cdots (90)$$

From this result it is seen that μ_k is a relaxation parameter similar to μ introduced for the spring dashpot elements. It represents the relaxation occurring in the kth cavity, during the decrease of excess pore pressure by discharge of fluid from that cavity towards its surroundings.

The character of $\hat{u}_k$ defined by equation (89) is a pore pressure experienced in the kth cavity as a weighted average of the pore pressures in the surrounding cavities. The weighting function is λ_{km}, the conductivity of the connecting channels.

The quantity $\hat{u}_k$ as defined by equation (89) will be called the *ambient excess pore pressure*. Substituting equations (88) and (89) in (87) gives

$$\bar{u}_k(s+\mu_k) = p+\mu_k\hat{u}_k \quad \dots \dots \dots \quad (91)$$

This expression indicates the relation between the transformed excess pore pressure in the kth cavity, $\bar{u}_k$, and the transformed ambient excess pore pressure, $\hat{u}_k$, which is a weighted average of the excess pore pressures in the adjacent cavities.

Settlement of a stochastic assembly of cavities and channels

The expression (91) permits determination of the settlement of a layer that is composed of an assembly of cavities. It will be assumed that the entire settlement consists of the reduction in volume of the cavities, and that the volume reduction ΔV_k of the kth cavity is equal to its compressibility ρ_k multiplied by the decrease in excess pore pressure $(p-u_k)$ after the moment of loading, so that

$$\Delta V_k = \rho_k(p-u_k) \quad \dots \dots \dots \quad (92)$$

The number of cavities, present in a volume L^3 will be N^3, with N a large number and L a length as defined in the sub-section on stochastic distribution of elements. These N^3 cavities will generate a volume reduction equal to

$$\Delta V = \sum_{k=1}^{N^3} \rho_k(p-u_k) \quad \dots \dots \dots \quad (93)$$

If N is large, a number N cavities gives a fair representation of the average, and so a summation over N^3 elements is equal to N^2 times the summation over N elements. Therefore equation (93) can be written as

$$\Delta V = N^2 \sum_{k=1}^{N} \rho_k(p-u_k) \quad \dots \dots \dots \quad (94)$$

Consider now a layer of small thickness Δz. This layer covers an area $\Delta A = L^3/\Delta z$, if its volume is L^3. When this layer is compressed in a manner such that lateral expansion is prohibited, the height reduction $\Delta\zeta$ of the layer is

$$\Delta\zeta = \Delta V/\Delta A = \Delta V(\Delta z/L^3)$$

or, using equation (94) the settlement is

$$\Delta\zeta = \Delta z(N^2/L^3) \sum_{k=1}^{N} \rho_k(p-u_k)$$

Transforming this with the Laplace operation gives $[(p/s)-\bar{u}_k]$ for $(p-u_k)$. So the transformed settlement $\Delta\bar{\zeta}(s)$ for a layer of thickness Δz is

$$\Delta\bar{\zeta}(s) = \Delta z(N^2/L^3) \sum_{k=1}^{N} \rho_k[(p/s)-\bar{u}_k]$$

If now the excess pore pressure in the kth cavity $\bar{u}_k$ is replaced by its ambient excess pore pressure $\hat{u}_k$ using equation (91) there results

$$\Delta\bar{\zeta}(s) = \Delta z \frac{N^2}{L^3} \sum_{k=1}^{N} \frac{\rho_k\mu_k}{s+\mu_k} \left[\frac{p}{s}-\hat{u}_k\right]$$

In this summation (p/s) is a constant, independent of k, because it was stated that the excess pore pressure at the moment of loading, $t=0$, is equal for all cavities. So (p/s) can be taken out of the summation, and the transformed settlement becomes

$$\Delta\hat{\zeta}(s) = \Delta z \frac{N^2}{L^3}\left[\left(\frac{p}{s}\sum_{k=1}^{N}\frac{\rho_k\mu_k}{s+\mu_k}\right) - \left(\sum_{k=1}^{N}\frac{\rho_k\mu_k}{s+\mu_k}\hat{a}_k\right)\right] \quad \ldots \quad (95)$$

The ambient excess pore pressure $\hat{a}_k$ is different for every cavity, because it is a weighted average of the surrounding excess pore pressures, with the conductivity of the connecting channels as weighting function. Make the assumption now that the correlation between $\hat{a}_k$ and the properties ρ_k, μ_k of the particular cavity can be disregarded.

It is not obvious that this is a reasonable assumption, since the conductivity of the adjacent channels λ_{km} is incorporated in both $\hat{a}_k$ and μ_k (see equations (88) and (89)). That the correlation will be weak follows from the fact that the excess pore pressure $\bar{u}_m$ in the adjacent cavities is practically independent of the conductivity of the connecting channel λ_{km} because that channel being only one of the ν channels emanating from the mth cavity contributes only a portion $(1/\nu)$ in the value of $\bar{u}_m$. So $\bar{u}_m$ can be considered nearly random with respect to the kth cavity.

Now rearrange the second summation in equation (95) such that all cavities with practically the same value for $[\rho_l\mu_k/(s+\mu_k)]$ are taken together, so that they form K subgroups, each containing ΔN cavities. Let ΔN still be a large number such that the $\hat{a}_k$ of a subgroup represents a fair sample of the total, which is possible because there is no correlation between $\hat{a}_k$ and ρ_k or μ_k. Then the summation of these ΔN values for $\hat{a}_k$ is ΔN times the average value $\hat{a}$, if the *average ambient excess pore pressure* $\hat{a}$ is defined as

$$\hat{a} = (1/N)\sum_{k=1}^{N}\hat{a}_k \quad \ldots \quad \ldots \quad (96)$$

According to the definition of the volume L^3 in the section on stochastic distribution of elements the stress condition in the layer with thickness Δz is nearly homogeneous. This means that the excess pore pressure in a continuous reference soil subjected to the same loading as the cavity channel assembly would have been practically a constant throughout this volume. The averaging procedure, applied to obtain $\hat{a}$, smoothes out the difference in excess pore pressures encountered in the individual pores, but does not obliterate pore pressure differences caused by consolidation.

It is now possible to extract the constant $\hat{a}$ from the last summation term in equation (95), because it can be removed from every subgroup. All subgroups give the same $\hat{a}$ because of the lack of correlation mentioned. The result is

$$\Delta\hat{\zeta}(s) = \Delta z \frac{N^2}{L^3}\left(\frac{p}{s}-\hat{a}\right)\left[\sum_{k=1}^{N}\frac{\rho_k\mu_k}{s+\mu_k}\right] \quad \ldots \quad \ldots \quad (97)$$

It may be remarked from this analysis that in general the summation over a product of two stochastic variables is equal to the average of one of these variables multiplied by the summation over the other, if there is no correlation between them. This property is used in the next sub-sections.

Similarity between settlement of cavity channel assembly and spring dashpot system

There is a great similarity between expression (97) for the transformed settlement of a cavity channel assembly and equation (65), the transformed settlement of a spring dashpot system whose retardation function obeys equation (4). The similarity appears by altering the height h of the block into the thickness Δz of the layer, and changing p/s into $[(p/s)-\hat{a}]$. A similar change was made to obtain equation (66) from equation (65) where, however, instead of $[(p/s)-\hat{a}]$ the effective stress $\bar{\sigma}'(s)$ was substituted. Since $\bar{\sigma}' = (p/s)-\bar{u}$ (see equation (70)) the similarity is complete if $\hat{a}$ can be identified with $\bar{u}$. For mathematical convenience this identification will be made, but this should not be done thoughtlessly, because there is a fundamental physical difference between $\hat{a}$ and $\bar{u}$. The quantity $\bar{u}$ stands for an excess pore pressure acting inside a grain skeleton with spring dashpot retardation at the grain contact points. This pore pressure is a deterministic quantity, whose computed value is the physical pore pressure present at the point of consideration. This pore pressure $\bar{u}$ does not interfere with the secular retardation mechanism.

On the other hand in a cavity channel system the pressures in the individual pores are stochastic. They may vary substantially between adjacent pores, and therefore it is impossible to determine the pore pressure at a point. Only a mathematical quantity $\hat{a}$ can enter into the computations as an average ambient excess pore pressure. This is the average of a complicated quantity, the ambient excess pore pressure, which interferes with the secular mechanism, and is also a stochastic variable. This means that it is not possible to assert the exact magnitude of it at a point, but only the average of the values that might occur there, weighted according to their frequency of occurrence.

The definition of the ambient excess pore pressure as given in the sub-section on physical mechanism of cavity-channel element indicates that $\hat{a}$ is an average over many cavities, in the same way as a pore pressure measured by a piezometer is an average. Such a device is usually large with respect to the individual pores, and therefore it cannot determine the pressure in a particular pore but senses some average of the pressures in the surrounding pores. It is plausible to assume that this average is weighted according to the conductivity of the connecting channels. A similar weighting procedure was obtained for $\hat{a}_k$, the ambient excess pore pressure. Since a piezometer is connected to many pores it senses some average; this average is somewhat different from $\hat{a}$ but may be related to it.

This complicated notion of an average ambient excess pore pressure is also reflected in the effective stress and hence in the transformed effective stress in the cavity channel system. Since this quantity will also involve an averaging it will be denoted by $\hat{\sigma}'$, and be defined as

$$\hat{\sigma}'(s) = (p/s) - \hat{a}(s) \quad \cdots \cdots \cdots \quad (98)$$

A physical meaning is not assigned to $\hat{\sigma}'$, but it is used as a mathematical entity which takes the place of $\bar{\sigma}'$ in the sub-section on basic equation of consolidation for retarded elements immersed in pore water. Since the similarity has now become complete the transition from equation (65) to equation (68) can be used and applied to equation (97) to obtain

$$\Delta\bar{\zeta}(s) = \hat{\sigma}'(s)\hat{a}g(s)\Delta z \quad \cdots \cdots \cdots \quad (99)$$

with

$$\hat{a}g(s) = (N^2/L^3) \sum_{k=1}^{N} \rho_k\mu_k/(s+\mu_k) = \hat{a}\int_0^\infty [\mu m(\mu)/(s+\mu)]d\mu \quad \cdots \quad (100)$$

Here $\hat{a}$ has the same meaning as given by equation (14), namely an average compressibility of the cavities. The frequency function $m(\mu)$ and the retardation parameter μ also have the same meaning.

This justifies the conclusion that it is not possible to decide from test results in the secondary period of consolidation which model gives a better description of a soil behaviour: the spring dashpot model or the cavity channel model. Both give a similar expression for the transformed compressibility $\hat{a}g(s)$.

Basic equation of consolidation for a stochastic assembly of cavities and channels

So far the settlement has been expressed in a form which accounts for the stochastic and time dependent compressibility if the effective stress $\hat{\sigma}' = (p/s) - \hat{a}$ is known. Since the load p is given, it remains to determine $\hat{a}$, the average ambient excess pore pressure. This can be done with the normal procedures used in the theory of consolidation, because, as is shown, the quantity $\hat{a}$ obeys a partial differential equation which is similar in structure to the basic equation governing Terzaghi's consolidation theory.

After the Laplace transformation, the basic equation of consolidation contains only

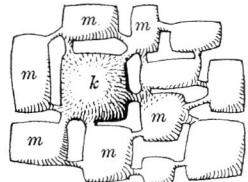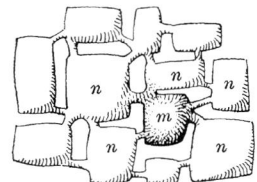

Fig. 10. The cavity k is the centre of ν surrounding cavities, labelled m. Every cavity m is again the centre of ν surrounding cavities, labelled n. The number ν may vary with the cavities. That k is again one of the n is disregarded in the text

derivatives with respect to the space co-ordinates, and in the one dimensional case it has the form

$$(k/\gamma_w)(d^2\bar{u}/dz^2) = a(s\bar{u}-p) \quad . \quad . \quad . \quad . \quad . \quad . \quad (101)$$

A comparable equation is obtained by starting from equation (86) which relates the transformed excess pore pressure, $\bar{u}_k$, in the kth cavity to $\bar{u}_m$, the transformed excess pore pressures in the surrounding cavities, labelled m. This equation is

$$\sum_{m=1}^{\nu} \lambda_{km}(\bar{u}_m - \bar{u}_k) = \rho_k(s\bar{u}_k - p) \quad . \quad . \quad . \quad . \quad . \quad (102)$$

The space co-ordinate will come in by considering the location of the mth cavity, which is one of the set of ν cavities surrounding the kth. Every mth cavity is in turn the centre of its own set of cavities, labelled n, which surround it and are connected to it by channels with conductivity λ_{mn} (see Fig. 10). The mth cavity then has an ambient excess pore pressure $\hat{a}_m$ which has a definition analogous to equation (89) and is related to $\bar{u}_m$ by an equation analogous to equation (91), being

$$\bar{u}_m(s+\mu_m) = p + \mu_m\hat{a}_m \quad . \quad . \quad . \quad . \quad . \quad . \quad (103)$$

By considering the orientation in space of the centres of these ambient excess pore pressures, $\hat{a}_m$, with respect to the kth cavity, the space derivative of the ambient excess pore pressures can be introduced.

The establishment of the basic equation of consolidation requires two steps. The first is to rearrange equation (102) in a manner such that the right hand side becomes comparable to the right hand side of equation (101). The second step is obtained by rearranging the left hand sides.

The *first step* starts by rearranging equation (102) in a manner such that a relation between the ambient excess pore pressures $\hat{a}_k$ and $\hat{a}_m$ is obtained, instead of the relation between $\bar{u}_k$ and $\bar{u}_m$, the actual pressures in the pores. Introducing equations (91) and (103) for $\bar{u}_k$ and $\bar{u}_m$ in (102) yields

$$p\sum_m \frac{\lambda_{km}}{s+\mu_m} + \sum_m \lambda_{km}\frac{\mu_m\hat{a}_m}{s+\mu_m} - \frac{p}{s+\mu_k}\sum_m \lambda_{km} - \frac{\mu_k\hat{a}_k}{s+\mu_k}\sum_m \lambda_{km} = (s\hat{a}_k - p)[\rho_k\mu_k/(s+\mu_k)]$$

Rearrangement gives, with equation (88),

$$\sum_m \frac{\lambda_{km}\mu_m}{s+\mu_m}\hat{a}_m - \hat{a}_k\sum_m \lambda_{km} = -p\sum_m \frac{\lambda_{km}}{s+\mu_m}$$

which can be simplified by adding the factor $s\hat{a}_k \sum_m (\lambda_{km}/s+\mu_m)$ to both sides. The result is then

$$\sum_m \left[\frac{\lambda_{km}\mu_m}{s+\mu_m}(\hat{a}_m - \hat{a}_k)\right] = (s\hat{a}_k - p)\sum_m \frac{\lambda_{km}}{s+\mu_m} \quad . \quad . \quad . \quad (104)$$

If an equation of the form (101) is to be obtained, the right hand side should contain the compressibility, which in the cavity channel case is $\hat{a}g(s)$ (see equation (100)). In order to produce this here, it is necessary to bring equation (104) into a shape which permits manipulation as was done with equation (95). This requires multiplication of both sides by a factor

$\left(\dfrac{\sum\limits_{m}\lambda_{km}}{s+\mu_k}\right)\Big/\left(\sum\limits_{m}\dfrac{\lambda_{km}}{s+\mu_m}\right)$ to yield

$$\left\{\sum_m\frac{\lambda_{km}\mu_m}{s+\mu_m}(\hat{a}_m-\hat{a}_k)\Big/\sum_m\frac{\lambda_{km}}{s+\mu_k}\right\}\frac{\sum\limits_m\lambda_{km}}{s+\mu_m}=(s\hat{a}_k-p)\frac{\sum\limits_m\lambda_{km}}{s+\mu_k}\qquad . \quad . \quad (105)$$

Summing over the volume L^3 which contains N^3 cavities gives, per unit volume, on the right hand side

$$\frac{N^2}{L^3}\left[\sum_{k=1}^{N}\left(\frac{\sum\limits_m\lambda_{km}}{s+\mu_k}s\hat{a}_k\right)-p\sum_{k=1}^{N}\left(\frac{\sum\limits_m\lambda_{km}}{s+\mu_k}\right)\right]$$

The similarity of this expression with the greater part of equation (95), (if equation (88) is used), shows that the analysis of the past two sub-sections is applicable to it up to equation (100).

So (N^2/L^3) times the summation over k from 1 to N brings the right hand side of equation (105) into the form $\hat{a}g(s)[s\hat{a}-p]$. That summation applied to the entire equation (105) finally gives

$$\frac{N^2}{L^3}\sum_{k=1}^{N}\left\{\sum_{m=1}^{v}\frac{\lambda_{km}\mu_m}{s+\mu_m}(\hat{a}_m-\hat{a}_k)\Big/\sum_{m=1}^{v}\frac{\lambda_{km}}{s+\mu_m}\right\}\frac{\sum\limits_{m=1}^{v}\lambda_{km}}{s+\mu_k}=\hat{a}g(s)[s\hat{a}-p]\quad . \quad (106)$$

The *second step* is to show that the left hand side of equation (106) represents something similar to the left hand side of equation (101), i.e. the second derivative of $\hat{a}$ multiplied by the permeability. In order to obtain this result use will be made of the property that the sum of a product of two non-correlated stochastic variables is equal to the product of the average over one of them and the sum of the other. Since $\hat{a}_k$ and $\hat{a}_m$ are not correlated to λ_{km}, μ_k or μ_m it is permissible to extract the average of $(\hat{a}_m-\hat{a}_k)$ from the summation over v and N on the left hand side of equation (106).

In order to bring in the space co-ordinate we introduce the centre of gravity of the volume occupied by the N^3 elements as the location to which the average ambient excess pore pressure is referred. In a soil body, where a one dimensional consolidation process is occurring in the z direction, this location is indicated by its z co-ordinate.

The assumption will be made, that the average ambient excess pore pressure $\hat{a}(z)$ is a gradually varying function of the co-ordinate z. Then, in view of the one dimensional character of the consolidation process the average ambient excess pore pressure in the neighbourhood of z_0 can be expressed by its Taylor expansion in z-direction only, as

$$\hat{a}(z)=\hat{a}(z_0)+(z-z_0)(d\hat{a}/dz)+\tfrac{1}{2}(z-z_0)^2(d^2\hat{a}/dz^2+\ldots)\qquad . \quad . \quad (107)$$

Let z_0 be the centre of gravity of the volume occupied by the set of N^3 cavities, labelled k. From this set select a subset of cavities which possess adjacent cavities m connected by channels of a length between l and $l+\Delta l$, and orientated with respect to the z direction at an angle between ψ and $\psi+\Delta\psi$ (*see* Fig. 11). The centre of gravity of the mth cavities belonging to this subset lies at a distance $l\cos\psi$ above z_0. If this subset contains enough cavities, the average ambient excess pore pressure of the corresponding mth cavities is representative for the $\hat{a}$ at that height. So the average of $(\hat{a}_m-\hat{a}_k)$ for this subset is with equation (107)

subset average $(\hat{a}_m-\hat{a}_k)=\hat{a}(z_0+l\cos\psi)-\hat{a}(z_0)$

$$=(l\cos\psi)(d\hat{a}/dz)+\tfrac{1}{2}(l\cos\psi)^2(d^2\hat{a}/dz^2+\ldots)\qquad . \quad . \quad (108)$$

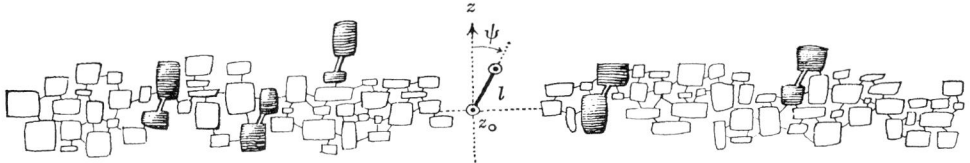

Fig. 11. From the set of cavities constituting a representative elementary volume, a subset is selected for which the distance between the cavities k and m is l in a direction ψ

Averaging over all the subsets means averaging over $(l \cos \psi)$ and $(l \cos \psi)^2$, since $(d\hat{a}/dz)$ and $(d^2\hat{a}/dz^2)$ are fixed quantities at the height z_0. In this averaging procedure it is assumed that l and ψ are not correlated and that the soil is isotropic so that the distribution of the channels is homogeneous with respect to direction. By this assumption the average of $(l \cos \psi)$ is the average of l multiplied by the average of $(\cos \psi)$. This is zero for $\cos \psi$ is an odd function if ψ ranges from 0 to π. So the term with $(d\hat{a}/dz)$ in equation (108) vanishes by taking the average of the subsets. The term with $(d^2\hat{a}/dz^2)$ does not vanish, because the average of $(\cos^2 \psi)$ over a sphere is

$$(1/4\pi) \int_0^\pi \cos^2 \psi \, 2\pi \sin \psi d\psi = \tfrac{1}{3} \quad \quad \cdots \cdots (109)$$

If $\hat{l}^2$ denotes the average of l^2 the last term of (108) becomes $(1/6)\hat{l}^2(d^2\hat{a}/dz^2)$ by averaging and this is the average of $(\hat{a}_m - \hat{a}_k)$ over the νN cavities of the summation in equation (106).
Therefore equation (106) can be written as

$$\frac{1}{6}\hat{l}^2\frac{d^2\hat{a}}{dz^2}\frac{N^2}{L^3}\sum_{k=1}^{N}\left\{\frac{\sum\limits_{m}^{\nu}[\lambda_{km}\mu_m/(s+\mu_m)]}{\sum\limits_{m}^{\nu}[\lambda_{km}/(s+\mu_m)]}\frac{\sum\limits_{m}^{\nu}\lambda_{km}}{(s+\mu_k)}\right\} = \hat{a}g(s)[s\hat{a}-p] \quad \cdots (110)$$

The part of the equation in the braces is a function of s whose dimensions are those of λ, because s and μ both have the dimension (1/time). Since λ is a conductivity with dimension [length⁵/force × time] multiplication by $(\hat{l}^2/L^3)$ brings the dimension to [length⁴/force × time] which is the same as the dimensions of the permeability term (k/γ_w) in equation (101).

Therefore, by analogy with the time dependent compressibility $\hat{a}g(s)$ introduced in equation (68), we can introduce here a time dependent permeability $(\hat{k}/\gamma_w)f(s)$, such that

$$\frac{\hat{k}}{\gamma_w} f(s) = \frac{1}{6}\hat{l}^2\frac{N^2}{L^3}\sum_{k=1}^{N}\left\{\frac{\sum\limits_{m}^{\nu}[\lambda_{km}\mu_m/(s+\mu_m)]}{\sum\limits_{m}^{\nu}[\lambda_{km}/(s+\mu_m)]}\frac{\sum\limits_{m}^{\nu}\lambda_{km}}{(s+\mu_k)}\right\} \quad \cdots (111)$$

where the time dependence is incorporated in $f(s)$. Substituting this in equation (110) gives finally

$$(\hat{k}/\gamma_w)f(s)[d^2\hat{a}/dz^2] = \hat{a}g(s)[s\hat{a}-p] \quad \cdots \cdots (112)$$

If again use is made of a coefficient of consolidation

$$\hat{c}_v = \hat{k}/\hat{a}\gamma_w \quad \cdots \cdots \cdots \cdots (113)$$

This equation becomes

$$\hat{c}_v[f(s)/g(s)](d^2\hat{a}/dz^2) = s\hat{a}-p \quad \cdots \cdots \cdots (114)$$

Equations (112) and (114) have the same structure as the classical consolidation equation (101).

Solution of boundary value problems and the consequence of the time dependent permeability

The settlement equation and the basic equation of consolidation for a cavity channel network (equations (99) and (112)) have the same structure as the corresponding equations for a spring dashpot assembly (equations (68) and (71)) as well as for the classical theory of consolidation without secular effects. The difference from the equations for spring dashpots is in the permeability, which for the cavity channel network is a function of s, being $(k/\gamma_w)f(s)$.

Solution of boundary value problems follows therefore the same lines as indicated in the sub-section on discussion of the consolidation equation. There it was mentioned that the geometric part of the problem remains identical to the solution for normal consolidation without secular effects. Only in the last operation of inversion will the result deviate from the known solutions.

Because the basic equations are similar, so also most of the results in equations (73) to (82) developed for the solution of the boundary value problem pertaining to a layer of thickness H, will be valid here. Apart from a few differences that are described below, the settlement for small t will be considered because the function $f(s)$ in the permeability will enable the fitting of test results at the beginning of consolidation.

The difference in the solution is primarily in $\hat{q}$, which instead of equation (74) will for a cavity channel system be defined as

$$\hat{q} = [sg(s)/f(s)\hat{c}_v]^{1/2} \qquad \ldots \ldots \ldots \quad (115)$$

This entails a few changes in equation (78). The denominators of the terms between braces in the summation become

$$1+s^+\left[\frac{g'(s^+)}{g(s^+)} - \frac{f'(s^+)}{f(s^+)}\right] \quad \text{and} \quad 1+s^-\left[\frac{g'(s^-)}{g(s^-)} - \frac{f'(s^-)}{f(s^-)}\right]$$

with

$$s^\pm g(s^\pm)/f(s^\pm) = (\hat{c}_v/H^2)[\pm\tfrac{1}{2}i\pi(2m-1)]^2$$

For small t and large s the solution is the same as the first of (81), which by use of $\hat{q}$ from equation (115) gives

$$\zeta(s) \approx (\hat{p}\hat{a}/s\hat{q})g(s) = \hat{p}\hat{a}[\hat{c}_v f(s)g(s)]^{1/2}s^{-3/2} \qquad \ldots \quad (116)$$

If again $g(s)$ is taken from (69) with $1/\theta_1 \ll s < \infty$ the result is

$$\zeta(s) \approx [\hat{c}_v\Gamma(1+\alpha)f(s)/\hat{p}\hat{a}\theta_o^\alpha]^{1/2}s^{-(3+\alpha)/2} \qquad \ldots \ldots \quad (117)$$

Test results show that the settlement for small t follows a law of the form $t^{(1-\epsilon)/2}$, when ϵ is positive. The transformed settlement is then a function of s of the form $s^{-(3-\epsilon)/2}$ for large s. For this $f(s)$ must be a function which contains s in the form $s^{\alpha+\epsilon}$.

Since $f(s)$ is already defined by the summation (111), it remains to establish that the summation can provide a function with s in the form $s^{\alpha+\epsilon}$. Since the form of equation (111) is rather involved, it is neither obvious from its structure that this possibility exists, nor is it easy to prove. Only the outlines are traced of a proof that was used to verify that $f(s)$ indeed satisfies the requirements.

The summation is over many kth cavities, and contains a quotient of two summations over the mth cavities that surround the kth cavities. If it is assumed that there is no correlation between the cavities k and m, the property of a product of non-correlated stochastic variables is used to write the right hand side of equation (111) as

$$\frac{1}{6}\hat{l}^2\left[\text{average of}\left\{\frac{\sum\limits_{m}^{v}[\lambda_{km}\mu_m/(s+\mu_m)]}{\sum\limits_{m}^{v}[\lambda_{km}/(s+\mu_m)]}\right\}\right]\frac{N^2}{L^3}\sum_{k}^{N}\frac{\rho_k\mu_k}{s+\mu_k} \qquad \ldots \quad (118)$$

where use is made of equation (88). From equation (100) we see that the last summation is $\hat{a}g(s)$, a function which contains s in the form $s^{-\alpha}$. This means that the average of the quotient between the braces should be of the form $s^{2\alpha+\epsilon}$.

For one value of k both numerator and denominator of the quotient consist of ν terms, each a quotient. If all these terms are multiplied by the product of the ν factors $(s+\mu_m)$ both numerator and denominator become polynomials of the order $(\nu-1)$ in s. By subtracting a term $[\sum_m^\nu \lambda_{km}/\sum_m^\nu (\lambda_{km}/\mu_m)]$ a reduced quotient remains whose numerator is s times a polynomial of order $(\nu-2)$ and that is one order lower than the denominator. To this reduced quotient Heaviside's expansion theorem can be applied, which allows us to split up the reduced quotient in s times a summation of $(\nu-1)$ terms of the form $C_p/(s+\mu_p)$. Since all λ_{km} and μ_m are positive, it can be shown that C_p is a positive quantity, related to λ_{km} and μ_m, and that the $(\nu-1)$ values μ_p lie in between the ν values μ_m.

The quotient of (118) between braces for one particular value of k can therefore be written as

$$\frac{\sum\limits_m^\nu \lambda_{km}}{\sum\limits_m^\nu (\lambda_{km}/\mu_m)} + s \sum_p^{\nu-1} \frac{C_p}{s+\mu_p} \quad . \quad . \quad . \quad . \quad . \quad . \quad (119)$$

The first term in (119) can be shown to be small with respect to the second for those values of s which correspond to the beginning of the consolidation process as follows. In the summation of (λ_{km}/μ_m) the terms with small μ_m dominate, and this means that the first term is of the order of ν times the smallest value of μ_m. The smaller values of μ produce the end value of the settlement and are reciprocal to the time that the final value sets in. Since also s is the reciprocal of the time considered, the first term of (119) is small compared with the second, if the end value is reached at a time large compared with the time when the consolidation process is at its starting period.

So, finally, the average of the quotient in the brace of equation (118) equals approximately $(\nu-1)$ times the average of $sC_p/(s+\mu_p)$ over many values of μ_p. Since the μ_p lie in between the actual values μ_m, their distribution is a fair representation of the actual distribution of the retardation parameters. Averaging over many cavities suggests the introduction of an integral of the form

$$s \int_0^\infty [C(\mu)/(s+\mu)]d\mu \quad . \quad . \quad . \quad . \quad . \quad . \quad (120)$$

This integral should yield a function of s of the form $s^{2\alpha+\epsilon}$, and this is possible since the factor s preceding it requires that the integral itself leads to a function $s^{-1+2\alpha+\epsilon}$. Since both α and ϵ are of the order 0·1 in the tests observed, this means that s is raised to a negative power.

As is mentioned above, the use of the theory of Stieltjes transforms to obtain a solution for $C(\mu)$ requires a negative power of s. The power $(-1+2\alpha+\epsilon)$ is therefore in agreement with the requirements of the theory and the solution gives for $C(\mu)$ a function of the form $\mu^{-1+2\alpha+\epsilon}$.

It is too long to develop here the relation between λ_{km} and μ_m which will give this $C(\mu)$. It turns out to be possible to find a suitable relation, but as it is complicated its presentation is omitted here. The purpose of establishing $C(\mu)$ was to see whether this theory is a realistic possibility, and it turned out to be so.

Conclusion

The basic equation of consolidation in the form of its Laplace transform for a cavity channel network is identical to the transform of the original Terzaghi equation, except that the coefficient of consolidation is itself a function of s, containing as component parts the compressibility and the permeability equal respectively to $\hat{a}g(s)$ and $(\hat{k}/\gamma_w)f(s)$. Solutions to boundary value problems with secular effects therefore follow the same lines as for consolidation without secular effect as long as only geometry is involved. Only the final inversion of the result is more elaborate because of the term $(\hat{k}/\hat{a}\gamma_w)[f(s)/g(s)]$. At the geometric stage of the solution, when integration is performed with respect to the co-ordinates, the results are identical to the solutions for normal consolidation. Solutions to Biot's three dimensional equations can also be used without modification.

For large values of time (t) the solution reduces to the case when $\hat{\sigma}'$ is practically equal to the load p, the situation treated in the third section for the spring dashpot assembly. During this secondary period of consolidation the average ambient excess pore pressure is practically zero, but nevertheless in a certain number of cavities, connected by badly conducting channels to the rest of the network, the excess pore pressure may still be appreciable. Release of water from these badly connected cavities gives the long term settlement in the secondary period of consolidation. During that period the cavity channel model behaves identically with the spring dashpot model, and therefore the frequency distribution of the retardation parameters of the constituent elements is the same for both models.

For small values of time (t) the solution reduces to a form which can be made to match test results by a proper choice of the distribution of conductivities of the channels.

This possibility, of arranging the cavity channel model in a manner such that it reproduces observed settlement behaviour of soils, indicates that the model contains the required properties. It is not a verification though that the model is the correct description of the physical mechanism at work. Other mechanisms may lead to identical effects, in the same way as the cavity channel network produced an identical behaviour to the spring dashpot assembly in the secondary period. The adoption of any mechanism can only be assumed to reflect reality if it has been established that the physics of the microstructure produces that particular mechanism.

It is surmised that in a consolidating soil the secular mechanism is basically of the same kind as the spring dashpot and cavity channel systems examined. This may be the case as well in the microstructure consisting of the assembly of individual particles as on the macro-scale, when cracks and channels run through a soil mass. Therefore the concepts developed in this study may serve the future study of micro- and macrostructure of soils.

ACKNOWLEDGEMENT

The Author is grateful for the help given by Professor R. E. Gibson in shaping the theory and the manuscript, by Ir H. W. Hoogstraten in considering the properties of certain integral transforms and by Ir A. Verruijt in overcoming difficulties at all stages of this work.

The basic ideas of this study were presented in a lecture in February 1966 at King's College, London.

REFERENCES

BARDEN, L. (1965). Consolidation of clay with non-linear viscosity. *Géotechnique* **15**, No. 4, 345–362.
DE JOSSELIN DE JONG, G. (1942). Kanaalstelselhypothese. Report, Lab. Grond. Mech. Delft, K 9153.
DE JOSSELIN DE JONG, G. & VERRUIJT, A. (1965). Primary and secondary consolidation of a spherical clay sample. *Proc. 6th Int. Conf. Soil Mech.*, Montreal, I, 254–258.
GIBSON, R. E. & LO, K. Y. (1961). *A theory of consolidation for soils exhibiting secondary compression.* Norwegian Geotechnical Inst. Publ., No. 41.
KEVERLING BUISMAN, A. S. (1936). Results of long duration settlement tests. *Proc. 2nd Int. Conf. Soil Mech.*, Cambridge, U.S.A., I, 103–105.
KEVERLING BUISMAN, A. S. (1938). Enkele grepen uit de grondmechanica en meer in het bijzonder uit het zettingsprobleem. *De Ingenieur*, No. 32, B 133–155.
KEVERLING BUISMAN, A. S. (1940). *Grondmechanica.* Delft: Waltman.
OLSEN, H. W. (1960). Hydraulic flow through saturated clays. *9th Natn. Clay Conf.*, Purdue University.
SCHIFFMAN, R. L., LADD, C. C. & CHEN, A. T. F. (1964). The secondary consolidation of clay. *Proc. Symp. Rheology Soil Mech.*, Grenoble, 273–298.

TAN, T. K. (1957).   Onderzoekingen over de rheologische eigenschappen van klei.   (Investigations on the
    rheological properties of clay.)   Dr. thesis Excelsior, 's Gravenhage.
TAYLOR, D. W. & MERCHANT, W. (1940).   A theory of clay consolidation accounting for secondary com-
    pression.   *J. math. Phys.* **19**, No. 3, 167–185.
TERZAGHI, K. (1925).   *Erdbaumechanik auf bodenphysikalischer Grundlage.*   Vienna: Deuticke.

APPENDIX

Some functions used in the text and their approximations for large and small values of their argument
are given.

Complete gamma function: $\Gamma(x)$

$$\Gamma(x) = \int_0^\infty \lambda^{x-1} \exp(-\lambda)d\lambda \qquad\qquad x > 0 \quad . \quad . \quad . \quad . \quad . \quad (121)$$

The complete gamma function is related to other functions by

$$x\Gamma(x) = \Gamma(1+x) = x! \quad . \quad . \quad . \quad . \quad . \quad . \quad . \quad . \quad (122)$$

$$\frac{\sin(x\pi)}{\pi} = \frac{1}{\Gamma(x)\Gamma(1-x)} = \frac{x}{\Gamma(1+x)\Gamma(1-x)} \quad . \quad . \quad . \quad . \quad (123)$$

Incomplete gamma function: $\Gamma\{x, y\}$

$$\Gamma\{x, y\} = \int_y^\infty \lambda^{x-1} \exp(-\lambda)d\lambda \quad . \quad . \quad . \quad . \quad . \quad . \quad (124)$$

Expansion of $\exp(-\lambda)$ in a series expression and integrating gives approximation for small y

$$\Gamma\{x, y\} = \Gamma(x) - y^x/x + y^{x+1}/1!(x+1+\ldots) \qquad y \ll 1 \quad . \quad . \quad . \quad (125)$$

Continued partial integration gives approximation for large y

$$\Gamma\{x, y\} = y^{x-1}\exp(-y)[1+(x-1)/y+(x-1)(x-2)/y^2+\ldots] \qquad y \gg 1 \quad . \quad (126)$$

Hypergeometric function: $_1F_1(p;q;z)$

$$_1F_1(p;q;z) = \frac{\Gamma(q)}{\Gamma(p)\Gamma(q-p)} z^{1-q}\int_0^z \lambda^{p-1}(z-\lambda)^{q-p-1}\exp(\lambda)d\lambda \quad . \quad . \quad . \quad (127)$$

Approximations

$$_1F_1(p;q;z) = 1+\frac{p}{q}z+\frac{p(p+1)}{q(q+1)}\frac{z^2}{2!}+\ldots \qquad z \ll 1 \quad . \quad . \quad . \quad (128)$$

$$_1F_1(p;q;z) = \frac{\Gamma(q)}{\Gamma(q-p)}(-z)^{-p}\left[1-\frac{p(p-q+1)}{z}+\ldots\right]+\frac{\Gamma(q)}{\Gamma(p)}z^{p-q}\exp(z)\left[1+\frac{(1-p)(q-p)}{z}+\ldots\right]$$

$$z \gg 1 \quad . \quad . \quad . \quad (129)$$

[Reprinted from The Journal of The American Ceramic Society, Vol. 40, No. 2. February, 1957.]

Verification of Use of Peak Area for the Quantitative Differential Thermal Analysis

by G. de Josselin de Jong

Research Department, Soil Mechanics Laboratory, Delft, Netherlands

The amount of reacting material in a sample investigated by differential thermal analysis can be determined from the peak area according to the Boersma equation, which accounts for heat flow through sample and thermocouple wires. This relation has been checked by experiment and it has been found that the use of different thermocouples for the calibrations may lead to variations of about 30%. The values obtained for heat transfer through a sample and thermocouple have been checked by comparison of computations and observation of base-line offset at the beginning of a run and exponential decay of the amount of heat dissipating out of a sample. It is shown that according to the equation it is not the total amount of reacting material that determines the peak area but merely the density of the material near the thermocouple. The sensitivity of the method for quantitative analysis is discussed in relation to the possible variations in the factors involved, namely, density and heat conductivity of the sample and heat transfer through the thermocouple.

I. Introduction

THE use of differential thermal analysis for quantitative measurements is of great importance and several workers have studied its possibilities. In this paper the writer considers in particular the method in which the peak area is used to indicate the quantity of reacting material in a sample.

In differential thermal analysis the temperature difference between an inert sample and the reacting material is recorded while the sample holder which contains both samples is heated at a constant rate. The peak area is the area enclosed by the base line and the curve of the differential temperature (θ) as recorded versus time (t). When the heat of reaction begins after t_1 and dissipates before t_2, the mathematical representation of the peak area is

$$\text{Peak area} = \int_{t_1}^{t_2} \theta \, dt$$

having the dimension of deg. sec.

That this peak area should be related to the amount of heat of reaction released during a test was experimentally established by Kracek et al.,[1] Alexander et al.,[2] Schafer and Russell,[3] Berg,[4] Grimshaw and Roberts,[5] and McLaughlin.[6]

Verification of this relation becomes possible if the heat dissipation occurring during a test is analyzed and if equations are available which describe this process explicitly. The first theory was advanced by Speil,[7] who deduced that

$$\text{Peak area} = \int_{t_1}^{t_2} \theta \, dt = wM/g\lambda \quad (1)$$

w = heat of reaction per gram of sample (cal./gm.).
M = total mass of sample (gm.).
g = geometrical factor (cm.) which accounts for temperature-gradient distribution in sample.
λ = heat conductivity of sample (cal./deg. cm. sec.).

By using this equation the unknown quantity of the heat of reaction per gram of sample, w, can be computed from the peak area if the factor $M/g\lambda$ is known. This factor is in fact the calibration value of the test device, but for brevity here the calibration factor ψ is introduced and equation (1) is rewritten as follows:

$$\psi = \frac{\text{peak area}}{w} = \frac{M}{g\lambda}$$

In the derivation of this equation the writer assumes that g has a constant value throughout the whole process of heat dissipation, an assumption not consistent with reality. Moreover, the value of g is not given in its physical components and therefore the equation is still incomplete for verification purposes, although it already shows the influence of the heat conductivity, λ, on the calibration value.

The correct derivation of the calibration factor was given by Kronig and Snoodijk.[8] They described in detail how the heat dissipates out of a sample, using the theory of heat conduction. From this, they determined the differential temperature rise originating in the center of the sample and brought about by the heat produced in various small zones within the sample; these individual contributions differ, depending on the place where they originate. By dissipation, the differential temperature will decrease in the course of time and tend toward zero, giving an area in the temperature-time graph which contributes to the total peak area.

Received March 4, 1956; revised copy received May 7, 1956.
The author is assistant director, Research Department, Soil Mechanics Laboratory.

[1] (a) F. C. Kracek, "Polymorphism of Potassium Nitrate," J. Phys. Chem., 34, 225–47 (1930).
(b) R. W. Goranson and F. C. Kracek, "Experimental Investigation of Phase Relations of $K_2Si_4O_9$ Under Pressure," ibid., 36 [3] 913–26 (1932); Ceram. Abstr., 11 [5] 325 (1932).
[2] L. T. Alexander, S. B. Hendricks, and R. A. Nelson, "Minerals Present in Soil Colloids: 11, Estimation in Some Representative Soils," Soil Sci., 48 [8] 273–79 (1939); Ceram. Abstr., 19 [7] 173 (1940).

[3] G. M. Schafer and M. B. Russell, "Thermal Method as a Quantitative Measure of Clay Mineral Content," Soil Sci., 53 [5] 353–64 (1942); Ceram. Abstr., 21 [9] 199 (1942).
[4] L. G. Berg, "Area Measurements in Thermograms for Quantitative Estimations and the Determination of Heats of Reaction," Compt. rend. acad. sci. U.R.S.S., 49, 648–51 (1945) (in English).
[5] (a) R. W. Grimshaw, E. Heaton, and A. L. Roberts, "Constitution of Refractory Clays: II, Thermal-Analysis Methods," Trans. Brit. Ceram. Soc., 44 [6] 76–92 (1945); Ceram. Abstr., 1946, April, p. 66.
(b) R. W. Grimshaw and A. L. Roberts, "Quantitative Determination of Some Minerals in Ceramic Materials by Thermal Means," Trans. Brit. Ceram. Soc., 52 [1] 50–67 (1953); Ceram. Abstr., 1954, April, p. 78d.
[6] R. J. McLaughlin, "Quantitative Mineralogical Analysis of Clay and Silt Fractions by Differential Thermal Analysis"; presented at meeting of the Mineralogical Society, London, March 26, 1953.
[7] Sidney Speil, "Applications of Thermal Analysis to Clays and Aluminous Minerals," U. S. Bur. Mines Rept. Invest., No. 3764, 36 pp. (1944); Ceram. Abstr., 23 [11] 200 (1944).
[8] R. Kronig and F. Snoodijk, "Determination of Heats of Transformation in Ceramic Materials," Appl. Sci. Research, A3 [1] 27–30 (1951); Ceram. Abstr., 1953, January, p. 17a.

42

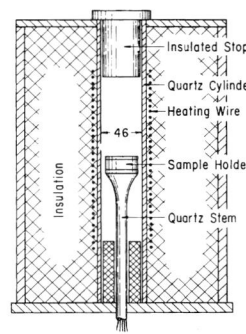

Fig. 1.

Test assembly show-
ing furnace, sample
holder, and thermo-
couple arrange-
ment.

Furnace with Sample Holder

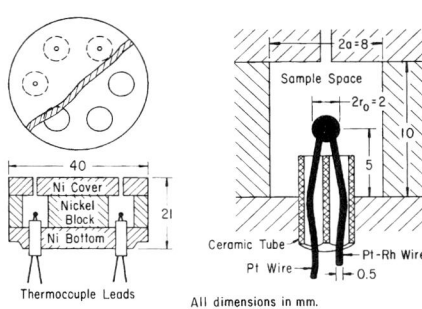

Sample Holder   Sample Space

During a test, the temperature at any place in the sample will pass the reaction temperature zone; all the elementary parts of the whole sample therefore will gradually produce the heat of reaction and their individual contributions to the peak area will gradually accumulate.

Mathematically, this means an integration versus time, and over the whole of the sample, to obtain their calibration factor, which is as follows:

For cylindrical symmetry

$$\psi = \rho a^2/4\lambda = M/4\pi h\lambda \qquad (2)$$

For spherical symmetry

$$\psi = \rho a^2/6\lambda = M/8\pi a\lambda \qquad (3)$$

ρ = density of sample (gm./cm.^{3}).
a = radius of sample holder (cm.).
h = height of sample (cm.).

De Bruyn and van der Marel,[9] among others, pointed out that these equations complete the equation obtained by Speil (equation (1)) because the geometrical factor g is shown to be $4\pi h$ or $8\pi a$, according to the equations for cylindrical or spherical symmetry.

Kronig and Snoodijk's equations (2) and (3) describe the temperature difference between the center of the reacting sample and the center of the inert sample, if the thermocouple which registers these temperatures is of very small dimensions.

Boersma[10] reaches an identical result elegantly avoiding the use of potential functions. This derivation enables him to introduce in a simple way the influence of heat loss through the thermocouple junction and wires. He obtains the following:

For cylindrical symmetry

$$\psi = \frac{\rho a^2}{4\lambda}\left\{\left[1 - \frac{r_0^2}{a^2}\left(1 + 2\ln\frac{a}{r_0}\right)\right]\Big/\left[1 + \frac{\Lambda}{\lambda}\ln\frac{a}{r_0}\right]\right\} \quad (4)$$

For spherical symmetry

$$\psi = \frac{\rho a^2}{6\lambda}\left\{\left[1 - \frac{r_0^2}{a^2}\left(3 - 2\frac{r_0}{a}\right)\right]\Big/\left[1 + \frac{\Lambda}{\lambda}\left(1 - \frac{r_0}{a}\right)\right]\right\} \quad (5)$$

r_0 = radius of thermocouple junction (cm.).
Λ = heat-transfer coefficient for thermocouple wires (cal./deg. cm. sec.).

The length and the cross-sectional area of the leads and their thermal conductivity are accounted for in Λ.

Equations (4) and (5) take into account all the physical phenomena of heat dissipation during a test and were used in

this laboratory to verify calibrations for other experimental work. Since agreement had been found, the different factors in these equations were considered separately and their effects were studied. For calibration, the method used was the method advanced by Barshad[11] and by de Bruyn and van der Marel,[9] who determined the peak area for chemical compounds with well-known heats of reaction.

Another way to obtain the calibration value directly from the differential thermal analysis curve was demonstrated by Vold.[12] She analyzes the exponential decay of the curve after the reaction has ceased and obtains therefrom the value of ψ. This value is determined mainly by heat transfer through the air surrounding the sample holder; the relaxation time therefore has such a large value that after the chemical reaction has ceased, the temperature relaxes over a range large enough to be interpretable.

In the apparatus used in the present tests, where the heat is drained away through the nickel block, which is a good conductor, the relaxation time is too short for such an analysis, as will be shown later.

In addition to calibration methods, the author studied the drift of the base line when the heating rate is suddenly changed because the equation describing this phenomenon contains a factor identical to the calibration factor ψ.

To verify these equations and study the effects of the different physical components, several tests were run under varying conditions. All the tests described in this paper were run with the apparatus constructed and described by de Bruyn and van der Marel[9] and which has been used in this laboratory since 1952. It was provided with an amplifier and a Brown recorder. The furnace contained six holes. Five samples could be analyzed simultaneously; the remaining hole was reserved for the inert reference sample (usually Al_2O_3 previously heated for a number of hours at about 1300 °C.).

Figure 1 shows the furnace with the nickel sample holder mounted on a quartz stem which guides the thermocouple

[9] C. M. A. de Bruyn and H. W. van der Marel, "Mineralogical Analysis of Soil Clays: I, Introduction and Differential Thermal Analysis," *Geol. en Mijnbouw*, [N.S.], **16**, 69–83 (1954); "II, Examples of Mineral Analysis by X-Ray Diffraction and Differential Thermal Analysis," *ibid.*, pp. 407–28.
[10] S. L. Boersma, "Theory of Differential Thermal Analysis and New Methods of Measurement and Interpretation," *J. Am. Ceram. Soc.*, **38** [8] 281–84 (1955).
[11] Isaac Barshad, "Temperature and Heat of Reaction Calibration of the Differential Thermal Analysis Apparatus," *Am. Mineralogist*, **37** [7 and 8] 667–94 (1952).
[12] M. J. Vold, "Differential Thermal Analysis," *Anal. Chem.*, **2** [6] 683–88 (1949); *Ceram. Abstr.*, **1950**, February, p. 31e.

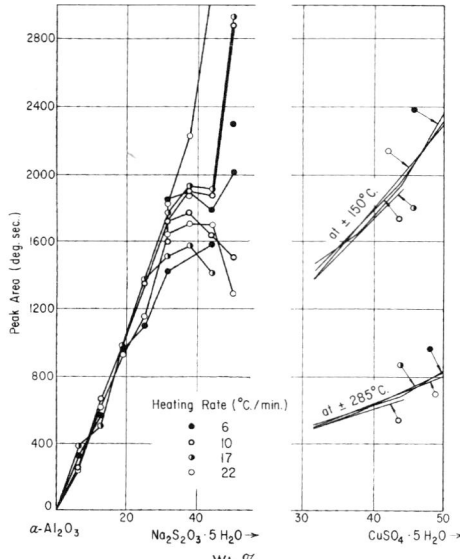

Fig. 2. Effect of heating rate on peak areas for dilutions of $Na_2S_2O_3 \cdot 5H_2O$ and $CuSO_4 \cdot 5H_2O$ in $\alpha\text{-}Al_2O_3$ (sample density, 1.1 gm. per cm.^{3}; total weight, 400 mg.).

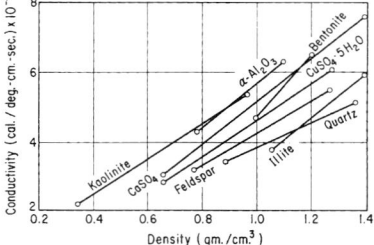

Fig. 3. Heat conductivity, λ, of sample powders of test materials at 20°C. packed under different conditions.

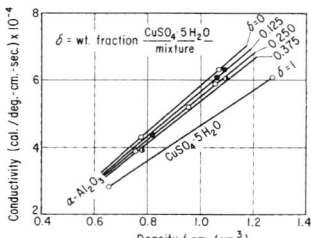

Fig. 4. Heat conductivity, λ, of dilutions of $CuSO_4 \cdot 5H_2O$ in $\alpha\text{-}Al_2O_3$ at 20°C. packed under different conditions.

leads down and out of the furnace. Enlarged diagrams of the sample holder show its dimensions and the placement of the thermocouples.

II. Applicability of Boersma Equation

(1) Heating Rate During Test

Equations (4) and (5) do not contain the heating rate as a factor which influences the calibration values. This is theoretically understood by reasoning as follows:

In the derivation of equations (2) and (3) and (4) and (5) the classical heat-conduction theory is employed, where use is made of the superposition principle and the assumption that the heat conductivity and heat capacity of the sample are constants during the process. The distribution of heat during the time of the test therefore should not influence the total peak area because each heat quantity has a fixed dissipation time and a contribution to the total peak area that is uninfluenced by the other contributions.

Speil *et al.*[13] found that the areas of the first endothermic reaction of kaolinite (near 600°C.) are reproducible to within 3% for heating rates varying from 5° to 20° per minute. This is in accordance with the theoretical considerations above and was verified in this laboratory by tests on $CuSO_4 \cdot 5H_2O$ and $Na_2S_2O_3 \cdot 5H_2O$ diluted with $\alpha\text{-}Al_2O_3$ (heating rates, 6° to 22° per minute). Figure 2 shows that the heating rate has no effect. Only for dilution ratios of the thiosulfate higher than 30% do the values of the peak areas vary considerably. This is explained by changes in the heat conductivity of the sample, as will be considered later, and is not associated with the different heating rates.

[13] Sidney Speil, L. H. Berkelhamer, J. A. Pask, and Ben Davies, "Differential Thermal Analysis, Its Application to Clays and Other Aluminous Minerals," *U. S. Bur. Mines Tech. Paper*, No. 664, 81 pp. (1945); *Ceram. Abstr.*, 24 [8] 153 (1945).

(2) Physical Components of Boersma Equation

For the normal test the dimensions of the sample are height = 10 mm., radius $a = 4$ mm. The thermocouple junction is a sphere with radius $r_0 = 1$ mm.; the thermocouple wires (Pt–Pt-Rh) are 0.5 mm. in diameter (see Fig. 1). These wires enter two holes in the ceramic tube which isolate them from the nickel bottom of the sample holder.

When, during an exothermic reaction,* heat is lost through the thermocouple wires, the heat must travel from the junction to the place where these wires touch the ceramic tube; the tube then conducts it toward the nickel sample holder. The length, l, of this travel through the wires therefore depends on the location of the wires in the holes and may vary with each renewal of the thermocouples. From Fig. 1 it is clear that this travel distance may be about 5 mm. The value of Λ, the heat-transfer coefficient, is, according to Boersma[10]:

For cylindrical symmetry

$$\Lambda = A\lambda_{pt}/2\pi lh$$

For spherical symmetry   (6)

$$\Lambda = A\lambda_{pt}/4\pi lr_0$$

A = cross-sectional area of thermocouple wires = 0.0039 sq. cm.
λ_{pt} = heat conductivity of Pt–Pt-Rh = 0.166 cal./deg. cm. sec.
l = travel distance = about 0.5 cm.
h = height of cylinder, which may be taken as $2r_0 = 0.2$ cm.

---

* During an endothermic reaction the heat flow is reversed without changing the considerations in the text.

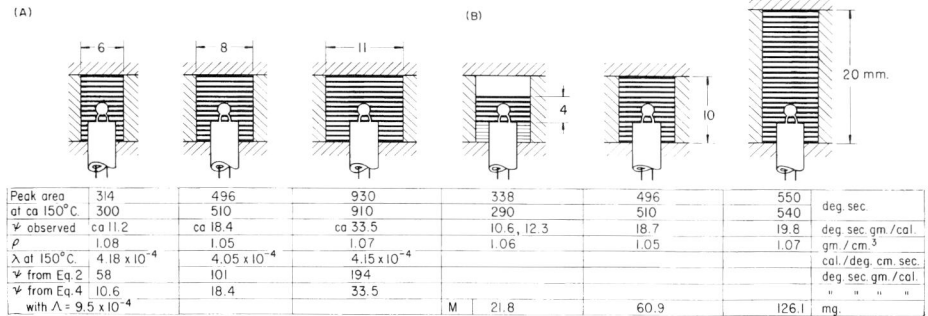

	(A)			(B)				
Peak area	314	496	930	338	496	550		deg. sec.
at ca I50°C.	300	510	910	290	510	540		"
ψ observed	ca II.2	ca I8.4	ca 33.5	10.6, I2.3	I8.7	I9.8		deg. sec. gm./cal.
ρ	1.08	1.05	1.07	1.06	1.05	1.07		gm./cm.^{3}
λ at I50°C.	4.18×10^{-4}	4.05×10^{-4}	4.15×10^{-4}					cal./deg. cm. sec.
ψ from Eq.2	58	101	194					deg. sec. gm./cal.
ψ from Eq.4	10.6	18.4	33.5					" " "
with $\Lambda = 9.5 \times 10^{-4}$				M	21.8	60.9	126.1	mg.

Fig. 5. Test results for $CuSO_4 \cdot 5H_2O$ diluted with $\alpha\text{-}Al_2O_3$ (weight ratio, 1 : 7). (A) Samples of different diameter and equal height, and (B) samples of different height and equal diameter.

Thus, for both cylinders and spheres, one obtains

$$\Lambda = 10.3 \times 10^{-4} \text{ cal./deg. cm. sec.}$$

This, however, is an approximate value because of the uncertainty about the length of l.

The cross-sectional area, A, of the thermocouple leads influences the value of Λ in a direct way. A reduction of the diameter of these wires would therefore reduce Λ appreciably. Because of the accompanying decrease of mechanical strength, the variation of this diameter could not be studied and only a diameter of 0.5 mm. was used.

The density, ρ, and the heat conductivity, λ, of the sample appear in equations (2) and (3) and (4) and (5). The density is easily evaluated if the packing has been carefully done and a homogeneous sample has been obtained. To determine the heat conductivity, λ, separate tests were carried out, as described in Appendix I (p. 49), in which a stationary heat flow through sample powders was studied. Figure 3 shows the results which were obtained on samples ranging in density from about 0.4 to 1.4 gm. per cm.^{3}. The temperature in the samples varied from 20° to 25°C. At higher temperatures the value of λ decreases, as shown by Kingery and McQuarrie[14]; it diminishes about 0.6 times when the temperature increases from 20° to 200°C. The variation of λ is nearly inversely proportional to T, the absolute temperature.

In Fig. 4 are shown the values of λ for different dilutions of $CuSO_4 \cdot 5H_2O$ in $\alpha\text{-}Al_2O_3$ and different densities valid at room temperature. The test results may be schematized to the following linear relation:

$$\lambda = [(6.96 - 1.2\delta) \rho - (1.09 - 1.12\delta)]10^{-4} \text{ cal./deg. cm. sec.} \quad (7)$$

where δ = weight fraction, i.e., weight of $CuSO_4 \cdot 5H_2O$ divided by weight of mixture.

For the interpretation of the differential thermal analysis calibration tests the values of λ for the sample were computed with this equation valid at room temperature and thereupon

---

[14] (a) W. D. Kingery and M. C. McQuarrie, "Thermal Conductivity: I, Concepts of Measurement and Factors Affecting Thermal Conductivity of Ceramic Materials," *J. Am. Ceram. Soc.*, **37** [2, Part II] 67–72 (1954).
(b) Malcolm McQuarrie, "Thermal Conductivity: V, High-Temperature Method and Results for Alumina, Magnesia, and Beryllia from 1000° to 1800°C.," *ibid.*, pp. 84–88.
(c) W. D. Kingery, "Thermal Conductivity: VI, Determination of Conductivity of Al_2O_3 by Spherical Envelope and Cylinder Methods," *ibid.*, pp. 88–90.

converted inversely proportional to the absolute temperature T in order to adapt them to the temperature of the reaction.

III. Determination of Calibration Factor

For the calibration experiments, $CuSO_4 \cdot 5H_2O$ was chosen. At the heating rate of 10° per minute as applied here this compound loses four H_2O molecules in the range 80° to 190°C. and the remaining one H_2O in the range 250° to 330°C. The loss of four H_2O molecules corresponds to about 220 cal. per gm., as was shown by de Bruyn and van der Marel.[9]

With the same thermocouple (No. 5), several tests were run with samples of different dimensions in order to verify the equations. Figure 5 shows the results of tests on samples of equal height and different diameter and of equal diameter and different height.

The table under the diagrams shows the results of tests on $CuSO_4 \cdot 5H_2O$ diluted with $\alpha\text{-}Al_2O_3$ (weight fraction $\delta = 1/8$). The peak areas given in the first two rows of the table were determined by a planimeter from the diagram where each square centimeter has the value of 49.7 deg. $\times$ sec. The involved heat of reaction per gram of sample was because of the weight fraction; $w = 220/8 = 27.5$ cal. per gm. for the release of four H_2O molecules near 150°C. By dividing the peak areas by this value of $w = 27.5$ cal. per gm., the calibration factor ψ (*third row*) was obtained.

Equations (2) and (4) were then applied to the samples of Fig. 5 (A) (samples 6, 8, and 11 mm. in diameter). The density, ρ, is given in the fourth row of the table, and with equation (7) it was possible to compute λ, the heat conductivity of the sample at 20°C. This factor was thereupon reduced inversely proportional to the absolute temperature of 150°C., giving the values in the fifth row of the table. The factor ψ for a cylinder is then easily computed, giving the values in the sixth and seventh rows. These values indicate that ignoring the heat loss through the thermocouple (equation (2)) gives values of ψ which are five or six times too large, whereas introducing the heat transfer through the thermocouple wires in the form of $\Lambda = 9.5 \times 10^{-4}$ cal. per deg. cm. sec. gives values which correspond to the values for samples of different diameter.

A comparison of the value of Λ obtained from these tests and the value predicted theoretically in the previous section shows that there is no great difference. They become equal if the value $l = 0.55$ cm. is adopted for the travel distance.

Figure 5 (B) shows the results of tests on samples 4, 10, and 20 mm. high. In these tests the density, ρ, was kept nearly constant, but the total mass, M, of the reacting material

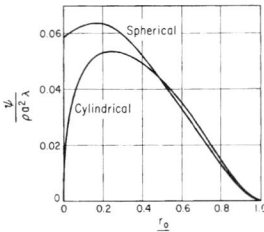

Fig. 6. Graph of calibration factor ψ according to Boersma's equations (4) and (5) for $\Lambda/\lambda = 2$ and for different ratios of thermocouple junction, r_0, versus sample diameter.

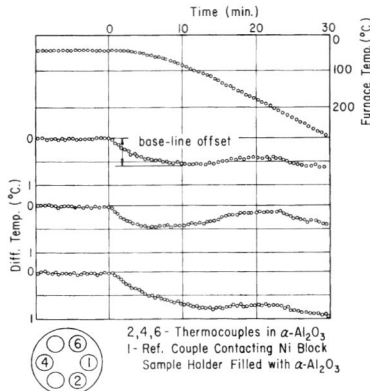

2,4,6 - Thermocouples in α-Al$_2$O$_3$
1 - Ref. Couple Contacting Ni Block
Sample Holder Filled with α-Al$_2$O$_3$

Fig. 7. Base-line offset for samples of α-Al$_2$O$_3$ for change in heating rate from 0° to 10°C. per minute.

varied from 21.8 to 126.1 mg. Note that the samples 10 and 20 mm. high show practically the same peak area, which indicates that it is not the total amount of the reacting material that counts but only the amount of the material present in a disk of unit height. This was predicted by equations (2) and (4) because the sample height in the cylindrical case cancels out and instead of M, the total mass, ρa^2 is encountered. Therefore if the sample height surpasses a certain limit, the surplus of material does not influence the peak area, apparently being out of the influence zone of the thermocouple (see also Barshad[11]). For the very small sample heights the peak area becomes too small because the geometry of the heat flow no longer resembles a part of an infinite cylinder.

From the graph in Fig. 6, which shows the variations of the calibration factor ψ when different diameters of thermocouple junction are used, it is deduced that a ratio $r_0/a = $ about 0.2 gives the largest peak areas. To decrease the diameter of the junction in order to avoid the effect of heat loss through the thermocouple leads is therefore meaningless. It is seen in Fig. 6 that the difference between spherical and cylindrical symmetry for $r_0/a = 0.2$ is not great; this study is therefore continued with the equations for cylindrical symmetry.

IV. Verification of λ and Λ by Base-Line Drift and Exponential Decay

To check the values of λ and Λ obtained from the calibration, tests were devised to verify these values by differential thermal analysis. These tests comprised the determination of base-line offset at the beginning of an experiment and exponential decay at the end of a chemical reaction.

(1) Base-Line Drift

In Appendix II (p. 49) it is derived that the base-line offset, $\Delta\theta$, corresponding to a change in the heating rate of V degrees per second is

$$\Delta\theta = V\rho c \left[\left(\frac{a^2 - r_0^2}{4} - \frac{r_0^2}{2} \ln \frac{a}{r_0} \right) \Big/ \left(\lambda + \Lambda \ln \frac{a}{r_0} \right) \right] \quad (8)$$

c = capacity of sample.

Here are encountered the same terms as in equation (4).

In Fig. 7 are shown the results of a test on samples of α-Al$_2$O$_3$ with $\rho c = 0.2$ cal. per cm.^{3} deg.; the reference couple was in contact with the nickel block. For the change of heating rate $V = 10°$ per minute ($a = 0.4$ cm., $r_0 = 0.1$ cm., $\lambda = 5.2 \times 10^{-4}$ cal. per deg. cm. sec., and $\Lambda = 9.5 \times 10^{-4}$ cal. per deg. cm. sec.) one obtains theoretically a base-line offset of $\Delta\theta = 1.55°$. This value for $\Delta\theta$ agrees with the

values determined experimentally. Without the thermocouple effect, this offset would have been five times larger.

This test could verify the calibrating properties of the test device for every test and would be useful especially at the end of a run. Its practical application fails, however, because of other uncontrollable drifting of the base line.

(2) Exponential Decay

The exponential decay at the end of a chemical reaction can be approximated by the first and dominating term of the series which represents the dissipation of heat out of a sample. This term prevails regardless of the place or time of the production of heat in the sample.

If the effect of the thermocouple is introduced, one obtains for the exponential part of this term, according to Carslaw and Jaeger,*

$$e^{-(\lambda\alpha^2 t/\rho c)}$$

with α the first root of

$$\frac{J_0(\alpha a)}{N_0(\alpha a)} = \frac{\lambda J_0(\alpha r_0) + \Lambda \alpha r_0 J_1(\alpha r_0)}{\lambda N_0(\alpha r_0) + \Lambda \alpha r_0 N_1(\alpha r_0)} \quad (9)$$

J_0 and J_1 = Bessel functions.
N_0 and N_1 = Neumann's functions of zero and first order, respectively.

Figure 8 shows the results of a test where corundum powder of about 60°C. was suddenly introduced into the sample holder. The relaxation time in this test turned out to be 5 to 10 seconds.

A computation of relaxation time with equation (9) gives a value of $\alpha = 9$ cm.$^{-1}$ with $a = 0.4$ cm., $r_0 = 0.1$ cm., $\lambda = 5.2 \times 10^{-4}$ cal. per deg. cm. sec., and $\Lambda = 9.5 \times 10^{-4}$ cal. per deg. cm. sec. and thus for the relaxation time about 5 seconds. This value agrees with the test value. If the thermocouple effect had been neglected, a value four times larger would have been obtained.

The shortest relaxation time observed for clay minerals is

---

* With $k_1 = 0$, $k_2 = 1$, $k_1' = -\lambda r_0/\Lambda$, $k_2' = 1$ (see H. S. Carslaw and J. C. Jaeger, Conduction of Heat in Solids, paragraph 126, page 278, formula (5), and paragraph 137, IV, page 307. Oxford University Press, London, 1947).

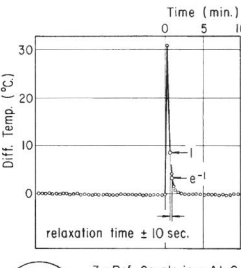

Fig. 8. Exponential decay as shown by the relaxation curve for the temperature in an α-Al$_2$O$_3$ sample that loses its heat toward the nickel block.

3 - Ref. Couple in α-Al$_2$O$_3$
5 - Sample Holder Suddenly Filled with Hot α-Al$_2$O$_3$

Fig. 10. Peak areas for CuSO$_4$·5H$_2$O diluted with α-Al$_2$O$_3$ (weight ratio, 1:7) as recorded by different thermocouples.

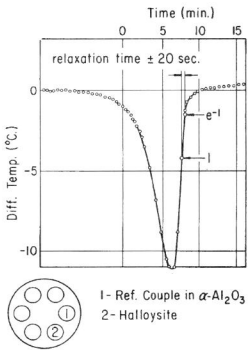

1 - Ref. Couple in α-Al$_2$O$_3$
2 - Halloysite

Fig. 9. Peak for halloysite at ±590°C. showing relaxation curve at end of the reaction. Exponential decay is slower than that shown in Fig. 8.

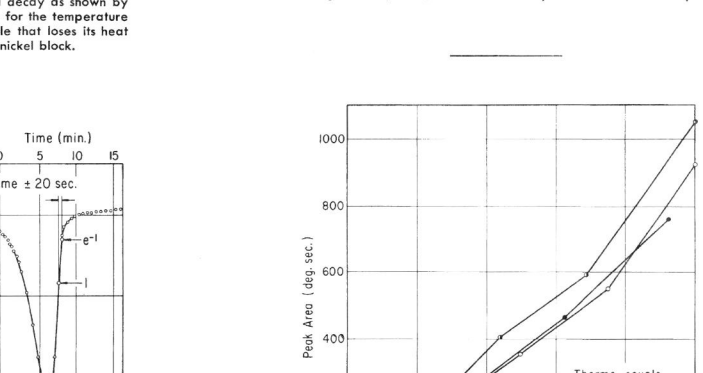

Fig. 11. Peak areas for illite and α-Al$_2$O$_3$ mixtures at about 600°C. (sample density, 1.1 gm./cm.^{3}; total weight, 400 mg.).

given by metahalloysite. According to Speil *et al.*[13] and to de Bruyn and van der Marel,[9] the endothermic reaction at about 600°C. ends abruptly because this mineral has an unordered structure. However, small quantities of kaolinite, which are always present, diminish the abruptness and the relaxation time is therefore always more than for the apparatus itself, as is revealed by Fig. 9.

Here it may be remarked that Vold's method of determining heat-transfer coefficients for a sample during differential thermal analysis as based on the relaxation period at the end of a thermal reaction may lead to unsatisfactory results when applied to an apparatus such as that considered here. The relaxation time of this apparatus is so short that no thermal reaction ends abruptly enough to show this relaxation devoid of the interference of an expiring reaction.

Neither base-line offset nor relaxation time is therefore sensitive enough to form a basis for peak-area interpretations, but they do verify that heat loss through the thermocouples is of the order of magnitude determined, so as to reduce the peak areas by a factor of about 5.

V. Sensitivity of Differential Thermal Analysis Apparatus for Quantitative Analysis

From the foregoing it is deduced that the factors that may vary in the different tests, and by this variation influence the calibration of the test device, are principally the density, ρ, of the sample, the heat conductivity, λ, of the sample, and the heat-transfer coefficient, Λ, for the thermocouple. The heating rate does not alter the calibrations if the heat conduction of the sample does not change during the reaction.

The heat transfer through the thermocouple, Λ, predominates over the heat conductivity of the sample, its influence being four to five times greater. In Fig. 10 the calibration of CuSO$_4$·5H$_2$O diluted with α-Al$_2$O$_3$ (1:7) is shown as measured with different thermocouples. A difference of about 15% between two thermocouples is observed.

Differences of about 30% are observed in Fig. 11 for several dilutions of illite in α-Al$_2$O$_3$; each thermocouple has a marked tendency to follow a calibration curve of its own. After replacement of the thermocouples, the results with

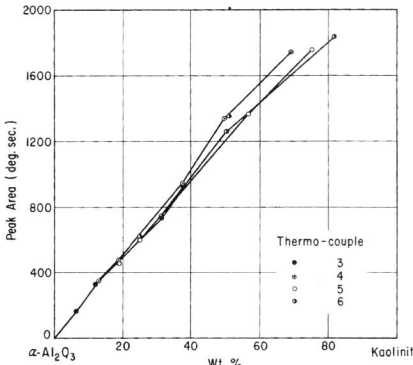

Fig. 12. Peak areas for kaolinite and α-Al₂O₃ mixtures at about 600°C. (sample density, 1.1 gm./cm.³; total weight, 400 mg.).

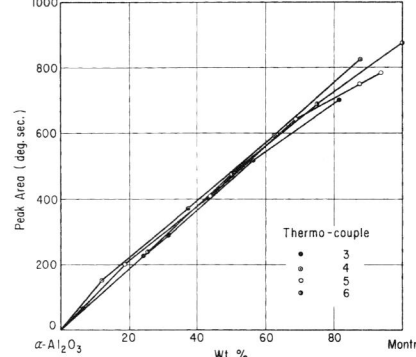

Fig. 13. Peak areas for montmorillonite and α-Al₂O₃ mixtures at about 720°C. (sample density, 1.1 gm./cm.³; total weight, 400 mg.).

kaolinite and montmorillonite diluted with α-Al₂O₃ (Figs. 12 and 13) show that the differences are less pronounced. Each renewal of the thermocouples changes Λ.

Several tests were run with thermocouple No. 5 without renewal, giving a standard deviation of 2.0 to 3.0% for 24 tests for a mean peak area of 1380 deg. × sec. (see Fig. 14).

The heat transfer through the samples varies from test to test because their heat conductivity may differ for several reasons. Figure 3 shows how λ differs for various materials; there the results of heat conduction tests at room temperature are given for α-Al₂O₃, CaSO₄, CuSO₄·5H₂O, a kaolinite, an illite, a bentonite, quartz, and feldspar. Figure 15 shows the results from some dilutions. It is worth noting that the kaolinite and α-Al₂O₃ and the dilution (1:1) of these materials show nearly the same λ for corresponding densities.

Tests with different dilutions of kaolinite and α-Al₂O₃ at equal density therefore have the same calibration value, resulting in a linear relationship between peak area and weight fraction. This is shown in Fig. 12 by the straight lines for each thermocouple. This proportionality for kaolinite mixtures has been found by several investigators. Grimshaw and Roberts,[5] de Bruyn and van der Marel,[9] and Wittels[15] have found the same proportionality for calcite.

Illite, however, has a λ differing considerably from the λ of α-Al₂O₃, as shown in Fig. 15. Therefore, the dilutions of illite and α-Al₂O₃ at equal density show no linear relationship (see Fig. 11) for each thermocouple, but a tendency toward greater peak areas for higher illite concentrations. This agrees with the theory, for more illite in the mixtures decreases the λ, and λ figures in the denominator of equation (2). The same is observed for copper sulfate (see Fig. 2). Figure 3 also reveals that the λ is nearly proportional to the density, ρ, in the region considered. The proportionality proposed by Arens[16] does not hold rigorously but is not far from reality. In the Boersma equation[10] ρ and λ appear as the fraction $\rho/[\lambda + \Lambda \ln (a/r_0)]$. The peak area therefore would be unaffected by density if $\Lambda \ln (a/r_0)$ were small compared with λ. In reality, however, $\Lambda \ln (a/r_0)$ is four or five

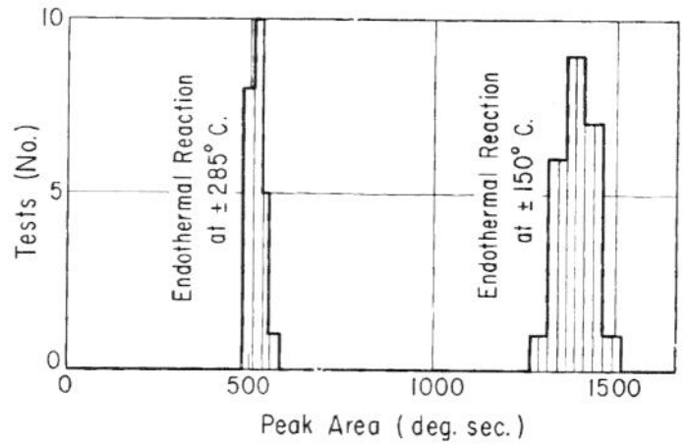

Fig. 14. Distribution of peak areas from 24 tests on CuSO₄·5H₂O with thermocouple No. 5 without renewal.

times the λ. The equalizing effect of λ on ρ is therefore partly eliminated. Tests in this respect proved that a higher density increased the peak area, although the difficulty of obtaining a constant density throughout the sample confused the results and made them inexact.

Another complication caused by the heat conductivity of a sample is the possibility of its changing during a test. This might be caused by losses of weight (see Norton[17] and Berg[4]) or by the appearance of a liquid phase (see Berg[4]). When the sample melts, the heat conductivity is increased owing to better contact between grains; the peak area is consequently reduced. On the other hand, loss of weight and especially loss of volume may decrease the heat conductivity because the sample shrinks and loses contact with wall or thermocouple; the peak area then is larger. The results of the tests with several concentrations of thiosulfate in α-Al₂O₃ as shown in Fig. 2 show that these effects begin to show only beyond dilution ratios of 30%.

Wittels[15] observed, when working with a vacuum differential thermal analysis apparatus, that the heating rate influences the peak area of calcite. Here again one may encounter the effect of the heat conductivity of the sample; in

[15] Mark Wittels, "Some Aspects of Mineral Calorimetry," *Am. Mineralogist*, **36** [9 and 10] 760–67 (1951).

[16] P. L. Arens, "Study on Differential Thermal Analysis of Clays and Clay Minerals"; doctor's thesis, Agricultural University of Wageningen, Netherlands, 1951.

[17] F. H. Norton, "Critical Study of Differential Thermal Method for Identification of the Clay Minerals," *J. Am. Ceram. Soc.*, **22** [2] 54–63 (1939).

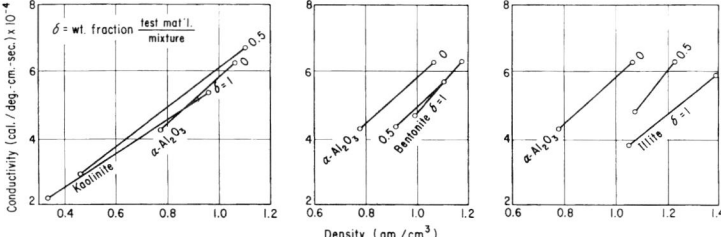

Fig. 15. Heat conductivity, λ, of dilutions of a kaolinite, a bentonite, and an illite in α-Al₂O₃ at 20°C. packed under different conditions.

this instance it decreases during the reaction. At higher heating rates the dissipation of previously produced heat, which at a lower heating rate could have dissipated under more favorable conditions, may be retarded.

Grimshaw and Roberts[5] propose to test samples when they are diluted with an inert material, e.g., α-Al₂O₃, in such a concentration that the heat-conducting properties of the mixture will equal those of the α-Al₂O₃. That it is possible to eliminate the effects of sintering and shrinking of the test material when the α-Al₂O₃ is preponderant is verified by the foregoing.

VI. Conclusion

The most serious errors in quantitative measurements with the differential thermal analysis apparatus are caused by possible differences in the heat transfer through the samples and through the thermocouple wires. Thus, for a dependable interpretation of the test results, it is necessary to take precautions to minimize this disturbing factor. This can be done by standardizing the packing method so that the same

density and the same volume will always be obtained and thus the heat flow near the thermocouple will follow the same geometrical pattern. Furthermore, the sample should be diluted with the reference inert material (usually α-Al₂O₃) to such a degree that the packing and heat-conducting properties will be determined principally by the diluter and will remain the same during the test regardless of shrinking, sintering, or liquefaction of the test material. Dilution, however, must not prevent the material investigated from giving enough thermal effect (see the dilution technique of Grimshaw and Roberts[5]). To minimize the effect of the thermocouple, it is necessary to calibrate the test device each time the thermocouple is renewed. This calibration may be done with standard chemicals of well-known heats of decomposition.

Acknowledgment

This investigation was undertaken in collaboration with H. W. van der Marel, Agricultural Experiment Station, Groningen, Netherlands. The author is indebted to H. Labrie and A. van der Wende for conducting the tests.

APPENDIX I

The heat conductivity of the sample powders was determined from the stationary heat flow through the sample toward the ellipsoidal wall of the container; the heat was generated by an electrically heated coil of confocal ellipsoidal shape. The wall of the container was kept at a constant temperature (20°C.) in a water bath and the temperature of the coil was registered by a thermistor (resistance with negative temperature coefficient). The coil was wound on a copper core so that the temperature of the surface of the inner ellipsoid was constant. The interior of this ellipsoid therefore had the same temperature as the boundary under stationary conditions as potential theory reveals, the region being simply connected and without heat-generating sources. The temperature registered by the thermistor therefore equaled that of the coil and the heat generated by the electric current moved outward only.

According to potential theory the temperature difference $\Delta\theta$ (°C.) between coil and container wall is related to the heat flow, Q (cal. per second), by the equation

$$\Delta\theta = 4\pi\lambda \cdot l \cdot Q \, (\cosh^{-1}\epsilon_1 - \cosh^{-1}\epsilon_2)$$

λ = heat conductivity of sample at 22.5°C.
$2l$ = focal distance.
ϵ_1 = eccentricity of inner ellipsoid.
ϵ_2 = eccentricity of outer ellipsoid.

The electric energy involved was $Q = 0.31$ watt, and the dimensions of the ellipsoids were $l = 1.25$ cm., $\epsilon_1 = 4.33$, and $\epsilon_2 = 1.18$. A calibrating control was made with gelatin, which has the same heat conductivity as pure water (1.37×10^{-4} cal. per deg. sec. cm.) but is without the disadvantage of heat loss through convection. The expected temperature difference of 5.2° was registered.

APPENDIX II

When the nickel block is heated at a rate V degrees per second, the temperature of the sample will lag behind, giving a constant temperature difference $\Delta\theta$ between the nickel block and the center of the sample, when the process has become stationary.

The heat per second that must enter the volume of the sample of height h, between the radii r_0 and r, is $\pi(r^2 - r_0^2)h\rho c V$. Part of this heat crosses the area $2\pi rh$ at a temperature gradient $\partial\theta/\partial r$ so that this quantity is, per second,

$$2\pi rh \cdot \lambda \cdot \partial\theta/\partial r$$

The other part flows through the thermocouple wires with cross-sectional area A, length l, and heat conductivity λ_{pl} at a rate of $\Delta\theta \cdot A\lambda_{pl}/l$.

Therefore one obtains, if $\Lambda = A\lambda_{pl}/2\pi lh$

$$\frac{\partial\theta}{\partial r} = V\frac{\rho c}{\lambda}\frac{r^2 - r_0^2}{2r} - \frac{\Lambda \cdot \Delta\theta}{\lambda \cdot r} \qquad (a)$$

with boundary conditions

$$\theta_{(r=a)} - \theta_{(r=r_0)} = \Delta\theta \qquad (b)$$

Integrating equation (a), one obtains

$$\theta = V\frac{\rho c}{\lambda}\left(-\frac{r_0^2}{2}\ln r + \frac{r^2}{4}\right) - \frac{\Lambda \cdot \Delta\theta}{\lambda}\ln r + c$$

and afterward by introduction of equation (b)

$$\Delta\theta = V\rho c\left[\left(\frac{a^2 - r_0^2}{4} - \frac{r_0^2}{2}\ln\frac{a}{r_0}\right)\Big/\left(\lambda + \Lambda\ln\frac{a}{r_0}\right)\right] \qquad (c)$$

A Capacitive Cell Apparatus

L'Appareil à Cellule Capacitive

by G. de Josselin de Jong, Ir., Sub-director, and E. C. W. A. Geuze, Professor, Delft Soil Mechanics Laboratory, Delft, Netherlands

Summary

The article describes a test device which permits the determination of the horizontal strain of cylindrical samples under loading conditions without touching the sample by means of an electrical capacitive method. Tests on samples to study at rest pressure, compression and shear separately are outlined.

The conception of shear resistance, according to the sample behaviour as determined by this test device, is discussed.

Sommaire

Cette communication décrit un dispositif d'essai permettant de déterminer la déformation transversale d'éprouvettes cylindriques comprimées axialement et transversalement, sans moyens mécaniques, mais en évaluant la variation de capacité électrique du système.

Au cours d'une série d'essais avec l'appareil, on étudie séparément la pression au repos, la compression et le cisaillement.

La signification de la résistance au cisaillement est discutée en fonction du comportement de l'échantillon, pendant les essais.

Introduction

The study of cylindrical soil samples under the influence of stresses as effected in test devices such as the Dutch cell or British triaxial apparatus becomes more interesting if the strains may be determined separately and without disturbing the applied stresses.

The capacitive cell apparatus described here permits the measurement of lateral deformation of the sample without touching it, so the horizontal stress applied by air pressure on the rubber envelope of the sample is not disturbed by the displacement determination. After a description of the test device and interpretation of the results, a discussion of typical test results in connection with the conception of shear resistance will be given.

Description of the Lateral Strain Determination

In Fig. 1 the apparatus is schematically represented. The cylindrical sample is loaded vertically by weights and horizontally by air pressure on the rubber envelope and the vertical movement is read on the dial gauge, these contrivances being of the conventional type.

The measurement of the horizontal strain is effected by using the sample surface as one of the two electrodes of a condenser, the metal housing being the other electrode.

If the sample dilates the distance between sample and outer wall will decrease, thus increasing the electrical capacity of this condenser because the capacity is inversely proportional to the distance between the electrodes. By determination of this electrical capacity the horizontal strain may be deduced.

In order to measure this capacity a reference condenser is installed in the base of the apparatus which can be made electrically equivalent to the sample-wall combination. In Fig. 1 this reference condenser is shown in a substitute circuit consisting of a fixed condenser C_p, which represents the main part of the capacity from the sample and connecting metallic parts, a resistance R_s as a substitute for the sample resistance and an adjustable condenser C_v which accounts for the remaining variable part of the sample-wall capacity.

To equalize this electrical substitute to the sample-wall capacity these two circuits are alternately connected to a zero indicator.

For this purpose use is made of an electronic device (type Boersma C.V.M. III, frequency ca. 1 Mc) which is sensitive to very small capacity variations: $0.002\ pF$ gives a visible deflection

on the Amp. scale. This sensitivity determines the possible accuracy for horizontal strain measurements. With a sample: height $(h) = 10$ cm, diameter $= 2r_s = 6.35$ cm and the dia-

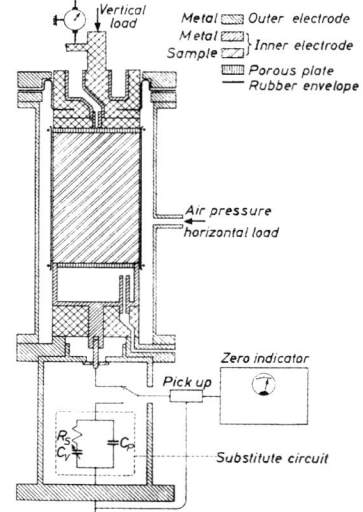

Metal ▨ Outer electrode
Metal ▨ ⎫ Inner electrode
Sample ▨ ⎭
▨ Porous plate
— Rubber envelope

Fig. 1   Schematic representation of test device
Schéma de l'appareil

meter of the inside of the housing wall $2r_w = 8.4$ cm the capacity of the coaxial condenser amounts to

$$C = \frac{h}{1.8 \log_e (r_w/r_s)} = 20\,pF$$

An increase of this capacity by $0.002\ pF$ corresponds to a horizontal strain of $\epsilon_H = 0.03$ per cent, being the accuracy of the measurement.

The apparatus was calibrated by the introduction of metal cylinders of different diameter and height. The experimental values for the capacity agreed with the values computed using formula 1: also, as a control, samples deformed in different stages were measured by an optical device of great accuracy again giving agreement with the capacitive measuring method.

Because of the rigid end plates the deformation of the sample in the horizontal direction will be larger in the middle if the friction between sample and end plates is large enough to prevent horizontal movement. By wetting the surface this friction is reduced and the barrel form is less pronounced, the disturbance being limited to small zones of the height of the sample, the rest deforming homogeneously and remaining cylindrical.

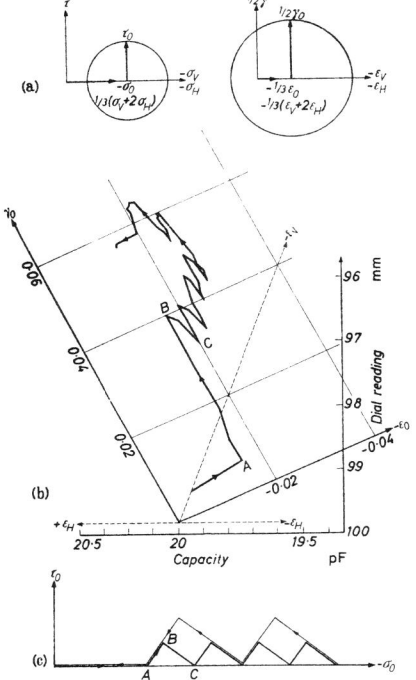

(a)

(b)

(c)

Fig. 2   Typical test result from capacitive cell apparatus
Résultats caractéristiques obtenus avec l'appareil

The capacity measuring system determines the summation of all horizontal displacements over the whole of the sample height and would therefore give the mean value if the capacity was proportional to the distance of the sample to the wall. However being inversely proportional to this distance, the parts that approach the wall most closely will dominate. The deformations studied did not give a strain greater than 5 per cent and if the sample had become a barrel with a wall in the form of a sinusoid, which is very exaggerated, the error in the mean value would amount to 0·05 per cent strain, which is a negligible part of the mean strain. So we can assume that the capacitive measuring system gives the mean horizontal strain. This may be compared with the vertical displacement obtained

from the dial gauge readings which is proportional to the mean vertical strain over the sample height.

When a sample loses water during the test this water is captured in the hollow end plates (see Fig. 1). By surrounding these water storage rooms with metal connected to the sample this water is enclosed by a Faraday cage and so does not influence the capacity of the condenser.

Representation of Test Results

The test results consist of the variations of four variables namely the principal stresses σ_V and σ_H and the principal strains ϵ_V and ϵ_H (the subscripts correspond to vertical and horizontal).

For the interpretation of the data however it is easier to consider compression and shear distinct from one another. We will therefore present the results in terms of the hydrostatic and deviational increments of stress and strain, σ_0, τ_0 and ϵ_0, γ_0, which are related to the principal stresses and strains by:

$$\sigma_0 = \tfrac{1}{2}(\sigma_V + \sigma_H) \qquad \tau_0 = \tfrac{1}{2}(-\sigma_V + \sigma_H)$$
$$\epsilon_0 = (\epsilon_V + \epsilon_H) \qquad \gamma_0 = (-\epsilon_V + \epsilon_H)$$

Thus the quantities σ_0 and ϵ_0 relate to compression only and τ_0 and γ_0 to shear, the last two being the radii of respectively stress and strain circles in Mohr's diagram (see Fig. 2a).

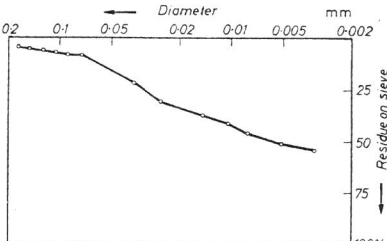

Fig. 3   Grain size distribution of test material
Granulométrie du matériau

In Fig. 2 b, and c a typical test result is shown in these coordinates. The strain diagram (Fig. 2b) is plotted from dial readings and capacity determinations. In the same diagram the coordinates for ϵ_0 and γ_0 are also given. Comparing the result in these ϵ_0 and γ_0 coordinates with the applied stresses in the $\sigma_0\tau_0$ diagram we observe a remarkable resemblance. (The negative sign indicates pressure in accordance with the notations in elasticity.)

This sample showed resiliency for the compression strain which could be accounted for by the presence of air in the clay. The air content amounted to about 4 per cent of the volume. The test duration was not long and the sample being very impervious to water movement, practically no consolidation was obtained during the test.

Some Tests Executed with the Capacitive Cell

To show the use of the apparatus we describe some typical tests on Gouda clay. The grain size distribution of this illite clay is given in Fig. 3. Liquid limit 45 per cent, plastic limit 18 per cent. The samples were moulded and cast with a water content of 34 per cent of dry weight and left for four days to regain their strength.

At-rest pressure test—The study of horizontal stress caused by horizontal confinement is facilitated by the apparatus as it permits horizontal strain measurement.

During the test the stresses are adjusted in such a way that ϵ_H remains zero. In Fig. 4 two tests are shown: one in which

53

the vertical stress is increased at a rate of 0·01 kg/cm² per minute and the horizontal stress is adjusted, and in the other the horizontal stress is increased gradually and the vertical stress is adjusted. Both test procedures give rise to a similar value for τ_0. The value of τ_0 depends on the rate of stress increase, but tends to a minimum value when one of the stresses is kept constant for a long time.

During this test the sample preserves the same diameter over its whole height. So deformation is homogeneous throughout

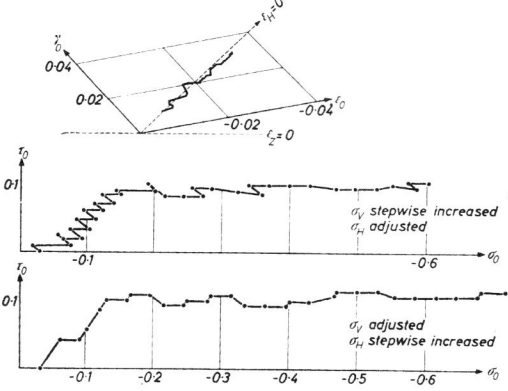

Fig. 4   Result of at-rest pressure test
Résultats de l'essai de compression au repos

the sample and also the stress distribution throughout its volume is homogeneous. This is the most favourable condition for the study of the sample.

Compression and Shear Combined

The curves of Fig. 5 represent the effect of compression and shear stresses applied simultaneously in various combinations.

For the study of shear properties we can leave the components of strain in ϵ_0 direction out of consideration and direct our attention to the movement of the strain in the γ_0 direction

by projection on the γ_0 axis. Another possibility is to load the sample in such a way that the compression strain is not varied. This is realized by increasing the vertical stress and at the same time decreasing the horizontal stress by an amount equal to one half the vertical change, thus keeping the sum of the three stresses constant.

Tests performed in this way proved that the strain consisted only of shear deformation and no compression component was induced.

Comparison of the shear deformation obtained in this way with that in cases where the compression was present showed no appreciable difference for this typical impervious clay. We will therefore in the subsequent discussion leave compression influences out of our considerations.

Shear Separated

Representation of shear influences is most effectively given in $\tau{-}\gamma$ curves. However, clay being a material with flow properties, the value of γ is not distinct for a certain value of τ as γ will increase gradually in course of time under the influence of the shear stress.

For a test where τ is increased in steps we therefore obtain a curve as indicated in Fig. 5a by the dotted line. The vertical parts of the curve are pursued in course of time and the heights of the steps in the graph therefore depend on the duration of each step.

The increase of γ with time is given in Fig. 5b. From this plot we determined the flow rate ($\partial\gamma/\partial t$) and incorporated this flow in the $\tau{-}\gamma$ curve by joining points of equal flow rate in Fig. 5a (full lines).

The family of flow rate curves is typical for a clay sample and different ways of testing reveal the same situation of these curves on the $\tau{-}\gamma$ plane as is shown in Fig. 5 (c, d, e). When τ is decreased during the test, lines of smaller flow-rate are passed indicating that the sample decreases its flow rate. Below a certain value this flow rate becomes smaller than zero, which means that the sample is resilient to shear strain. The family of negative flow rate curves however is very narrow, just like the band of positive flow rates which is characteristic for the repeated increase of τ (see thin lines in Fig. 5a). When the previous maximum value of τ is reassumed this thin band joins the primary family of flow rate curves whose situation in the $\tau{-}\gamma$ field is not seriously affected.

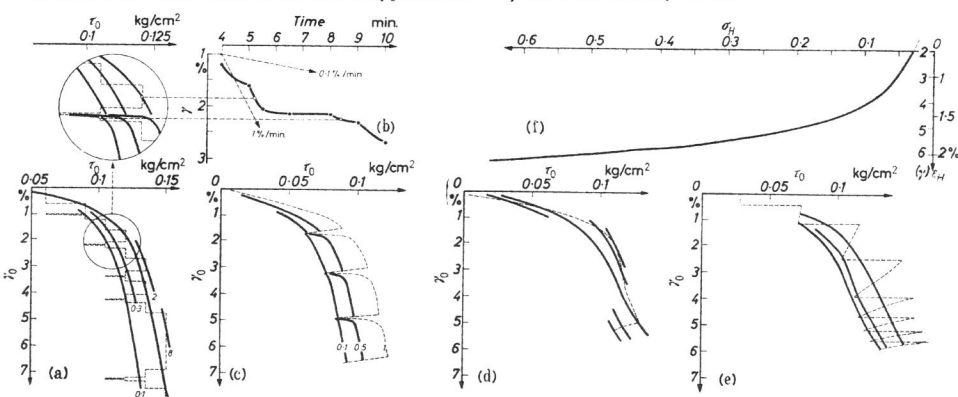

Fig. 5   Comparison of families of $\tau{-}\gamma{-}D$ curves as obtained from different test procedures
(a) Capacitive cell, (c) triaxial test with unloadings, (d) cell test continuously loaded, (e) cell test loaded in steps, (f) rigidity of cell apparatus
Comparaison des familles de courbes $\tau{-}\gamma{-}D$ obtenues par différentes méthodes expérimentales

54

An attractive way of representing this behaviour is the block diagram shown in Fig. 6 where the vertical coordinate is the flow rate $D = \partial\gamma/\partial t$.

When the soil is in the condition represented by the point A on the slope of the curved plane the material will yield with a velocity indicated by the height above zero level. Starting from this point A different paths may be followed depending on the conditions imposed by the test procedure.

A horizontal path AB is pursued if the flow rate is kept constant. This condition is realized by the triaxial test procedure when there is no consolidation during the test.

The path AC corresponds to a constant value of the shear stress τ and is obtained in the test described above.

The path AD of steepest descent is followed if a decrease of τ is accompanied by an increase of γ. This happens in an apparatus such as the Dutch-cell where the horizontal displacement of the sample is restricted by an increase of horizontal stress. When the vertical stress is kept constant this signifies

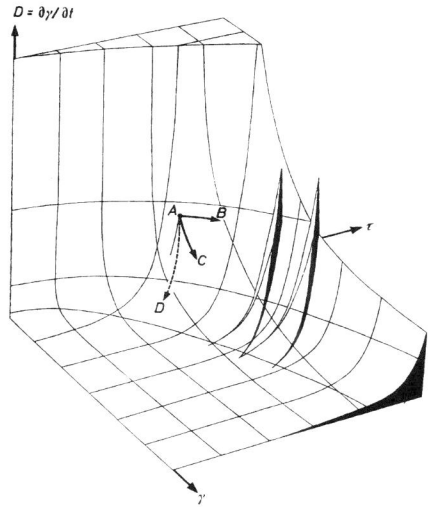

Fig. 6   Block diagram for $\tau-\gamma-D$ relation
Graphique à trois dimensions des relations $\tau-\gamma-D$

a decrease of τ. The rigidity of the apparatus determines the direction of the descent path which is more nearly parallel to the $\tau-D$ plane for greater rigidity.

In Fig. 6 is demonstrated, by the sharp pointed zig-zags, the path followed by the sample behaviour for a repeated loading as executed in the Dutch-cell test. The primary sheet is immediately rejoined, there being no serious deviations. The conception of this $\tau-\gamma-D$ sheet is of great help in the interpretation of admissible shear resistance.

Study of Shear Resistance

The resistance offered by the soil to shear is of primary importance when the strength of soil constructions is to be judged. For if the soil stresses are divided into compression and shear, it is well known that soil can withstand every compression stress at the expense of certain deformations, but that the shear stress, exceeding a certain value, causes unlimited deformations, which bring about a break down of the construction.

However in certain instances it is not the total failure that is of interest but the deformations which may assume intolerable values, causing overloading in other parts of the construction.

In a test of short duration (in comparison to the life of civil engineering construction) the flow properties of the soil under loading conditions have to be considered because continual flow may cause the unwanted deformations. As a criterion for admissibility of shear stress the flow rate has to be introduced.

In a test such as the triaxial test where only one velocity is induced we obtain an insight of the situation of one contour of D, but the slope of the sheet does not enter into the test result. This slope however determines how much the observed τ has to be reduced in order to obtain an admissible value. Preferable therefore is a test which by exploration of the sheet slope gives this information.

Such a test is a Dutch-cell test. The rigidity of the cell apparatus in the horizontal direction is represented in Fig. 5f which shows how the horizontal pressure increases as the sample dilates laterally. As the vertical stress is increased in steps τ also increases in steps, but during the time the vertical stress is left constant the sample strains and a gradual increase of E_H occurs, thus decreasing τ. This produces the oblique lines in the $\tau-\gamma$ plane shown in Fig. 5e where a test result is represented. All oblique parts together form the rigidity line of the apparatus being parallel to successive segments of Fig. 5f. The corresponding movements in the $\tau-\gamma-D$ sheet of Fig. 6 follow lines such as AD, which turn out to be nearly perpendicular to the contour lines ($D = $ constant) and so follow the direction of greatest slope on the sheet. Requirements for a good test performance are that each loading step should last long enough for observation of the small velocities and that the loading steps should be small enough not to disturb the sample.

The cell is constructed so rigidly that although a great many steps of loading and unloading are executed and a broad band of the family of velocities is explored, the total shear strain γ is still limited and shows no large values.

This reveals the intention of the cell tests to obtain information about the flow properties of the soil in an equilibrium state, that is with small deformations.

Conclusion

The capacitive cell apparatus permits the study of different conditions of stress and strain, and by measuring the vertical and horizontal strains separately compressive and shear strains may be obtained. The relation between shear stress and shear strain is only uniquely determined if the flow rate is considered. The family of curves for different flow velocities in the $\tau-\gamma$ plane seems to be characteristic of the sample and is not shifted seriously when loadings are repeatedly applied if the total shear strain does not exceed too great a value.

The exploration of the velocities in different stages of strain is most effectively obtained by an apparatus such as the Dutch-cell apparatus which, by rigidity against strain, limits the deformations of the sample even when large velocities have been applied.

The authors wish to thank Mr Boersma for suggesting the use of a capacity method for the measurement of lateral strain, Mr Winkel for carrying out the tests and Mr Van der Beld and Mr 't Hart for their assistance in the development of the apparatus.

References

GEUZE, E. C. W. A. and TAN, T. K. (1950). The shearing properties of soils. *Géotechnique*, **2**, 141
— (1953). *General Report. Proc. 3rd International Conference on Soil Mechanics and Foundation Engineering*, Vol. 2, p. 317
Proc. of the Conference on the Measurement of Shear Strength of Soils in Relation to Practice

55

ÉTUDE PHOTO-ÉLASTIQUE D'UN EMPILEMENT DE DISQUES

par G. DE JOSSELIN DE JONG et A. VERRUIJT

(UNIVERSITE TECHNIQUE, DELFT, PAYS - BAS)

RESUME

Des essais exécutés sur un empilement de disques, constituant un modèle de milieu pulvérulent, sont décrits. Les disques ont été fabriqués avec un matériau photo-élastique, ce qui permet d'étudier les contraintes à l'intérieur des disques. Il est montré comment l'analyse des essais conduit à la détermination des forces de contact entre les disques, aussi bien en grandeur qu'en direction. Les conditions d'équilibre des disques individuels sont vérifiées à l'aide d'une épure des forces.

I. — INTRODUCTION

Les propriétés mécaniques des milieux pulvérulents, comme le sable, sont souvent étudiées à l'aide d'un modèle analogique constitué d'un assemblage bidimensionnel de rouleaux ou de disques cylindriques. DANTU (1957) a été le premier à appliquer la photo-élasticité à l'étude d'un tel assemblage. Il a exécuté des recherches qualitatives sur la répartition des forces intergranulaires, aussi bien dans un empilement régulier que dans un empilement quelconque. Il résulte des recherches de DANTU, ainsi que de celles d'autres chercheurs (WAKABAYASHI, 1957 ; DE JOSSELIN DE JONG, 1960 ; WEBER, 1966), que la distribution des forces intergranulaires dans un milieu pulvérulent n'est pas du tout homogène, mais qu'il existe des chaînons de disques (ou de rouleaux) qui transmettent la plupart des forces, tandis qu'un grand nombre de disques reste non chargé.

Les recherches rappelées ci-dessus ont toutes été effectuées en utilisant de la lumière polarisée circulaire, les forces intergranulaires restant si petites qu'à l'intérieur des disques chargés, on ne constate qu'un effet photo-élastique dit du premier ordre. Alors, les disques ne transmettant aucune force ne se distinguent pas du fond obscur, tandis que les disques transmettant une certaine force deviennent lumineux.

En utilisant un empilement de billes de verre entre deux parois de verre parallèles, dont les vides ont été remplis par un liquide de même indice de réfraction que le verre, on peut observer la transmission des forces dans un milieu tridimensionnel. L'assemblage est transparent pour un certain type de lumière monochromatique. On a l'impression (DANTU, 1957 ; WARABAYASHI, 1957) que les lignes formées dans un tel milieu par l'illumination des billes chargées sont dirigées selon la direction de la contrainte principale majeure. Il est remarquable que, parfois, on voie aussi un réseau de lignes orthogonales noires, c'est-à-dire un réseau de lignes à peu près dirigées selon la direction de la contrainte principale mineure (DE JOSSELIN DE JONG, 1960).

— 73 —

Par les méthodes décrites ci-dessus, il est impossible de déterminer individuellement les forces inter-granulaires. Le présent article décrit une méthode, qui, en utilisant CR 39, un matériau photo-élastique plus sensible que le verre utilisé auparavant, donne des possibilités d'analyse plus prononcées. L'assemblage consiste en disques assez minces soutenus latéralement par deux panneaux de Perspex. En agrandissant les photographies des disques, il a été possible d'en étudier les arabesques photo-élastiques plus en détail. Le résultat est une détermination des forces individuelles agissant entre les constituants d e l'assemblage. On a pu déterminer ces forces en valeur absolue aussi bien qu'en direction.

Après avoir déterminé toutes les forces de contact, il a été possible de construire un diagramme d e forces qui, étant fermé, prouve que toutes les conditions d'équilibre sont satisfaites. Quelques exemples de résultats expérimentaux et d'épures dérivées des photographies sont présentés ci-dessous.

II. — CONSEQUENCES PHOTO-ELASTIQUES DU CHARGEMENT DES DISQUES

Ce paragraphe sera consacré à l'effet photo-élastique d'un disque circulaire chargé à sa périphérie par des contraintes exercées par les disques voisins. Ces résultats seront utilisés pour indiquer comment o n peut analyser la distribution des forces dans un assemblage de disques.

Considérons d'abord un seul disque en contact avec un certain nombre de disques voisins. Si les dimensions des plans de contact sont suffisamment petites, on peut considérer les forces agissant au bord du disque comme des forces concentrées. Pour assurer l'équilibre du disque, il est nécessaire que la résultante de toutes ces forces soit nulle et que ces forces ne produisent pas de moment résultant.

Bien que la distribution des contraintes à l'intérieur d'un disque chargé par des forces concentrées agissant à sa périphérie puisse être calculée par les méthodes de la théorie d'élasticité, il est instructif de commencer par des considérations approximatives. Près du point d'application d'une force concentrée, l a distribution des contraintes s'approchera de celle d'un plan semi-infini chargé par une force concentrée. Ce dernier cas est un problème classique de la théorie d'élasticité, résolu par FLAMANT (1892) pour le cas spécial d'une force dirigée perpendiculairement au bord, et généralisé par BOUSSINESQ (1892) pour le cas d'une force oblique.

De la solution de ce problème, on peut déduire que les isochromes (courbes à différence constante des deux contraintes principales) sont des cercles. Ces cercles passent par le point d'application de la force concentrée et leurs centres se trouvent sur une ligne dans le prolongement de la force (FROCHT, 1948).

On peut comparer cette approximation avec les résultats obtenus par PORITSKY (1950), qui a résolu rigoureusement le problème de deux disques élastiques en contact. Puisque le point de contact, dans ce cas, s'est étendu sur une petite surface, les isochromes ne sont plus des cercles, mais la déviation est a s s e z petite. La figure 1, ci-contre, présente, dans le cas particulier où la tangente de l'angle de la force a v e c la normale au plan de contact est égale à 1/3, les isochromes données par PORITSKY et les isochromes empruntées à FROCHT.

Il ressort de la figure 1 que les deux systèmes d'isochromes sont à peu près identiques. Cela justifie l'application au problème des deux disques en contact, de certains résultats obtenus pour une force agissant sur un plan semi-infini. En particulier, le fait mentionné ci-dessus que les centres des cercles isochromatiques sont situés dans le prolongement de la force, conduit à la détermination du diamètre commun d e s

— 74 —

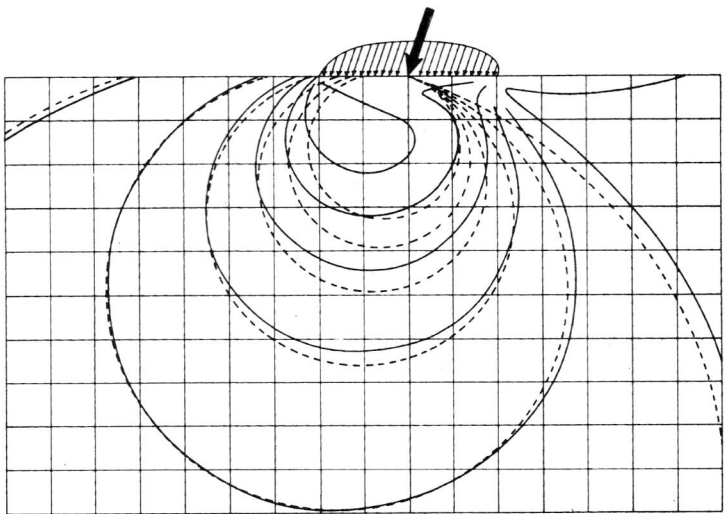

Figure 1 - Isochromes pour le cas d'une force concentrée agissant sur un demi-plan, comparées aux isochromes pour le cas de deux disques en contact. D'après FROCHT (1948), resp. PORITSKY (1950).

cercles isochromatiques pour en déduire la direction de la force. Les lignes photographiées sont exactement les isochromes et il n'est pas difficile de localiser la direction du diamètre commun.

La détermination de la valeur absolue de la force transmise d'un disque à l'autre peut être obtenue à l'aide d'une formule donnée par FROCHT (1941). Si l'expérience est exécutée sur champ clair, la différence des contraintes principales, $\sigma_1 - \sigma_2$, le long d'une ligne isochromatique est donnée par :

$$\sigma_1 - \sigma_2 = \frac{n\lambda}{Ct}, \tag{1}$$

où : λ est la longueur d'onde de la lumière,
  t l'épaisseur du modèle,
  C la constante optique du matériau,
  n l'ordre de l'isochrome (égal à un nombre entier au centre des franges lumineuses).

D'autre part, il résulte de la linéarité des propriétés mécaniques du matériau que $\sigma_1 - \sigma_2$ est proportionnel à la valeur absolue P de la force. Alors, on peut écrire :

$$\sigma_1 - \sigma_2 = P f(x, y), \tag{2}$$

où f(x, y) est une fonction exprimant la forme géométrique des isochromes. Cette fonction, en général, est également dépendante d'un certain nombre de paramètres liés au mode d'application de la force (par exemple : l'angle d'inclinaison de la force).

— 75 —

Les formules (1) et (2) expriment une relation linéaire entre la force P et l'ordre de l'isochrome n. Cette relation permet de comparer les résultats de deux essais photo-élastiques. En effet, on peut déterminer la valeur absolue de la force dans l'un des deux essais, si l'on connaît la force dans l'autre essai, par comparaison des ordres des lignes isochromatiques en un certain point.

Il est connu (FROCHT, 1948) que, pour le cas d'une force concentrée agissant sur un plan semi-infini, la fonction f(x, y) est donnée par la formule :

$$f(x, y) = \frac{2}{\pi t d} \qquad (3)$$

Dans cette fraction, t est encore l'épaisseur du modèle, et d est le diamètre du cercle passant par le point (x, y) et le point d'application de la force, le centre de ce cercle étant situé dans l'axe de la force. Lorsqu'on compare les résultats de deux essais (désignés respectivement par les indices 1, 2), il résulte des formules (1), (2) et (3) que :

$$\frac{n_1 \lambda}{C t} = \frac{P_1}{\pi t d_1} \quad , \quad \frac{n_2 \lambda}{C t} = \frac{P_2}{\pi t d_2} \quad ,$$

d'où l'on obtient :

$$\frac{P_2}{P_1} = \frac{n_2}{n_1} \frac{d_2}{d_1} \qquad (4)$$

Si l'on dispose des résultats d'un essai, dont on connaît la force P_1 et l'ordre n_1 d'une isochrome de diamètre d_1, on peut facilement calculer la force P_2 d'un second essai avec la formule (4). Il suffit de mesurer le diamètre d_2 d'une isochrome de l'ordre n_2. De ce que la fonction f(x, y) est donnée par la formule (3) quelle que soit la direction de la force P, il résulte qu'il n'est même pas nécessaire que les forces P_1 et P_2 aient même direction.

Naturellement, la mesure du diamètre d'une isochrome circulaire (ou à peu près circulaire) ne présente pas de difficulté sérieuse, mais la méthode permettant de déterminer l'ordre d'une telle isochrome n'est pas évidente. Considérons pour cela un disque soumis à l'influence de quelques forces concentrées, appliquées à sa périphérie. Il a été montré par FROCHT (1948) que, dans le cas d'un disque chargé par deux forces diamétralement opposées, toutes les contraintes s'annulent au bord du disque. Alors, la différence des contraintes principales s'annule aussi dans ce cas, et, par conséquent, le bord du disque constitue l'isochrome d'ordre zéro.

Des calculs assez simples, analogues aux calculs de FROCHT, montrent que, dans le cas d'un disque chargé par deux forces colinéaires, opposées et dirigées selon une corde arbitraire AB (figure 2),

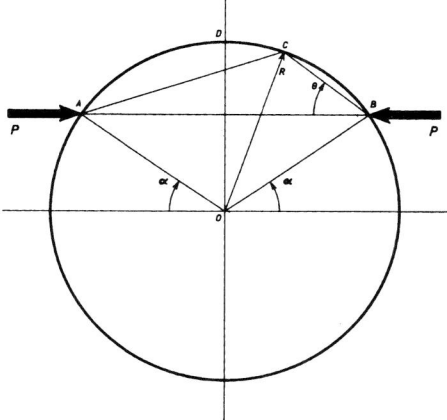

Figure 2 - Disque chargé par deux forces colinéaires et opposées.

la différence des contraintes principales en un point C du bord, déterminé par l'angle θ de la ligne BC avec la corde AB est donnée par la formule :

$$(\sigma_1 - \sigma_2)_C = \frac{P}{\pi R t} \; \frac{\sin 2\alpha}{\sin \alpha + \sin(\alpha + 2\theta)} \; , \qquad (5)$$

où :
                    R est le rayon du disque,
                    α l'angle de la direction OB avec la direction AB,

Les angles α et θ sont considérés positifs s'ils sont orientés comme indiqué sur la figure.

En posant $\alpha = 0$ dans la formule (5), on obtient la relation obtenue par FROCHT, $\sigma_1 - \sigma_2 = 0$, valable pour un couple de forces diamétrales.

Au point D, situé symétriquement par rapport à A et B, où $2\theta = \frac{\pi}{2} - \alpha$, on a :

$$(\sigma_1 - \sigma_2)_D = \frac{P}{\pi R t} \; \frac{\sin 2\alpha}{1 + \sin \alpha} \; . \qquad (6)$$

En utilisant la formule générale (1), n étant maintenant un nombre pas nécessairement entier, on obtient :

$$n_D = \frac{PC}{\pi R \lambda} \; \frac{\sin 2\alpha}{1 + \sin \alpha} \; . \qquad (7)$$

Cette formule montre que, dans le cas présent, l'ordre isochromatique ne s'annule pas au bord. Néanmoins, cela n'empêche pas l'analyse, parce que la valeur de n_D est, en pratique, limitée pour les raisons suivantes.

L'angle α est également l'angle de la force P avec la normale au plan de contact. Il ne peut donc dépasser l'angle de frottement, qui est d'environ 20° pour le matériau utilisé dans les essais. Lorsque l'angle α varie entre $-20°$ et $+20°$, la quantité $\sin 2\alpha / (1 + \sin \alpha)$ varie entre $-0,98$ et $+0,48$.

Dans les essais les forces P sont au maximum de l'ordre de 300 N et agissent sur des disques de 1,5 cm de rayon. Pour le matériau utilisé (CR 39), la constante C est d'environ 3×10^{-7} cm²/N. La longueur d'onde λ étant d'environ 600×10^{-7} cm, la quantité $PC/(\pi R \lambda)$ est toujours inférieure à 0,3.

La formule (7) conduit donc à la conclusion que la valeur absolue de n_D, c'est-à-dire de l'ordre iso-chromatique en un point du bord du disque équidistant des points d'application des forces, est toujours inférieure à 0,3. Ainsi, puisque le nombre entier le plus proche de la valeur de n_D est toujours zéro, il est à prévoir que la recherche de l'isochrome d'ordre zéro ne présentera pas beaucoup de difficulté dans le cas d'un disque chargé par deux forces dirigées selon une corde arbitraire. On notera que ce type de sollici-tation est le cas le plus général de système équilibré composé de deux forces seulement.

En extrapolant ce résultat, on peut admettre qu'il est probable que les contraintes à la périphérie d'un disque chargé par plusieurs forces concentrées (cf. la figure 3, au verso, qui présente un certain nombre de disques chargés de cette façon) seront également très petites, et que l'isochrome d'ordre zéro sera située à proximité du bord du disque.

$$- 77 -$$

Figure 3 - Détail de la photographie d'un empilement chargé, montrant les isochromes à l'intérieur des disques.

Quand l'isochrome d'ordre zéro est repéré, il est facile de déterminer l'ordre d'une autre isochrome (par exemple, une isochrome située près du point d'application d'une force), puisque l'on sait que l'ordre croît ou décroît d'une unité en passant d'une isochrome à la suivante. L'ordre n_2 et le diamètre d_2 d'une telle isochrome étant ainsi déterminés, on peut alors, par la formule (4), calculer la force P_2 transmise au point de contact.

III. — ANALYSE DES FORCES SUR LES DISQUES

Comme nous l'avons vu, la direction d'une force de contact est déterminée par le lieu des centres des cercles, et sa valeur absolue est déterminée par le diamètre et l'ordre d'une courbe isochromatique. Alors, toutes les forces transmises à l'intérieur de l'assemblage sont connues en direction aussi bien qu'en valeur absolue.

Il est ainsi possible de vérifier que les résultats satisfont aux conditions d'équilibre des disques individuels. Le procédé de vérification va être décrit à l'aide des figures 3, 4 et 5. Les figures 3 et 4 montrent le détail d'un certain empilement chargé, et la figure 5 représente le diagramme des forces y correspondant.

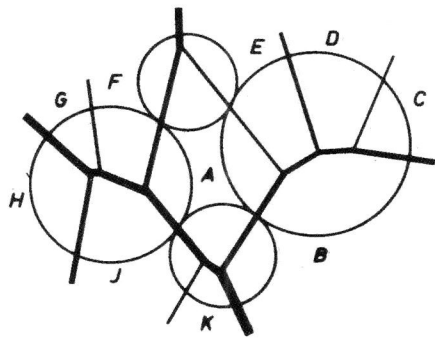

Figure 4 - Les quatre disques au centre de la figure 3 avec leurs forces de contact. L'épaisseur des lignes est proportionnelle à la valeur absolue des forces.

Pour que l'équilibre d'un disque soit satisfait, il faut que les forces agissant sur ce disque constituent un polygone fermé. Par exemple, l'équilibre du disque situé à l'extrême gauche de la figure 4 est satisfait, puisque le chemin AJHGFA de la figure 5, chemin composé des forces AJ, JH, HG, GF et FA, est un polygone fermé. Les points d'intersection des forces de la figure 5 (par exemple, les points A et J), correspondent sur la figure 4 à des régions. Ces régions ont été désignées par les mêmes caractères (A et J). La frontière commune à deux régions est la ligne d'action de la force, correspondant dans la figure 5 au segment joignant les points A et J.

Ces figures donnent également la possibilité de vérifier l'équilibre des moments des disques. Pour ce faire, les directions des forces de contact (déduites de la figure 3 et déjà utilisées pour construire la figure 5) ont été tracées sur la figure 4 à partir des points de contact correspondants. S'il s'agit d'un disque chargé par trois forces, il est nécessaire que les trois directions passent par un même point.

Dans le cas d'un disque chargé par plus de trois forces, par exemple quatre, il faut plusieurs étapes. D'abord, on compose la résultante de deux forces (quelconques parmi les quatre) et on construit sa ligne d'action qui passe par l'intersection de ces deux forces. Ensuite, on vérifie qu'elle passe par l'intersection des deux forces restantes. De la même façon, on vérifie l'équilibre des moments d'un disque chargé par cinq forces ou davantage encore. La figure 4 montre que l'équilibre des moments de tous les disques est effectivement satisfait.

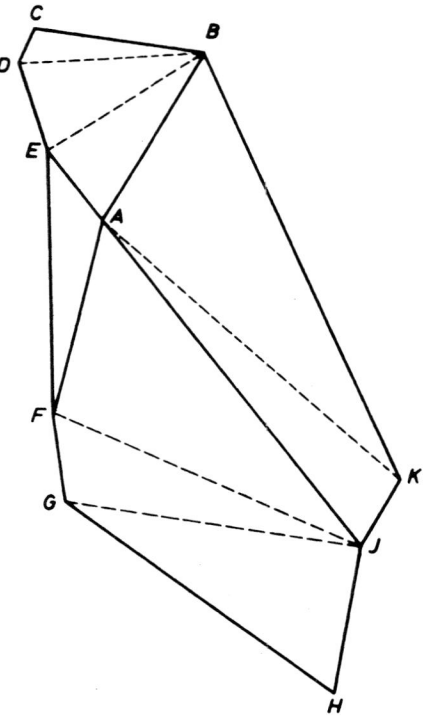

Figure 5 - Diagramme des forces correspondant aux Fig. 3 et 4, illustrant l'équilibre des disques.

IV. — DESCRIPTION DES ESSAIS

L'appareil utilisé pour les essais est composé d'une « cuvette » formée par deux panneaux de Perspex, d'environ 30×30 cm², parallèles et placés à une distance de 0,8 cm.

Entre les panneaux se trouve placé l'assemblage de disques circulaires, fabriqués à partir du matériau photo-élastique CR 39. La répartition détaillée de ces disques était la suivante : 6 avaient un diamètre de 4,0 cm ; 7 de 3,5 cm ; 16 de 3,0 cm ; 33 de 2,5 cm ; 33 de 2,0 cm ; 36 de 1,8 cm ; 28 de 1,5 cm et 31 de 1,0 cm.

Ces disques sont disposés au hasard dans la cuvette, à l'exception de chacun des quatre coins de l'assemblage carré, où l'on a mis, pour des raisons pratiques, un grand disque. Pour éviter l'apparition, au cours de la fabrication, de contraintes initiales aux bords des disques, il a fallu opérer avec beaucoup de soin.

L'assemblage peut être chargé horizontalement et verticalement par des plaques rigides de même épaisseur que les disques. La plaque inférieure est fixée rigidement ; les trois autres plaques sont libres de

— 79 —

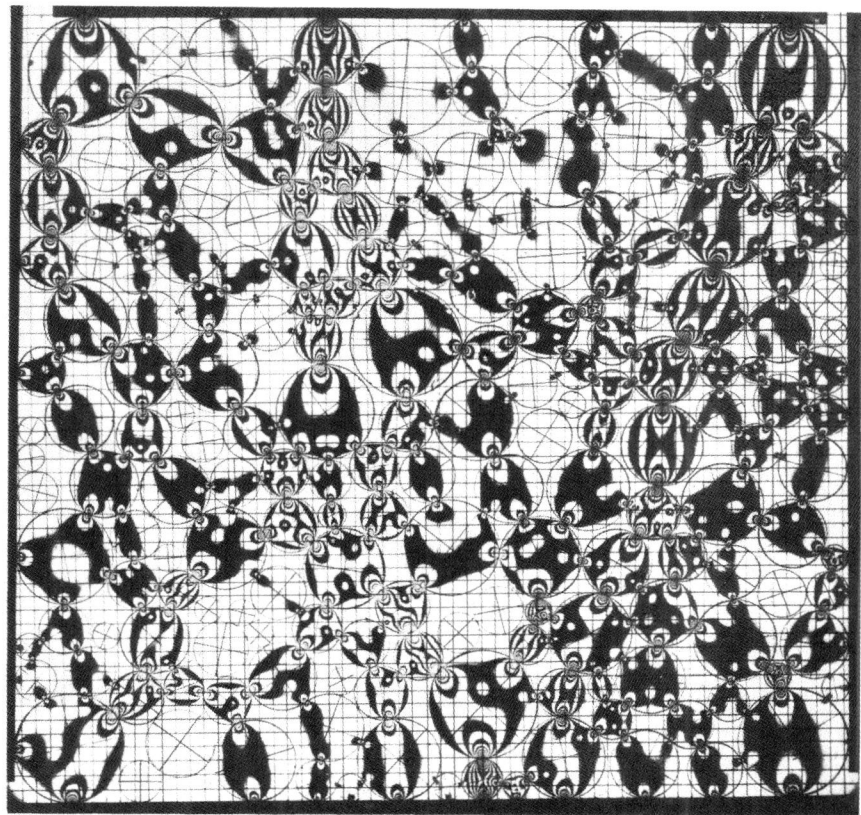

Figure 6 - Photographie d'un empilement chargé. Force verticale : 1380 N. Force horizontale : 540 N.
Le détail de la figure 3 se trouve légèrement au-dessous du centre.

se déplacer dans leur plan. Les plaques latérales sont chargées par des forces purement horizontales, égales et opposées. La plaque supérieure subit une force purement verticale.

La cuvette est placée dans un appareil photo-élastique de type normal, composé successivement d'une source de lumière, d'un polariseur, de deux lames quart-d'onde et d'un analyseur. Le modèle se trouve entre les deux lames quart-d'onde.

Le modèle étant trop grand pour que l'on puisse en obtenir une photographie intégrale en une seule fois, il a été placé dans un bâti mobile qui permet de déplacer le modèle chargé. Au total, quatre prises de vue sont nécessaires pour obtenir l'image photographique de tout l'assemblage.

— 80 —

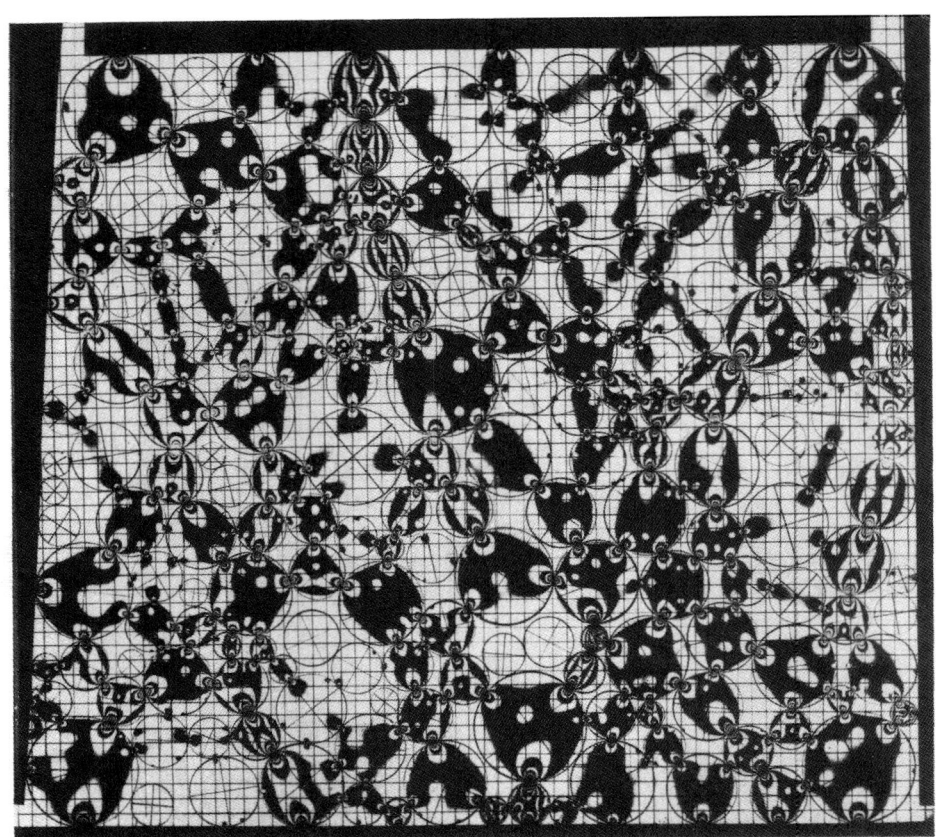

Figure 7 - Photographie d'un empilement chargé. Force verticale : 1440 N. Force horizontale : 480 N.

La figure 6 ci-contre présente le résultat d'un essai où la force verticale était égale à 2,556 f o i s la force horizontale. Le système de lignes orthogonales superposées est un réseau tracé sur l'une des d e u x parois de la cuvette. La distance mutuelle de ces lignes est de 0,5 cm.

La figure 7, ci-dessus, montre le résultat d'un autre essai sur le même assemblage, la seule différence avec l'essai de la figure 6 étant que la force verticale est augmentée de 4,348% et que la force horizontale est réduite de la même quantité. Ceci veut dire que le déviateur des contraintes, déjà assez grand d a n s l'essai de la figure 6, a encore été un peu augmenté. En comparant les deux photos, on observe que l'assemblage a un peu changé, du fait de la rupture de quelques plans de contact. D'autres plans de contact ont été créés durant la transition d'un système d'équilibre à l'autre.

— 81 —

V. — LES FORCES

A partir des photographies du modèle chargé, telles que celles des figures 6 et 7, les forces de contact peuvent être déterminées de la manière décrite plus haut. On voit qu'aux bords de tous les disques, il y a une certaine région qui ne se distingue pas du fond clair. A cette région, l'ordre isochromatique zéro a été assigné, eu égard aux considérations ci-dessus. Le résultat d'un étalonnage exécuté sur un disque de 5 cm de diamètre, chargé par une force de 300 N., a été utilisé comme base de comparaison.

Les forces ainsi déterminées ont été utilisées pour construire les épures des forces. Les épures correspondant aux photographies 6 et 7 sont représentées sur les figures 8 et 9. L'expérience nous a montré qu'il n'est pas trop difficile d'obtenir une épure fermée. Les corrections qu'il est parfois nécessaire d'apporter aux valeurs des forces, sont assez petites et l'on peut légitimement les attribuer aux erreurs inévitables pouvant provenir de l'exécution des essais, des relevés de mesure (détermination du diamètre d'une isochrome), des écarts d'arrondi, etc... Il n'est pas impossible non plus que ces corrections soient partiellement dues à de petites forces de frottement entre les disques et les parois de la cuvette.

Les figures 10 et 11 montrent le réseau des lignes de force. L'épaisseur des lignes a été choisie proportionnelle à la valeur absolue de ces forces. De ces figures, il résulte que les forces agissant sur un seul disque réalisent l'équilibre des moments, car partout trois forces passent par un même point.

Le fait que les forces déterminées à partir des photographies de l'assemblage obéissent aux conditions d'équilibre des disques individuels, indique que la méthode utilisée pour déterminer les forces de contact

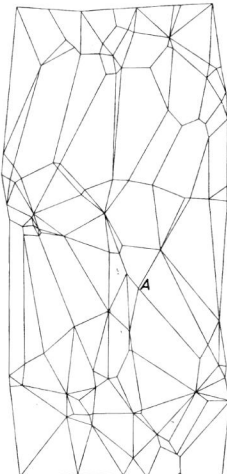

Figure 8 - Epure des forces correspondant à l'empilement de la Fig. 6. La lettre A correspond à celle de la Fig. 5

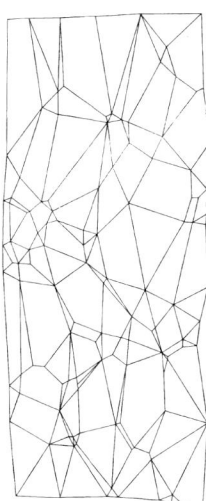

Figure 9 - Epure des forces correspondant à l'empilement de la Fig. 7.

est correcte. De même, on constate que le rapport de la hauteur totale d'une épure à sa largeur correspond bien au rapport de la force totale verticale à la force horizontale. Cependant les dimensions absolues des épures dépassent en général les valeurs des charges extérieures d'environ 10 %. Ceci doit être le résultat d'une erreur systématique, par exemple d'une différence entre les propriétés mécaniques des disques de l'empilement et celles du disque de comparaison. Cette erreur a été compensée par une modification des échelles des épures.

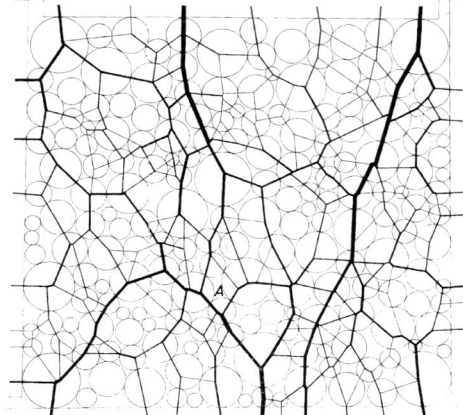
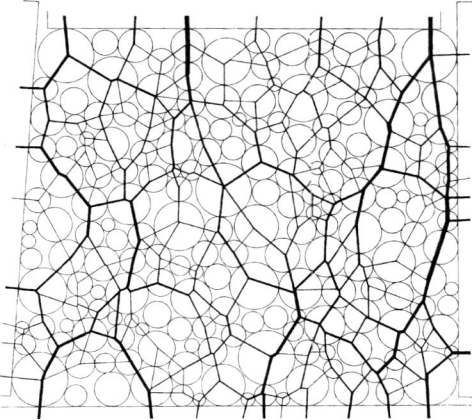

Figure 10 - Empilement de la Fig. 6 avec les forces de contact. L'épaisseur des lignes est proportionnelle à la valeur absolue des forces. La lettre A correspond à celle de la Fig. 4.

Figure 11 - Empilement de la Fig. 7 avec les forces de contact. L'épaisseur des lignes est proportionnelle à la valeur absolue des forces.

VI. — DISCUSSION DES RESULTATS

Normalement, la description du comportement mécanique des matériaux est basée sur les principes de la mécanique des milieux continus. Pour cela, le matériau est remplacé, en imagination, par un matériau fictif, dont toutes les propriétés sont des variables continues. La transmission des forces dans un tel matériau continu se décrit à l'aide de la conception du tenseur des contraintes, et les changements de la géométrie sont décrits, au cas le plus simple, par un autre tenseur du deuxième ordre, le tenseur des déformations. Les composantes de ce dernier tenseur sont obtenues par dérivation partielle des déplacements. Le comportement mécanique du matériau est décrit en choisissant, entre les tenseurs des contraintes et des déformations, une relation fonctionnelle telle que l'on obtienne la meilleure correspondance avec des résultats expérimentaux.

Le comportement mécanique d'un empilement de disques possède un caractère tout-à-fait différent de celui d'un matériau continu. La transmission des forces de contact, aussi bien que les déplacements relatifs des disques, qui prennent la place, respectivement, du tenseur des contraintes et du tenseur des déformations d'un milieu continu, sont des phénomènes essentiellement discontinus.

La transmission des forces est discontinue parce qu'elle ne s'effectue, d'un disque à un autre, que par les points de contact. D'autre part, chaque déformation est produite par un glissement, qui n'a lieu qu'en quelques points de contact, pourvu que la déformation ne soit pas excessive. Ces quelques points de contact

— 83 —

Soil Mechanics and Transport in Porous Media

disparaissent et d'autres sont créés, tandis que la plupart des points de contact est conservée. La géométrie du système de disques change, et la description de ce changement possède un caractère discontinu à cause de la rigidité des disques.

Il est très difficile de remplacer ce comportement très compliqué d'un empilement de disques par des phénomènes continus caractérisés par des tenseurs de contraintes et de déformations.

On peut essayer de caractériser la transmission des forces par un tenseur de « contraintes moyennes ». Les composantes de ce tenseur peuvent être obtenues en traçant une ligne droite à travers l'empilement, et en divisant la résultante des forces transmises par le matériau à travers cette ligne, par la longueur de celle-ci. En général, la valeur d'une telle composante dépend de la longueur de la ligne à travers les disques, et oscille autour d'une valeur moyenne. Si ces oscillations sont trop fortes, il est inepte d'introduire ce tenseur comme mesure de la transmission des forces. C'est seulement lorsqu'on peut indiquer des longueurs d e lignes correspondant à des composantes assez constantes qu'il est raisonnable de considérer le tenseur de contraintes moyennes comme mesure des forces intergranulaires. On appelle « volume élémentaire représentatif » le plus petit volume qui contient des lignes à composantes raisonnablement constantes.

Un procédé algébrique pour le calcul des composantes d'un tenseur de contraintes moyennes a été donné par WEBER (1966).

On peut obtenir une idée qualitative de la variation de la contrainte moyenne avec la longueur de la ligne élémentaire à l'aide de l'épure des forces. Pour cela, il faut noter que l'empilement et l'épure sont géométriquement duaux. Un certain nœud, jonction de trois lignes d'action de forces dans l'empilement, correspond, dans l'épure, au triangle formé par ces trois forces. D'autre part, chaque nœud de l'épure, réunion de plusieurs forces, correspond, dans l'empilement, au domaine polygonal limité par les l i g n e s d'action de ces forces. Par exemple, le point A de la figure 8 (page 82) correspond au domaine A de la figure 10 (page 83). Ainsi, une ligne arbitraire, tracée à travers l'empilement et passant par un certain nombre de ces domaines, se projette sur l'épure, en un chemin qui suit les nœuds correspondant a u x domaines. Si la projection d'une ligne droite, tracée à travers l'empilement, consiste en un chemin tortueux dans l'épure, cela indique que les forces agissantes sur la ligne droite sont distribuées d'une façon inhomogène et, en ce cas, on ne peut pas représenter, avec suffisamment de précision, les forces concentrées par une contrainte moyenne.

L'information donnée par les figures 8, 9, 10 et 11 est suffisante pour vérifier la conclusion que les chemins obtenus dans l'épure comme projections de lignes droites à travers l'empilement sont très tortueux, surtout s'il s'agit de lignes éloignées du bord. Les tortuosités consistent en déviations par rapport à une ligne droite dans l'épure, déviations qui sont d'un ordre de grandeur égal aux dimensions de l'épure. Ceci indique que les valeurs des composantes du tenseur des contraintes moyennes dépendent de la longueur de la ligne considérée, pour toutes les longueurs inférieures à la dimension de l'empilement. C'est pour cela qu'en ce cas-ci, le tenseur des contraintes moyennes n'est pas une bonne mesure pour la transmission des forces. Puisque la distribution des forces aux bords de l'empilement était bien homogène, il s'ensuit q u e le nombre de disques, que doit contenir une région élémentaire pour l'on puisse parler d'un tenseur des contraintes moyennes représentatif, sera au moins égal au nombre de disques présents dans les e s s a i s décrits ici, c'est-à-dire environ 200. Ceci confirme une remarque de WEBER (1966).

Pour décrire les déformations, il faut partir des déplacements des disques individuels. Ces déplacements peuvent être assemblés dans un hodographe, qui possède une dualité avec l'empilement de disques semblable à la dualité qui existe entre l'épure des forces et les lignes d'action des forces dans l'empilement. Par conséquent, il est possible de construire la projection sur l'hodographe d'une ligne quelconque tracée à travers l'empilement, ce qui donne le déplacement relatif des extrémités de cette ligne. La construction d'un tenseur des déformations pour un point de l'empilement exige la division des déplacements relatifs par

— 84 —

la longueur de la ligne considérée, pour un certain nombre de lignes passant par ce point. Puisque l'empilement ne contient pas assez de disques, et puisqu'il est assez difficile de mesurer les déplacements des
disques avec suffisamment de précision, les résultats que l'on peut obtenir des photographies de l'empilement
ne sont pas d'une très grande valeur. C'est pour cela que nous ne donnerons pas ici de considérations plus
détaillées.

Des difficultés essentielles se produisent lorsque l'on veut déduire théoriquement, en partant du
système élémentaire de deux disques en contact, le comportement mécanique d'un empilement de disques,
exprimé en termes des tenseurs de contraintes et de déformations. Actuellement, nous ne disposons pas
d'un procédé mathématique qui permette de prévoir le mouvement d'un empilement de disques sous l'influence
d'une variation des charges extérieures, en partant du comportement de l'unité élémentaire (deux disques en
contact). Il est néanmoins possible de décrire, qualitativement, les effets qui se produisent dans l'empilement de la façon suivante.

Dans un empilement en équilibre, les forces de contact sont distribuées de telle façon que l'angle de
la force avec la direction normale au plan de contact ne dépasse l'angle de frottement en aucun des points
de contact. Lorsque les charges extérieures varient, les angles d'inclinaison des forces varient, et dès
qu'en un certain point cet angle atteint l'angle de frottement, ce point quitte le régime élastique. En ce
point de contact des déformations irréversibles auront lieu, c'est-à-dire que les deux disques glisseront l'un
par rapport à l'autre, ou encore que le contact entre les deux disques disparaîtra complètement. Il est
possible que l'empilement retrouve son équilibre après une déformation irréversible très petite, mais il est
également possible que la déformation irréversible devienne si grande, que la redistribution des forces
qu'elle entraîne, amène un autre point de contact à passer au régime irréversible, etc... L'empilement ne
présentera des déformations importantes que si une telle déformation irréversible a lieu en au moins un point
de contact. Il est à noter que des nouveaux points de contact peuvent être créés pendant que l'empilement
tend vers sa nouvelle position d'équilibre. Il est possible aussi que l'empilement ne parvienne pas à une
nouvelle position d'équilibre, mais que le décrochement des points de contact constitue une réaction en
chaîne. Dans ce cas, on dit que l'empilement a dépassé l'équilibre limite.

La description mathématique du comportement de l'empilement décrit ci-dessus ne nous est pas possible.
La difficulté primordiale à surmonter nous semble être la description géométrique des changements de structure de l'empilement. Par ce fait, nous nous abstiendrons d'un exposé plus détaillé de ces difficultés.

VII. — CONCLUSION

Il a été montré que, par l'emploi des méthodes de la photo-élasticité, il était possible de déterminer,
avec suffisamment de précision, les forces de contact dans un empilement de disques. Les photographies
permettent de reconstruire les déplacements relatifs des disques, qui constituent les déformations géomé-
triques de l'empilement. La mesure de ces déformations a déjà été faite par beaucoup de chercheurs, à
l'aide de l'étude d'un empilement de rouleaux en aluminium. Cependant, des expériences sur un empilement
de rouleaux ne donnent aucune information sur les forces à l'intérieur de l'empilement. Nous avons l'impression que des recherches, telles que celles décrites ici, sont nécessaires pour comprendre le comportement
mécanique d'un empilement de disques.

BIBLIOGRAPHIE

J. BIAREZ. — Contribution à l'étude des propriétés mécaniques des sols et des matériaux pulvérulents, Thèse de Doctorat, Grenoble, 240 pp., Louis-Jean, Gap, 1962.

J. BIAREZ et K. WIENDIECK. — Remarque sur l'élasticité et l'anisotropie des matériaux pulvérulents, C.R. Acad. Sc. Paris, 254, p. 2712-2714, 1962.

J. BIAREZ et K. WIENDIECK. — La comparaison qualitative entre l'anisotropie mécanique et l'anisotropie de structure des milieux pulvérulents, C.R. Acad. Sc. Paris, 256, p. 1217-1220, 1963.

J. BOUSSINESQ. — Des perturbations locales que produit au-dessous d'elle une forte charge, etc, C.R. Acad. Sc. Paris, 114, p. 1510-1516, 1892.

P. DANTU. — Contribution à l'étude mécanique et géométrique des milieux pulvérulents, Comptes Rendus du IVème Congrès de Mécanique des Sols et des Fondations, p. 144-148, Londres, 1957.

G. DE JOSSELIN DE JONG. — Foto-elastisch onderzoek van korrelstapelingen, L.G.M. Mededelingen, 4, p. 119-134, 1960.

G. DE JOSSELIN DE JONG. — Statics and kinematics in the failable zone of a granular material, Doctor's thesis, Delft, 119 pp., Waltmar Delft, 1959.

M. FLAMANT. — Sur la répartition des pressions dans un solide rectangulaire chargé transversalement, C.R. Acad. Sc. Paris, 114, p. 1465-1468, 1892.

M.M. FROCHT. — Photoelasticity, vol. I, 411 pp., New-York, Wiley, 1941.

M.M. FROCHT. — Photoelasticity, vol. II, 505 pp., New-York, Wiley, 1948.

T. WAKABAYASHI. — Photoelastic method for determination of stress in powdered mass, Proc. 7 th Japan National Congress for Applied Mechanics, p. 153-158, Tokyo, 1957.

J. WEBER. — Recherches concernant les contraintes intergranulaires dans les milieux pulvérulents. Application à la rhéologie de ces milieux. Cahiers du Groupe Français de Rhéologie, n° 3, t. I, p. 161-170, 1966.

*

*   *

Les auteurs adressent leurs remerciements à Monsieur J.C. Baas, assistant au Laboratoire STEVIN de l'Université, qui a effectué et élaboré les essais.

Selected Works of G. de Josselin de Jong   *161*

PART II

FLOW AND TRANSPORT IN POROUS MEDIA

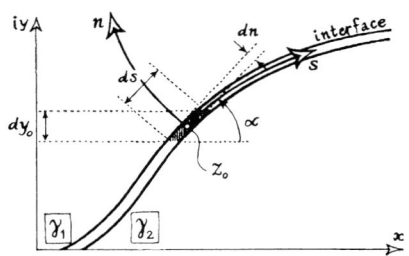

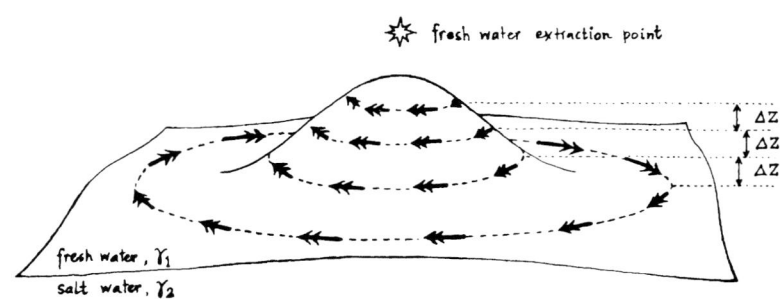

4

Flow and Transport in Porous Media

4.1 Introduction to Flow and Transport in Porous Media

When studying a certain problem, Gerard de Josselin de Jong's way of approach was to create a clear visual concept of the mechanism of the underlying physical process. Each time this was the most important step in his research. He could make no progress without it. In the early days of his career only quite primitive two dimensional visualization techniques, basically Hele - - Shaw models, were available to see what actually goes on inside a porous medium. Nowadays, advanced three dimensional computer aided techniques have been developed to study flow processes in porous media. An example is given in Figure 1. It shows snap shots of a fluorescent dye which was injected into a homogeneous porous medium, occupied by a single fluid moving at constant average velocity. The tracer, initially present in a small ball, clearly develops into an ellipsoid which is elongated in the direction of flow and symmetric in the transverse directions. Also note that the ellipsoids are perturbed by the micro structure of the porous material.

Jos must have had this picture in mind when he developed the concepts of tracer dispersion in the mid fifties. In his Dijon (1957) and AGU (1958) papers he was the first to present a quantitative analysis of the observed longitudinal and transverse dispersion. As a first characteristic step he visualized a porous medium as a network of randomly oriented tubes (canals) of fixed length. Of course he supported this concept by adding a number of artistic impressions of pore systems in the paper. Further he introduced a probability distribution - a choice of path - in the analysis. This second critical step asserts that a particle arriving at a junction between tubes has a probability to move in a certain direction. He supposed that this probability is equal to the ratio of water flowing in that direction. The combination of these two steps leads to a scattering mechanism which resembles a Brownian motion with a super-imposed convection in mean flow direction. Following the classical work of Chandrasekhar he then developed a probability density function describing the probability that a particle, after N steps, arrives in a given infinitesimal volume during a given infinitesimal time interval. Once this probability density function is known, he computed the longitudinal (flow direction) and transverse standard deviations and thereby quantifying the longitudinal and transverse dispersivities in terms of the system parameters.

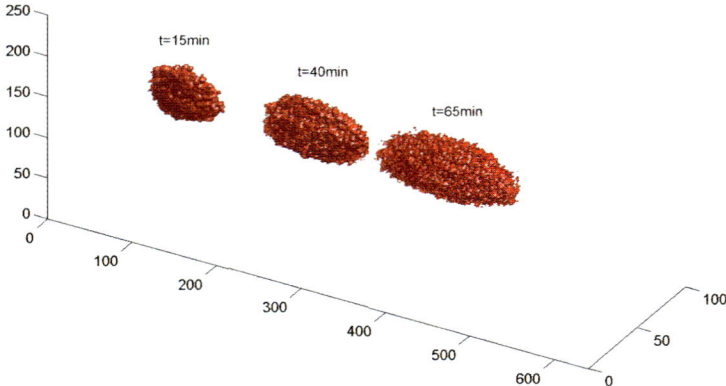

Figure 4.1: *Dispersion in a porous medium. The tracer experiment was carried out by Michael Röhr of the Institute for Environmental Physics of the University of Heidelberg (2001). We acknowledge Professor K. Roth for allowing us to use this material.*

From his approach it is clear that he considers tracer transport in porous media as a discrete process. Unlike many of his colleagues in the field, Jos was never very much in favor of a partial differential equation of convection - diffusion type for the tracers. He describes transport in terms of integrals based on probability density functions describing the local phenomena.

Much later, in his Socorro (1970) report, he considered tracer dispersion again. In this unpublished work he introduced the concept of Elementary Conveyer Unit (ECU) as opposed to the well-known and much used Representative Elementary Volume (REV). An ECU carries the direction of flow and is characterized by repetition and by independence. The latter means that a tracer or fluid particle has a choice of path at the exit of the elementary unit, which is independent of the situation at the entrance of the unit. As a result, again a scattering mechanism results. Using ECU's as elementary units he considered in the Socorro report tracer dispersion in fissured rock and he was able to give a complete description of the dispersion mechanism.

Salt-water intrusion arises naturally in many coastal regions, and in particular in a sub sea-level country as The Netherlands. As a consequence a certain tradition was established in the Dutch hydrological community, in which various kinds of fresh-salt groundwater models were developed and studied. One of the issues in the fifties was the existence or non-existence of a potential. It was known that single phase flow of constant density can be considered as potential flow, in the sense that the specific discharge is proportional to the gradient of a potential (Darcy Law). Also interface models, with an abrupt transition between the densities of fresh and salt groundwater, can be handled by a (pseudo) potential. But the general case of a smoothly varying density could not be put in potential form, in spite of several severe attempts.

In his AGU (1960) paper on 'singularities in multiple fluid flow', Jos was one of the first to give a proper mathematical formulation of variable density fluid flow in porous media. In this paper he derived his famous stream function equation

$$\Delta \Psi + \frac{\kappa}{\mu} \frac{\partial \gamma}{\partial x} = 0$$

for two dimensional incompressible flow. He gave an equivalent formulation in terms of pressure as well. In particular he pointed out that a sharp interface approach leads to a vortex distribution along the interface. This explains quite elegantly the interface shear flow and the resulting motion of the fluid bodies. Further he introduced the complex potential as a tool to obtain the specific discharge in a more direct way. In this way he was able to formulate and solve initial boundary value problems for variable density and for variable viscosity fluid flow. In particular he treated the case of an initially vertical fresh-salt interface in a horizontal porous layer (aquifer): he computed the corresponding discharge and was able to make a quantitative statement about the evolving rotating interface. Later, Euromech (1981), he used some of these results to give a justification of the Dupuit-Dietz approximation in the case of flat, horizontally extended interfaces.

Since Polubarinova-Kochina it is known that many two dimensional, stationary groundwater flow problems can be solved by the hodograph method. In this method one introduces a complex potential and a complex discharge and one studies the relation between the original two dimensional flow domain, the complex potential plane and the complex discharge plane (the hodograph). In particular fresh-salt interface problems, with stagnant salt water below the moving fresh water, can be treated by the hodograph because the interface translates simply into a circle in the complex discharge plane. Bear and Dagan (1964) observed that the presence of wells or drains leads to many-valuedness of the hodograph. A specific problem of this nature, a drain intercepting part of the fresh water discharging into the sea, was considered by Jos in his WWR (1965) paper. There he solved this extremely complicated problem by finding a way of handling the many-valuedness of the hodograph. This paper is one of his best in the mathematical as well as artistic sense. Beautiful drawings were made to support the idea of the construction and the transformations involved.

Sofar we only discussed theoretical issues. However going through his papers, one finds that each theoretical concept is verified by an experiment. The dispersion paper comes with salinity breakthrough curves to verify the longitudinal spreading and in the discussion that followed he showed two dimensional results to support the idea of transversal spreading. The multiple fluid paper shows a Hele-Shaw experiment of the rotating interface and finally in the hodograph paper a Hele-Shaw experiment confirms his elaborated construction of the flow pattern and the position of the stagnation point.

Jos had no aim to solve so-called real life problems, in which many phenomena and complexities are mixed. He was also not an isolated theoretician. He had the rare talent to select a characteristic feature out of the complex reality, to create an innovative idea which was then confirmed by experiment. That is true science and that is what we have to teach our students.

JOURNAL OF GEOPHYSICAL RESEARCH   VOLUME 65, No. 11   NOVEMBER 1960

Singularity Distributions for the Analysis of Multiple-Fluid Flow through Porous Media

G. DE JOSSELIN DE JONG[1]

Institute of Engineering Research
University of California, Berkeley, California

Abstract. In Part 1 the simultaneous flow of fluids of different properties is treated by substituting these fluids by one hypothetical fluid and applying singularities at those points where the properties of the actual fluids change. Their magnitude is chosen so that the specific discharges in the hypothetical fluid are everywhere identical to the specific discharges in the actual fluids. The flow in the hypothetical fluid can be determined by potential theory from the transformed boundary conditions and the influence of the singularities.

For the determination of the discharge a stream function is used which contains singularities in the form of vortices. For the determination of the fluid pressures a multiple-fluid potential is defined which contains singularities in the form of source and sink distributions. The stream and the potential functions each combine with auxiliary, many-valued functions to form complex potentials. These permit solutions in the form of one integral in complex variables, valid for any point in the entire field, irrespective of the fluid present. The solution for the transition zone between fluids as well as the abrupt interface is elaborated.

In Part 2 the two-dimensional example of an infinite, confined aquifer with an initial vertical interface between two fluids of different specific weight is elaborated, giving as a result the movement of the fluids in the entire field at the first moment and a first approximation for the rotation of the interface around the center as a function of time.

These results are verified by a parallel plate model and an electric resistance model. In the latter model the vortices are replaced by sources for the tracing of streamlines and by source-sink combinations forming doublets for the potential lines.

NOTATION

a		Time dependent coefficient describing inclination of interface in example
b	$[L]$	Breadth of stream channel
c	$[L]$	Half height of aquifer
d	$[L]$	Slot width of parallel plate model
f	$[L]$	Thickness of two-dimensional aquifer
f_e	$[L]$	Thickness of two-dimensional electric resistance model
$i = \sqrt{-1}$		Imaginary unit
$\mathbf{i}$	$[Amp\ L^{-2}]$	Electric current density vector
k	$[L^2]$	Specific permeability of aquifer
m		Integer number
n	$[L]$	Coordinate perpendicular to the interface
p	$[FL^{-2}]$	Pressure in the fluids
q	$[LT^{-1}]$	Specific discharge
q_s, q_n, q_x, q_y	$[LT^{-1}]$	Specific discharge components in s, n, x, y directions
q_c, q_d	$[LT^{-1}]$	Continuous, discontinuous components of q on interface
$\bar{q} = q_x - iq_y$	$[LT^{-1}]$	Complex specific discharge
s	$[L]$	Coordinate along the interface
t	$[T]$	Time
$\bar{v}$	$[LT^{-1}]$	Mean velocity of the fluid
x	$[L]$	Horizontal coordinate

---

[1] Since August 1960 at Technische Hogeschool, Delft, The Netherlands.

3739

y	$[L]$	Vertical coordinate
$z = x + iy$	$[L]$	Complex coordinate
A, dA	$[L^2]$	Area
A, B, C, D, E, F		Points on boundary
E	$[Volt]$	Electric potential
E^*	$[Volt]$	Electric potential in current supply system
I	$[Amp]$	Electric current
Im		Imaginary part of complex expression
J, K		Intersection points of interface and boundary
M		Center point of interface in example
N		Image of M in ζ plane
P		Arbitrary running point
P_ζ		Arbitrary running point in ζ plane
Q	$[L^3 T^{-1}]$	Discharge in aquifer
S	$[L]$	Contour or path of line integral
V, W		Unspecified functions of x and y
α		Inclination of interface to horizontal
β		Auxiliary variable
γ	$[FL^{-3}]$	Specific weight of fluid
δ''	$[L]$	Doublet distance
ϵ		Porosity of aquifer
$\zeta = \xi + i\eta$		Complex coordinate of transformation
κ	$[T^{-1}]$	Strength of source per unit area
λ''	$[LT^{-1}]$	Strength of doublet per unit area
μ	$[FTL^{-2}]$	Dynamic viscosity of fluid
ρ	$[Ohm\ L]$	Specific electric resistance
ψ		Exterior angle $BP_\zeta E$
ω	$[T^{-1}]$	Strength of vorticity in x, y plane per unit area
Θ	$[L^2 T^{-1}]$	Multiple fluid potential
Λ''	$[L^4 T^{-1}]$	Doublet discharge
Φ	$[L^2 T^{-1}]$	Auxiliary function
Ψ	$[L^2 T^{-1}]$	Specific discharge stream function
$\Omega = \Phi + i\Psi$		Complex specific discharge potential

Subscripts

0	Point, where a singularity is present, or point of interface containing a singularity
I	Part of complex potential accounting for singularities
II	Part of complex potential accounting for boundary conditions
1, 2	Pertaining to fluid 1, 2 (1 is light, 2 is heavy)
μ	Potential functions convenient for the study of viscosity influences with electric resistance analogy

Superscripts

'	Inversion for electric resistance model (tracing of stream lines)
''	Doublets in electric resistance model (tracing of potential lines)

PART 1. DERIVATION

Introduction. In the study of ground-water flow a theory of the interaction between different fluids needs to be developed so that any arbitrary boundary value problem may be solved rigorously. Many hydrologists have devoted their attention to this subject, because in several fields of hydrology important problems are created by the presence of two fluids. The solutions obtained, however, are mainly limited to cases which yield certain geometrical approximations. It is the purpose of this paper to describe a method which permits the determination of the behavior of two or more fluids, for any form of distribution over a field with arbitrary boundary conditions.

Badon Ghijben [1888] and afterwards *Herzberg* [1901] stated the hydrostatic equilibrium for a fresh water lens floating on top of salt water in a porous aquifer. Since then several authors have considered the more relevant case that either one of the two fluids or both are in movement. The first to have given a correct account of how the movement of the fluids influences the behavior of the interface seems to have been *Lorentz* [1913]. His discussion of the upconing of salt water under a well was later investigated experimentally by *Muskat* [1937]. *Hubbert* [1940] established a general description of the flow of two fluids at either side of a steady interface between fluids. By application of these considerations to a stationary oil deposit above an underlying body of flowing water, the tilt of the interface was derived for a one-dimensional flow system [*Hubbert*, 1953].

Edelman [1940] determined the shape of the interface between moving fresh water and stationary salt water in a dune area adjacent to the sea by use of a graphical flow net analysis. A mathematical solution was obtained for the case in which vertical flow components could be ignored, adopting the Dupuit-Forchheimer assumption. *Todd and Huisman* [1959] applied Edelman's method for determining the influence of recharge and pumping operations to limit overdraft in the drinking water supply area of Amsterdam.

If only one fluid is in motion, the hodograph method can be applied because the interface is circular. Applications of this technique for particular solutions have been carried out by *Harder, Simpson, Lau, Hotes, and McGauhey* [1953], *Kidder* [1956a, b], *Glover* [1959], and *Henry* [1959]. When both fluids are moving, the interface, when mapped into the hodograph plane, takes the form of an unpredictable curve, so that the mapping procedure cannot be applied. An approximate method based on the Dupuit-Forchheimer assumptions was developed for one case by *Dietz* [1953].

Model studies provide the only means of studying the location and movement of interfaces for general problems involving movement of both fluids. A parallel plate model has been employed for this purpose by *Santing* [1951], and sand models have been used by *Harder, Simpson, Lau, Hotes, and McGauhey*, [1953] and *Keulegan* [1954].

None of the publications indicated above contains a mathematical method by which the correct computation of the movement of both fluids as derived from given boundary conditions can be made. The first work which suggested a solution to this problem was an unpublished study by Edelman in 1957, 'Grondwaterstroming van een niet homogene vloeistof.' In this study he introduced a concept which proved inspiring for the development of the present theory: the replacement of the two fluids with their own characteristics by one hypothetical fluid with the same properties over the entire field. In this hypothetical fluid a row of sources coinciding with the position of the interface takes care of the change in properties of the two fluids. Edelman determined the source distribution in such a way that the velocity field created by these sources is equal to the real velocity distribution in one of the two fluids. The velocities in the other fluid can be obtained from the velocities created by the sources by the addition of a fictitious velocity. This concept was developed only for a horizontal interface, and the absence of sources along a vertical interface was inferred without further proof.

The present work [*de Josselin de Jong*, 1959] was based on this concept of replacing the two different fluids by one hypothetical fluid and of introducing the different fluid properties by singularities along the interface. In determining the character of the singularities, however, the aim was a singularity distribution that would directly create the actual velocity distribution in both fluids. By this approach the treatment of the two fluids becomes equivalent, and because of this equivalency the extension to more fluids can be made without further complications. The present theory shows that it is possible to meet these requirements with two kinds of singularities, vortices or sources and sinks. The choice of the kind to use depends on the objective of the study. If the objective is the determination of the discharges, it is convenient to introduce vortices. If the pressure in the fluids is to be determined, it is more convenient to use the sources and sinks. Furthermore, in the present study any inclination of the interface and the case of a gradual transition zone are considered, as well as the introduction of boundary conditions.

The important advantage of the singularities is the possibility of solving any boundary value problem involving two or more fluids by application of the established methods of potential theory for the solution of boundary value problems of one fluid. Although the singularity method is completely general with regard to differences of fluid properties, the present work deals mostly with the influence of differences in density only. The effect of differences in viscosity can also be represented in terms of singularities, as indicated, but the procedure is not amenable to mathematical treatment. How an electric analogy can be used to account for viscosity differences in connection with singularities for the density differences will be shown at the end of Part 2.

Physical assumptions and basic equations. This study is concerned with the behavior of miscible fluids filling the pores of a porous medium. The distribution of the fluids over the aquifer is given as initial information, and the value of the specific weight γ and the dynamic viscosity μ is known as a function of position.

The geometry of the pore space is the same, irrespective of the prevailing fluid; therefore, the specific permeability k is identical over the entire field. Both fluids obey Darcy's law while flowing through the porous medium, as all velocities remain in the laminar region.

For the derivation of the formulas the transition zone between two fluids will be considered. Such a zone exists between two miscible fluids. To simplify the mathematical treatment of a special case, which will serve as a demonstration of the method, the transition zone will be reduced to an abrupt interface. Such an interface will not occur in reality, because of effects of dispersion and miscibility. Changes in permeability created by entrapment of fluids and pressure jumps resulting from surface tension differences, both occurring with immiscible fluids, will be ignored in this study. For the sake of simplicity only the two-dimensional case will be treated here, because the third dimension adds no important differences.

Two basic assumptions are inherent in the following derivation. One is that there shall be continuity of incompressible fluids; thus, no

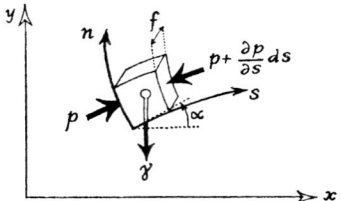

Fig. 1.　Coordinate system.

gaps can exist in either fluid or at the interface. The second is that the pressure distribution in the interior of the system shall be continuous. This condition, in which the occurrence of steps in pressure is avoided, is dictated by equilibrium conditions.

It is convenient to consider only those variables which possess magnitudes to be determined independently of the fluid properties. The variables are pressure in the fluid p and specific discharge q (q is ϵ times the mean velocity $\bar{v}$ of the liquid, where ϵ is porosity).

To relate p and q, the forces which drive the fluid through the ground will be equated to the resistance offered by the pore system. Consider in a vertical section of a two-dimensional aquifer of thickness f, a small cube ($ds\,dn\,f$), where s is arc length along an arbitrary but fixed curve, n is the perpendicular to that curve, and α is the direction to the horizontal x coordinate (Fig. 1).

The force acting on the fluid in the s direction consists of a pressure component

$$-(\partial p/\partial s)\ ds\ dn\ f$$

and a gravity component

$$-\gamma(\partial y/\partial s)\ ds\ dn\ f$$

By Darcy's law this force is counteracted by a resistance offered by the pore system, which is proportional to q_s, the s component of q. This requires a driving force of magnitude

$$+(\mu/k)q_s\ ds\ dn\ f$$

The equation for these three forces, after division by the elementary volume, is

$$(\mu/k)q_s\ =\ -(\partial p/\partial s)\ -\ \gamma(\partial y/\partial s) \qquad (1)$$

Similarly, in the n direction

$$(\mu/k)q_n\ =\ -(\partial p/\partial n)\ -\ \gamma(\partial y/\partial n) \qquad (2)$$

The two requirements mentioned above can be written in terms of the variables q and p. The condition of continuity requires that

$$\mathrm{div}\ q = (\partial q_s/\partial s) + (\partial q_n/\partial n) \qquad (3)$$

The condition of equilibrium requires that p be a single-valued function containing no jumps. This may be written in two ways:

$$(\partial^2 p/\partial s\ \partial n) - (\partial^2 p/\partial n\ \partial s) = 0$$

$$\text{or}\quad \oint_s (\partial p/\partial s)\ ds = 0 \qquad (4)$$

The contour integral applies to any closed contour S.

Solution with vortices. If the objective is the determination of the discharge q, it is convenient to introduce a stream function Ψ, whose derivatives are the components of q according to

$$(\partial \Psi/\partial n) = -q_s, \qquad (\partial \Psi/\partial s) = +q_n \qquad (5)$$

Ψ is merely a function which generates the specific discharge without the material properties γ or μ being involved. Therefore, its influence on the fluid motion is equivalent to whatever fluid is present and may therefore be considered to be related to the hypothetical fluid.

Introduction of (5) into the continuity equation (3) shows that

$$-(\partial^2 \Psi/\partial n\ \partial s) + (\partial^2 \Psi/\partial s\ \partial n) = 0 \qquad (6)$$

which implies that Ψ is a single-valued function without jumps.

The condition of equilibrium may be used in the contour integral form of (4). Writing (1) as a similar contour integral results in

$$-\oint_s (\mu/k) q_s\ ds = \oint_s (\partial p/\partial s)\ ds$$

$$+ \oint_s \gamma (\partial y/\partial s)\ ds$$

Substituting for q_s with (5) and using (4) reduces this to

$$\oint_s (\mu/k)(\partial \Psi/\partial n)\ ds = \oint_s \gamma (\partial y/\partial s)\ ds \qquad (7)$$

The contour integrals in this equation may be replaced by surface integrals by use of Green's theorem, which in two dimensions has the form

$$\oint_s V \frac{\partial W}{\partial n}\ ds$$

$$= -\iint_A \left(\frac{\partial V}{\partial x} \frac{\partial W}{\partial x} + \frac{\partial V}{\partial y} \frac{\partial W}{\partial y} \right) dx\ dy$$

$$- \iint_A V\ \nabla^2 W\ dx\ dy$$

The surface integral extends over the whole area A enclosed by the contour S. If the quantities V and W, which represent any two functions of x and y, are applied to the Ψ term of equation 7 with $V = \mu$ and $W = \Psi$,

$$\oint_s \frac{\mu}{k} \frac{\partial \Psi}{\partial n}\ ds$$

$$= -\frac{1}{k} \iint_A \left(\frac{\partial \mu}{\partial x} \frac{\partial \Psi}{\partial x} + \frac{\partial \mu}{\partial y} \frac{\partial \Psi}{\partial y} \right) dx\ dy$$

$$- \frac{1}{k} \iint_A \mu\ \nabla^2 \Psi\ dx\ dy \qquad (8)$$

To apply Green's theorem to the γ term of (7) it is necessary to convert $\partial y/\partial s$ into an expression involving the partial derivation of n. As s and n are perpendicular in such a way that n makes the same angle with y as x makes with s, $\partial y/\partial s$ is equal to $-\partial x/\partial n$. Therefore the γ term of (7) becomes, with $V = \gamma$ and $W = x$,

$$-\oint_s \gamma \frac{\partial x}{\partial n}\ ds$$

$$= +\iint_A \left(\frac{\partial \gamma}{\partial x} \frac{\partial x}{\partial x} + \frac{\partial \gamma}{\partial y} \frac{\partial x}{\partial y} \right) dx\ dy$$

$$+ \iint_A \gamma\ \nabla^2 x\ dx\ dy \qquad (9)$$

This reduces to $\iint_A (\partial \gamma/\partial x)\ dx\ dy$ because $(\partial x/\partial x) = 1$ and $(\partial x/\partial y) = \nabla^2 x = 0$.

Combining (7), (8), and (9) results in, for any small area A, i.e., for every point of the interior,

$$\nabla^2 \Psi = -\frac{k}{\mu} \frac{\partial \gamma}{\partial x} - \frac{1}{\mu} \left(\frac{\partial \mu}{\partial x} \frac{\partial \Psi}{\partial x} + \frac{\partial \mu}{\partial y} \frac{\partial \Psi}{\partial y} \right) \qquad (10)$$

This is known in potential theory to represent a singularity of vorticity ω.

$$\omega = -\frac{k}{\mu} \frac{\partial \gamma}{\partial x} - \frac{1}{\mu} \left(\frac{\partial \mu}{\partial x} \frac{\partial \Psi}{\partial x} + \frac{\partial \mu}{\partial y} \frac{\partial \Psi}{\partial y} \right) \qquad (11)$$

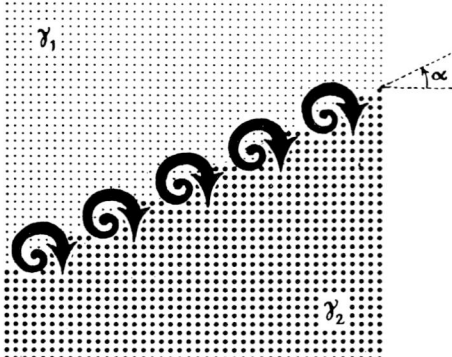

Fig. 2. Vortices on interface between two fluids ($\gamma_1 < \gamma_2$), clockwise rotating in points where γ increases in x direction.

The first term on the right side represents the tendency of an interface between two fluids of different specific weight to rotate toward the horizontal position in such a way that the heavier fluid underlies the lighter fluid. This is clarified in Figure 2, where the heavier fluid γ_2, represented by the darker area, is to the right of the lighter fluid γ_1. Here $(\partial\gamma/\partial x)$ is positive. At all points of the transition zone between the two fluids, vorticity of negative sign exists, which, according to potential theory, imposes clockwise rotation.

The second term on the right side of (11) describes the vorticity caused by differences in viscosity of the two fluids. For simplicity in the subsequent treatment it is convenient to consider only those cases in which the gravity term dominates the viscosity term. A comparison of their respective contribution to the vorticity may be made by writing (11) as

$$\omega = -\frac{k\gamma}{\mu}\frac{\partial \ln \gamma}{\partial x} - \left(\frac{\partial \ln \mu}{\partial x} q_y - \frac{\partial \ln \mu}{\partial y} q_x\right)$$

In the study of problems of sea water intrusion the properties of fresh water (fluid 1) and salt water (fluid 2) are given approximately by the ratios

$$\gamma_2/\gamma_1 = 1.025 \qquad \mu_2/\mu_1 = 1.070$$

The components of the gradients $(\partial \ln \gamma/\partial x)$, $(\partial \ln \mu/\partial x)$, $(\partial \ln \mu/\partial y)$ are therefore of the same order of magnitude, and the relative importance of the terms depends on the magnitude of $\mathbf{q}$ with respect to the value of $(k\gamma/\mu)$.

Now $(k\gamma/\mu)$ is the specific discharge for a head gradient of 1. In general, such high discharge velocities are not present; in problems of sea water intrusion, gradients in head will be of the order of $(\gamma_2 - \gamma_1)/\gamma_1$, which are small with respect to 1. Therefore, an analysis in which the viscosity differences are ignored is justified in such problems. Because, however, the magnitude of the terms also depends on the orientation of the interface with respect to the horizontal and the vector $\mathbf{q}$, it should be verified that their relative importance is not changed because of the sine of the angles involved.

In those cases in which the variation of μ may be disregarded, equation 11 reduces to

$$\nabla^2\Psi = -(k/\mu)(\partial\gamma/\partial x) \qquad (12)$$

Since the distribution of γ is known, the right-hand side of (12) is a known function of position and this relation for Ψ is of the type called Poisson's equation, whose solution in terms of vortices is known from potential theory. The solution can be written as

$$\Psi = -\iint_A (k/2\pi\mu)(\partial\gamma/\partial x) \ln r \, dA \qquad (13)$$

where r is the distance between the point of consideration and the point where the vortex of strength $-(k/\mu)(\partial\gamma/\partial x)$ is present. Integration has to be effected over all vortices, i.e., the entire area A, where $(\partial\gamma/\partial x) \neq 0$.

Ψ is regular in the entire field and also at a point containing a vortex. This is true because, although $\ln r$ becomes infinite for r approaching zero, the product $\ln r \cdot dA$ reduces to zero, since dA is of the order of r^2.

The magnitude of the vortex at a point of the field depends only on the change in properties of the fluids present at that point. The number of fluids involved in the problem is of no importance, and the method is therefore suitable for computing the flow of several fluids and the zones of gradual transition between them.

Treatment of vortices by complex potential. To adapt this flow problem to a treatment by complex variables the stream function Ψ may be combined with an auxiliary function Φ, so that

$$\frac{\partial\Phi}{\partial s} = \frac{\partial\Psi}{\partial n} = -q_s; \quad \frac{\partial\Phi}{\partial n} = -\frac{\partial\Psi}{\partial s} = -q_n \quad (14)$$

The condition of continuity (3) then gives

$$\nabla^2\Phi = 0 \qquad (15)$$

and from the condition of equilibrium, which finally resulted in (12), it follows that

$$\partial^2\Phi/\partial s\ \partial n - \partial^2\Phi/\partial n\ \partial s = \nabla^2\Psi$$
$$= -(k/\mu)(\partial\gamma/\partial x) \qquad (16)$$

Equation 16 shows that Φ is not single-valued and therefore has no particular physical meaning. However, the complex combination, to be called 'complex specific discharge potential,'

$$\Omega = \Phi + i\Psi \qquad (17)$$

is convenient because the solution (13) can be written as a function of the complex variables $z = x + iy$, as follows:

$$\Omega = (-ik/2\pi\mu)$$
$$\cdot \iint_A (\partial\gamma/\partial x)_{z_0}\ \ln\ (z - z_0)\ dA \qquad (18)$$

Here z is the point of consideration and z_0 the point containing the vortex. The integration extends over all points z_0, where $\partial\gamma/\partial x \neq 0$.

The expression (18) gives the correct value for Ψ, since $\ln\ (z - z_0)$ can be separated into real and imaginary parts as $[\ln r + i\theta]$, where r is the distance between z and z_0 and θ the angle with the horizontal. Therefore, $\ln r$ appears in the imaginary part of Ω in the same way as in (13). The real part of Ω consists of the angles θ and is therefore many-valued. This many-valuedness, however, only reflects on the auxiliary function Φ.

From Ω the movement of the fluid is obtained as the complex specific discharge $\bar{q}$ by taking the derivative with respect to z, because

$$\bar{q} = q_x - iq_y = -\frac{\partial\Phi}{\partial x} - i\frac{\partial\Psi}{\partial x}$$
$$= -\frac{d\Omega}{dz}\cdot\frac{\partial z}{\partial x} = -\frac{d\Omega}{dz} \qquad (19)$$

The kind of fluid present at the point z is irrelevant in this computation.

Although Φ is many-valued, its derivatives are single-valued and constitute no difficulties for the determination of $\bar{q}$. Since the discharge is the objective of the study, the awkwardness of the auxiliary function Φ is of no importance in

the analysis. The expression (18) for Ω and its derivative are therefore usable instruments in the analysis and give correct answers for every point in the entire field, even though the point contains vorticity.

Vortices on an abrupt interface. An abrupt interface may be considered to be the degeneration of a transition zone between two fluids of specific weight γ_1 and γ_2, respectively. The gradient of γ in the zone tends to infinity as the thickness of the zone approaches zero in such a way that in the limit

$$\lim_{dn\to 0} (\partial\gamma/\partial n)\ dn = -(\gamma_2 - \gamma_1) \qquad (20)$$

Let s be the coordinate along the interface, and n the normal; then the elementary length ds of the interface corresponds to an area $dA = dsdn$ (shaded part of Fig. 3). The contribution to the complex potential created by this part of the interface is

$$d\Omega = (-ik/2\pi\mu)(\partial\gamma/\partial x)_{z_0}\ \ln\ (z - z_0)\ ds\ dn$$

where z_0 is the complex coordinate of the center of the area.

If α is the angle of ds with the horizontal x axis, $(\partial\gamma/\partial x)$ is equal to $-\sin\alpha\ (\partial\gamma/\partial n)$. By use of (20) the complex potential becomes

$$d\Omega = (-ik/2\pi\mu)(\gamma_2 - \gamma_1)$$
$$\cdot[(\sin\alpha)_{z_0}\ \ln\ (z - z_0)\ ds]$$

The term between brackets varies along the interface, so that the complex potential originated by the whole interface is given by

$$\Omega = (-ik/2\pi\mu)(\gamma_2 - \gamma_1)$$
$$\cdot\int_S (\sin\alpha)_{z_0}\ \ln\ (z - z_0)\ ds \qquad (21)$$

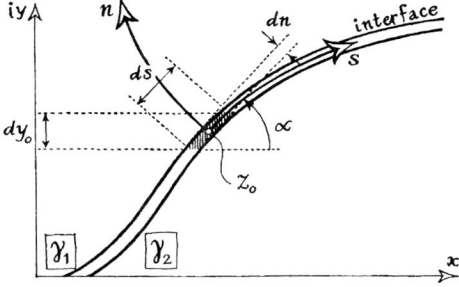

Fig. 3.   Interface of infinitesimal thickness dn.

The integral is a line integral running along a line S, which coincides with the interface; the distance ds represents an incremental length along this line and z_0 is any point on the interface. The interface can be represented as a vortex sheet composed of vortex lines. The intersection of this interface with the x, iy plane, which is perpendicular to it, is the line S. As $(\sin \alpha)_{z_0} \, ds$ is equal to dy_0, the vertical component of ds (see Fig. 3), the strength of the vortex contained in ds is equal to

$$\omega \, ds = (k/\mu)(\gamma_2 - \gamma_1) \, d\mu_0 \qquad (22)$$

and the complex specific discharge potential may be simplified to

$$\Omega = (-ik/2\pi\mu)(\gamma_2 - \gamma_1)$$
$$\cdot \int_S \ln \, (z - z_0) \, dy_0 \qquad (23)$$

Treatment of a boundary value problem. The foregoing introduction of the stream function Ψ permits the solution of boundary value problems. Besides the knowledge of conditions along the boundary, the position of the interface between the fluids also has to be known in order to specify Ψ over the entire field. Then the motion of the fluids can be derived for a time interval short enough to neglect the displacement of the interface. From the motion of the fluids the movement of the interface, and subsequent positions, each having a different flow pattern, can be determined.

For the computation of Ψ at a given moment it is convenient to treat the influence of the conditions along the boundaries and the velocity created by the interface separately. The superposition principle allows the addition of the particular solutions afterwards.

The boundary conditions are generally given in terms of pressures or discharges. Let s be a coordinate running along the boundary with n perpendicular to it.

If the boundary condition consists of discharge, the component q_n is known, and, from equation 5, $(\partial\psi/\partial s)$ is determined. Integration along s gives the value Ψ for every point along such a boundary.

If the boundary condition consists of pressure, it is possible to determine q_s along that boundary with (1), since p, γ, y are known functions of the coordinate s. The knowledge of q_s, with (5), gives $(\partial\Psi/\partial n)$.

Once the boundary conditions are translated in terms of the stream function Ψ, the solution of the problem can proceed as follows. The actual solution Ψ is broken down into two parts, to be called I and II, respectively.

$$\Psi = \Psi_I + \Psi_{II} \qquad (24)$$

Part I takes care of the vortices

$$\nabla^2\Psi_I = -(k/\mu)(\partial\gamma/\partial x) \qquad (25)$$

with the solution (13). From this solution the values of Ψ_I and $(\partial\Psi_I/\partial n)$ can be computed along the boundaries. In general, these values will not coincide with the boundary values Ψ or $(\partial\Psi/\partial n)$ required in the problem. The differences $(\Psi - \Psi_I)$ and $[(\partial\Psi/\partial n) - (\partial\Psi_I/\partial n)]$ then constitute new boundary conditions which can be satisfied by part II of the solution.

The quantity Ψ_{II} must satisfy the deficiencies of part I at the boundaries and

$$\nabla^2\Psi_{II} = 0 \qquad (26)$$

over the entire field. The determination of Ψ_{II} constitutes a boundary value problem, with a unique solution to be obtained by known methods of potential theory.

The addition of (25) and (26) shows that the superposition $(\Psi_I + \Psi_{II})$ satisfies the requirement (12). It will prove convenient to use image vortices of opposite sign in order to create values of Ψ_I = constant along impermeable boundaries so that, for part II of the solution also, these boundaries remain impermeable.

The shear flow at an abrupt interface. A sheet with singularities is known to contain discontinuities. Instead of showing this by evaluating the principal Cauchy value of the integral expression in (23) as the interface is approached from either side, a shorter presentation may be given, using equation 12 as a starting point. Gauss's gradient theorem applied to Ψ for a small rectangle $ds\,dn$ straddling the interface gives, for a vanishingly small width of dn,

$$\iint \nabla^2\Psi \, ds \, dn = \int (\partial\Psi/\partial n_1) \, ds_1$$
$$+ \int (\partial\Psi/\partial n_2) \, ds_2 \qquad (27)$$

where n_1 and n_2 are the normals to the faces ds_1 and ds_2 pointing toward the exterior of the rectangle. These coordinates are related to n

with a direction (Fig. 3) according to

$$+n = +n_1 \qquad +n = -n_2$$

From the surface integral of (12) and the results of the abrupt interface analysis it follows that

$$\int \left[\left(\frac{\partial \Psi}{\partial n} \right)_1 - \left(\frac{\partial \Psi}{\partial n} \right)_2 \right] ds = -\frac{k}{\mu} \iint \frac{\partial \gamma}{\partial x} \, ds \, dn$$

$$= -\frac{k}{\mu} (\gamma_2 - \gamma_1) \int \sin \alpha \, ds$$

As this applies to any portion of the line s, it may be written, using (5), as

$$q_{s_1} - q_{s_2} = (k/\mu)(\gamma_2 - \gamma_1) \sin \alpha \qquad (28)$$

This relation shows that there is a discontinuity in discharge parallel to the interface. It is a function of the inclination α and the difference in specific weights of the fluids on either side.

The discontinuity is restricted to the discharge parallel to the interface. Perpendicular to it the discharge is continuous, because otherwise a gap in the fluids would occur and would violate the continuity principle; hence

$$q_{n_1} = q_{n_2} \qquad (29)$$

The continuous part of the discharge, however, is not restricted to the normal direction. Superimposed on the discontinuous discharge there may be a continuous component parallel to the interface.

The discontinuous component of the discharge (q_d in Fig. 4) is created by the vortex at the point of consideration on the interface. The continuous component q_c, which can have any direction with respect to the interface, is created by all the vortices outside the point of consideration and outside the boundary conditions.

The normal component q_n is responsible for an advance of the interface, which, for complete displacement of one fluid by the other, is equal to the mean velocity q_n/ϵ.

Solution with sources and sinks. In problems of irrotational flow the pressure in the fluid can be determined from the potential function, which is defined in the same way as Φ is defined by (14). Since, however, in this case, Φ is many-valued, it is an unsuitable function for the determination of the pressure. If the pressures are the objective of the study, it is therefore

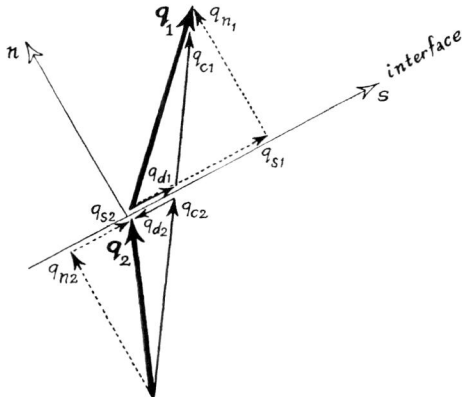

Fig. 4. Discharge discontinuity at an interface.

convenient to introduce a quantity which has the character of Φ but does not have its many-valuedness. Let this quantity be called Θ, the multiple fluid potential, and be defined as

$$(\mu/k)\Theta = p + \gamma y \qquad (30)$$

where the values of p, γ, y are those of the point under consideration. Since all these quantities are single-valued also, Θ is a single-valued function of position in the entire field.

Taking the derivative in an arbitrary direction s, it follows for $\mu = $ constant that

$$(\mu/k)(\partial\Theta/\partial s) = (\partial p/\partial s) + \gamma(\partial y/\partial s)$$
$$+ y(\partial\gamma/\partial s)$$

Application of (1) reduces this to

$$(\mu/k)q_s = -(\mu/k)(\partial\Theta/\partial s) + y(\partial\gamma/\partial s)$$

which may be written

$$(\mu/k)\mathbf{q} = -(\mu/k) \, \mathrm{grad} \, \Theta + y \, \mathrm{grad} \, \gamma \qquad (31)$$

since s is any direction. A comparison with (14) shows that the relation between Θ and Φ is

$$\mathrm{grad} \, \Phi = \mathrm{grad} \, \Theta - (k/\mu)y \, \mathrm{grad} \, \gamma \qquad (32)$$

where the last term takes care of the many-valuedness, so that Θ can be single-valued in the entire field, including the region of vorticity. At the points, where grad γ is zero, i.e., the points of no vorticity, the quantities Φ and Θ are the same.

Application of the condition of continuity (3) to (31) gives

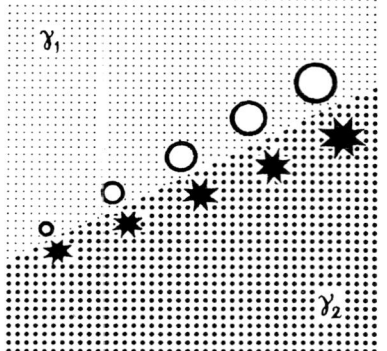

Fig. 5. Doublets on interface between two fluids ($\gamma_1 < \gamma_2$). Doublet strength increases with height of position.

$$\text{div } q = -\nabla^2 \Theta$$

$$+ (k/\mu) \text{ div } [y \text{ grad } \gamma] = 0 \qquad (33)$$

This is a Poisson's equation showing singularities in Θ of magnitude $(k/\mu) \text{ div } [y \text{ grad } \gamma]$. Since Θ has the character of a potential in the hypothetical fluid, the singularities may be termed sources for negative and sinks for positive values of their magnitude.

Application of the condition of equilibrium, by using the first expression of (4) and the condition that the function (γy) is single-valued, gives

$$\partial^2 \Theta / \partial s \, \partial n = \partial^2 \Theta / \partial n \, \partial s$$

which confirms the single-valuedness of Θ. The solution for Θ can be written in analogy to (13):

$$\Theta = (k/2\pi\mu) \iint_A \text{div } [y \text{ grad } \gamma] \ln r \cdot dA \quad (34)$$

Again Θ can be combined with an auxiliary function having the character of a stream function in order to form a complex potential. In this case the stream function is many-valued and unsuitable for the determination of the discharges, but it reduces to Ψ for points of no vorticity. The sources and sinks can be expressed in terms of the complex potential by replacing $\ln r$ in (34) by $\ln(z - z_0)$.

Doublets on an abrupt interface and boundaries. That the expression div $[y \text{ grad } \gamma]$ reduces to a doublet layer if the transition zone reduces to an abrupt interface (Fig. 5), is recognized as

follows. Consider a line n perpendicular to the interface pointing from the γ_2 fluid to the γ_1 fluid (Fig. 3). At the point of intersection with the interface, y is a constant y_0, and the source distribution is $y_0 \text{ div grad } \gamma = y_0 \nabla^2 \gamma$. Since γ varies in the n direction only, $\nabla^2 \gamma$ is equal to $\partial^2 \gamma / \partial n^2$. Let the interface consist of two infinitesimal bands of width dn at a mutual distance δ''. Let $(\partial \gamma / \partial n)$ in the region between the bands be constant and equal to $-(\gamma_2 - \gamma_1)/\delta''$, and let $(\partial^2 \gamma / \partial n^2) \, dn$ in the two bands be equal to $\pm (\gamma_2 - \gamma_1)\delta''$. Then the value of γ will indeed decrease from γ_2 to γ_1 in the region of the interface.

The quantity (34) then becomes

$$\Theta = (k/2\pi\mu) \iint y_0 (\partial^2 \gamma / \partial n^2)(\ln r) \, dn \, ds$$

$$= [k(\gamma_2 - \gamma_1)/2\pi\mu \, \delta'']$$

$$\cdot \int_s y_0 [+\ln r_1 - \ln r_2] \, ds \quad (35)$$

where r_1 and r_2 are the distances from the point of consideration to the two sides of the interface. This expression describes a source and a sink at either side of the mutual distance δ'', their strength being equal to $d \kappa'' = [k (\gamma_2 - \gamma_1)/\mu\delta''] y_0 ds$. If δ'' is allowed to shrink to zero, the $[\ln r]$ term becomes $[\delta''/r]$ cos ϕ, ϕ being the angle between n and the radius r to the point under consideration. This is known to be the potential of a doublet oriented perpendicularly to the interface. The expression (35) therefore represents a doublet layer, containing a doublet of strength.

$$d\lambda'' = \delta'' \cdot d\kappa'' = [k(\gamma_2 - \gamma_1)/\mu] y_0 \, ds \quad (36)$$

for each line segment ds of the interface.

For the determination of boundary value problems the solution can again be broken down into two parts, Θ_I and Θ_{II}, where Θ_I accounts for the sources and sinks and Θ_{II} satisfies the deficiences at the boundaries. For boundaries with known pressure, the value of Θ can be computed with (30). For boundaries with discharge information, the value of q_n is known, and from (31) the value of $(\partial \Theta / \partial n)$ can be determined, since $(\partial \gamma / \partial n)$ can be obtained from the γ distribution, which is assumed to be available as initial information. Since either Θ or $(\partial \Theta / \partial n)$ is known along the boundaries, Θ_I

and Θ_{II} can be evaluated with the usual methods of potential theory.

PART 2. MATHEMATICAL EXAMPLE AND EXPERIMENTAL VERIFICATION

Introduction. The application of the foregoing basic results may be demonstrated by an example of a two-fluid system enclosed by impermeable boundaries on all sides. By this disposition the movement of the fluids is due exclusively to the weight differences of the fluids and can be computed by use of the singularity method.

This example is treated mathematically, with the result that the movement of every point in the entire field is given. Since the distribution of the movements is the objective here, the treatment with vortices will be used. From the movements obtained as the solution the initial displacement of the interface can be derived. The subsequent rotation of the interface in course of time cannot be computed rigorously, but an approximate evaluation of the rotation of the center part is developed.

As the singularities permit the application of electric resistance analog models, the potential and streamlines for the same example were determined by the method of electric analogy. In addition to this verification, parallel plate model tests were run in order to obtain photographs of the streamlines traced by particles in the fluids. Both verifications are treated.

Given conditions for the mathematical example. The example consists of a two-fluid system contained in a confined infinite aquifer of thickness $2c$ (Fig. 6). The interface has, at time $= 0$, a vertical position BME, and it rotates in a clockwise direction from there under the influence of the differences in specific weight of the fluids at either side (fluid 1 is considered to be lighter than fluid 2).

From the two parts of the solution (equation 24), the first part, corresponding to the vortices along the interface, will include images which provide a flow parallel to the boundaries. The second part of the solution then has to satisfy $\nabla^2 \Psi_{II} = 0$ in the interior and $\Psi_{II} = $ constant along all the boundaries. From potential theory it is known that Ψ_{II} must be constant if it is to satisfy these conditions. If the arbitrary value zero is assigned to the constant, the solution consists only of the first part, and the suffix I may be omitted in the subsequent treatment.

Movement of the entire field at $t = o$. Instead of computing Ω from the vortices in the $z = x + iy$ plane of Figure 6, it is more convenient first to introduce a transformation in order to obtain a simpler arrangement of the images than to work with the infinite series necessitated in the z plane. By the transformation function,

$$z = (2c/\pi) \ln \zeta \qquad (37)$$

the infinite strip of Figure 6, is mapped into the half plane corresponding to positive real values of $\zeta = \xi + i\eta$ (Fig. 7). The boundaries are mapped into one vertical line, the η axis AF. The interface is transformed into a half circle BME with radius 1. Fluid 1 occupies the region of the half plane inside the circle and fluid 2 the rest of the half plane outside the circle. The images are located on the left half circle BNE and have the opposite sign, in order to make AF an impermeable boundary.

The subscript zero will indicate that the interface is concerned. A line segment dz_0 of the interface contains, according to (22), a vortex of strength

$$(k/\mu)(\gamma_2 - \gamma_1)\, dy_0 \qquad (38)$$

Therefore the vortex strength associated with the line segment dy_0 has to be applied in the transformed field to the line segment $d\zeta_0$ which is the image of dy_0. Since $x_0 = 0$ is the equation for the interface, a line segment dz_0 is equal to $dx_0 + idy_0 = idy_0$. Application of the transformation (37) gives

$$dz_0 = i\, dy_0 = (2c/\pi)\zeta_0^{-1}\, d\zeta_0 \qquad (39)$$

which represents dy_0 as a function of ζ_0. The strength of the vortex associated with the line segment $d\zeta_0$ of the transformed interface is therefore, according to (38) and (39), $[2ck (\gamma_2 - \gamma_1)/i\pi\mu\zeta_0]d\zeta_0$. The complex specific discharge potential in the ζ plane is the integral of the influences of these elementary vortices, each of which is a contribution to the potential of magnitude $d\Omega = (i/2\pi) \ln (\zeta - \zeta_0)$ times the vortex strength in ζ_0. This gives

$$\Omega(\zeta)_{\text{(vortex part)}} = -\frac{i}{2\pi}\frac{2ck(\gamma_2 - \gamma_1)}{i\pi\mu}$$
$$\cdot \int_{EMB} \ln (\zeta - \zeta_0)\frac{d\zeta_0}{\zeta_0} \qquad (40)$$

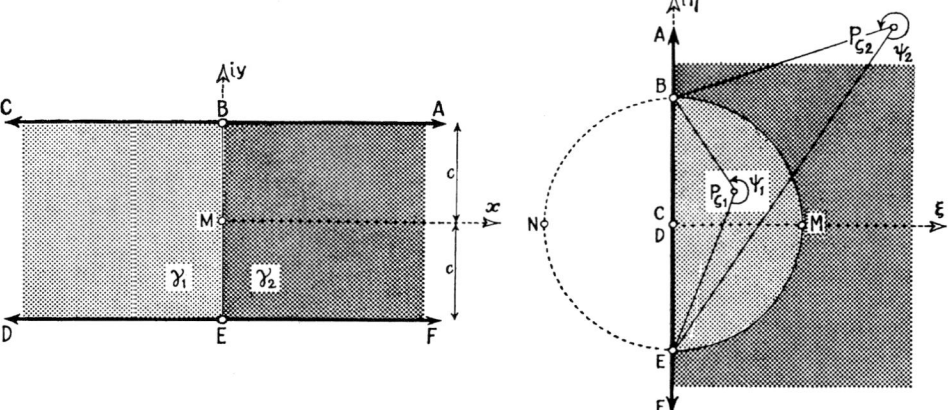

Fig. 6. Infinite confined aquifer in z plane.   Fig. 7. Transformed infinite confined aquifer in ζ plane.

The integration extends over the values of ζ_0 along the half circle EMB, counterclockwise.

The image vortices have opposite sign and the path of integration is the half circle BNE, counterclockwise. If the direction of integration is inverted, a second sign inversion is obtained, giving a value for $\Omega(\zeta)$ of the image part, which has the same sign as (40). The result after addition is

$$\Omega(\zeta)_{total} = \Omega(\zeta)_{vortices} + \Omega(\zeta)_{images}$$

$$= -\frac{ck(\gamma_2 - \gamma_1)}{\pi^2 \mu}$$

$$\cdot \left[\int_{EMB} \ln (\zeta - \zeta_0) \frac{d\zeta_0}{\zeta_0} \right.$$

$$\left. + \int_{ENB} \ln (\zeta - \zeta_0) \frac{d\zeta_0}{\zeta_0} \right] \qquad (41)$$

This integral has no solution in closed form, and, since the aim of the computation is the movement of the fluid, the specific discharge might as well be introduced at this moment because it facilitates the evaluation of the integral. The specific discharge, according to (19), is obtained by a differentiation, and, since it is permissible here to carry this out under the integral sign, (41) yields

$$\bar{q} = -\frac{d\Omega(\zeta)}{d\zeta} \frac{d\zeta}{dz} = -\frac{d\Omega(\zeta)}{d\zeta} \frac{\pi}{2c} \zeta$$

$$= +\frac{k(\gamma_2 - \gamma_1)}{2\pi\mu} \left[\int_{EMB} \frac{\zeta \, d\zeta_0}{(\zeta - \zeta_0)\zeta_0} \right.$$

$$\left. + \int_{ENB} \frac{\zeta \, d\zeta_0}{(\zeta - \zeta_0)\zeta_0} \right] \qquad (42)$$

Now integration with respect to ζ_0 gives the solution

$$\bar{q} = +\frac{k(\gamma_2 - \gamma_1)}{2\pi\mu} \{ [\ln \zeta_0 - \ln (\zeta - \zeta_0)]_{EMB}$$

$$+ [\ln \zeta_0 - \ln (\zeta - \zeta_0)]_{ENB} \} \qquad (43)$$

The real part, the x component of the specific discharge, is obtained from $\ln |\zeta_0|$ and $\ln |\zeta - \zeta_0|$, corresponding to the limits B and E. As the interface circle has radius 1, $\ln |\zeta_0| = 0$ and

$$q_x = -\frac{k(\gamma_2 - \gamma_1)}{\pi\mu} \cdot \ln \left| \frac{BP_\zeta}{EP_\zeta} \right| \qquad (44)$$

remains. Here P_ζ is the point in the ζ plane corresponding to the point P of consideration in the z plane. The loci of points P_ζ for which (BP_ζ/EP_ζ) is a constant are circles in which the points E and B are inverse points. These circles for equal values of q_x are shown in Figure 8. The imaginary part, the y component of the specific discharge, is obtained from angle (ζ_0) and angle $(\zeta - \zeta_0)$. The first of these two is the angle swept by the vector ζ_0, with starting point at the origin, as its end point traces the half circle EMB and ENB. The angles are $+\pi$ and

$-\pi$, respectively, which cancel each other.

There remains, therefore, the angle $(\zeta - \zeta_0)$, which is the angle swept by the vector with **starting point** on the half circle and end point **at the point** P_ζ. The total value of this angle depends on the location of P_ζ. If P_ζ, is located in fluid 1, the paths of integration BME and BNE pass on either side, giving $(-\psi_1)$ and $(2\pi - \psi_1)$ for the angle swept, respectively, where ψ_1 is the exterior angle $EP_\zeta B$ (Fig. 7). Therefore, the vertical component in fluid 1 will be

$$q_{v_1} = -\frac{k(\gamma_2 - \gamma_1)}{2\pi\mu} [-\psi_1 + (2\pi - \psi_1)]$$

$$= +\frac{k(\gamma_2 - \gamma_1)}{\mu} \cdot \frac{(\psi_1 - \pi)}{\pi} \quad (45)$$

If P_ζ, is located in fluid 2, the paths of integration both **pass** on the left side, and the angle swept is for both $(2\pi - \psi_2)$. The vertical component in fluid 2 will therefore be

$$q_{v_2} = -\frac{k(\gamma_2 - \gamma_1)}{\mu} \cdot \frac{(2\pi - \psi_2)}{\pi} \quad (46)$$

On the interface the angles ψ_1 and ψ_2 both have the value $3\pi/2$, so the difference of the parallel discharge at either side of the interface is

$$(q_{v_1} - q_{v_2})_{(\text{interface})} = (k/\mu)(\gamma_2 - \gamma_1) \quad (47)$$

which is the value expected for the shear flow **at a vertical interface** $(\alpha = \frac{1}{2}\pi)$ according to (28).

The expressions (45) and (46) show that on the lines of equal value for q_v the angles ψ_1 or ψ_2 are constants; these lines, therefore are arcs of circles through the points B and E (Fig. 8).

Transformed to the z plane these lines for equal q_x and q_y form the pattern of Figure 9. For each point of the aquifer the specific discharge vector can be determined from the components q_x and q_y, as is shown in Figure 10, where arrows represent the flow in magnitude and direction.

The advance of the interface dx_0/dt at the moment $t = 0$ is equal to the component of the mean velocity perpendicular to the interface; i.e., in this case, the x component of the specific discharge divided by the porosity. From (44) it follows, therefore, that

$$\frac{dx_0}{dt} = \frac{q_x}{\epsilon} = \frac{k(\gamma_2 - \gamma_1)}{\epsilon\pi\mu} \ln \left| \frac{EP_{\zeta_0}}{BP_{\zeta_0}} \right|$$

$$= \frac{k(\gamma_2 - \gamma_1)}{\epsilon\pi\mu} \ln \left| \frac{\zeta_0 + i}{\zeta_0 - i} \right|$$

By use of $\zeta_0 \exp (\pi z_0/2c) = \exp (i\pi y_0/2c)$ this reduces to

$$\frac{dx_0}{dt} = \frac{k(\gamma_2 - \gamma_1)}{\epsilon\pi\mu} \ln \left\{ \cot \left[\frac{\pi(c - y_0)}{4c} \right] \right\} \quad (48)$$

Figure 11 shows this velocity as a function of y_0, an S-shaped curve which is representative of the shape of the interface a small time after $t = 0$.

Displacement of the interface at $t > 0$. As the interface turns, the shape continues to be in the form of an S which is tangent to the impermeable boundaries. For all subsequent positions a different flow pattern applies, because the vortices, which generate the flow pattern, shift position with the interface.

The determination of the advancement of the interface necessitates the computation of the specific discharge component perpendicular to the interface. Also this component changes with the shifting interface. Since an exact mathematical solution of this displacement was not obtained, a first approximation is given here, treating the rotation in the center point of a straight-lined interface.

Let the inclination of the interface with the horizontal be $\alpha = \text{arc cot } a$, so that the interface equation is

$$x_0 = y_0 \cot \alpha = ay_0 \quad (49)$$

Then a line segment dz_0 of the interface is equal to

$$dz_0 = dx_0 + i\, dy_0 = (a + i)\, dy_0 \quad (50)$$

A comparison with (39) shows the only difference to be the replacement of i by a factor $(a + i)$. The vortex strength associated with a line segment $d\zeta_0$ becomes, by similar reasoning,

$$+[2ck(\gamma_2 - \gamma_1)/(a + i)\pi\mu\zeta_0]\, d\zeta_0 \quad (51)$$

The image interface is given by the equation $(m = \text{integer})$

$$x_0 = a[(4m + 2)c - y_0]$$

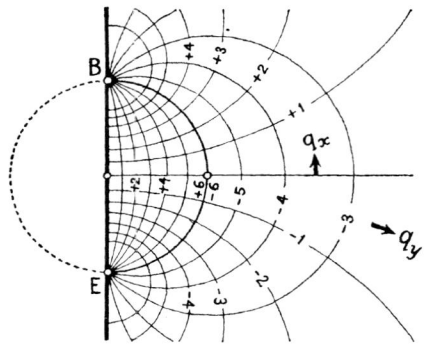 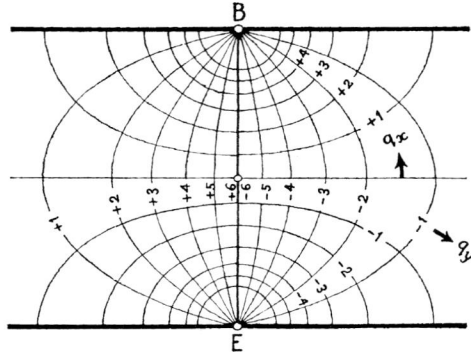

Fig. 8. Lines of constant q_x and q_y in ζ plane. Fig. 9. Lines of constant q_x and q_y in z plane.
Values indicate $12\,q_x\,\mu/k(\gamma_2-\gamma_1)$ and $12\,q_y\,\mu/k(\gamma_2-\gamma_1)$, respectively.

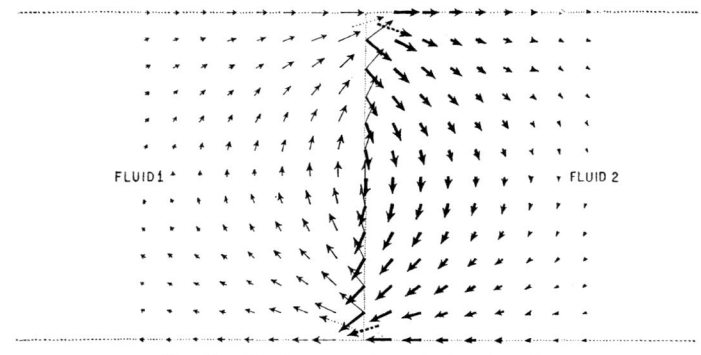

Fig. 10. Discharge vectors at time $t = 0$.

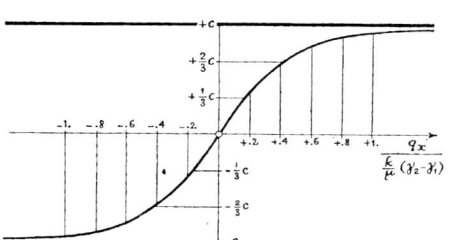

Fig. 11. Distribution of discharge perpendicular
to interface.

and a line segment dz_0 of the image interface is

$$dz_0 = dx_0 + i\,dy_0 = -(a - i)\,dy_0$$

Since the image vortices have opposite signs, their strength associated with a line segment $d\zeta_0$ of the transformed image interface is

$$+[2ck(\gamma_2 - \gamma_1)/(a - i)\pi\mu\zeta_0]\,d\zeta_0 \qquad (52)$$

Elaboration of the complex specific discharge potential, as demonstrated previously, gives the complex specific discharge in a form similar to (43):

$$\bar{q} = \frac{k(\gamma_2 - \gamma_1)}{2\pi\mu}$$

$$\cdot\left\{\frac{i}{a + i}\,[\ln\zeta_0 - \ln(\zeta - \zeta_0)]_{EMB}\right.$$

$$\left.-\frac{i}{a - i}\,[\ln\zeta_0 - \ln(\zeta - \zeta_0)]_{ENB}\right\} \qquad (53)$$

where the minus sign for the second term between braces is due to the path inversion from BNE to ENB.

The derivation $(d\bar{q}/dz)$ represents the complex deformation of the fluid and the imaginary part of $[\epsilon^{-1}(d\bar{q}/dz)\,e^{2i\alpha}]$, the clockwise rotation of a line of fluid particles oriented at an angle

α with the horizontal. If computed for the point $z = 0$, it represents the rotation of the interface at the center point M of the confined layer. This rotation is equal to the decrease of the interface inclination α with time, or

$$- d\alpha/dt = \epsilon^{-1} \text{ Im } [(d\bar{q}/dz)e^{2i\alpha}]_{z=0} \quad (54)$$

By use of (53) the result for $z = 0$ ($\zeta = 1$) is

$$\frac{d\bar{q}}{dz} = \frac{d\bar{q}}{d\zeta}\frac{d\zeta}{dz} = \frac{k(\gamma_2 - \gamma_1)}{4\mu c}$$

$$\cdot \left\{ -\frac{i}{a+i}\frac{1}{1-\zeta_0} + \frac{i}{a-i}\frac{1}{1-\zeta_0} \right\}_E^B$$

For the inclined interface the point B corresponds to $z_0 = (a + i)c$ and the point E to $z_0 = -(a + i)c$. Therefore, the integration limits for ζ_0 are, respectively,

$$\zeta_0 = e^{\pi(a+i)c/2c} = ie^{\pi a/2c} \text{ and } \zeta_0 = -ie^{-\pi a/2c}$$

and

$$\frac{d\bar{q}}{dz} = \frac{k(\gamma_2 - \gamma_1)}{2\mu c(a^2 + 1)} \left[\frac{\sinh(\pi a/2c) - i}{\cosh(\pi a/2c)} \right]$$

By introduction of $\beta = \text{arc cot } [\sinh(\pi a/2c)]$ and $a = \cot\alpha$ the expression is simplified to

$$d\bar{q}/dz = [k(\gamma_2 - \gamma_1)/2\mu c] \cdot [\sin^2 \alpha e^{-i\beta}]$$

This value for $d\bar{q}/dz$ used in equation (54) gives

$$d\alpha/dt = -[k(\gamma_2 - \gamma_1)/2\epsilon\mu c]$$

$$\cdot \sin^2 \alpha \sin(2\alpha - \beta) \quad (55)$$

The right side of (55) being a function of α only, the relation between α and t may be written as

$$t[\epsilon\mu c/k(\gamma_2 - \gamma_1)]$$

$$= -\int_{(1/2)\pi}^{\alpha} [1/2 \sin^2 \alpha \cdot \sin(2\alpha - \beta)] \, d\alpha \quad (56)$$

Because of the intricate function relating β to α, a solution of the integral is not available in closed form. The curve of Figure 12 was obtained by numerical step-by-step evaluation of (56), and it represents as the final result of this analysis the decrease of the interface inclination at the center point M with time in a first approximation.

Verification with parallel plate model. To verify the mathematical solution a parallel plate

model test was carried out, with the results as represented in the photographs (Figs. 13 and 14). These pictures show the center part of a 15 $\times$ 60 cm Lucite model, with a slot width of $d = 1$ mm. Since in the beginning the movement is primarily concentrated in the center part, the deviation between this model of limited length and the infinite confined aquifer treated theoretically can be disregarded.

The fluids at either side of the originally vertical interface were, to the left, a glycerine with $\gamma_1 = 1.23$ g/cm³ and, to the right, 60 to 40 per cent mixture of glycerine and phosphoric acid with $\gamma_2 = 1.40$ g/cm³. The viscosity for both fluids at 24°C was $\mu = 0.005$ g sec/cm². Since ϵ, the porosity, is 1 for a parallel plate model, the value of the half-aquifer height c is 7.5 cm, and the specific permeability is $k = d^2/12 = 8.3 \times 10^{-4}$ cm², the factor associated with t in (56) has the value

$$\epsilon\mu c/k(\gamma_2 - \gamma_1) = 265 \text{ sec}$$

The white particles showing in the photographs are flakes of gold leaf, which by their large surface-thickness ratio had a small settling velocity with respect to the motion of the fluid itself. During the 90-sec exposure time the flakes traced lines which are representative of the fluid motion. Figure 13 is therefore the experimental representation of the specific discharge vectors shown in Figure 10. The resemblance of both patterns is apparent.

Figure 13 covers the period $0 < t < 90$ sec, Figure 14 the period $120 < t < 210$ sec. According to the graph (Fig. 12), the inclination of the interface is expected to be approximately $90° > \alpha > 80°$ for Figure 13 and $77° > \alpha > 68°$ for Figure 14. In the pictures the area swept by the interface can be observed as a lighter band, covered by particle tracings of crossing directions. These cross paths are created by the shear flow at the interface and are also present in the analytically obtained flow pattern of Figure 10. In both figures the S-shaped interface predicted by Figure 11 is distinguishable, and the inclination at the center follows the computed values approximately.

Electric analogy. In the electric models, which are particularly appropriate for the plotting of potential lines and streamlines [*Malavard*, 1956], it is impossible to simulate differences of specific weight directly. However,

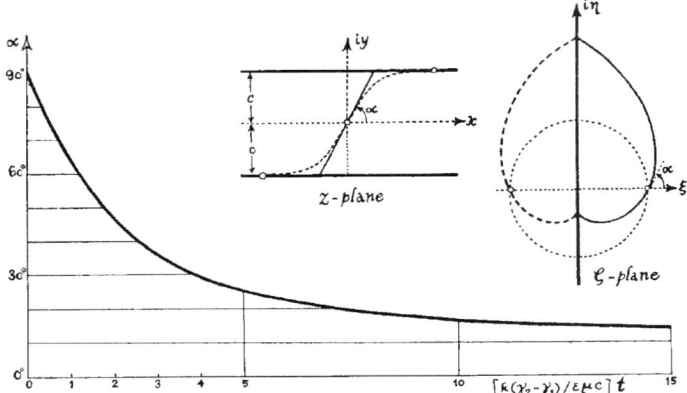

Fig. 12. Slope of rectilinear interface as a function of time.

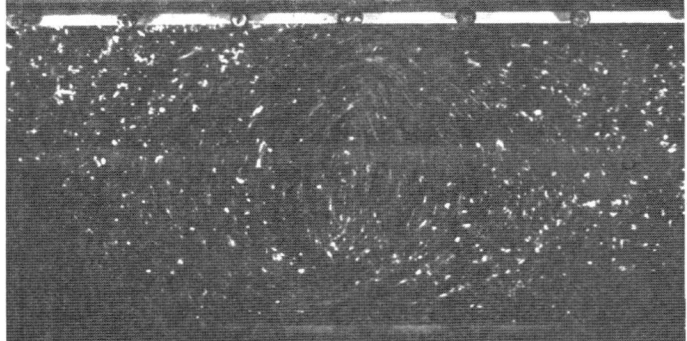

Fig. 13. Photograph of parallel plate model test, with gold leaf flakes tracing streamlines; left side light fluid, right side heavy fluid; exposure time $0 < t < 90$ sec.

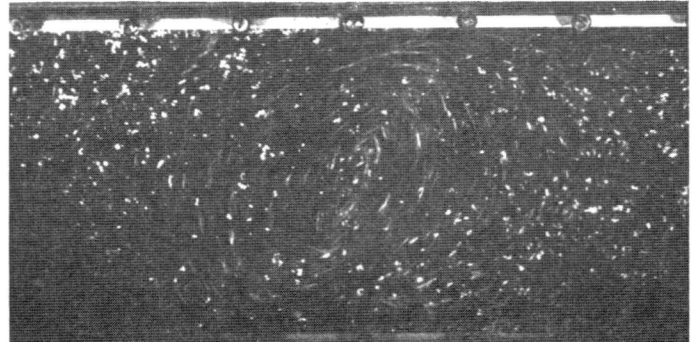

Fig. 14. Photograph of parellel plate model test of Figure 13 at a later stage; exposure time $120 < t < 210$ sec.

by the replacement of the fluids with different properties by the hypothetical fluid and the introduction of property changes by singularities, a concept is obtained which has electric analogies.

In the electric resistance model, where lines of equal electric potential can be traced, sources and sinks can be simulated by electrodes which discharge or withdraw electric current from the model. In this model the electric resistance body with homogeneous specific resistance can represent the hypothetical fluid.

If the multiple fluid potential Θ is to be determined, the strength of the source or sink in the area dA must be proportional to (k/μ) div $[y \text{ grad } \gamma]dA$, according to (33). Lines of equal electric potential are then lines of equal Θ, and they are called potential lines. For a gradual transition zone the source-sink distribution is continuous. Exact simulation by a continuous electrode arrangement would give the correct potential lines in the entire field, including the transition zone. In practice, a discrete number of electrodes will be used, giving an eventual distortion of the potential lines. At an abrupt interface the source-sink distribution reduces to a line of doublets whose strength per line element ds, according to (36), is equal to

$$d\lambda'' = [k(\gamma_2 - \gamma_1)/\mu]y_0 \, ds \qquad (57)$$

If the stream function Ψ is to be determined, vortices must be simulated. Actually, vortices have no analogy in the electric resistance model, but this presents no difficulties because the inverted model must be used for tracing the streamlines. With the inversion, where streamlines are simulated by electric potential lines, the vortices are inverted to sources or sinks. For the gradual transition zone the strength of the source in an area dA is $- (k/\mu) (\partial\gamma/\partial x) dA$, according to (12).

For the abrupt interface the strength for a line segment ds corresponding to an elevation increment dy_0 is, according to (22),

$$\kappa' \, dy_0 = (k/\mu)(\gamma_2 - \gamma_1) \, dy_0 \qquad (59)$$

The correct streamlines would be obtained in the entire field, including a gradual transition, if the exact arrangement of continuous electrodes could be applied.

Since Θ is equal to Φ in the region outside the zone containing the singularities, and since Φ

and Ψ satisfy the Cauchy-Riemann equations (see equation 14), the Θ and Ψ lines form an orthogonal, equilateral grid in this outside region. Inside the gradual transition zone, Θ and Ψ no longer form squares because Θ is unequal to Φ, and therefore the Cauchy-Riemann relations no longer hold between Θ and Ψ.

For completeness it may be mentioned that vortices are encountered in the theory of magnetism, because any wire conducting an electric current creates a magnetic field around it, which, according to the law of Biot and Savart, obeys the same formulas as the potential around the vortex filament. To make a model in which the magnetic analogy is used is difficult from a practical standpoint. The application of boundary conditions, the determination of the flux, and the direct current necessary for the elimination of eddy current losses all create unattractive complications. The elaboration of this type of model is therefore omitted in this study.

Verification with an electric resistance analogy model. In the previous section it was demonstrated how streamlines and potential lines can be determined by means of an electric resistance analogy model. A test by this means was conducted for the vertical interface situation investigated previously by analytic and parallel plate model procedures. Conducting paper was used to simulate the porous medium. Instead of applying the continuous sink and doublet distribution as required theoretically for an abrupt interface, a limited number (10 for the half height c) was used.

For the streamline determination in the inverted model, each sink, according to (59), requires a discharge of

$$Q' = \kappa'f(c/10) = kf(\gamma_2 - \gamma_1)c/10\mu \qquad (60)$$

where f is the thickness of the aquifer.

The electric current I flowing through the electric model of thickness f_e is considered representative of the discharge flowing through the aquifer of thickness f. Ohm's law for the current flowing through a stream channel of breadth b, perpendicular to the potential gradient (dE/ds), with ρ the specific electric resistivity, gives $I = - (bf_e/\rho) (dE/ds)$.

With (5) the discharge through a breadth b in the hydraulic case is $Q = - bf (\partial\Psi/\partial n)$. Because of the inversion d/ds is similar to d/dn,

supplied with a potential difference of $\frac{1}{24}$ E*

and the relation between model and aquifer is

$$(f_eE/\rho)\,[\text{amperes}] \rightarrow f\Psi\,[\text{cm}^3/\text{sec}] \qquad (61)$$

The intervals for the streamlines $\Delta\Psi$ were chosen at the arbitrary value of

$$\Delta\Psi = k(\gamma_2 - \gamma_1)c/20\mu \qquad (62)$$

By (61), these were simulated by electric potential intervals of magnitude ΔE, according to

$$(f_e/\rho)\,\Delta E \rightarrow f\,\Delta\Psi = [k(\gamma_2 - \gamma_1)cf/20\mu] \qquad (63)$$

A 500-volt a-c source (E^*) was used to supply current to the point sources. With a 500-kΩ resistance, the current at each source was $E^*/500,000$ [amp] which, according to (60), simulates $k\,(\gamma_2 - \gamma_1)\,cf/10\,\mu$. The resistance of the conducting paper in the model was negligible with respect to the 500 kΩ. Because the conducting paper used in the model has a specific resistivity per unit area of $(\rho/f_e) = 1625\ \Omega$, the electric potential intervals to be traced were

$$\Delta E = \tfrac{1}{2} \times 1625 \times E^*/500,000$$
$$= 1.625 \times 10^{-3}E^*$$

A potential divider with 1000 subdivisions was E^*. The intervals ΔE of the streamlines were represented by 39 subdivisions of the potential divider.

For the determination of the potential lines the continuous distribution of doublets was replaced by ten doublets of equal strength. Since, according to (59), the doublet strength increases with height y_0, the strength distribution is triangular. The ten doublets were therefore placed at heights

$$[2c/3\sqrt{10}]\{[(m-1)^{1/2} + m^{1/2}]$$
$$- [(m-1)^{-1/2} + m^{-1/2}]^{-1}\}$$

where m represents the integers from 1 to 10, their positions corresponding to the centers of gravity of the 10 equal subdivisions of the triangle. Each doublet had a strength

$$\Lambda'' = (kf/10\mu)(\gamma_2 - \gamma_1)\int_0^c y_0\,dy_0$$
$$= (kf/20\mu)(\gamma_2 - \gamma_1)c^2 \qquad (64)$$

The doublets were represented by a source and a sink with discharge Q'' at a mutual distance $\delta'' = c/40$, and their strength was $\Lambda'' = Q''c/40$. To meet the requirement (64), the discharges of the sources and sinks were given by

$$Q'' = 2kf(\gamma_2 - \gamma_1)c/\mu \qquad (65)$$

The required current for the sources was obtained by replacing the 500-kΩ resistance by a smaller one of 250 kΩ. For the sinks a counter-phase potential E^* of equal magnitude with a 250-kΩ resistance was used. So $E^*/250,000$ [amp] was equivalent to $2kf\,(\gamma_2 - \gamma_1)\,c/\mu$.

To obtain potential lines at similar intervals as in (62)

$$\Delta\Theta = k(\gamma_2 - \gamma_1)c/20\mu$$

the electric potential intervals ΔE had to be taken as

$$\Delta E = (\rho/f_e)(f\,\Delta\Theta/Q'')(E^*/250,000)$$
$$= 1.625 \times 10^{-4}E^*$$

In Figure 15 the full lines are streamlines and the dotted lines are the potential lines obtained with the electric resistance model. This corresponds to the half-aquifer height, and the streamline pattern is therefore comparable with the upper half of Figure 13. The abrupt change of direction of the streamlines created by the shear flow is again visible.

Because of the replacement of continuous sources and doublet distributions by discrete sources and source-sink combinations, the pattern is distorted along the interface. In Figure 15 the distortion of the streamlines by the discrete sources is shown to the left; the potential distortion by the discrete doublets is shown in Figure 15 to the right.

Outside this region the lines form squares.

Difference in viscosity. If, besides specific weight differences, variations of viscosity also have to be considered, the vortex strength depends on the gradient of Ψ according to (10). This excludes mathematical treatment along the lines developed, but the electric resistance model offers a possibility for determination of the flow characteristics. A slightly different stream function Ψ_u and corresponding multiple-fluid potential Θ_u must be introduced. Their definition is given by

$$\Theta_u = kp + k\gamma y$$
$$\mu q_s = -(\partial\Theta_u/\partial s) + ky(\partial\gamma/\partial s) = -(\partial\Psi_u/\partial n)$$
$$\mu q_n = -(\partial\Theta_u/\partial n) + ky(\partial\gamma/\partial n) = +(\partial\Psi_u/\partial s)$$

Elaboration of the conditions of equilibrium then gives

$$\nabla^2 \Psi_u = -k(\partial\gamma/\partial x)$$

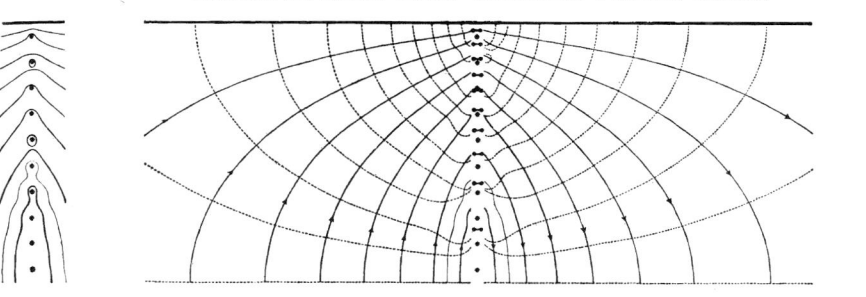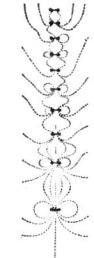

Fig. 15. Result of electric resistance model test, showing lines for constant Ψ (full lines) and constant Θ (dotted lines).

an expression devoid of the μ term. The condition of continuity remains div q = 0.

In the electric resistance model Ohm's law gives

$$\rho \mathbf{i} = -\operatorname{grad} E$$

where $\mathbf{i}$ is the current density vector, and ρ, the specific resistance, is now a variable function of position. The relation between hydraulic and electric quantities may now be as follows:

For the tracing of potential lines

$$\Theta_\mu \to E; \quad [\mathbf{q} - (ky/\mu)\operatorname{grad}\gamma] \to \mathbf{i}; \quad \mu \to \rho$$

For the tracing of streamlines

$$\operatorname{grad}\Psi_\mu \to -\mathbf{i}; \quad q_n \to (\partial E/\partial s);$$

$$q_s \to (-\partial E/\partial n); \quad \mu \to 1/\rho$$

Application of the conditions of continuity and equilibrium to the hydraulic system gives the following conditions for the electric quantities:

For the case of potential lines

$$-k \operatorname{div}\left[(y/\mu)\operatorname{grad}\gamma\right] \to \operatorname{div}\mathbf{i}$$
$$[\partial^2\Theta/\partial s\,\partial n - \partial^2\Theta/\partial n\,\partial s] = 0$$
$$\to [\partial^2 E/\partial s\,\partial n - \partial^2 E/\partial n\,\partial s] = 0$$

For the case of stream lines

$$\operatorname{div}\mathbf{q} = 0 \to [\partial^2 E/\partial s\,\partial n - \partial^2 E/\partial n\,\partial s] = 0$$
$$-k(\partial\gamma/\partial x) \to \operatorname{div}\mathbf{i}$$

These equations show that E is a single-valued function. The values of div $\mathbf{i}$ indicate a source and sink distribution which differs but slightly from the previous results. In the first case the specific resistance must be directly proportional to the viscosity, in the second case inversely

proportional. As the total resistance in the electric model consists of ρ multiplied by the thickness, it is sometimes more convenient to use a fluid of constant resistivity as conducting material and to vary the layer thickness of the fluid to meet the differences in viscosity.

CONCLUSION

This investigation has shown that it is possible to determine the flow pattern in a porous medium which is saturated by several fluids of different specific weight by the use of singularities. All the fluids are replaced by one hypothetical homogeneous fluid. The motion of the hypothetical fluid will be identical to the movements of the different fluids, for which it is substituted, if singularites are applied at all the points where the original fluids show change in specific weight.

It is proved to be convenient to give the singularities the character of vortices, if the motion of the fluids is to be determined. If the emphasis is on the pressure determination, it is better to use a distribution of sources and sinks. In the case of the vortices, a stream function is the basic variable; in the case of sources and sinks, it is a multiple-fluid potential. Outside the region of the singularities the potential and stream functions obey the Cauchy-Riemann equations; inside they are not orthogonal.

The magnitude of the singularities is related to the gradients in specific weight. If the initial position of the interface is known, the action of the singularities on the fluids contained within boundaries with known conditions can be determined from potential theory. The solution is given in the form of an integral for which the region of integration is that part of the aquifer where weight differences are present.

In most practical cases the interface or the mixing zone will have a form which prohibits a mathematically convenient calculation of the integral in closed form. There are, however, methods available for approximation of the integrals involved to any desired degree.

The method requires a knowledge of the position of the interface at a certain moment, and, from that and the known boundary conditions, the subsequent motion of the boundary can be computed from the specific discharge perpendicular to the interface. The displacement of the interface creates a new position of the singularities, with specific discharge changed accordingly, from which a gradual or step-wise movement of the interface can be computed.

A special feature of the singularity distribution is the possibility it offers for the use of the electric resistance model, which is especially useful because it makes possible the direct tracing of potentials and streamlines. Actually, weight differences of fluids have no analogy in electric resistance models, but the hypothetical fluid with homogeneous properties and the singularities applied to this fluid can be simulated by the use of sources and sinks with different distributions for the determination of both potential lines and streamlines.

By introducing a small change in the definition of the potential and stream function, the influence of viscosity differences may also be studied with an electric resistance analogy. Then similar sources and sinks take care of the specific weight variations, but differences in the specific resistance of the model are needed to simulate the viscosity variations.

Acknowledgment. This study was sponsored by the Water Resources Center of the University of California and was conducted in the Hydraulic Laboratory at Berkeley. For the opportunity to study this subject, for various suggestions in the course of the investigation, and for valuable help in the preparation of the manuscript I am especially indebted to Professor David K. Todd, University of California. In the laboratory phases of the study Dr. J. Bear, Technion, Israel, offered his experienced assistance.

REFERENCES

Badon, Ghijben, W., Nota in verband met de voorgenomen putboring nabij Amsterdam, *Ingenieur, Utrecht*, p. 21, 1888–1889.

Cooper, H. H., Jr., A hypothesis concerning the dynamic balance of fresh water and salt water in a coastal aquifer, *J. Geophys. Research, 64,* 461–467, 1959.

Dietz, D. N., A theoretical approach to the problem of encroaching and bypassing edgewater, *Koninkl. Ned. Akad. Wetenschap., Proc., Ser. B, 56,* 83–92, 1953.

Edelman, J. H., Strooming van Zoet en Zout Water, *Rappt. 1940 in zake de watervoorziening van Amsterdam,* Bylage 2, 1940.

Glover, R. E., The pattern of fresh-water flow in a coastal aquifer, *J. Geophys. Research, 64,* 457–459, 1959.

Harder, J. A., T. R. Simpson, L. K. Lau, F. L. Hotes, and P. H. McGauhey, Laboratory research on sea water intrusion into fresh ground-water sources and methods of its prevention, *Final Rept., Sanitary Eng. Research Lab.,* Univ. Calif., Berkeley, 68 pages, 1953.

Henry, H. R., Salt water intrusion into fresh-water aquifers, *J. Geophys. Research, 64,* 1911–1919, 1959.

Herzberg, B., Die Wasserversorgung einiger Nordseebader, *J. Gasbeleuchtung and Wasserversorgung, 44,* 815–819, 842–844, 1901.

Hubbert, M. K., The theory of ground-water motion, *J. Geol., 47,* 785–944, 1940.

Hubbert, M. K., Entrapment of petroleum under hydrodynamic conditions, *Bull. Am. Assoc. Petrol. Geologists, 37,* 1954–2026, 1953.

de Josselin de Jong, G., Vortex theory for multiple phase flow through porous media, *Contrib. 23, Water Resources Center,* Univ. Calif., Berkeley, 80 pp., 1959.

Keulegan, G. H., An example of density current flow in permeable media, unpublished report of the Natl. Bur. Standards, U. S., June 1954.

Kidder, R. E., Flow of immiscible fluids in porous media, *J. Appl. Phys., 27,* 867–869, 1956*a*.

Kidder, R. E., Motion of the interface between two immiscible liquids of unequal density in a porous solid, *J. Appl. Phys., 27,* 1546–1548, 1956*b*.

Lorentz, H. A., Grondwaterbewegung in de nabijheid van bronnen, *Ingenieur, Utrecht, 28,* 24–26, 1913 (for English translation see: *Lorentz's Collected Papers, 4,* 59–66, Nyhoff, the Hague, 1937).

Malavard, L. C., The use of rheoelectrical analogies in aerodynamics, *AGARDograph 18,* NATO Advisory Group of Aeronaut. Research and Develop., Paris, 175 pp., 1956.

Muskat, M., *The Flow of Homogeneous Fluids through Porous Media,* McGraw-Hill Book Co., New York, 763 pages, 1937.

Santing, G., Modèle pour l'étude des problèmes de l'écoulement simultane des eaux souterrains douces et salées, *Assemblée Générale de Bruxelles, Assoc., Intern. d'Hydrologie Scientifique, 2,* 184–193, 1951.

Todd, D. K., and L. Huisman, Ground water flow in the Netherlands coastal dunes, *J. of the Hydrol. Div., Proc. Am. Soc. Civil Engrs., 85,* no. HY 7, 63–81, 1959.

(Manuscript received June 28, 1960.)

Moiré Patterns of the Membrane Analogy for Ground-Water Movement Applied to Multiple Fluid Flow

G. DE JOSSELIN DE JONG

Civil Engineering Department, Technological University
Delft, the Netherlands

Introduction. In the membrane analogy for flow through porous media, contour lines of a deflected membrane represent either streamlines or equipotential lines. The membrane was used by *Prandtl* [1903] to solve torsion problems. *Hansen* [1952] gave a description of the application of the analogy to solve the flow patterns resulting from systems of sources and sinks. Multiple fluid flow through porous media may be treated by considering a suitable distribution of sources and sinks [*de Josselin de Jong*, 1960] and therefore the membrane analogy is also applicable to this problem.

The object of this letter is to point out that such problems may be solved by a moiré method, since this technique provides a convenient procedure for establishing contour lines of a deflected membrane. The procedure is a slight modification of the method initiated by *Ligtenberg* [1955], who superimposed two photographic exposures of a loaded and an unloaded model. Ligtenberg used the reflection of the grid as seen in the mirrored model surface to obtain a moiré pattern which is a measure of surface rotation. The modified procedure differs in that a grid pattern is projected on the nonreflecting model and the moiré pattern obtained by the superimposed images is a measure of the normal displacement. *Otto* [1954] mentioned this kind of experiment in connection with the study of hanging roofs.

Membrane-moiré test setup. The membrane is a thin rubber sheet placed vertically to eliminate its own weight and stretched uniformly by a unit force *S*. Sinks or sources are simulated by concentrated point loads *P* normal to the plane of the membrance as shown in Figure 1.

A horizontal beam of parallel light rays, at an angle of incidence α with the normal to the unloaded membrane, is projected through a grid so as to throw a shadow image on the membrane. The grid consists of a ruling of parallel black lines on a glass plate, oriented in such a way that the lines are vertical.

When the membrane is given a small displacement, *w*, normal to its plane, the shadow image of the grid is displaced. If the axis of the camera is placed normally to the unloaded membrane the apparent normal displacement, *u*, of the projected grid lines is

$$u = w \tan \alpha$$

With *b* the pitch of the shadow lines in the unloaded condition, a displacement of *n* spacings corresponds to a deflection w_n equal to

$$w_n = nb \cot \alpha$$

If the spacing of the grid lines is such that the pitch is twice the width of the lines, photographic superposition will produce complete exposure in those regions where *w'* is given by

$$w' = (n + \tfrac{1}{2})b \cot \alpha \tag{1}$$

therefore these regions are black in the negative and show up as white bands in the positive prints.

These bands indicate the desired contour lines for the deflected membrane. An example of the results which can be obtained by this method is given in Figure 2. This figure shows the streamlines for a confined aquifer filled with two fluids of different specific weight at the moment the fluid motion starts from an abrupt vertical interface. Since this is the same problem as treated previously by the author (1960), the upper half of Figure 2 can be compared with the full lines of Figure 15 of the article quoted, wherein the results of the electric resistance model are represented.

Computation of analogous quantities. The dimensions of the test setup were determined as follows. The relation between the deflection *w* and the normal load per unit area *p* of a membrane uniformly stretched by a unit force *S* is

3625

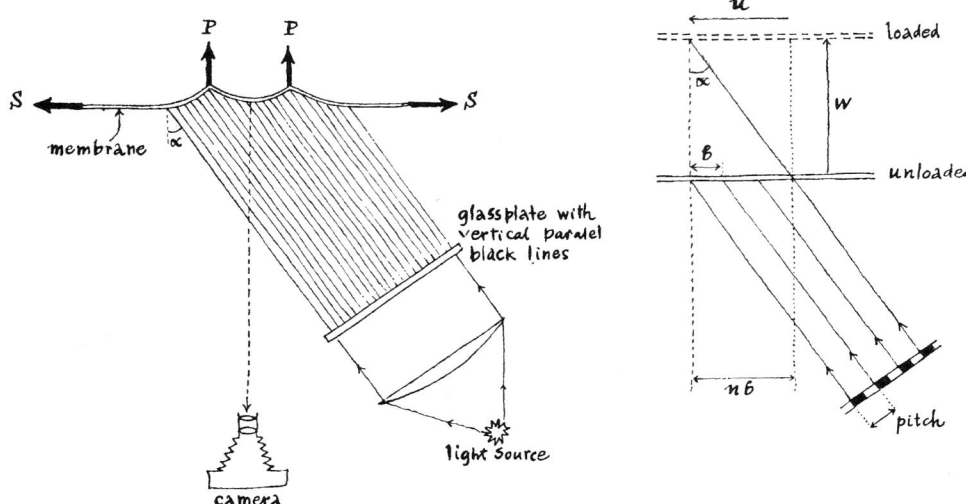

Fig. 1. Top view of test arrangement and detail of displaced membrane.

given by the well-known equation

$$\nabla^2 w = -p/S \qquad (2)$$

[e.g. see *Timoshenko and Goodier*, p. 269, 1951.] This equation is analogous to the basic equation for multiple fluid flow given by the author in the reference quoted. Equating (10) to (11) from this reference gives

$$\nabla^2 \Psi = \omega \qquad (3)$$

where Ψ is the specific discharge stream function and ω is the vorticity. Therefore, a vortex of strength ωdA acting in the region dA can be represented by a distribution $(-p/S)dA$ over the region dA of the membrance. The force distribution p may be approximated by a concentrated force, $P = -pdA$, applied at the centroid of the region dA. The vortex strength for an abrupt interface is given by equation 22 of the article quoted in corrected form:

$$\omega \, dA = \omega \, ds \, dn = (k/\mu)(\gamma_2 - \gamma_1) \, dy_0$$

From (2) and (3) the relationship between analogous quantities is therefore established:

$$\frac{w}{\Psi} = \frac{-p/S}{\omega} = \frac{-p \, dA/S}{\omega \, dA}$$

$$= \frac{P/S}{(k/\mu)(\gamma_2 - \gamma_1) \, dy_0}$$

For the electric resistance model the streamline interval was arbitrarily selected as (eq. 62)

$$\Delta \Psi = (k/\mu)(\gamma_2 - \gamma_1)(c/20)$$

To permit comparison of test results the contour interval for the moiré test was therefore chosen to be

$$\Delta w = \Delta \Psi \frac{P/S}{(k/\mu)(\gamma_2 - \gamma_1) \, dy_0} = \frac{Pc}{20 S \, dy_0}$$

Since nine point loads were used, for a half aquifer height c in the membrane, dy_0 equals $c/9$, and

$$\Delta w = 9P/20S \qquad (4)$$

In order that Δw correspond to one interval, n is 1 in (1), and Δw of (4) should be equal to $b \cot \alpha$. Thus it follows that

$$P = (20/9) Sb \cot \alpha$$

Using a stretching force $S = 200$ g/cm, a projected grid width $b = 0.22$ cm and an angle of incidence $\alpha = 45°$, point loads of magnitude $P = 98$ g were required. Since the theory is only valid for small values of w a correction had to be made for regions of deviation.

Figure 3 gives the pattern for the instantaneous streamlines corresponding to a subsequent position of the interface occurring at a time $t = \frac{1}{2}\epsilon\mu c/$

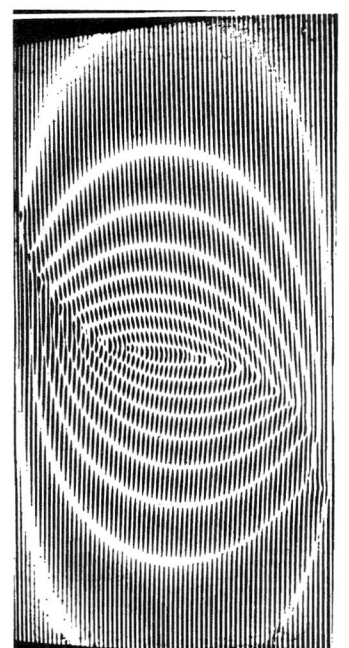

Fig. 3. Moiré pattern of streamlines for the displaced interface.

Fig. 5. Comparison of moiré pattern and parallel plate test for the displaced interface.

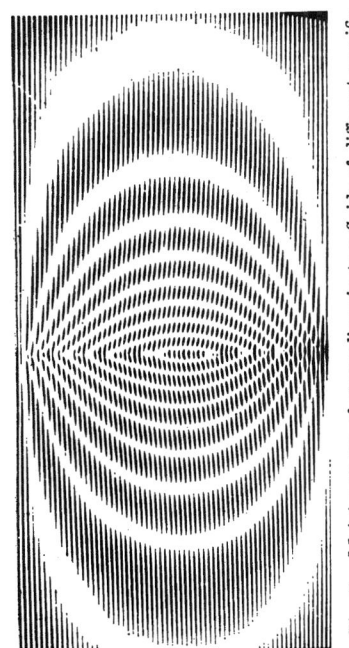

Fig. 2. Moiré pattern of streamlines in two fluids of different specific weight filling a confined aquifer and separated by a vertical interface.

Fig. 4. Comparison of moiré pattern and parallel plate test for the vertical interface.

$k(\gamma_2 - \gamma_1)$ sec, after the initial vertical position of the interface given in Figure 2. The position of the interface at a subsequent instant may be approximated from the streamline pattern for the previous instant, assuming steady-state flow during the time interval.

In the membrane analogy it is then necessary to shift the point loads to the new position of the interface. For a complete investigation of the movement of the fluids, a succession of photographs representing the instantaneous motion in the entire field at each of the time intervals selected is required.

The membrane moiré results of Figures 2 and 3 are comparable to the parallel plate test results given in Figures 13 and 14 of the article quoted. Their agreement is demonstrated by the superposition shown in Figures 4 and 5.

Equipotential lines may also be obtained by the moiré method. Point load moment couples are then applied to the membrane to represent the doublets in the analogy.

The principal advantage of the membrane moiré method is the convenient way of obtaining a photograph of streamlines and equipotentials. This procedure eliminates the need for point to point plotting of contour lines, which is required with the electrical resistance analogy.

REFERENCES

Hansen, V. E., Complicated well problems solved by membrane analogy, *Trans. Am. Geophys. Union, 33,* 912–916, 1952.

de Josselin de Jong, G., Singularity distributions for the analysis of multiple-fluid flow through porous media, *J. Geophys. Research, 65*(11), 3739–3758, 1960.

Ligtenberg, F. K., The moiré method—A new experimental method for the determination of moments in small slab models, *Proc. Soc. Exptl. Stress Anal., 12*(2), 1955.

Otto, F., *Das Hängende Dach,* Ullstein A. G., Berlin, p. 22, 1954.

Prandtl, L., Zur Torsion von prismatischen Stäben, *Physik. Z., 4*(26b), 1903.

Timoshenko, S., and J. N. Goodier, *Theory of Elasticity,* 2nd ed., p. 269, 1951.

Todd, D. K., *Groundwater Hydrology,* John Wiley & Sons, New York, (chap. 14, p. 316: 'Membrane Models'), 1959.

A Many-Valued Hodograph in an Interface Problem

G. DE JOSSELIN DE JONG

Civil Engineering Department, Technological University
Delft, the Netherlands

Abstract. The hodograph method for determining patterns of flow of groundwater in a coastal aquifer with a drain requires the treatment of double-sheeted hodograph planes. This many-valuedness does not prohibit the use of Schwarz-Christoffel analysis if the hodograph domain is simply connected. In the two-fluid case of fresh water flowing over stationary salt water the hodograph is simply connected, and the hodograph method is shown to give a solution. This solution was verified by a test that showed a stable interface in the predicted position.

INTRODUCTION

In solving several problems of two-dimensional groundwater flow encountered in a coastal region, *Bear and Dagan* [1964] used the hodograph method successfully to cope with the difficulties created by the freshwater-saltwater interface. The hodographs treated in that study were all single-valued. There are, however, cases of practical importance where the hodograph is many-valued. An example is a drainage well located in the vicinity of the seacoast to recover part of the fresh water that flows out to the sea.

In their progress report *Bear and Dagan* [1962] suggested the possibility of many-valuedness of the hodograph in that case. This drainage problem will be treated here to show how to deal with hodographs when they are many-valued.

Many-valuedness occurs when the same specific discharge vector is encountered in several different places of the aquifer. *Polubarinova-Kochina* [1952] shows the shape of a double-valued hodograph in the case of seepage through a dam, without further mathematical elaboration (see 1962 translation, p. 46).

To show the character of a double-valued hodograph first in a simplified form, the one fluid case of water flowing around a wall while a drain intercepts part of that water will be treated here (Figure 1a). Double-valuedness is a consequence of the inflection points of the streamlines in the region *CSGFD*.

The two-fluid case of fresh water flowing over a stationary salt water region with a drain

shows a similar pattern of the streamlines (see Figure 2a). The interface has inflection points in the region of upconing created by the drain.

ONE-FLUID CASE

The one-fluid case (Figure 1a) can be solved without the use of a hodograph, because it is possible to map the complex potential $\Omega = \Phi + i\Psi$ (Figure 1c) directly on the $z = x + iy$ plane, which is a section of the aquifer (Figure 1a).

The potential Φ is zero along the seabottom, i.e., the positive x axis, the line GE up to infinity. The stream function Ψ is zero along the line CS, which beyond S separates in the two lines SA_2 and SG. The discharge of the drain is Q_1, so that the value of Ψ is $-Q_1$ along A_1B.

The real axis of z, the line BE, is a line of symmetry.

With these specifications the region of the Ω plane corresponding to the lower half of the z plane can be drawn as in Figure 1c. Only the location of the stagnation point S is unknown. The situation of the drain is known and given by

$$z_A = ae^{-i\pi} = -a$$

Since the flow region is bounded in the Ω plane by straight lines, mapping can be affected by the Schwarz-Christoffel formula. Because this formula maps on the upper half of a plane, the z plane is turned into the t plane, represented in Figure 1b, by multiplying z with $e^{i\pi} = -1$. The relation between Ω and t can

543

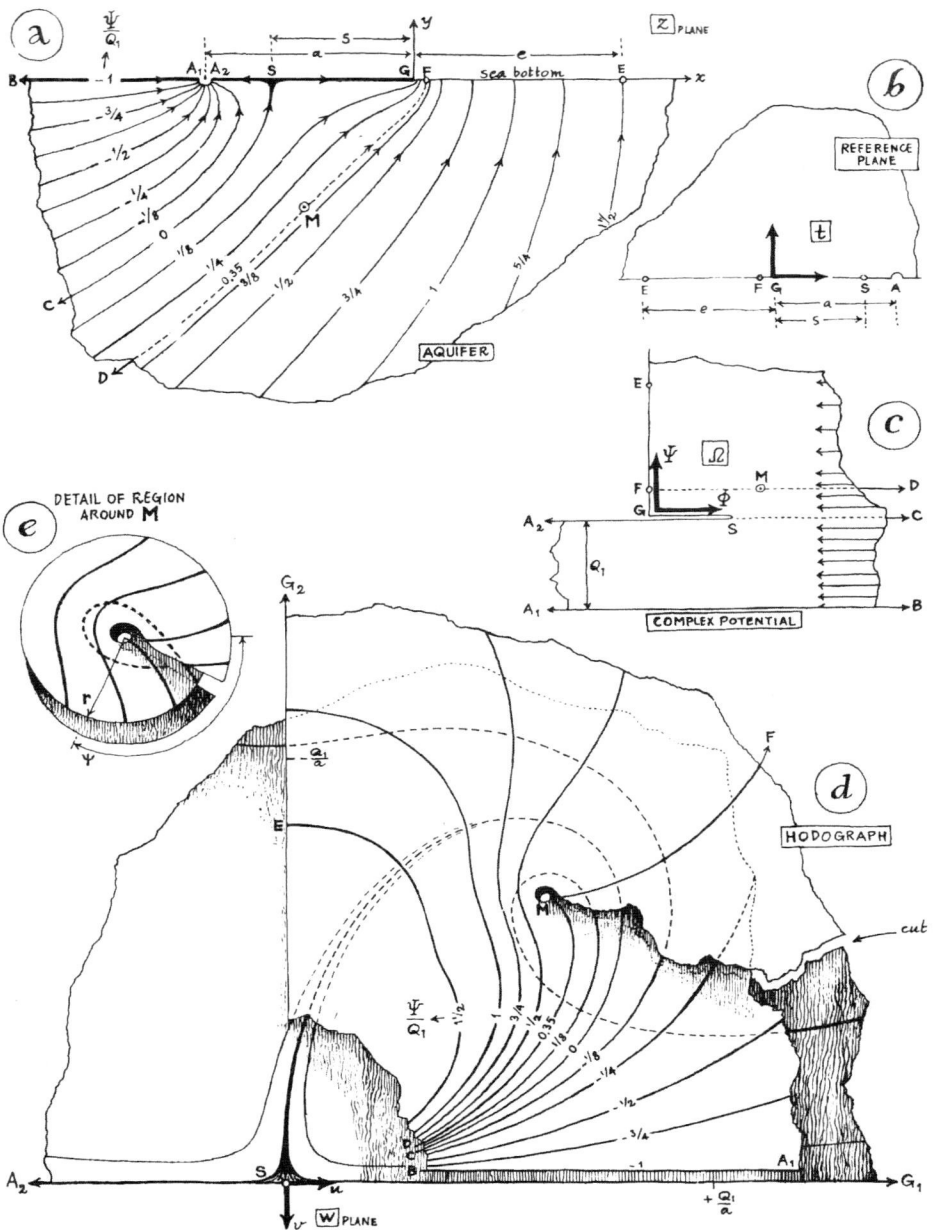

Fig. 1. One-fluid case of a drain at A intercepting water flowing underneath the impermeable line BG toward the sea bottom. All figures are drawn for the case $a = e$, $Q_E = 2Q_1$.

Soil Mechanics and Transport in Porous Media

then be written as

$$\Omega(t) = \alpha_1 \int^t \frac{(\lambda - s)\, d\lambda}{(\lambda - a)\lambda^{1/2}} + \beta$$

$$= \alpha_1 \int^t \frac{d\lambda}{\lambda^{1/2}} + \alpha_1 \int^t \frac{(a - s)}{(\lambda - a)} \frac{d\lambda}{\lambda^{1/2}} + \beta$$

$$= 2\alpha_1 t^{1/2} + \alpha_1 \frac{(a - s)}{a^{1/2}}$$

$$\log \left(\frac{t^{1/2} - a^{1/2}}{t^{1/2} + a^{1/2}} \right) + \beta_1$$

Introduction of the boundary conditions

on $A_1 B$: $\mathrm{Im}\,[\Omega(t)] = -Q_1$

for $\mathrm{Re}\,(t) > a$ $\mathrm{Im}\,(t) = 0$

at G: $\Omega(t) = 0$ for $t = 0$

permits evaluation of the constants

$$\alpha_1 = Q_1 a^{1/2}/\pi(a - s)$$

$$\beta_1 = -iQ_1$$

Turning t into $ze^{i\pi}$ finally gives the complex potential as a function of z,

$$\Omega(z) = \frac{Q_1}{\pi} \left[2i \frac{(az)^{1/2}}{(a - s)} + \log \left(\frac{ia^{1/2} + z^{1/2}}{ia^{1/2} - z^{1/2}} \right) \right] \tag{1}$$

In (1) only s is unknown. If the discharge of the drain Q_1 is kept constant, the value of s changes, if the strength of the flow field from infinity to the boundary GE is varied.

Without drain the flow around a wall is given by the complex potential

$$\Omega(z) = iQ_E(z/e)^{1/2} \tag{2}$$

where Q_E is the discharge through the line GE with length e. For large values of z the two potential fields (1) and (2) will tend to coincide if

$$(a - s) = 2Q_1(ae)^{1/2}/(\pi Q_E) \tag{3}$$

This defines s.

In Figure 1 all drawings correspond to the case $a = e$ and $Q_E = 2\,Q_1$.

Hodograph in one-fluid case. The hodograph **is** represented in Figure 1d, where the endpoints **of** the specific discharge vectors **q** are mapped

in the complex w plane. Here w is defined as

$$w = u + iv = q_x - iq_y \tag{4}$$

By this definition the relation between Ω and w is

$$w = -d\Omega/dz \tag{5}$$

because $\mathbf{q} = -\mathrm{grad}\,\Phi$ and the harmonic functions Φ and Ψ obey the Cauchy-Riemann equations.

To obtain the complex specific discharge w (5) is applied to (1)

$$w = -i\,\frac{Q_1}{\pi}\,\frac{(z + s)}{(a - s)(z + a)}\left[\frac{a}{z}\right]^{1/2} \tag{6}$$

The w plane is drawn in Figure 1d with its positive imaginary axis v downward, because then according to (4) the specific discharge vector $\mathbf{q}$ in the hodograph plane has the same direction as the corresponding vector in the z plane.

In the representation of Figure 1d, the map turns out to be double-sheeted, as can be seen by following a path that encircles the lower half of the z plane once, by starting in A_1 to the left of the drain encircling the drain counter-clockwise to A_2, then going over SGF to E, encircling the aquifer at infinity clockwise from E to B, and returning to A_1.

The endpoints of the corresponding specific discharge vectors then trace the following path. In A_1 of z this vector points to the right, so that A_1 is mapped on the positive u axis. Encircling the drain the vector endpoint turns over a large circle counterclockwise to A_2. The point S being a stagnation point, the velocity is zero there and its map in the w plane is the origin. In the point G the velocity is infinitely large and changes direction pointing to the right in G_1 and upward in G_2, so that the vector endpoint traces $\frac{1}{2}\pi$ of a large circle.

Since all along $G_2 E$ the velocity is upward and decreasing in strength, in the hodograph the endpoint of the vector follows the negative v axis toward the origin. The origin of the hodograph corresponds to the infinity of the z plane $EDCB$, because the velocity reduces to zero there. From B to A_1 the velocity vector points to the right, so that BA_1 is mapped along the positive u axis of the hodograph plane.

If the path $BA_1A_2SGFEDCB$ is traveled in the aquifer (Figure 1a), the total flow region is encircled once and lies at the right-hand side of that path. In the hodograph (Figure 1d) the relevant part of the plane lies to the left of the traveler, because there is a reflection by drawing the positive v axis downward. Figure 1d shows that the hodograph path makes a double loop, and therefore the encircled region is double-sheeted.

Because the single-valued z plane is turned into a double-sheeted w plane, the w plane must contain a branch point of order 1. This point M can be found because, as will be shown presently, a point of w is a branch point of order 1 if the first derivative vanishes there and the second derivative is not zero. In the symbols

$$\begin{cases} w_M' = (dw/dz)_{z=z_M} = 0 \\ w_M'' = (d^2w/dz^2)_{z=z_M} \neq 0 \end{cases} \quad (7)$$

The validity of this statement can be shown by expanding w in a Taylor series around w_M as

$$w = w_M + w_M'(z - z_M)$$
$$+ \tfrac{1}{2}w_M''(z - z_M)^2 + \cdots$$

Because of (7) this reduces to

$$w - w_M = \tfrac{1}{2}w_M''(z - z_M)^2 + \cdots \quad (8)$$

If $(z - z_M) = \rho e^{i\varphi}$ with ρ a small, constant length, the point z describes a small circle with radius ρ around z_M. A complete revolution is traced when φ increases from 0 to 2π. The point w then traces approximately a small circle with radius $r = \tfrac{1}{2}|w_M''|\rho^2$ (see Figure 1e), which is a detail of Figure 1d. If we call $w - w_M = re^{i\psi}$, then according to (8) ψ is equal to $2\varphi + \arg(w_M'')$. If therefore φ increases from 0 to 2π, the angle ψ increases by 4π, indicating that the point w_M is encircled twice if the point z_M is encircled once.

A branch of higher order, say n, is obtained if all derivatives up to $w_M^{(n)}$ vanish and $w_M^{(n+1)}$ is not zero. For more information about branch points the reader is referred to the textbooks on function theory. A detailed description of branch points is, for instance, given by *Nehari* [1952], pages 78 and 152. This description will be helpful in understanding the representation in Figure 1d, which may be difficult to visualize since it is impossible to build a correct paper model of such surfaces.

From (6), $w_M' = 0$ if

$$w_M' = i\,\frac{Q_1}{\pi}\,\frac{[z_M^2 + (3s - a)z_M + as]}{2(a - s)(z_M + a)^2}$$
$$\left[\frac{a}{z_M^3}\right]^{1/2} = 0$$

This is the case for

$$z_M^2 + (3s - a)z_M + as = 0$$

giving the roots

$$z_M + a = \tfrac{1}{2}(a - s)^{1/2}$$
$$\cdot [3(a - s)^{1/2} \pm (9s - a)^{1/2}] \quad (9)$$

It can be verified by direct computations that w_M'' is not zero for z_M given by (9). Therefore z_M as defined by (9) is mapped into the branch point of the w plane. If DF is the streamline passing through the point M in the z plane given by z_M, then only the streamlines between DF and CSG have inflection points. In the hodograph (Figure 1d) these streamlines make loops around the branch point (see, for instance, the streamline for $\Psi = \tfrac{1}{8}Q_1$).

The $\pm$ sign in (9) reflects the symmetry with respect to the z axis. Since the lower half of the z plane is considered here, only the $-$ sign is needed.

Because of the branch point the hodograph is not only double-valued but also double-connected. The location of the branch point in w cannot be inferred beforehand, and therefore it is impossible to indicate the cut (as shown in Figure 1d) that makes the hodograph simply connected. In the two-fluid case this difficulty is avoided because the hodograph is itself simply connected.

TWO-FLUID CASE

In the two-fluid case the presence of a drain that intercepts part of the fresh water that flows toward the sea will cause upconing of the interface between the fresh water and salt water. The flow pattern in the aquifer for this case is shown in Figure 2a. All pictures in Figure 2 represent the values of the test described in the last section. In Figure 2e, however, the points F, M_2, M_1, and S are shifted somewhat in the horizontal direction in order to make them discernible.

Figure 2a shows that in the aquifer the drain is located at a distance a_0 from the seashore. The interface between the moving fresh water

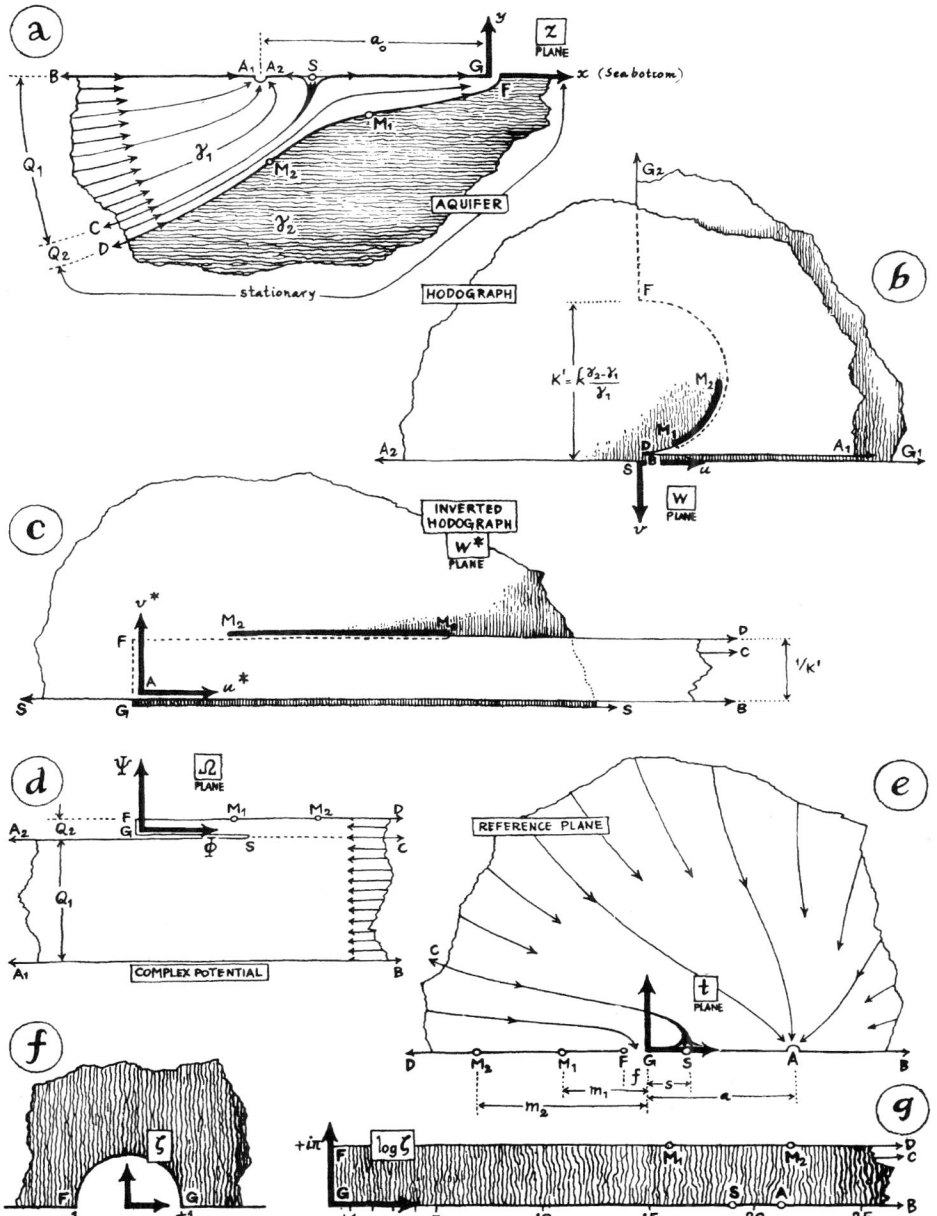

Fig. 2. Two-fluid case of a drain at A intercepting fresh water flowing underneath the impermeable line BG toward the sea bottom over stationary salt water. All figures are drawn for the case $Q_1/K'a_0 = 0.304$, $Q_2/K'a_0 = 0.0416$ except 2e, which is out of scale.

and the underlying stationary salt water (horizontal shading) is the line DM_2M_1F. This line has two inflection points, M_1 and M_2. The hump between these two points can be considered as the upconing.

If the salt water is stationary, the interface is known to map in the hodograph plane as a circle with diameter

$$K' = k(\gamma_2 - \gamma_1)/\gamma_1$$

with k = hydraulic conductivity, γ_2 = specific weight of salt water, and γ_1 = specific weight of fresh water. The hodograph is represented in Figure 2b. It is left to the reader to verify that the path $BA_1A_2SGFM_1M_2DCB$ that encircles the region of the aquifer occupied by the fresh water is mapped in the hodograph by the path $BA_1A_2SG_1G_2FM_1M_2DB$. The areas A_1A_2 and G_1G_2 in the hodographs are circles with large radii.

In the same way as in Figure 1 the relevant region of the aquifer lies to the right of the traveler and in the hodograph to the left of the traveler. The points FM_1M_2D from the interface lie on the circle with diameter K'. Since M_1 and M_2 are inflection points on the interface in the aquifer, these points are turning points on the circle of the hodograph. This makes, for example, the line FM_1M_2 a slot in the sheet that is shaded vertically in Figure 2b. There is a narrow bridge between the points SM_1 that connects the region FG_2G_1 to A_2. From A_2 over the large circle to A_1 and farther toward B there is an overlap encircling the slot at M_2. This overlap ends in the white beak that contains the points B and D. The relevant region in the hodograph is double-sheeted in such a way that it cannot be cut out of one piece of paper. By gluing two pieces together it is possible to make a model that is simply connected. It is therefore possible to map the hodograph on a half infinite plane, for example, the t plane (Figure 2e), which is also simply connected.

The complex potential is shown in Figure 2d. The discharges Q_1 and Q_2 through the drain and the sea bottom FG, respectively, are considered known. The location of the stagnation point S is as yet unknown.

The shape of the interface in the z plane can be computed by mapping the pertinent domains in both the Ω and w planes (Figures 2b and 2d) on the upper half of the same t plane (Figure

2e). By using relation (5), z can then be expressed as a function of the parameter t, which forms a solution because for every point of z the value of Ω and w can be computed through the intermediary of t.

Complex Potential in the Two-Fluid Case

The mapping of Ω on the upper half of the t plane is obtained by the Schwarz-Christoffel formula as follows:

$$\Omega(t) = \alpha_2 \int^t \frac{(\lambda - s)}{(\lambda - a)} \frac{d\lambda}{\lambda^{1/2}(\lambda + f)^{1/2}} + \beta$$

$$= \alpha_2 \int^t \frac{d\lambda}{\lambda^{1/2}(\lambda + f)^{1/2}}$$

$$+ \alpha_2 \int^t \frac{(a - s)\, d\lambda}{(\lambda - a)\lambda^{1/2}(\lambda + f)^{1/2}} + \beta$$

$$\Omega(t) = \alpha_2[\log \zeta - (a - s)a^{-1/2}$$

$$\cdot (a + f)^{-1/2}\mathcal{L}(t, f, a)] + \beta_2 \qquad (10)$$

where

$$\begin{cases} \zeta = [t^{1/2} + (t + f)^{1/2}]^2/f \\ \mathcal{L}(t, f, a) = 2 \log [t^{1/2}(a + f)^{1/2} \\ \quad + (t + f)^{1/2}a^{1/2}] - \log f(a - t) \end{cases} \qquad (11)$$

The value of the constants α_2, β_2 and a relation between the as yet unknown values of a, f, s can be obtained from the boundary conditions.

From (11) it can be verified that for t on the real axis the imaginary parts of $\log \zeta$ and $\mathcal{L}(t, f, a)$ have the following values:

$$\begin{cases} \text{Im } [\log \zeta] = 0 \quad \text{for } 0 < t < \infty \\ \text{Im } [\log \zeta] = i\pi \quad \text{for } -\infty = \infty e^{i\pi} \\ \qquad\qquad\qquad\qquad < t < -f = fe^{i\pi} \\ \text{and} \qquad\qquad\qquad\qquad\qquad (12) \\ \text{Im } [\mathcal{L}(t, f, a)] = 0 \quad \text{for } 0 < t < a \\ \text{Im } [\mathcal{L}(t, f, a)] = i\pi \quad \text{for } a < t < \infty \\ \qquad\qquad \text{and} \\ \qquad\qquad -\infty < t < -f \end{cases}$$

The boundary conditions for Ω can be taken

from Figure 2d to be for t on the real axis

$$\begin{cases} \text{on } A_1B: & \text{Im } [\Omega(t)] = -iQ_1 \\ & \text{for} \quad a < t < \infty \\ \text{on } FD: & \text{Im } [\Omega(t)] = +iQ_2 \\ & \text{for} \quad -\infty < t < -f \quad (13) \\ \text{at } G: & \Omega(t) = 0 \\ & \text{for} \quad t = 0 \end{cases}$$

Introduction of (12) and (13) in (10) permits evaluation of α_2 and β_2 and deduction of the following relation between a, f, s and the discharges:

$$(Q_1 + Q_2)/Q_1 = a^{1/2}(a + f)^{1/2}/(a - s) \quad (14)$$

The final expression for $\Omega(t)$ is then

$$\Omega(t) = \pi^{-1}[(Q_1 + Q_2) \log \zeta$$

$$- Q_1 \mathcal{L}(t, f, a)] \quad (15)$$

Hodograph in Two-Fluid Case

To apply the Schwarz-Christoffel formula for the mapping of the hodograph on the t plane it is convenient to transform the w plane into the w^* plane by the inversion

$$w^* = 1/w \quad (16)$$

because the boundaries then become straight lines. As is seen in Figure 2c particularly, the circle DM_2M_1F maps into a line parallel to the real axis, at a distance $1/K'$ above it.

The mapping of w^* on t is then given by the expression

$$w^* = \alpha_3 \int^t \frac{(\lambda + m_1)(\lambda + m_2)\,d\lambda}{(\lambda - s)^2\lambda^{1/2}(\lambda + f)^{1/2}} + \beta$$

which after integration becomes

$$w^* = \alpha_3 \left[\log \zeta - \frac{(m_1 + s)(m_2 + s)\,t^{1/2}(t + f)^{1/2}}{s(s + f)(t - s)} \right. $$
$$\left. + \frac{(m_1 + s)(m_2 + s)(s + \tfrac{1}{2}f) - (m_1 + m_2 + 2s)s(s + f)}{s^{3/2}(s + f)^{3/2}} \mathcal{L}(t, f, s) \right] + \beta_3 \quad (17)$$

The boundary conditions are now from Figure 2c:

$$\begin{cases} \text{on } GS \text{ and } SB: & \text{Im } [w^*(t)] = 0 \\ & \text{for} \quad 0 < t < \infty \\ \text{in } A \text{ and } G: & w^*(t) = 0 \\ & \text{for} \quad t = a \quad \text{and} \quad t = 0 \quad (18) \\ \text{on } FD: & \text{Im } [w^*(t)] = 1/K' \\ & \text{for} \quad -\infty < t < -f \end{cases}$$

According to (12) the imaginary part of $\mathcal{L}(t, f, s)$ changes by π if t passes the point S. The first of (18) indicates that such a jump is not required. Therefore the coefficient of $\mathcal{L}(t, f, s)$ in (17) must vanish, giving the condition

$$(m_1 + s)(m_2 + s)(s + \tfrac{1}{2}f)$$
$$= (m_1 + m_2 + 2s)s(s + f) \quad (19)$$

Since both $\log \zeta$ and $t^{1/2}$ vanish for $t = 0$, the condition in G indicates that $\beta_3 = 0$. In the point A, t is equal to a and the value of ζ is

$$\zeta_a = [a^{1/2} + (a + f)^{1/2}]^2/f \quad (20)$$

The condition in A is fulfilled if

$$\log \zeta_a = \frac{(m_1 + s)(m_2 + s)a^{1/2}(a + f)^{1/2}}{s(s + f)(a - s)} \quad (21)$$

With (14), (19), (21) it is possible to express s, m_1, m_2 as functions of Q_1, Q_2, a, and f as follows:

$$s = a - [Q_1/(Q_1 + Q_2)][a(a + f)]^{1/2}$$

$$m_1 + s = \tfrac{1}{2}\Lambda\{(s + \tfrac{1}{2}f) - [(s + \tfrac{1}{2}f)^2 - 4s(s + f)\Lambda^{-1}]^{1/2}\}$$

$$m_2 + s = \tfrac{1}{2}\Lambda\{(s + \tfrac{1}{2}f) + [(s + \tfrac{1}{2}f)^2 - 4s(s + f)\Lambda^{-1}]^{1/2}\} \quad (22)$$

where

$$\Lambda = [Q_1/(Q_1 + Q_2)] \log \zeta_a \quad (23)$$

Finally the condition along FD makes $\alpha_3 =$

$1/\pi K'$. Therefore the expression for w^* reduces to

$$w^* = (1/\pi K')\left[\log \zeta \right.$$
$$\left. - \frac{(a - s)t^{1/2}(t + f)^{1/2}}{(t - s)a^{1/2}(a + f)^{1/2}} \log \zeta_a \right] \quad (24)$$

Relation between z and t. The two expressions (15) and (24) permit computing z as a function of t. Writing (5) in a slightly different way and introducing (16) gives

$$dz = -w^{-1}\, d\Omega = -w^*\, d\Omega \quad (25)$$
$$z = -\int^t w^*(d\Omega/dt)\, dt + \beta$$

From (15) we obtain

$$\frac{d\Omega}{dt} = \left[\frac{Q_1 + Q_2}{\pi} \right.$$
$$\left. + \frac{Q_1}{\pi} \frac{a^{1/2}(a + f)^{1/2}}{(t - a)} \right] \frac{1}{t^{1/2}(t + f)^{1/2}}$$

Insertion of this result and (24) in (25) gives with (14)

The expression (27) can be further adjusted so that the points A and G in z correspond to these points in t. In order that G lie in the origin of z, β_4 should be zero, because ζ becomes 1 and $\log \zeta$ vanishes for $t = 0$.

Let the distance GA in the aquifer be a_0; then the condition that A coincides in both planes is $z = -a_0$ for $t = a$. This condition gives

$$a_0 = \frac{Q_2}{2\pi^2 K'}(\log \zeta_a)^2 +$$
$$\frac{Q_1}{\pi^2 K'}\left\{ -\log\left[\left(\frac{a + f}{a}\right)^{1/2}\left(\frac{\zeta_a^2 - 1}{\zeta_a^2}\right) \right] \log \zeta_a \right.$$
$$\left. + (\pi^2/6) - 2Li_2(1/\zeta_a) + Li_2(1/\zeta_a^2) \right\} \quad (29)$$

If for the aquifer the distance a_0, the discharges Q_1, Q_2, and the value for K' are given, it is possible to determine a/f with (29), since it is the only unknown there. It is not feasible to give an explicit expression for a/f except when it is very large. (All the cases mentioned by *Bear and Dagan* [1962] have very large a/f values). From (29) it follows that for $a/f \gg 1$ there results

$$z = -[(Q_1 + Q_2)/\pi^2 K']\int^t \log \zeta\, d(\log \zeta) + [Q_1/\pi^2 K']\left\{ \log \zeta_a \int^t (t - a)^{-1}\, dt \right.$$
$$\left. - \int^t [a(a + f)/t(t + f)]^{1/2}(t - a)^{-1}\, \log \zeta\, dt \right\} + \beta \quad (26)$$

Integration of this expression gives

$$z = -\frac{Q_1 + Q_2}{2\pi^2 K'}[\log \zeta]^2 + \frac{Q_1}{\pi^2 K'}\left\{ \log \frac{a - t}{a} \log \zeta_a + \log\left[\frac{1 - \zeta\zeta_a}{1 - (\zeta/\zeta_a)} \right] \log \zeta \right.$$
$$\left. - Li_2(\zeta/\zeta_a) + Li_2(1/\zeta_a) + Li_2(\zeta\zeta_a) - Li_2(\zeta_a) \right\} + \beta_4 \quad (27)$$

where $Li_2(\sigma)$ stands for the dilogarithm defined by

$$Li_2(\sigma) = -\int_0^\sigma \lambda^{-1}\, \log(1 - \lambda)\, d\lambda \quad (28)$$

The terms consisting of dilogarithms in (27) are derived from the last integral of (26). Since the integration of this term is not so easily verified, a further justification of the result is given in the appendix. The other terms of (26) have a more elementary character.

$$\log \zeta_a \approx \log(4a/f)$$
$$\approx \pi[(2K'a_0 - \tfrac{1}{3}Q_1)/Q_2]^{1/2} \quad (30)$$

This relation is represented in Figure 3.

The region of applicability of Figure 3 is limited, because the analysis developed is only valid if the interface in the aquifer contains inflection points M_1 and M_2. From (22) it is seen that the expressions for m_1 and m_2 contain roots. These roots are only real if

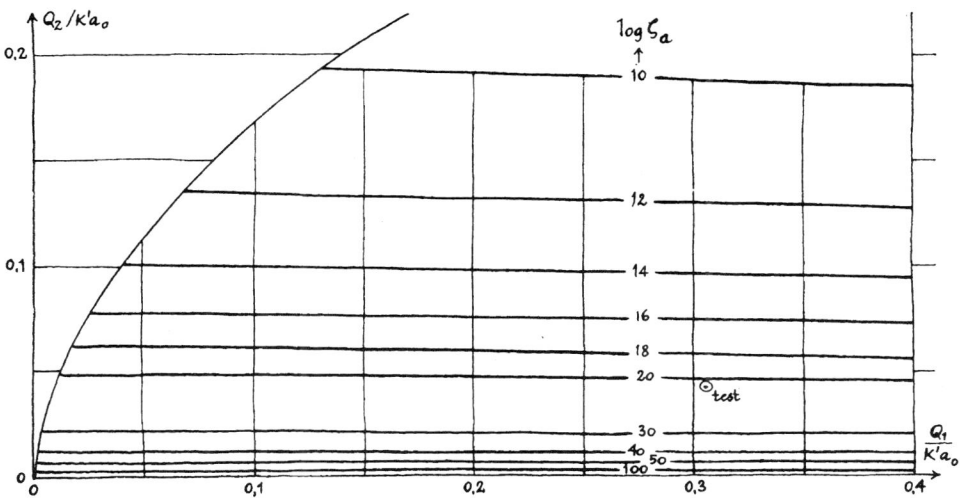

Fig. 3. Plot of log ζ_a as a function of Q_1 and Q_2 according to equation 30.

$$(s + \tfrac{1}{2}f)^2 > 4s(s + f)\Lambda^{-1} \qquad (31)$$

If (31) is not satisfied, the roots are imaginary and there are no inflection points. Then the hodograph contains a branch point as shown in the one-fluid case. If (31) is not satisfied, the formulas are not valid because the Schwarz-Christoffel analysis as used here is not applicable if a branch point is involved. The condition (31) sets an upper bound to the ratio Q_2/Q_1 which has the form

$$\frac{Q_2}{Q_1} < \left[\frac{(s + \tfrac{1}{2}f)^2}{4s(s + f)} \log \zeta_a \right] - 1 \qquad (32)$$

This line limits the zone of validity in Figure 3 to the left.

The region of applicability in Figure 3 is also limited at the bottom. This condition is visualized by considering the case that the discharge through the drain Q_1 is kept constant while the supply from infinity decreases, so that Q_2 diminishes and eventually reduces to zero. During this process a moment will come when the interface becomes unstable and the drain starts to discharge salt water.

By diminishing Q_2 the points M_2 and M_1 will move along the circle in the hodograph in a counterclockwise direction. Instability occurs for that part of the interface that maps in the hodograph on points of the circle that lie to the

left of the vertical. This is understood if it is realized that a velocity vector pointing to the upper left means an interface with fresh water below salt water.

When the slot (M_2M_1) moves over the top of the circle, M_2 is the first to pass the vertical. For instability not to occur the criterion is therefore that the real part of w at M_2 remains positive. This implies in the inverted hodograph that

$$\text{Re}\,[w^*(M_2)] > 0 \qquad (33)$$

The criterion (33) can only be represented in a simplified form if $a \gg f$, $K'a_0 \gg Q_1$, and $Q_1 \gg Q_2$, as is true for the region of Figure 3. In that case the requirement for the lower bound is

$$1 + \pi q + \log \pi q - [Q_1/12K'a_0] > 0 \qquad (34)$$

with

$$q = (2Q_2K'a_0)^{1/2}/Q_1$$

On the scale of Figure 3 the lower limit is so close to the horizontal axis that it is not to be distinguished from it.

Shape of the interface for $a/f \gg 1$. For large values of a/f the formulas describing the interface can be given in a more compact form. The interface is the line DM_2M_1F in the z plane (see Figure 2a), which will be represented by $z_0 =$

$x_0 + iy_0$. This line corresponds to a part of the negative real axis in the t plane. Let t_0 be defined by $t = t_0 e^{i\pi}$; then t_0 is a positive quantity along DM_2M_1F.

For points to the left of M_2 the value of t_0 is large with respect to f (because m_2 is of the order a). Therefore the quantity ζ_0 defined by $\zeta = \zeta_0 e^{i\pi}$ is then approximately $\zeta_0 \approx 4t_0/f$. Inserting these quantities in (27) and separating real and imaginary parts of z_0 yields

the fluids was $\nu = 0.34$ cm²/sec. From these values K' can be deduced to have been $K' = [(\gamma_2 - \gamma_1)/\gamma_1](gb^2/12\nu) = 2.54$ cm/sec. The drain distance a_0 was 37.5 cm. The discharge through the drain (per unit width) was $Q_1 = 29.0$ cm³/sec. The amount of fluid flowing through FG per unit width was $Q_2 = 3.97$ cm³/sec.

With these specifications it can be deduced that $a/f \gg 1$ and $\log \zeta_a = 21.2$ (see Figure 3).

$$\pi^2 K' \text{ Re } (z_0) = \pi^2 K' x_0$$

$$= Q_1\{\tfrac{1}{2}[\log(\zeta_0/\zeta_a)]^2 - \log(\zeta_0/\zeta_a)\log[(\zeta_0 + \zeta_a)/\zeta_a] + \tfrac{1}{2}Li_2(\zeta_a^2/\zeta_0^2) - Li_2(\zeta_a/\zeta_0) + (\pi^2/6)\}$$

$$+ Q_2\{-\tfrac{1}{2}[\log \zeta_0]^2 + \tfrac{1}{2}\pi^2\} \qquad (35)$$

$$\pi^2 K' \text{ Im } (z_0) = \pi^2 K' y_0 = -\pi\{Q_1 \log[(\zeta_0 + \zeta_a)/\zeta_a] + Q_2 \log \zeta_0\} \qquad (36)$$

These expressions can be simplified for $\zeta_0 \gg \zeta_a$. In the formula for x_0 the dilogarithms can be approximated by $Li_2(1/x) \approx 1/x$ for $x \gg 1$ and they then drop out. The remaining terms permit elimination of ζ_0 by squaring y_0 and comparing the terms obtained with $2(Q_1 + Q_2)x_0/K'$. The following relation between x_0 and y_0 is the result:

$$y_0{}^2 = -2(Q_1 + Q_2)x_0/K'$$
$$- Q_1Q_2[\log \zeta_a]^2/(\pi K')^2$$
$$+ (Q_1 + Q_2)(3Q_2 + Q_1)/3K'^2$$

By using the approximation (30) for $\log \zeta_a$ this reduces to

$$y_0{}^2 = -2(Q_1 + Q_2)x_0/K' + [2Q_1{}^2 + 4Q_1Q_2$$
$$+ 3Q_2{}^2 - 6Q_1K'a_0]/3K'^2 \qquad (37)$$

This is the shape of the interface for the quantity of water $(Q_1 + Q_2)$ coming from infinity and streaming around a wall with its edge located at a distance

$$[Q_1/(Q_1 + Q_2)][a_0 + (Q_1 + 2Q_2)/6K'] \quad (38)$$

to the left from the actual edge.

Test result. To verify the above formulas a test was run in a parallel plate model, with a slot width of $b = 2.1$ mm. Figure 4 is a photograph of the resulting flow pattern. The two fluids were resin oil ($\gamma_1 = 0.864$ gf/cm³) and glycerine ($\gamma_2 = 1.071$ gf/cm³). The viscosity of

From (22) it then follows that $s = 0.12a$, $m_1 = 0.00645a$, $m_2 = 2.0a$. Inserting these values in (27) gives the location of points S, M_1, and M_2 in the z plane:

$$z(s) = -29.3 \quad \text{cm}$$
$$z(M_1) = -20.02 - 8.05i \quad \text{cm} \qquad (39)$$
$$z(M_2) = -36.44 - 14.9i \quad \text{cm}$$

From (24) the inclination of the interface at M_1 and M_2 can be computed, because the angle of the interface with the horizontal is equal to the angles M_1AB and M_2AB in the inverted hodograph (Figure 2c). These angles can be obtained from the complex values for w^* at M_1 and M_2, which are

$$w^*(M_1) = (4.85 + i)/K'$$
$$w^*(M_2) = (1.28 + i)/K' \qquad (40)$$

The streamlines obtained from the photograph of Figure 4 are reproduced in Figure 5. In this drawing the points surrounded by a square give the computed location of the points S, M_1, and M_2 according to (39). The computed inclination of the interface at M_1 and M_2 according to (40) is represented by the dashed lines through these points. The agreement between test result and theory can be considered satisfactory. The point S coincides with the stagnation point that separates the water flowing to the well, A, and the water flowing to the sea, GF. The points M_1

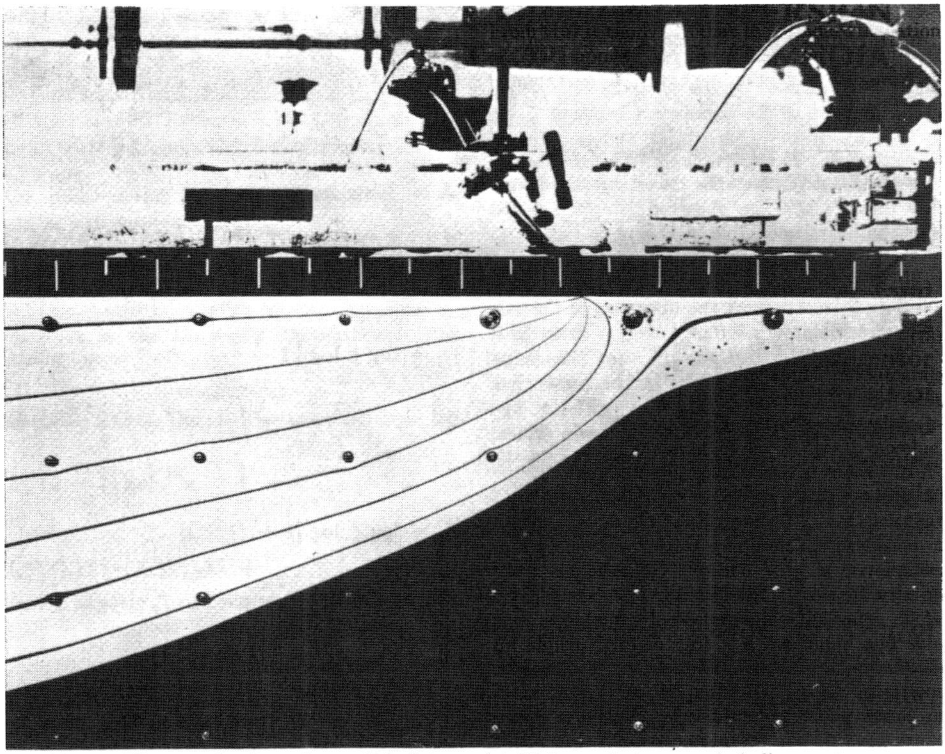

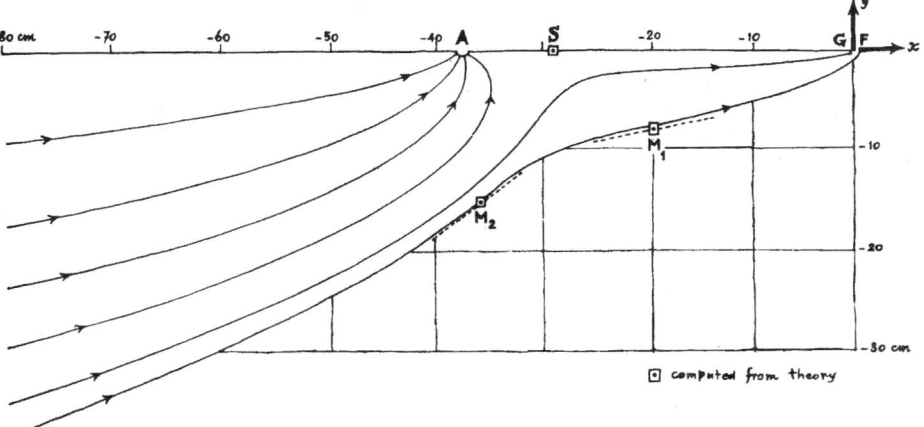

Fig. 4. Photograph of test in parallel plate model. Distances between scale indices on top are 5 and 10 cm.

Fig. 5. Plot of interface and streamlines of test in parallel plate model. Verification between theory and experiments consists of the stagnation point *S* and the inflection points M_1, M_2 with the local inclinations of the interface.

and M_2 coincide with the inflection points, and the dashed lines are practically tangent to the observed interface. The interface to the left of M_2 is approximated by the parabola (37) which by application of the parameters becomes

$$y_0{}^2 + 25.9\kappa_0 + 740 = 0$$

The deviation between the shape computed with this formula and the observed interface is too small to be reproducible in Figure 5.

From the agreement between test results and theory, it may be deduced that the equations developed do represent the upconing of the salt water-fresh water interface under a drain in a coastal area. The test showed that the shape of the interface was stable, in the sense that small disturbances disappeared without appreciably changing the location of the interface.

APPENDIX

Equation 26 requires the integration

$$J = \int_0^t \left[\frac{a(a+f)}{t(t+f)}\right]^{1/2}$$
$$\cdot \frac{\log\{[t^{1/2} + (t+f)^{1/2}]^2/f\}\, dt}{(t-a)} \quad (A1)$$

Consider the variable ζ introduced in (11). The following relations between t and ζ exist:

$$t = f[(\zeta - 1)^2/4\zeta]$$
$$dt = f[(\zeta^2 - 1)/4\zeta^2]\, d\zeta$$
$$[t(t+f)]^{1/2} = f[(\zeta^2 - 1)/4\zeta]$$

The domain of ζ corresponding to the domain of t is given in Figures 2f and 2g. The result of substituting ζ in (A1) is

$$J = \int_1^\zeta \frac{4[a(a+f)/f^2]^{1/2}\log \zeta\, d\zeta}{\{\zeta^2 - [2 + 4(a/f)]\zeta + 1\}} \quad (A2)$$

Using the constant ζ_a introduced by the relation (20), $[2 + 4(a/f)]$ is equal to $[\zeta_a + (\zeta_a)^{-1}]$. The denominator of the integrand in (A2) can therefore be written as

$$(\zeta - \zeta_a)[\zeta - (\zeta_a)^{-1}]$$

Because $[\zeta_a - (\zeta_a)^{-1}] = 4[a(a+f)/f^2]^{1/2}$, the integral J falls apart into two pieces

$$J = J_1 + J_2$$
$$= \int_1^\zeta \frac{\log \zeta\, d\zeta}{(\zeta - \zeta_a)} - \int_1^\zeta \frac{\log \zeta\, d\zeta}{[\zeta - (\zeta_a)^{-1}]} \quad (A3)$$

Consider J_1 and substitute $\mu = \zeta/\zeta_a$; this integral then becomes

$$J_1 = -\int_{1/\zeta_a}^{\zeta/\zeta_a} \frac{(\log \mu + \log \zeta_a)\, d\mu}{(1 - \mu)}$$

Partial integration of this integral gives

$$J_1 = \log \mu \log (1 - \mu)|_{1/\zeta_a}^{\zeta/\zeta_a}$$
$$+ \log \zeta_a \log (1 - \mu)|_{1/\zeta_a}^{\zeta/\zeta_a}$$
$$- \int_{1/\zeta_a}^{\zeta/\zeta_a} \mu^{-1} \log (1 - \mu)\, d\mu$$
$$= \log \zeta \log \left(1 - \frac{\zeta}{\zeta_a}\right)$$
$$- \int_0^{\zeta/\zeta_a} \mu^{-1} \log (1 - \mu)\, d\mu$$
$$+ \int_0^{1/\zeta_a} \mu^{-1} \log (1 - \mu)\, d\mu$$

$$J_1 = \log \zeta \log [1 - (\zeta/\zeta_a)]$$
$$+ Li_2(\zeta/\zeta_a) - Li_2(1/\zeta_a)$$

A similar operation applied to T_2 gives as a final result for (A3)

$$J = \log \left[\frac{1 - (\zeta/\zeta_a)}{1 - \zeta\zeta_a}\right] \log \zeta + Li_2(\zeta/\zeta_a)$$
$$- Li_2(1/\zeta_a) - Li_2(\zeta\zeta_a) + Li_2(\zeta_a) \quad (A4)$$

This result is used in (27).

In these expressions the notation $Li_2(\sigma)$ stands for the dilogarithm as indicated in (28). This function is tabulated for the real positive values of σ in the region $0 < \sigma < +1$ (see for instance Lewin [1958]), which also indicates how to deal with complex values of σ.

In this analysis only real values of σ are considered. For values of σ outside the region $0 < \sigma < +1$ the following recurrence formulas exist:

$$Li_2(\sigma) = -Li_2(1/\sigma) + (\pi^2/3) + i\pi \log \sigma$$
$$- \tfrac{1}{2}[\log \sigma]^2 \quad \text{for} \quad 1 < \sigma < \infty \quad (A5)$$

$$Li_2(\sigma) = \tfrac{1}{2}Li_2(\sigma^2) - Li_2(-\sigma)$$
$$\text{for} \quad -1 < \sigma < 0 \quad (A6)$$

$$Li_2(\sigma) = -\tfrac{1}{2}Li_2(1/\sigma^2) + Li_2(-1/\sigma)$$
$$- (\pi^2/6) - \tfrac{1}{2}[\log (-\sigma)]^2$$
$$\text{for} \quad -\infty < \sigma < -1 \quad (A7)$$

The expression (A5) differs from the value given by Lewin with respect to the sign of the term $i\pi \log \sigma$, as a consequence of an alteration applied in this analysis to the definition of the dilogarithm. At variance with Lewin, the domain of validity of σ is limited to $-\pi \leq \arg (1 - \sigma) < +\pi$. These limits are chosen so that the positive real axis of ζ can correspond to $\arg \zeta = 0$.

Acknowledgments. For verification and expansion of the mathematics I am indebted to A. Verruyt for his invaluable assistance. The verification test was skillfully run by K. Vervoort and G. Zandstra in the University Laboratory of Soil Mechanics at Delft.

REFERENCES

Bear, J., and G. Dagan, The transition zone between fresh and salt waters in coastal aquifers, Progress Report 1, The steady interface between two immiscible fluids in a two dimensional aquifer, 123 pp., *Technion-Israel Institute of Technology, Hydraulic Lab., P.N. 3/62*, December 1962.

Bear, J., and G. Dagan, Some exact solutions of interface problems by means of the hodograph method, *J. Geophys. Res., 69,* 1563–1572, 1964.

Lewin, L., *Dilogarithms and Associated Functions,* 353 pp., MacDonald, London, 1958.

Nehari, Z., *Conformal Mapping,* 396 pp., McGraw-Hill Book Company, 1st ed., 1952.

Polubarinova-Kochina, P. I., *Theory of Groundwater Movement,* translated from the Russian by Roger J. M. Dewiest, 613 pp., Princeton University Press, 1962.

(Manuscript received October 23, 1964.)

9

Generating Functions in the Theory of Flow through Porous Media

G. de Josselin de Jong
Civil Engineering Department
Delft University of Technology
Delft, the Netherlands

1. Introduction

In this study the flow of fluids through porous media is considered for the case in which fluid and porous medium are inhomogeneous. The properties that may differ from place to place in the field are the density and viscosity of the fluid and the intrinsic permeability of the medium. These inhomogeneities are responsible for rotations in the flow pattern. Since the Darcian flow of fluids through porous media is usually irrotational, it was considered instructive in this presentation to elaborate especially the character of these rotations.

For the determination of the distribution of the pressure and the specific discharge vector it is convenient to make use of potential and streamfunctions. These functions may be called generating functions, because their gradients generate the specific discharge vector. For incompressible homogeneous fluids these functions obey Laplace equations of the type

$$\nabla^2 \mathscr{F} = 0$$

377

where $\mathscr{F}$ is the relevant function. The solution of a problem is obtained by selecting from the functions that obey this basic equation those that satisfy the boundary conditions.

In this study, functions will be determined that in a similar way generate the specific discharge vector by differentiation if inhomogeneities are present. The basic equations obeyed by these generating functions will turn out to be Poisson equations. These are equations of the type

$$\nabla^2 \mathscr{F} = \mathscr{G}$$

where $\mathscr{G}$ is a function of place.

If the value of $\mathscr{G}$ is known everywhere, it is possible to solve $\mathscr{F}$ in a region enclosed by boundaries along which either the value of $\mathscr{F}$ or its derivative normal to the boundary is known. The solution of the Poisson equation will be given in the form of an integral which has to be extended over all points where $\mathscr{G}$ has a nonvanishing value. This method of solving may be called the *solution by singularities*. In addition, solutions of the Laplace equation have to be applied to satisfy the remaining boundary conditions.

If $\mathscr{G}$ is an unknown function (for instance, if $\mathscr{G}$ depends on $\mathscr{F}$), it is not possible to use the solution by means of the singularities mentioned. It may be possible that other kinds of singularities are able to cope with Poisson equations where $\mathscr{G}$ is a linear function of $\mathscr{F}$. This possibility is not investigated here and only functions obeying a Poisson equation with known values for $\mathscr{G}$ are considered.

The generating functions appropriate for flow of fluids with varying properties may be called *multiple fluid functions*. It will be required that the multiple fluid functions are single valued in the entire field irrespective of how many fluids are involved. The multiple fluid functions will be chosen such that they resemble as closely as possible the potential and streamfunctions for the single fluid case.

The multiple fluid functions developed here are essentially the same as those described in a previous article (de Josselin de Jong, 1960). In that paper Poisson's equations were established for the streamfunction Ψ and a multiple fluid potential, Θ. Yih (1961) arrived at the same result with respect to Ψ. Knudsen (1962) showed that the pressure obeys a Poisson equation of a kind very similar to that for Ψ.

Lusczynski (1961) suggested several types of heads for multiple fluid flow. De Wiest (1964, 1965) reviewed this work and selected H_{ip}, the type called "point water head" by Lusczynski, as the most suitable physical quantity to use. It will be shown here that Lusczynski's "point water head" H_{ip} is directly related to Θ. By elaborating the second-order differential equation governing H_{ip} and Θ, it turns out, however, that H_{ip} obeys a Poisson equation that is less convenient to solve than the relation governing Θ.

The name chosen by de Josselin de Jong (1960) to indicate Θ may be confusing because Θ is not a potential in the sense that grad Θ is proportional to the specific discharge vector. Because there is a rotation in the fluid with varying density, the specific discharge cannot be derived from a potential. Until a better name is suggested, the term *multiple fluid potential* will be used here, with the stipulation that it is not a velocity potential in the usual sense.

In Sections 2–8, the case of a two-dimensional aquifer is treated. In Section 9, formulas are given for the three-dimensional case.

The density is considered variable throughout the entire chapter. The viscosity of the fluids and the permeability of the porous medium are taken constant, except in Section 8. In that section, an electric analogy is described which provides a means of treating variability of these other properties.

All fluids will be considered to be incompressible. This means that fluid elements will preserve their density during displacement. The variability of density considered is the difference in density between fluid elements at different places in the field. No surface tension will be assumed to exist between fluids of different properties. The porous medium is isotropic.

The multiple fluid functions generate the fluid displacement at a given instant. It is assumed in this analysis that the distribution of the fluids and their properties is known at that moment. From the multiple fluid functions it is then possible to determine the displacement of fluid particles during a small time interval. By taking into account the manner in which the properties of the fluids are convected by this displacement, the distribution of the properties after that time interval can be determined. This creates a new situation which requires new multiple fluid functions. It is not the objective of this study to pursue what is happening at successive time intervals, but only to describe the instantaneous flow situation.

2. Basic Properties of the Flow

Since the fluids are assumed to be incompressible, conservation of mass requires that the specific discharge vector $\mathbf{q}$ obeys the continuity equation

$$\text{div } \mathbf{q} = \frac{\partial q_x}{\partial x} + \frac{\partial q_z}{\partial z} = 0 \tag{1}$$

The second requirement is Darcy's law which can be written in terms of an equilibrium of forces. When the forces acting on the fluid per unit volume of the fluid are equated, the equilibrium equation results:

$$(\mu/k)\,\mathbf{q} = -\text{grad}\,p - \rho\,\text{grad}\,F \tag{2}$$

The left side of (2) expresses the force per unit volume of fluid necessary to drive a specific discharge $\mathbf{q}$ through the porous medium according to Darcy's law. In (2) μ is the dynamic viscosity, k is the intrinsic permeability, and ρ is the density of the fluid. The two terms on the right of (2) represent the two forces that can work on the fluid and produce the driving action. The first term is obtained from the gradient of the pressure p. The second term represents a body force which can be derived from a potential F acting per unit mass of the fluid.

If only gravity works, F is equal to gz, in which z is measured vertically upwards. The equilibrium equation becomes in this case

$$(\mu/k)\,\mathbf{q} = -\text{grad}\,p - \rho g\,\text{grad}\,z \tag{3}$$

Equations (1) and (2) are the basic relationships sufficient for the establishment of the multiple fluid functions. They are relations describing the fluid behavior in a point. Because (3) describes a point relationship it is irrelevant for the last term whether the density is a constant or a variable in the field.

Since this point was questioned, it may be helpful to derive this term by integrating the forces acting on the fluid in an infinitesimally small circular disc with radius r_1 (Fig. 1).

Let ξ be a coordinate running in the direction of grad ρ and let ζ be a coordinate perpendicular to ξ. Further, let the subscript zero in the expressions ρ_0, $(\partial\rho/\partial\xi)_0$,..., etc., indicate the value of these quantities in the center of the circle. Then the density in the circle is approximated by its Taylor series:

$$\rho = \rho_0 + \xi(\partial\rho/\partial\xi)_0 + \tfrac{1}{2}\xi^2(\partial^2\rho/\partial\xi^2)_0$$
$$+ \xi\zeta(\partial^2\rho/\partial\xi\,\partial\zeta)_0 + \tfrac{1}{2}\zeta^2(\partial^2\rho/\partial\zeta^2)_0 + \cdots \tag{4}$$

The force acting on a unit of mass in the gravity field is $-g$ grad z. In order to find the total force on the fluid element, the mass of that fluid amount has to be evaluated by integration over the volume of the disc. If f is the thickness of the disc and n the porosity, this gives

$$M = \int_0^{r_1} \int_0^{2\pi} \rho \cdot nf \cdot dr \cdot r\,d\theta \tag{5}$$

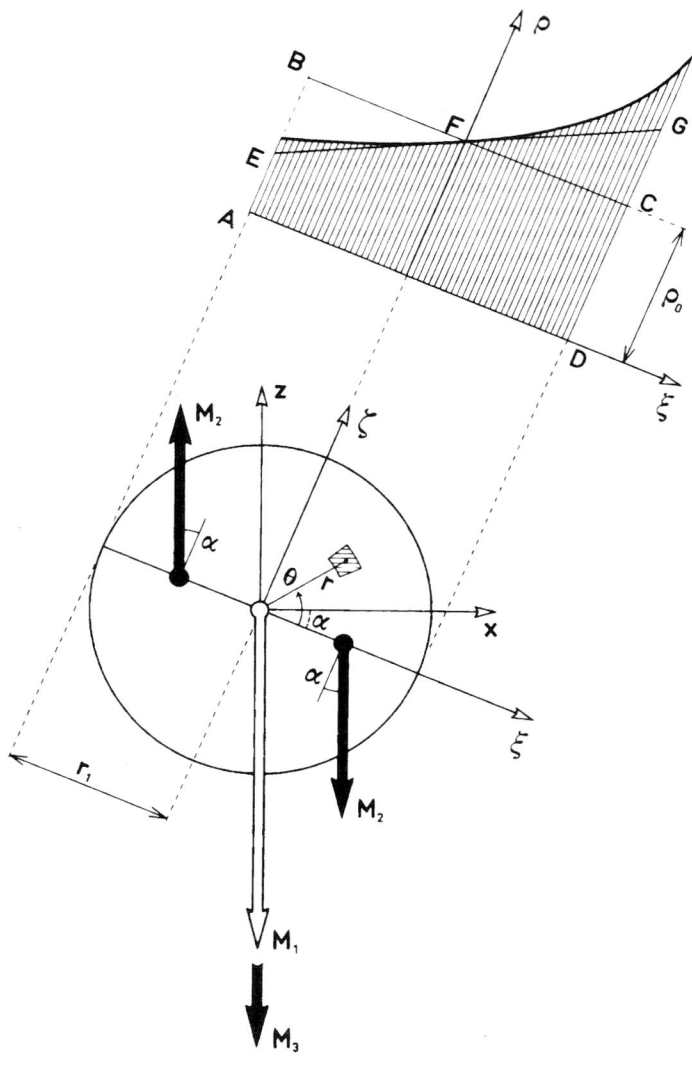

FIG. 1.

where ρ can be replaced by the series in (4). Let the series be split in three parts, giving for the integral three contributions M_1, M_2, M_3, respectively, such that M_1 is obtained from the first term in the series, M_2 from the second term, and M_3 from the third, fourth, and fifth terms together.

The result is

$$M_1 = \rho_0 nf \cdot \pi r_1^2, \qquad M_2 = 0$$

$$M_3 = [(\partial^2\rho/\partial\xi^2)_0 + (\partial^2\rho/\partial\zeta^2)_0]\, nf \cdot \tfrac{1}{4}\pi r_1^4 \tag{6}$$

Since M_3 is of the order r_1^4, its value may be disregarded with respect to M_1 which is of order r_1^2, because in the limit r_1 tends to zero.

To obtain the force per unit volume of the fluid, (6) has to be multiplied by g grad z and divided by $nf \cdot \pi r_1^2$, the volume of the fluid in the disc. Thus, the force on the infinitesimal element finally becomes $-\rho_0 g$ grad z as indicated in (3). The second term from the series expression for ρ does not contribute to a force on the element because M_2 is zero.

In the diagram for ρ in Fig. 1, $(\partial\rho/\partial\xi)_0$ is the inclination of the line EFG, the tangent to the curve for ρ. The second term, $\xi(\partial\rho/\partial\xi)_0$, covers the triangles BFE and CFG, which being similar and opposite give cancelling forces. However, they contribute a clockwise torque which can be written as

$$T = g \cos\alpha \int_0^{r_1} \int_0^{2\pi} \xi \cdot \xi(\partial\rho/\partial\xi)_0\, nf\, dr \cdot r\, d\theta$$

where α is the angle between ξ and x. By evaluating this integral it is found that

$$T = g \cos\alpha(\partial\rho/\partial\xi)_0 \cdot \tfrac{1}{4}\pi r_1^4 \cdot nf$$

Since $\cos\alpha = (\partial\xi/\partial x)$, the torque can be written as

$$T = \tfrac{1}{4}\pi r_1^4 g(\partial\rho/\partial x)_0\, nf \tag{7}$$

This torque is responsible for the rotational properties of the flow fields, as will be shown in Section 6.

The expressions (1) and (2) are therefore point relationships which only require ρ_0, the value of ρ at the point considered. The variability of ρ with location only contributes if higher order differential equations are established. This actually happens in the subsequent treatment.

3. Single-Valuedness of the Functions

For the multiple fluid functions to be applicable in the analysis it is of importance that they be single-valued because otherwise they have no physical meaning and may become misleading in computations.

The single-valuedness of a function Φ is established if according to Young's theorem,

$$\partial^2\Phi/\partial x\,\partial z = \partial^2\Phi/\partial z\,\partial x \tag{8}$$

This theorem can be proved by starting from the contour integral expression

$$\int_C (\partial\Phi/\partial s)\,ds = 0 \tag{9}$$

which indicates that for every closed contour C, the value of Φ will return to its original value. After changing to x, z coordinates, and introducing $M = \partial\Phi/\partial x$, $N = \partial\Phi/\partial z$, this contour integral can be converted into a surface integral by use of Green's theorem

$$\int_C (\partial\Phi/\partial s)\,ds = \int_C [(\partial\Phi/\partial x)\,dx + (\partial\Phi/\partial z)\,dz]$$

$$= \int_C (M\,dx + N\,dz)$$

$$= \iint_A [(\partial N/\partial x) - (\partial M/\partial z)]\,dx\,dz$$

$$= \iint_A [(\partial^2\Phi/\partial z\,\partial x) - (\partial^2\Phi/\partial x\,\partial z)]\,dA = 0 \tag{10}$$

Since (9) applies for every contour C in the field, (10) applies for every area A enclosed by that contour. By reducing the contour to as small a circle as one wishes, the last integral of (10) indicates that (8) applies to every point in the field.

Wherever expression (8) will occur in the sequel, it will be concluded that single-valuedness of the relevant function is established.

4. Multiple Fluid Functions for μ and k Constant, ρ Variable

Of the multiple fluid functions, the stream function Ψ resembles most the conventional stream function. It will be defined in two-dimensional flow as

$$q_x = -(\partial\Psi/\partial z), \qquad q_z = +(\partial\Psi/\partial x) \tag{11}$$

This means that the specific discharge vector has the magnitude of the gradient of Ψ and runs perpendicular to grad Ψ such that the angle $\frac{1}{2}\pi$ from $\mathbf{q}$ to grad Ψ is traced clockwise (Fig. 2).

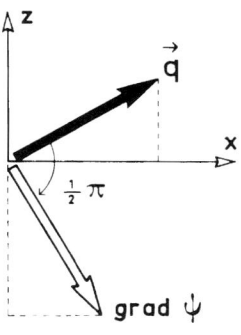

FIG. 2.

Application of the continuity Eq. (1) gives

$$-(\partial^2 \Psi / \partial z\ \partial x) + (\partial^2 \Psi / \partial x\ \partial z) = 0 \tag{12}$$

which shows that Ψ is a single-valued function irrespective of the properties of the fluids involved.

More information about Ψ may be obtained with the equilibrium Eq. (2). This equation written in its x, z components gives

$$(\mu/k)\ q_x = -(\partial p/\partial x), \qquad (\mu/k)\ q_z = -(\partial p/\partial z) - \rho g \tag{13}$$

From the physical viewpoint the pressure p must be single-valued because it is a physical scalar quantity which has one value in every point of the field. Therefore, the relation

$$-(\partial^2 p/\partial x\ \partial z) + (\partial^2 p/\partial z\ \partial x) = 0 \tag{14}$$

is satisfied everywhere in the field.

The identity (14) can be used to eliminate p from (13). Since (μ/k) is a constant, differentiating Eqs. (13) with respect to z and x, respectively, and subtracting gives

$$(\mu/k)[(\partial q_x/\partial z) - (\partial q_z/\partial x)] = -(\partial^2 p/\partial x\ \partial z) + (\partial^2 p/\partial z\ \partial x) + [\partial(\rho g)/\partial x]$$
$$= \partial(\rho g)/\partial x \tag{15}$$

Applying expressions (11) to this result gives

$$-(\partial^2 \Psi/\partial z^2) - (\partial^2 \Psi/\partial x^2) = (k/\mu)[\partial(\rho g)/\partial x]$$

or

$$\boxed{\nabla^2 \Psi = -(k/\mu)[\partial(\rho g)/\partial x]} \tag{16}$$

This relation, obtained previously (de Josselin de Jong, 1960), permits computation of Ψ with the aid of singularities of rotational type if the distribution of ρ is known. This will be shown in Section 5.

Applying the continuity Eq. (1) to (13) gives

$$(\mu/k) \operatorname{div} \mathbf{q} = -(\partial^2 p/\partial x^2) - (\partial^2 p/\partial z^2) - [\partial(\rho g)/\partial z] = 0$$

or

$$\nabla^2 p = -\partial(\rho g)/\partial z \tag{17}$$

This relation for p was obtained by Knudsen (1961). The two Poisson Eqs. (16) and (17) are very similar because they contain the derivatives of the density in the x and z directions, respectively. For p, solutions of boundary value problems can be obtained by use of singularities in a similar way as used for Ψ.

In a previous article (de Josselin de Jong, 1960), a function was introduced which can be considered the equivalent of the potential function for the flow of homogeneous fluids. This function Θ was called the *multiple fluid potential*, although it is not a potential in the sense that grad Θ is proportional to the discharge vector. This function Θ is related to pressure p and elevation z by the expression

$$(\mu/k) \Theta = p + \rho g z \tag{18}$$

Here ρ is the density of the fluid at the point of consideration. Since p, ρ, and z are single valued, Θ is also single valued. From (18) it follows that the relation between Θ and H_{ip}, the *point water head* as defined by Lusczynski (1961), is

$$(\mu/k) \Theta = \rho_i g H_{ip} \tag{19}$$

Taking the gradient of (18) gives, since (μ/k) is constant,

$$(\mu/k) \operatorname{grad} \Theta = \operatorname{grad} p + z \operatorname{grad}(\rho g) + (\rho g) \operatorname{grad} z \tag{20}$$

A comparison with (3) shows that this expression contains one term more than the expression for $(\mu/k)\mathbf{q}$. Therefore, Θ is not a regular potential.

By use of the equilibrium Eq. (3), the pressure can be eliminated to give

$$(\mu/k) \mathbf{q} = -(\mu/k) \operatorname{grad} \Theta + z \operatorname{grad}(\rho g) \tag{21}$$

Taking the divergence of the expression in order to establish continuity gives

$$(\mu/k) \operatorname{div} \mathbf{q} = 0 = -(\mu/k) \nabla^2 \Theta + \operatorname{div}[z \operatorname{grad}(\rho g)] \tag{22}$$

This finally gives for Θ the Poisson equation

$$\boxed{\nabla^2 \Theta = +(k/\mu) \operatorname{div}[z \operatorname{grad}(\rho g)]} \tag{23}$$

The Poisson equation (23) shows that Θ can be solved by use of singularities.

Since H_{ip}, the "point water head," is so closely related to Θ it might be expected that H_{ip} also obeys some convenient Poisson equation. Elaboration along the same lines as above gives

$$\nabla^2 H_{ip} = (\rho g)^{-1}\{\operatorname{div}[z \operatorname{grad}(\rho g)] - 2 \operatorname{grad} H_{ip} \cdot \operatorname{grad}(\rho g) - H_{ip} \nabla^2(\rho g)\}$$

Since the right-hand side contains H_{ip}, this equation is inconvenient in the analysis. H_{ip} is therefore less suitable then Θ.

The results of this section are summarized in Table 1.

TABLE 1

Multiple Fluid Functions

Multiple fluid functions:	Ψ	p	$\Theta = (k/\mu)(p + \rho g z)$
$\dfrac{\mu}{k} q_x =$	$-\dfrac{\mu}{k}\dfrac{\partial \Psi}{\partial z} =$	$-\dfrac{\partial p}{\partial x} =$	$-\dfrac{\mu}{k}\dfrac{\partial \Theta}{\partial x} + z\dfrac{\partial(\rho g)}{\partial x}$
$\dfrac{\mu}{k} q_z =$	$+\dfrac{\mu}{k}\dfrac{\partial \Psi}{\partial x} =$	$-\dfrac{\partial p}{\partial z} - \rho g =$	$-\dfrac{\mu}{k}\dfrac{\partial \Theta}{\partial z} + z\dfrac{\partial(\rho g)}{\partial z}$
Single-valuedness established by:	$\operatorname{div} \mathbf{q} = 0$	Physical necessity	Physical necessity
Poisson equation established by:	Elimination of p	$\operatorname{div} \mathbf{q} = 0$	$\operatorname{div} \mathbf{q} = 0$
Gives:	$\nabla^2\Psi = -\dfrac{k}{\mu}\dfrac{\partial(\rho g)}{\partial x}$	$\nabla^2 p = -\dfrac{\partial(\rho g)}{\partial z}$	$\nabla^2\Theta = +\dfrac{k}{\mu}\operatorname{div}[z \operatorname{grad}(\rho g)]$
Type of singularity	Vortex	Source or sink	Source–sink dipole

5. Velocity of the Flow Field

The Poisson equation (16) indicates that the flow contains rotation in those points where the gradient of the fluid density has a horizontal component. Although reasons can be found in textbooks why this equation represents vorticity, it is helpful for the subsequent treatment to show this here.

Consider the case that in a circular area A_2 with radius r_2 (Fig. 3), the following relation holds

$$\nabla^2 \Psi = 2\omega, \qquad 0 < r < r_2 \tag{24}$$

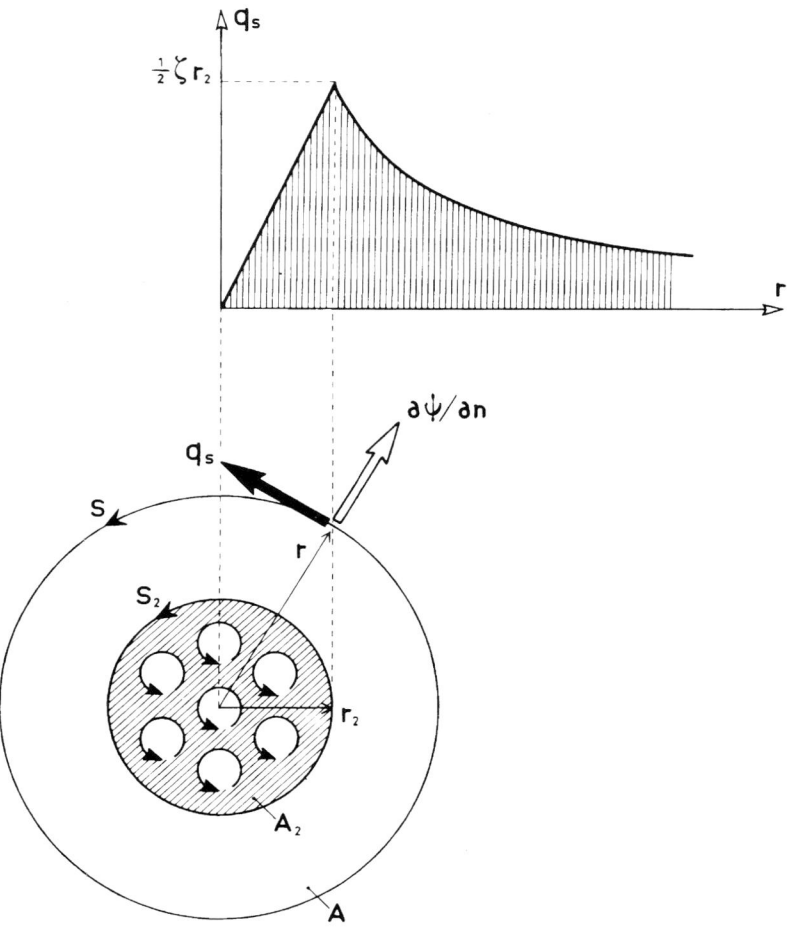

Fig. 3.

with ω = constant, and that outside A_2 it is

$$\nabla^2 \Psi = 0, \qquad r_2 < r < \infty \qquad (25)$$

Let S be a circular contour, concentric with A_2 and with a radius r that can be either larger or smaller than r_2.

Consider first the case where r is smaller than r_2, Then only (24) is relevant. Integration of (24) over the area A gives

$$\iint_A \nabla^2 \Psi \, dA = \iint_A 2\omega \, dA = 2\omega \iint_A dA = 2\omega \pi r^2 \qquad (26)$$

By Gauss' theorem the first integral can be written

$$\iint_A \nabla^2 \Psi \, dA = \int_S (\partial \Psi / \partial n) \, dS \qquad (27)$$

In this expression n is the outer normal to the circle S, which is related to the radius r in such a manner that $(\partial \Psi / \partial n) = (\partial \Psi / \partial r)$. Because the system is axial symmetric, $(\partial \Psi / \partial r)$ is constant over the circle S, and (27) becomes

$$\iint_A \nabla^2 \Psi \, dA = (\partial \Psi / \partial r) \int_S dS = (\partial \Psi / \partial r) \, 2\pi r \qquad (28)$$

Combination of (26) and (28) gives

$$(\partial \Psi / \partial r) = \omega r, \qquad 0 < r < r_2 \qquad (29)$$

From Fig. 2 it is seen that $(\partial \Psi / \partial r)$ represents a discharge q_S tangential to the circle S. Hence, (29) can be written

$$q_S = \omega r, \qquad 0 < r < r_2 \qquad (30)$$

and this indicates that ω in the region A creates a tangential flow circulating around this region with a discharge velocity proportional to the distance to the center. ω is called the *vorticity*.

The second case to consider is r larger than r_2. Since according to (25), the value of $\nabla^2 \Psi$ is zero outside A_2, integration of $\nabla^2 \Psi$ over the area A now only consists of a contribution over the area A_2. This gives

$$\iint_A \nabla^2 \Psi \, dA = \iint_{A_2} \nabla^2 \Psi \, dA = 2\omega \pi r_2^2 \qquad (31)$$

Again, the first integral of (31) can be elaborated by the Gauss theorem as (27) to yield a result (28) with r now the radius larger than r_2. By combining (28) and (31), there results

$$\partial\Psi/\partial r = \omega(r_2^2/r)$$

which shows that the discharge circulating around A_2 is

$$q_S = \omega(r_2^2/r), \qquad r_2 < r < \infty \tag{32}$$

This expression indicates that the discharge outside the region A_2, which contains the vorticity, circulates around that region with a velocity which is inversely proportional to the distance from the center.

The distribution of the specific discharge as given by the Eqs. (30) and (32) is represented in Fig. 3. In that figure it is seen that on the boundary of the region A_2 containing the vorticity, the circulating discharge q_s is continuous. This follows from (30) and (32) which give the same value for q_s if r is made equal to r_2. The continuity of q_s is a direct consequence of the above analysis, which shows that a discontinuity can only exist in a place where ω is infinite.

From (16) and (24) it follows that the vorticity is equal to

$$\omega = -\tfrac{1}{2}(k/\mu)[\partial(\rho g)/\partial x] \tag{33}$$

This expression shows that ω is only infinite if $\partial\rho/\partial x$ is infinite. This is the case if ρ changes abruptly, as is encountered at a sharp interface between two different fluids. Since this aspect was treated extensively by de Josselin de Jong (1960), its treatment will be omitted here. Only gradually changing densities (grad ρ = finite) will be taken into account in the sequel.

A further conclusion from (33) is that every small region, where the density varies in the horizontal x direction, contributes a vorticity similar to the region A_2 in the example. Around every elementary region ΔA_2 containing density variation, a circulating discharge Δq_S is created which obeys Eq. (32). The influence of the density variations in the entire field is thus obtained by adding vectorially all elementary discharges Δq_S created by all elementary areas ΔA_2. By choosing the elementary areas ΔA_2 infinitesimally small, the discharge Δq_S in the area itself obeying Eq. (30) can be disregarded.

If this reasoning is correct it is also necessary that the discharge (30) be obtained in this way by integrating discharges of the character (32) as generated by infinitesimally small vortices over the region A_2.

This may seem contradictory because it requires that the result obtained for the region outside A_2 (which differs from the solution inside), when applied to the region inside, actually gives that different inside solution. A lengthy but straightforward evaluation of the integral involved shows, however, that the solution (30) is indeed obtained. The verification is an entertaining exercise in elliptic integrals.

It may seem to be a difficulty that the discharge according to (32) becomes infinite for $r = 0$. This means that the vortex at the point of consideration contributes an infinitely large discharge. However, this difficulty disappears since the magnitude of the vortex at a point, being proportional to the area, becomes infinitely small of the same order. Therefore, the integral remains finite. This exorbitant behavior at a point containing vorticity is the reason we apply the word *singularity*. Using the solution (32) for the discharges around points containing vorticity and adding those in an integral is called the *integral solution with singularities*.

Instead of adding vectorially elementary discharges, it is mathematically more convenient to add the elementary streamfunctions which generate them. This gives an integral expression for Ψ which is explained in Section 6.

6. Solutions with Singularities

If a small region of the size $dx_0\, dz_0$ around the point x_0, z_0 contains vorticity $\omega(x_0, z_0)$, the discharge is rotating around that point according to (32) with a magnitude

$$q_S = 2\omega(x_0, z_0)\, dx_0\, dz_0/2\pi r \qquad (34)$$

where r is the distance from the point of vorticity. Written in terms of the streamfunction Ψ, this is

$$(\partial\Psi/\partial r) = 2\omega(x_0, z_0)\, dx_0\, dz_0/2\pi r$$

which by integration gives

$$\Psi = (1/\pi)\, \omega(x_0, z_0) \ln r\, dx_0\, dz_0 + \text{constant} \qquad (35)$$

The constant is not a function of θ, because $q_r = 0$ since the discharge only rotates around the point of vorticity. Thus, $\partial\Psi/\partial\theta = 0$. Since we are only interested in the derivative of Ψ, the constant is irrelevant and can be taken as zero without loss of generality.

Because all the regions containing vorticity contribute to Ψ in this manner, the total expression for Ψ as created by all the vortices is the integral

$$\Psi(x, z) = (1/\pi) \int\int \omega(x_0, z_0) \ln r \, dx_0 \, dz_0 \tag{36}$$

where

$$r = [(x - x_0)^2 + (z - z_0)^2]^{1/2} \tag{37}$$

In the multiple fluid case where density variations create vorticity according to (33), the expression for Ψ becomes

$$\Psi_{\mathrm{I}}(x, z) = (-k/2\pi\mu) \int\int [\partial(\rho g)/\partial x]_0 \ln r \, dx_0 \, dz_0 \tag{38}$$

The surface integral for Ψ_{I} covers the entire region between the boundaries containing fluids. This expression for Ψ_{I} forms the vortex part of the streamfunction.

In general, the values of Ψ_{I} on the boundaries differ from the values required by the boundary conditions. These differences form a new boundary value problem to be satisfied by a harmonic streamfunction Ψ_{II} obeying the Laplace equation $\nabla^2 \Psi_{\mathrm{II}} = 0$. The function Ψ_{II} forms the second part of the solution. The final solution to the boundary value problem is $\Psi = \Psi_{\mathrm{I}} + \Psi_{\mathrm{II}}$.

This way of solving the problem is known in potential theory as the solution of Poisson's equation by singularities. In this case the Poisson equation for Ψ_{I} is (16) which has (38) as its solution.

In a similar way, a solution for the pressure p or the multiple fluid potential Θ can be given because they obey similar Poisson equations i.e., (17) and (23). This gives

$$p_{\mathrm{I}} = (-1/2\pi) \int\int [\partial(\rho g)/\partial z]_0 \ln r \, dx_0 \, dz_0 \tag{39}$$

$$\Theta_{\mathrm{I}} = (+k/2\pi\mu) \int\int [z \, \nabla^2(\rho g) + \partial(\rho g)/\partial z]_0 \ln r \, dx_0 \, dz_0 \tag{40}$$

Also in these cases, additional functions p_{II} and Θ_{II} obeying Laplace's equation must be introduced to complete the solution and to satisfy all boundary conditions.

An alternative way of decomposing the solution of Poisson's equation for boundary conditions is pointed out by De Wiest (1969). Instead of $\ln r$ which can be considered as a Green's function for the infinite domain, Greens functions can be introduced which have zero conditions along the boundaries of the domain considered in each specific problem. (The function or its normal derivative is zero along the boundary.) The differences with the required boundary values are then those boundary values themselves, creating a boundary value problem with a solution satisfying Laplace's equation.

Since, however, a boundary value problem has to be solved anyway, the method proposed here seems less laborious because it circumvents the construction of a Green's function, which in contradistinction to ln r is different for every boundary geometry and for every point in the field.

We are now ready to compare the torque, caused by density variation in a circular area as determined in Section 4, to the driving force necessary for the circulating discharge caused by the vorticity. If, however, the vorticity is limited to the circular area, the available torque is not sufficient to support the discharge outside the area. This outside discharge requires a torque of infinite magnitude.

This discrepancy is a consequence of the vorticity, being caused by density variations. In other instances vortices exist which can be limited to isolated regions. In this case it is impossible to visualize an isolated region containing a density gradient. Isolation means that it is surrounded by a field which, up to infinity, has the same density. The gradient actually means that the field contains fluids of different density. If this difference exists up to infinity, this entails also a torque which is infinitely large.

In order to circumvent this difficulty and to obtain a realistic situation, the case of a circular aquifer of finite dimensions and surrounded by an impermeable boundary will be considered (see Fig. 4). The radius is r_3, the thickness f, and the porosity n.

For the analysis to remain simple the density will be taken to vary linearly over the field according to

$$\rho = \rho_0 + \xi \rho_0' \tag{41}$$

where ρ_0' is a constant.

For this situation, $(\partial \rho / \partial x)$ is constant over the field and equal to

$$\frac{\partial \rho}{\partial x} = \frac{\partial \rho}{\partial \xi} \cdot \frac{\partial \xi}{\partial x} = \rho_0' \cos \alpha \tag{42}$$

From (33), the vorticity is

$$\omega = -\tfrac{1}{2}(k/\mu)[\partial(\rho g)/\partial x], \qquad 0 < r < r_3$$

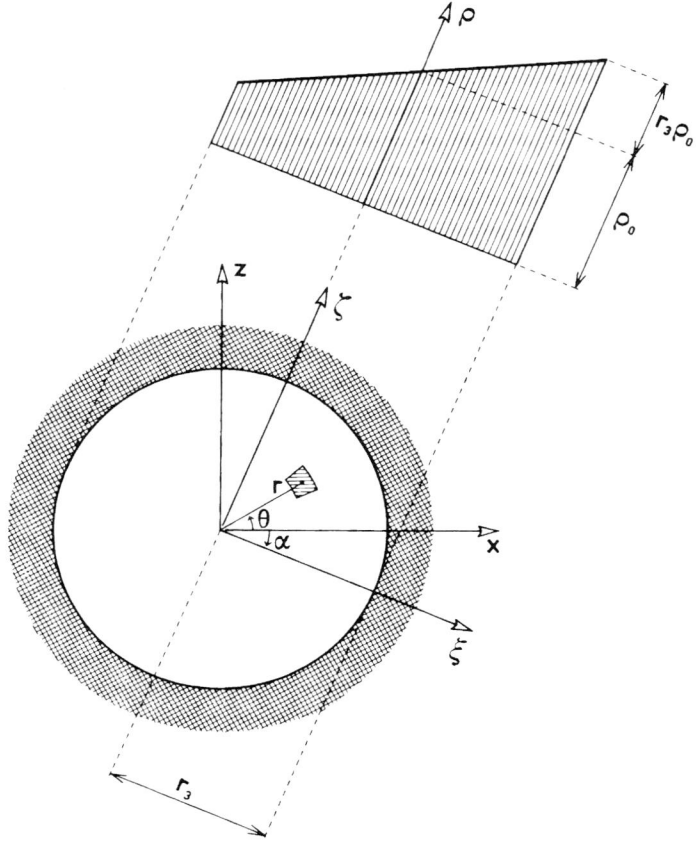

FIG. 4.

which is also constant. As explained above, an integration of the influence according to (32) of all the vortices in the entire region gives the solution similar to (30)

$$q_s = \omega r = -\tfrac{1}{2}(k/\mu)[\partial(\rho g)/\partial x]\, r \tag{43}$$

Since this gives directly a discharge perpendicular to the radius, this solution satisfies the boundary conditions which have actually been chosen in this example in such a way that the discharge obeys the simple formula (43). Therefore, the second part of the solution is zero.

Since in Fig. 3, q_s is positive for counter clockwise rotation, (43) gives a clockwise rotation for positive ρ_0'. In order to show that the clockwise torque (7) is apt to create this circulating flow, the torque will be computed here by considering the required forces.

The force necessary to drive fluid through the ring between r and $r + dr$ is

$$(\mu/k)\, q_s n 2\pi r \, dr \, f$$

Hence the total torque clockwise in the circular region is

$$T = \int_0^{r_3} r \left(\frac{\mu}{k}\right) \frac{1}{2}\left(\frac{k}{\mu}\right) \frac{\partial(\rho g)}{\partial x}\, rn2\pi r f \, dr$$

$$= \tfrac{1}{4}\pi r_3{}^4 \left[\frac{\partial(\rho g)}{\partial x}\right] nf \tag{44}$$

A comparison with (7) shows that if $r = r_3$, the two torques are equal.

From this analysis it follows that the rotational flow within the circular area is possible because a torque created by density variations acts on the fluid of this region. This is comparable with the general concept that forces of the boundaries cannot create rotations in the interior of the fluid. Vorticity can only occur in those places where body forces act in the form of torques on the fluid.

The geometry of Fig. 4 gives a flow pattern, which by its simplicity permits us to predict the rotation of the fluid at all subsequent times. The discharge is such that the fluid rotates as a rigid body. Therefore, regardless of dispersion and diffusion, the fluid conserves its density distribution (41) if the coordinate ξ is allowed to rotate clockwise with the fluid. This means that the angular velocity is equal to

$$\frac{d\alpha}{dt} = \frac{-q_s}{nr} = \frac{1}{2}\left(\frac{k}{\mu}\right)\left[\frac{\rho_0' g \cos \alpha}{n}\right]$$

Integration gives for $\alpha = 0$, $t = 0$,

$$\tan(\tfrac{1}{4}\pi + \tfrac{1}{2}\alpha) = \exp[\tfrac{1}{2}(k/\mu)(\rho_0' g/n)\, t] \tag{45}$$

7. Multiple Fluid Potential for Variable μ, ρ, and k

Equation (46) is not correct. Since (H) is defined by (18) to be

$$(H) = \frac{k}{\mu}p + \frac{k}{\mu}\rho g z \tag{46}$$

we find with (21)

$$\vec{q} = -\left(\frac{k}{\mu}\right) \operatorname{grad}\left(\frac{\mu}{k}(H)\right) + \left(\frac{kz}{\mu}\right) \operatorname{grad}\,(\rho g), \tag{47}$$

therefore the continuity equation should be

$$\text{div } \vec{q} = 0 = -\text{div}\left[\left(\frac{k}{\mu}\right)\text{grad}\left(\frac{\mu}{k}\Theta\right)\right] + \text{div}\left[\left(\frac{kz}{\mu}\right)\text{grad }(\rho g)\right]. \qquad (48)$$

It is impossible to build a Poisson equation in Θ from this expression and therefore also it is impossible to construct an integral expression for Θ of the form (48).

which represents the rotational part of the solution.

For Ψ and p, the equations are

$$\nabla^2\Psi = -\frac{k}{\mu}\frac{\partial(\rho g)}{\partial x} - \frac{k}{\mu}\text{grad}\left(\frac{\mu}{k}\right)\cdot\text{grad }\Psi \qquad (49)$$

$$\nabla^2 p = -\frac{\partial(\rho g)}{\partial z} - \text{grad}\left(\frac{\mu}{k}\right)\cdot[\text{grad }p + \rho g\text{ grad }z] \qquad (50)$$

Both these expressions contain Ψ and p, respectively, on the right sides. Therefore, the solution cannot be evaluated with the integral expression because the integrand still contains unknown terms.

A possible way to obtain a solution is by electric analogy. [See Malavard (1956).] Although this manner of solving a multiple fluid problem may only be of academic value, it will be given here for the sake of completeness.

8. Electric Analogy for Variable μ, ρ, and k

8.1. POTENTIAL LINES

If the objective is to determine the pressure in the fluid, the electric potential can be identified with the multiple fluid potential multiplied by (μ/k). This slightly revised function will be called Θ^*.

The analogy between fluid and electric quantities is

$$\Theta^* = (p + \rho g z) \leftrightarrow E \qquad (51)$$

$$(\mu/k)\,\mathbf{q} - z\,\text{grad}(\rho g) \leftrightarrow \rho_e \mathbf{i}_e \qquad (52)$$

$$(\mu/k) \leftrightarrow \rho_e \qquad (53)$$

in which the electric quantities are: E, the electric potential, ρ_e, the specific electric resistance, and $\mathbf{i}_e$, the specific electric current vector.

Because p, z, μ, ρ, and k are single-valued functions, Θ^* is single-valued and therefore, the single-valuedness of E is assured. In the electrical model the current will flow according to Ohm's law, which can be written as

$$\rho_e \mathbf{i}_e = -\operatorname{grad} E \tag{54}$$

which holds also if ρ_e is variable in the field. For the fluids this means that the following relation is supposed to hold

$$(\mu/k)\,\mathbf{q} - z\operatorname{grad}(\rho g) = -\operatorname{grad} p - z\operatorname{grad}(\rho g) - (\rho g)\operatorname{grad} z \tag{55}$$

A comparison with (3) shows that this is indeed required.

Application of the continuity equation for the fluids can be effected after dividing (52) by (μ/k) at the left side and by ρ_e at the right side. Taking the divergence then gives

$$\operatorname{div} \mathbf{q} - \operatorname{div}(kz/\mu)\operatorname{grad}(\rho g) \leftrightarrow \operatorname{div} \mathbf{i}_e \tag{56}$$

Requirement (1) indicates that $\operatorname{div} \mathbf{i}_e$ must be made equal to the electric analogy of $-\operatorname{div}(kz/\mu)\operatorname{grad}(\rho g)$. Since $\operatorname{div} \mathbf{i}_e$ is the amount of electric current that has to be injected into the electric flow field, it means that in those places where ρ or k or μ and ρ vary, current has to be fed into the model.

In an electric model, only electric potential can be observed or for that matter, equal potential lines may be drawn. From these lines the values of the pressure can be found by means of (51).

8.2. STREAMLINES

If the objective is to determine the streamlines in the fluid, the electric model must be used in an inverse way such that electric potential lines are related to the streamlines of the fluid. The inversion of streamlines and potential lines is obtained by inverting the boundary conditions. But besides that, in the interior, the inversion also has its consequences because the analogy of the resistance must be made reciprocal.

The analogy between fluid and electric properties will now be taken as

$$q_x \leftrightarrow -(\partial E/\partial z), \qquad q_z \leftrightarrow +(\partial E/\partial x) \tag{57}$$

$$\mu/k \leftrightarrow 1/\rho_e \tag{58}$$

Continuity of the fluids requires that

$$-(\partial^2 E/\partial z\, \partial x) + (\partial^2 E/\partial x\, \partial z) = 0$$

which indicates that E is single-valued.

Ohm's law applied to the electric system gives with (57), (58), and (13) the following relation between the specific electric current and the specific discharge of the fluids

$$i_{e_x} = -\frac{1}{\rho_e}\frac{\partial E}{\partial x} \leftrightarrow -\frac{\mu}{k}q_z = +\frac{\partial p}{\partial z} + \rho g$$

$$i_{e_z} = -\frac{1}{\rho_e}\frac{\partial E}{\partial z} \leftrightarrow +\frac{\mu}{k}q_x = -\frac{\partial p}{\partial x} \tag{59}$$

Taking the divergence of i_e and using the single-valuedness of p then gives

$$\text{div } i_e \leftrightarrow [\partial(\rho g)/\partial x] \tag{60}$$

From (60) it follows that also in this case an injection in the electric flow field is necessary with a magnitude of the electric current equivalent to $[\partial(\rho g)/\partial x]$. Because of the inversion, injection in this case has the effect of vorticity.

From (57) it follows directly that lines of equal electric potential will trace streamlines.

9. Variable Density Flow in Three Dimensions

For three dimensions, Poisson equations similar to the two dimensional case can be derived. For the pressure p, this was done by Knudsen (1962). Writing the equilibrium equation (3) in vector notation gives

$$(\mu/k)\, \mathbf{q} = -\nabla p - \rho g \mathbf{k} \tag{61}$$

where the vector $\mathbf{k}$ indicates the unit vector in the z direction, vertically upward. Continuity applied to this expression if μ and k are constant, leads to the following result in view of (1),

$$(\mu/k)\, \nabla \cdot \mathbf{q} = -\nabla^2 p - \nabla\rho \cdot g\mathbf{k} = 0$$

or

$$\nabla^2 p = -g(\partial\rho/\partial z) \tag{62}$$

For Θ, the Poisson equation is already stated in general coordinates by (23) giving

$$\nabla^2\Theta = +(k/\mu)\,\nabla \cdot [z\,\nabla\rho] \tag{63}$$

For the streamfunction it is more complicated since the transition to three dimensions requires the introduction of two streamfunctions. How this can be done was described by Yih (1957), who showed that the discharge can be written as

$$\mathbf{q} = (\nabla\Psi) \times (\nabla\chi) \tag{64}$$

Since the establishment of the appropriate Poisson equations for Ψ and χ was too difficult to achieve, it will be shown here that the discharge itself obeys a Poisson's equation which permits a direct computation with the solution by singularities.

Application of the curl operation to (61) gives, if μ and k are constant,

$$(\mu/k)\,\nabla \times \mathbf{q} = -\nabla \times \nabla p - \nabla \times \rho g\mathbf{k} \tag{65}$$

Expansion of the first term to the right side gives

$$\begin{aligned}
\nabla \times \nabla p = {}&\mathbf{i}[(\partial^2 p/\partial y\,\partial z) - (\partial^2 p/\partial z\,\partial y)]\\
&+ \mathbf{j}[(\partial^2 p/\partial z\,\partial x) - (\partial^2 p/\partial x\,\partial z)]\\
&+ \mathbf{k}[(\partial^2 p/\partial x\,\partial y) - (\partial^2 p/\partial y\,\partial x)]
\end{aligned}$$

For physical reasons p is single-valued, hence by use of (10), all three vectors are zero and the term vanishes. Thus, (65) becomes

$$(\mu/k)\,\nabla \times \mathbf{q} = -\nabla \times \rho g\mathbf{k} = g[\mathbf{i}(\partial\rho/\partial y) - \mathbf{j}(\partial\rho/\partial x)] \tag{66}$$

Since $\nabla \times \mathbf{q}$ represents a vortex, this expression shows that the axis of the vortex lies in the horizontal plane and has the direction of lines for ρ constant in that plane. Further, since $\mathbf{i}(\partial\rho/\partial x) + \mathbf{j}(\partial\rho/\partial y)$ is the horizontal component of the density gradient, the expression (66) shows that the vortex strength is proportional to this horizontal component of grad ρ.

The curl operation on (66) gives

$$(\mu/k)\,\nabla \times (\nabla \times \mathbf{q}) = -\nabla \times (\nabla \times \rho g\mathbf{k})$$

The first term can be written as

$$\nabla \times (\nabla \times \mathbf{q}) = \nabla(\nabla \cdot \mathbf{q}) - \nabla^2\mathbf{q}$$

Continuity requires that $\nabla \cdot \mathbf{q}$ be zero so there remains

$$\nabla^2\mathbf{q} = (k/\mu)\,\nabla \times (\nabla \times \rho g\mathbf{k}) \tag{67}$$

This is Poisson's equation for **q**. Solutions of this equation can be given in the form of the integral

$$\mathbf{q}_I = +(k/4\pi\mu) \int_V \nabla \times (\nabla \times \rho g\mathbf{k})\, r^{-1}\, d\forall \tag{68}$$

This integral describes the part of the discharge vector that is created by the vortices caused by density variations. A second part, $\mathbf{q}_{II}$, that is irrotational has to be added to satisfy the boundary conditions. The integrand of (68) contains the density variations because the development of the curls gives

$$\nabla \times (\nabla \times \rho g\mathbf{k}) = +(\partial^2\rho/\partial x\, \partial z)\,\mathbf{i} + (\partial^2\rho/\partial y\, \partial z)\,\mathbf{j}$$
$$\mp [(\partial^2\rho/\partial x^2) \mp (\partial^2\rho/\partial y^2)]\,\mathbf{k}$$

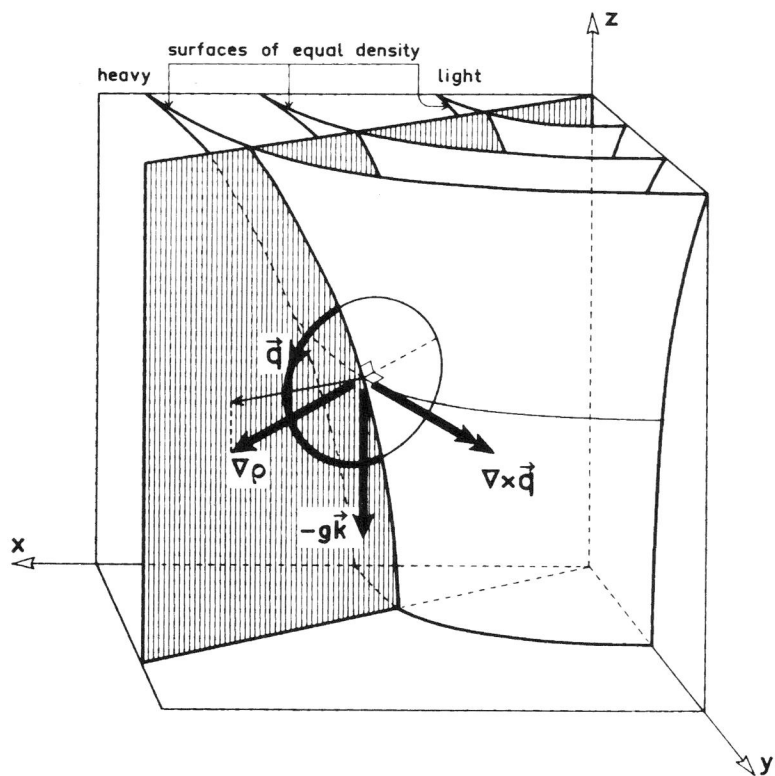

FIG. 5.

The behavior of the discharge is represented in Fig. 5 where a three-dimensional picture is given of surfaces of equal density. In the region where the density varies continuously, many of these sheets can be visualized as lying perpendicular to the density gradient. According to (66), the discharge vortex created by the density gradient has its axis in the direction of horizontal lines on these sheets.

The vortex axis, which is horizontal, is perpendicular to the vertical direction of gravity; it lies in the isopicnic (of equal density) sheet and is perpendicular to the gradient of density. Therefore, the vortex axis is perpendicular to the plane through the vertical and the density gradient (shaded area in Fig. 5).

The discharge from the vortex rotates in a plane perpendicular to the vortex axis. Hence, finally, the discharge vortex in a point is rotating in the plane through the vertical and the density gradient in that point. The strength of the vortex is proportional to the horizontal component of the density gradient.

REFERENCES

de Josselin de Jong, G. (1960). Singularity distributions for the analysis of multiple fluid flow through porous media, *J. Geophys. Res.* **65**, 3739–3758.

De Wiest, R. J. M. (1964). Dispersion and Salt water Intrusion, *Groundwater, J. of NWWA* **2**, 39–40.

De Wiest, R. J. M. (1965). "Geohydrology," pp. 305–313. Wiley, New York.

De Wiest, R. J. M. (1969). Green's functions in the flow through porous media, this volume, Chapter X.

Knudsen, W. C. (1962). Equations of fluid flow through porous media—incompressible fluids of varying density, *J. Geophys. Res.* **67**, 733–737.

Lusczynski, N. J. (1961). Head and Flow of Groundwater of variable Density, *J. Geophys. Res.* **66**, 4247–4256.

Malavard, L. C. (1956). The use of rheoelectrical analogies in aerodynamics, *AGARDograph* 18, NATO Advisory Group of Aeronautical Res. and Devel., Paris.

Yih, C. S. (1957). Fonctions de courant dans les écoulements à trois dimensions, *La Houille Blanche* **12**, 439-444.

Yih, C. S. (1961). Flow of a nonhomogeneous fluid in a porous medium, *J. Fluid Mech.* **10**, pp. 133–141.

Vortex Theory for Multiple Fluid in Three Dimensions

G. de Josselin de Jong

Geotechnical Laboratory
Department of Civil Engineering
Delft University of Technology
Keverling Buismanweg 1
2628 CL Delft, The Netherlands

Delft Progr. Rep., **4** (1979) pp. 87-102

Received: 1 IV 79

The specific discharge in a pore fluid with specific weight, γ, that is variable in space, is a rotational flow possessing vorticity in those regions, where γ varies in horizontal direction. The vortices have horizontal axes parallel to lines of constant γ. They create a circulating flow in the aquifer around them. On a sharp interface the contourlines of equal height form vortex lines, that enclose reentrant vortex ribbons of constant strength. Formulas are given for the specific discharge in an arbitrary point of the aquifer, created by the vortices in a triangular part of an interface. These relations are suitable for determining the displacement of an interface in time. Finally the specific discharge in a point of the interface, as created by the vortices in a small circular region of the interface around that point, is demonstrated to consist of a shear flow only, similar to the shear flow occurring in the two dimensional case.

Introduction

The vortex theory, for the simultaneous flow of fluids with different specific weight through porous media, was reviewed in this journal (ref. 3). for the two dimensional case. In an adjoining article *Haitjema* (Ref. 1) published the possibilities offered by a computer program based on the two dimensional vortex theory. The extension towards three dimensions of this work turned out to be involved and to require more than trivial elaborations. Therefore it was considered appropriate to present the three dimensional vortex theory here in some detail, starting from the basic equations, allthough the basis of the theory was already published previously (Ref. 2).

For collateral reading, chapter 5 of the book "Hydrodynamics" by *Lamb* is recommended.

1. Basic equations for pressure and stream functions

Darcy's law and continuity

The fluid flow through the pores of an aquifer is governed by Darcy's law and the requirement of continuity. *Darcy's law* can be obtained by considering equilibrium of forces acting on the pore fluid. The specific weight of the fluid, γ, and the gradient of the pressure, p, produce forces, that drive the fluid through the pores. These driving forces equalize the resistance force of magnitude $(\mu/\kappa)\vec{q}$, ecountered by the fluid when a specific discharge $\vec{q}$ is flowing through the pores.

Expressed in terms of the components q_x, q_y, q_z, of the vector $\vec{q}$ the pertinent relations are (see f.i. *Verruijt*, 1970)

$$\left.\begin{aligned}
q_x &= - (\kappa/\mu)(\partial p/\partial x) \\
q_y &= - (\kappa/\mu)(\partial p/\partial y) \\
q_z &= - (\kappa/\mu)\{(\partial p/\partial z) + \gamma\}
\end{aligned}\right\} \tag{1.1}$$

where κ is the intrinsic permeability of the aquifer and μ is dynamic viscosity of the fluid. The x, y coordinates are in the horizontal plane and z is vertical upwards. In the following treatment, κ and μ will be considered to be constants throughout the aquifer, but γ is variable in space.

The requirement of *continuity* is expressed by

$$(\partial q_x/\partial x) + (\partial q_y/\partial y) + (\partial q_z/\partial z) = 0 \tag{1.2}$$

Poisson equation for pressure

By introducing Darcy's law (1.1) into (1.2), the specific discharge components can be eliminated and there results, if (κ/μ) is a constant,

$$\nabla^2 p = - (\partial\gamma/\partial z) \tag{1.3}$$

This relation is due to *Knudsen* (1962) and shows that the fluid pressure p obeys a Poisson type equation. Such an equation describes the distribution of the potential created by sources of strength, $(\partial\gamma/\partial z)$. Every volume dV, where γ varies with height, contributes to the distribution of p. This contribution has in an infinite medium the magnitude $(1/4\pi R)(\partial\gamma/\partial z)dV$, where R is the distance between the volume dV and the point, where p is considered. The total result is an integral over the entire region, where γ is variable,

$$p = (1/4\pi) \iiint (\partial\gamma/\partial z) \ R^{-1} \ dV \tag{1.4}$$

Since this value of p occurs in an infinite medium, adjustments p_1 necessary to comply with boundary conditions have to be added. These adjustments obey the usual Laplace equation

$$\nabla^2 p_1 = 0 \tag{1.5}$$

The solution (1.4) may be of interest in problems, where the emphasis is on pressure. The specific discharge vector can then be obtained from the gradient of the pressure by use of (1.1). In the case, where the displacements of saline pore water or other contamminants are of principal importance, it is convenient to ~~have the~~ ~~disposal~~ of a method to determine the specific discharge directly. This is possible, since the solution for discharge can be written in the form of integrals similar to (1.4). These integrals are obtained by considering stream functions that obey Poisson equations of the type (1.3).

Stream functions

In the two dimensional case only one stream function is sufficient for describing the flow behaviour. In the three dimensional case it is necessary to introduce three stream functions, called F, G, H here. These stream functions generate the specific discharge components as follows

$$\left. \begin{aligned} q_x &= (\partial H/\partial y) - (\partial G/\partial z) \\ q_y &= (\partial F/\partial z) - (\partial H/\partial x) \\ q_z &= (\partial G/\partial x) - (\partial F/\partial y) \end{aligned} \right\} \tag{1.6}$$

The partial derivatives of F, G, H not mentioned in (1.6) are still free and are assumed to satisfy the additional requirement

$$(\partial F/\partial x) + (\partial G/\partial y) + (\partial H/\partial z) = 0 \tag{1.7}$$

The functions F, G, H are mathematically acceptable, only if they are single valued in the entire region. Then their mixed derivatives are commutative. In particular we have

$$\left. \begin{aligned} \partial^2 F/\partial y \partial z &= \partial^2 F/\partial z \partial y \\ \partial^2 G/\partial z \partial x &= \partial^2 G/\partial x \partial z \\ \partial^2 H/\partial x \partial y &= \partial^2 H/\partial y \partial x \end{aligned} \right\} \tag{1.8}$$

The equalities (1.8) are sufficient to ensure continuity, as can be verified by introducing (1.6) into (1.2). For the equivalence of singlevaluedness and the commutative property of partial derivatives see f.i. the appendix of the previous article (*de Josselin de Jong*, 1977).

Vorticity vector

At this point it is appropriate to introduce the vorticity vector $\vec{\Omega}$ with components Ω_x, Ω_y, Ω_z in x, y, z directions. These components are related to the specific discharge components q_x, q_y, q_z by

$$\left. \begin{aligned} 2\Omega_x &= (\partial q_z/\partial y) - (\partial q_y/\partial z) \\ 2\Omega_y &= (\partial q_x/\partial z) - (\partial q_z/\partial x) \\ 2\Omega_z &= (\partial q_y/\partial x) - (\partial q_x/\partial y) \end{aligned} \right\} (1.10)$$

(In *Lamb's* notation: $2\Omega_x = \xi$; $2\Omega_y = \eta$; $2\Omega_z = \zeta$.)
The components Ω_x, Ω_y, Ω_z correspond to a rotation around the respective coordinate axis in the direction of a right handed screw. The components are drawn as double pointed arrows in fig. 1 and their mode of rotation is shown by the circular arrows.

The components of vorticity can be expressed in terms of the stream functions by eliminating q_x q_y q_z from (1.10) by use of (1.6). This gives

$$\left. \begin{aligned} 2\Omega_x &= (\partial^2 G/\partial x\partial y) - (\partial^2 F/\partial y^2) - (\partial^2 F/\partial z^2) + (\partial^2 H/\partial x\partial z) \\ 2\Omega_y &= (\partial^2 H/\partial y\partial z) - (\partial^2 G/\partial z^2) - (\partial^2 G/\partial x^2) + (\partial^2 F/\partial y\partial x) \\ 2\Omega_z &= (\partial^2 F/\partial z\partial x) - (\partial^2 H/\partial x^2) - (\partial^2 H/\partial y^2) + (\partial^2 G/\partial z\partial y) \end{aligned} \right\} (1.11)$$

These expressions can be simplified by use of the additional requirement (1.7). By taking the x-derivative of (1.7) it follows that

$$(\partial^2 G/\partial y\partial x) + (\partial^2 H/\partial z\partial x) = - (\partial^2 F/\partial x^2) \tag{1.12}$$

Since the stream functions F, G, H are single valued everywhere in the aquifer, their mixed derivatives are commutative. So (1.12) can be used in the first of (1.11) to elinminate G and H. By similar operations on the other two of (1.11) there results

$$\left. \begin{aligned} 2\Omega_x &= - \nabla^2 F \\ 2\Omega_y &= - \nabla^2 G \\ 2\Omega_z &= - \nabla^2 H \end{aligned} \right\} (1.13)$$

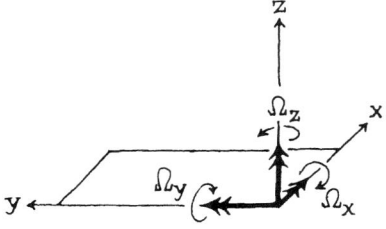

Fig. 1. Components of vorticity and their mode of rotation.

Poisson equation for the stream functions

The equations (1.13) have the character of Poisson equations. Their solution can be given in the form of integrals if the vorticity components Ω_x, Ω_y, Ω_z have a known magnitude. Actually this is the case, because their value can be obtained by eliminating q_x, q_y, q_z from (1.10) by use of Darcy's law (1.1). Since the pressure p is a physical quantity, that has only one value in a point, p is single valued and its mixed derivatives are commutative. So we have

$$
\left.
\begin{aligned}
\partial^2 p/\partial x \partial y &= \partial^2 p/\partial y \partial x \\
\partial^2 p/\partial y \partial z &= \partial^2 p/\partial z \partial y \\
\partial^2 p/\partial z \partial x &= \partial^2 p/\partial x \partial z
\end{aligned}
\right\} \quad (1.14)
$$

Therefore introducing (1.1) in (1.10) also results in eliminating p and there remains

$$
\left.
\begin{aligned}
2\Omega_x &= - \ (\kappa/\mu)(\partial\gamma/\partial y) \\
2\Omega_y &= + \ (\kappa/\mu)(\partial\gamma/\partial x) \\
2\Omega_z &= 0
\end{aligned}
\right\} \quad (1.15)
$$

Therefore the stream functions F, G, H obey Poisson type equations of the form

$$
\left.
\begin{aligned}
\nabla^2 F &= - \ 2\Omega_x = + \ (\kappa/\mu)(\partial\gamma/\partial y) \\
\nabla^2 G &= - \ 2\Omega_y = - \ (\kappa/\mu)(\partial\gamma/\partial x) \\
\nabla^2 H &= - \ 2\Omega_z = 0
\end{aligned}
\right\} \quad (1.16)
$$

Since the right hand sides are known quantities, the solution of F, G, H can be written in the form of integrals similar to the expression (1.4) for p. These solutions are

$$
\left.
\begin{aligned}
F &= (1/4\pi) \iiint 2\Omega_x, \ R^{-1} \ dV' = - \ (\kappa/4\pi\mu) \iiint (\partial\gamma/\partial y') \ R^{-1} \ dV' \\
G &= (1/4\pi) \iiint 2\Omega_y, \ R^{-1} \ dV' = + \ (\kappa/4\pi\mu) \iiint (\partial\gamma/\partial x') \ R^{-1} \ dV' \\
H &= (1/4\pi) \iiint 2\Omega_z, \ R^{-1} \ dV' = 0
\end{aligned}
\right\} \quad (1.17)
$$

2. Physical interpretation

Local coordinate system

According to (1.15) the specific discharge has vorticity (and therefore rotates) in those points of the aquifer, where the specific weight γ varies in x and/or y direction. Everywhere else, where γ is a constant, the flow is irrotational. In order to specify the character of the motion it is appropriate to introduce a local orthogonal, right handed coordinate system n, s, t,

Fig. 2. Local n, s, t coordinate
system oriented normal and
tangent to the γ-constant plane.

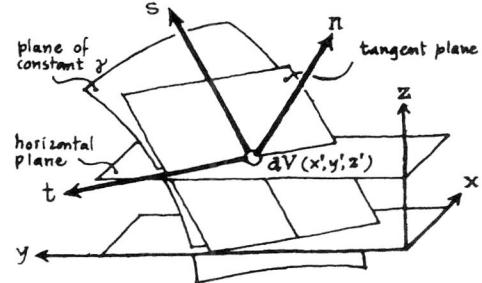

tangent to the plane of constant γ fig. 2. The origin of the n, s, t
coordinates is in the centre of an infinitesimal volume dV containing γ-
variation and therefore vorticity.

Let n be normal to the plane of constant γ, pointing upwards. Then n has
the direction of gradient γ. Since however the lighter fluid will be above the
heavier, the positive direction of n corresponds to γ decrease and therefore
with $-\nabla\gamma$. The coordinates s, t are orthogonal to n and ly both in the γ-
constant plane. We then have

$$\partial\gamma/\partial n = - \nabla\gamma ; \qquad \partial\gamma/\partial s = 0 ; \qquad \partial\gamma/\partial t = 0. \tag{2.1}$$

The coordinates s, t are chosen in such a manner, that t is horizontal and
s is in the direction of steepest ascent. The angle between s and its
projection on the horizontal plane is α, while the angle between this
projection and the x-axis is β, see fig. 3. With x', y', z' the coordinates
of the centre of dV, the transformation relations between the coordinates
x, y, z and n, s, t are

$$
\left.
\begin{aligned}
(x - x') &= - n \sin\alpha \cos\beta + s \cos\alpha \cos\beta - t \sin\beta \\
(y - y') &= - n \sin\alpha \sin\beta + s \cos\alpha \sin\beta + t \cos\beta \\
(z - z') &= + n \cos\alpha + s \sin\alpha
\end{aligned}
\right\} \tag{2.2}
$$

$$
\left.
\begin{aligned}
n &= - (x - x') \sin\alpha \cos\beta - (y - y') \sin\alpha \sin\beta + (z - z') \cos\alpha \\
s &= + (x - x') \cos\alpha \cos\beta + (y - y') \cos\alpha \sin\beta + (z - z') \sin\alpha \\
t &= - (x - x') \sin\beta + (y - y') \cos\beta
\end{aligned}
\right\} \tag{2.3}
$$

Direction of vorticity

Using the chain rule

$$(\partial\gamma/\partial x) = (\partial\gamma/\partial n)(\partial n/\partial x) + (\partial\gamma/\partial s)(\partial s/\partial x) + (\partial\gamma/\partial t)(\partial t/\partial x) \tag{2.4}$$

and similarly for y, it is found by use of (2.1)(2.3) that

$$
\left.
\begin{aligned}
(\partial\gamma/\partial x) &= + \nabla\gamma \sin\alpha \cos\beta \\
(\partial\gamma/\partial y) &= + \nabla\gamma \sin\alpha \sin\beta \\
(\partial\gamma/\partial z) &= - \nabla\gamma \cos\alpha .
\end{aligned}
\right\} \tag{2.5}
$$

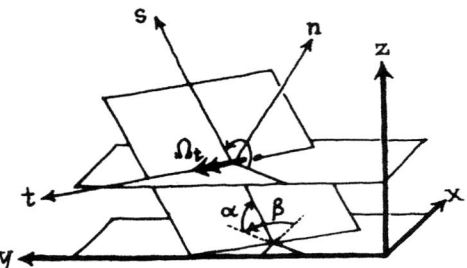

Fig. 3. Orientation of n, s, t coordinates by angles α and β. Vorticity vector parallel to t.

Application of this result to (1.15) gives for the components of the vorticity

$$2\Omega_x = - (\kappa/\mu)\ \nabla\gamma\ \sin \alpha\ \sin \beta$$
$$2\Omega_y = + (\kappa/\mu)\ \nabla\gamma\ \sin \alpha\ \cos \beta$$
$$2\Omega_z = 0 .$$

$\left.\right\}$ (2.6)

The components of vorticity $\Omega_n\ \Omega_s\ \Omega_t$ in the n, s, t directions can be found by application of similar transformation formulas as (2.3) giving

$$\Omega_n = - \Omega_x\ \sin \alpha\ \cos \beta - \Omega_y\ \sin \alpha\ \sin \beta + \Omega_z\ \cos \alpha$$
$$\Omega_s = + \Omega_x\ \cos \alpha\ \cos \beta + \Omega_y\ \cos \alpha\ \sin \beta + \Omega_z\ \sin \alpha$$
$$\Omega_t = - \Omega_x\ \sin \beta + \Omega_y\ \cos \beta$$

$\left.\right\}$ (2.7)

and this results with (2.6) in

$$2\Omega_n = 0\ ;\qquad 2\Omega_s = 0\ ;\qquad 2\Omega_t = (\kappa/\mu)\ \nabla\gamma\ \sin \alpha. \qquad (2.8)$$

This result indicates, that the vorticity in a point created by a specific weight gradient in the pore fluid, has its axis tangent to the intersection line of a horizontal plane and the plane of constant γ. This vorticity is represented by the double pointed arrow in fig. 3. Its mode of rotation is such as to revolve the γ-constant plane towards a horizontal position, a physically plausible action.

Sharp interface

A sharp interface between two fluids, of specific weight γ_1, γ_2 respectively, can be considered to be a zone of thickness h of gradually decreasing γ from γ_1 to γ_2, in the limit that h is reduced to zero. The zone contains sheets of planes for γ-constant, that are locally parallel and in the limit are sqeezed together into one plane.

Consider an infinitesimal area dA of the interface. This area corresponds to a volume hdA of the zone of thickness, h. In this volume the vorticity is according to (2.8) and the total strength of the vortex acting is

$$2\Omega_t\ dV = (\kappa/\mu)\ \nabla\gamma\ \sin \alpha\ hdA \qquad (2.9)$$

Since γ is assumed to decrease gradually within the zone we have

$$\nabla\gamma = (\gamma_2 - \gamma_1)/h \tag{2.10}$$

and so the vortex acting in an area dA of a sharp interface has the magnitude

$$2\Omega_t \, dV = (\kappa/\mu)(\gamma_2 - \gamma_1) \sin \alpha \, dA \tag{2.11}$$

The direction of the vortex corresponds to the positive t-coordinate, which is locally tangent to the line of intersection between planes of constant γ and the horizontal plane. In this case the planes of constant γ form a stack, that is sqeezed into the sharp interface plane. So the vortex has its axis parallel to the horizontal contourlines of the interface plane.

A sharp interface plane between two fluids of different density actually is a vortex sheet. The horizontal contour lines are the vortex lines on that sheet fig. 4. The ribbon cut out by two adjacent horizontal contour lines forms a reentrant vortex tube. Let the height distance between the contour lines be Δz. Then Δz is an constant for these two contour lines and the width of the ribbon is $\Delta z/\sin \alpha$. So unit length of the ribbon has an area $dA = \Delta z/\sin \alpha$, and therefore the total strength of all vortices per unit length of ribbon is with (2.11)

$$2\Omega_t dV = (\kappa/\mu)(\gamma_2 - \gamma_1) \, \Delta z \tag{2.12}$$

This is a constant and so (2.12) shows that the ribbon cut out from a sharp interface by two horizontal contour lines with constant height difference, is a reentrant vortex tube with constant strength.

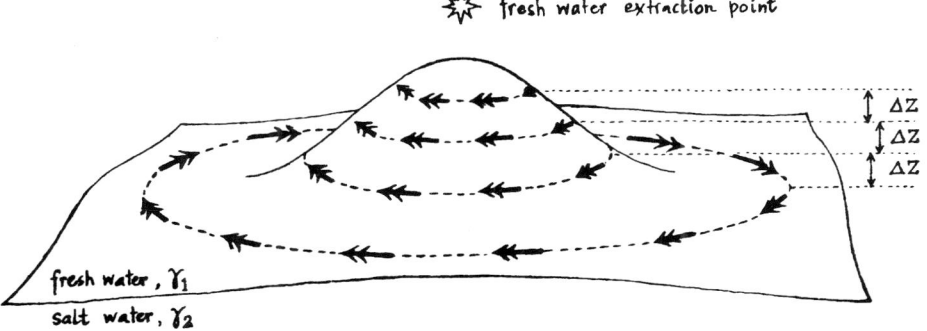

Fig. 4. The contourlines on a sharp interface are vortex lines. The ribbons between contourlines are vortex tubes with constant strength.

3. Circulation of the pore fluid

Specific discharge created by isolated vortex

The vortices in the regions, where γ varies, produce a rotation of the fluid in those regions. Besides that, each vortex induces a motion of the fluid around its region of action up to infinity. This motion consists of a circulation of the fluid around the axis of the vortex. The character of this motion can be shown by deducing the specific discharge created in a point x, y, z outside the region of vorticity from the solutions (1.17) for the stream functions.

The solutions (1.17) consist of integrals over the entire region of the aquifer, where γ variations occur. Consider the contributions dF, dG, dH to the stream functions F, G, H produced by the infinitesimal volume dV', only. These contributions are, taking account of (1.15) which specifies that $\Omega_{z'} = 0$,

$$
\left.
\begin{aligned}
dF &= (1/4\pi)\ 2\Omega_{x'}\ R^{-1}\ dV' \\
dG &= (1/4\pi)\ 2\Omega_{y'}\ R^{-1}\ dV' \\
dH &= 0 ,
\end{aligned}
\right\} \quad (3.1)
$$

with

$$
R = \{(x - x')^2 + (y - y')^2 + (z - z')^2\}^{\frac{1}{2}}. \qquad (3.2)
$$

The quantity R is the distance between the point x', y', z', where the vortex $\vec{\Omega}$ is located, and the point x, y, z, where the specific discharge is to be computed.

Introducing (3.1) into the relations (1.6), the contributions dq_x, dq_y, dq_z to the specific discharge components q_x, q_y, q_z in the point x, y, z are found to be

$$
\left.
\begin{aligned}
dq_x &= (1/4\pi)\ \{+ 2\Omega_{y'}\ (z - z')\}\ R^{-3}\ dV' \\
dq_y &= (1/4\pi)\ \{- 2\Omega_{x'}\ (z - z')\}\ R^{-3}\ dV' \\
dq_z &= (1/4\pi)\ \{- 2\Omega_{y'}\ (x - x') + 2\Omega_{x'}\ (y - y')\}\ R^{-3}\ dV'.
\end{aligned}
\right\} \quad (3.3)
$$

These relations are simplified by transforming them into n, s, t directions. By substituting the relation (2.8) for Ω_t into the equations (2.6) it is found that

$$
\left.
\begin{aligned}
2\Omega_{x'} &= - 2\Omega_t \sin \beta \\
2\Omega_{y'} &= + 2\Omega_t \cos \beta
\end{aligned}
\right\} \quad (3.4)
$$

Further the terms $(z - z')$ etc. can be transformed by use of (2.2).

So (3.3) becomes

$$
\left.\begin{aligned}
dq_x &= (dV'/4\pi R^3)\ 2\Omega_t\ (n \cos \alpha \cos \beta + s \sin \alpha \cos \beta] \\
dq_y &= (dV'/4\pi R^3)\ 2\Omega_t\ (n \cos \alpha \sin \beta + s \sin \alpha \sin \beta] \\
dq_z &= (dV'/4\pi R^3)\ 2\Omega_t\ (n \sin \alpha - s \cos \alpha).
\end{aligned}\right\} \quad (3.5)
$$

From these the components of $\vec{q}$ in n, s, t directions are found by using relations derived from (2.3)

$$
\left.\begin{aligned}
dq_n &= - dq_x \sin \alpha \cos \beta - dq_y \sin \alpha \sin \beta + dq_z \cos \alpha \\
dq_s &= + dq_x \cos \alpha \cos \beta + dq_y \cos \alpha \sin \beta + dq_z \sin \alpha \\
dq_t &= - dq_x \sin \beta + dq_y \cos \beta
\end{aligned}\right\} \quad (3.6)
$$

which give with (3.5)

$$
\left.\begin{aligned}
dq_n &= (dV'/4\pi R^3)\ (-2\Omega_t \cdot s) \\
dq_s &= (dV'/4\pi R^3)\ (+2\Omega_t \cdot n) \\
dq_t &= 0
\end{aligned}\right\} \quad (3.7)
$$

Form of flow pattern

The result (3.7) represents the specific discharge created around a vortex of strength $2\Omega_t\ dV'$ with its axis in t-direction. The specific discharge vector $\vec{dq}$ lies in a plane perpendicular to the axis of the vortex, because dq_t is zero see fig. 5. Further, the vector $\vec{dq}$ is perpendicular to the line, that projects the point of consideration x, y, z on the t-axis. So the flow is a circulating motion around the t-axis as executed by points on the rim of a wheel. Actually all points on a sphere around the vortex have specific discharges, that have their direction and magnitude coinciding with the motions of points on the sphere, when it rotates as a rigid body around the axis of the vortex. This applies to all concentric spheres revolving with angular velocities that decrease proportional to R^{-3}.

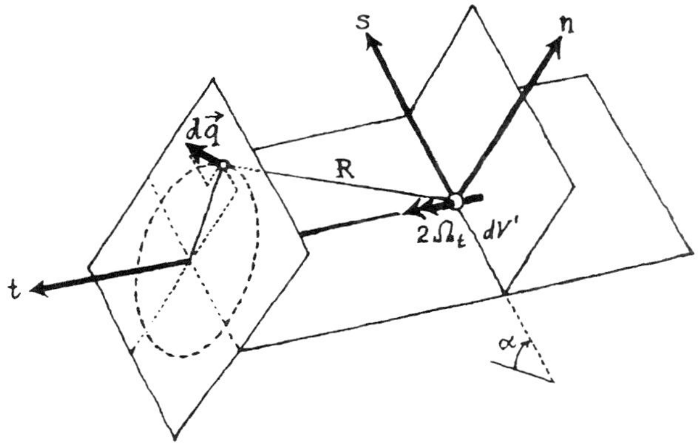

Fig. 5. An isolated vortex creates a circulating flow perpendicular to its axis.

Completion of the solution

The result (3.7) is derived from the relations (3.1), that are solutions of the relevant Poisson equations (1.16), which for the case of an isolated vortex $2\vec{\Omega}$ dV' have the form

$$\nabla^2 (dF) = \begin{cases} -2\Omega_x, & dV' \quad \text{within } dV' \\ 0 & \text{outside } dV' \end{cases}$$

$$\nabla^2 (dG) = \begin{cases} -2\Omega_y, & dV' \quad \text{within } dV' \\ 0 & \text{outside } dV' \end{cases} \Bigg\} (3.8)$$

$$\nabla^2 (dH) = 0 \qquad \text{everywhere}$$

In order to constitute the solution to the problem also the equation (1.7) has to be satisfied, which in this case has the form

$$\partial (dF)/\partial x + \partial (dG)/\partial y + \partial (dH)/\partial z = 0 \qquad (3.9)$$

It can be shown (see f.i. *Lamb*, pg. 209) that in order to satisfy also (3.9), it is necessary to combine isolated vortices into reentrant vortex tubes, whose surfaces consist of closed vortex lines. In fact this situation is encountered in the case of a sharp interface, that is subdivided into vortex ribbons by horizontal contour lines. So the solutions (3.7) are complete only after integration over an entire interface. It is possible to show, that also in the case of gradually changing γ the entire region of γ variations can be subdivided into reentrant vortex tubes, with constant vortex strength in their cross sections. So also in that case the solution is complete only after integration over the entire region of varying γ.

4. Motion of a sharp interface

Subdivision in triangles

The solution (3.7) with (2.11) is suitable to determine the motion of a sharp interface. Let the interface position be given at a certain instant. Then a surface element dA contains a vortex of known strength and direction. The direction is parallel to the contour line in dA and its strength is proportional to sin α , with α the local inclination angle of the interface. The displacement of the interface during a time step is obtained by determining the specific discharge in points of the interface surface as an integration of the individual contribution of all area's dA.

For an arbitrary shape of the interface, this is executed by choosing a number of nodal points and schematizing the gradually curved interface by a

diamondshaped surface consisting of flat triangles between nodal points.
From the coordinates of the nodal points, the angles α and β are determined that
specify the local coordinate system n, s, t. Since the triangles are flat,
α and β are constants for each triangle. Let g be the vortex strength per unit
area, such that according to (2.11)

$$g = (\kappa/\mu)(\gamma_2 - \gamma_1) \sin \alpha. \tag{4.1}$$

Then also g is a constant for each triangle.

Each triangle contributes a specific discharge $\vec{q}$ with components q_n, q_s, q_t
in a point P, whose coordinates in the local n, s, t coordinates of the triangle
are n_p, s_p, t_p. The origin of the local coordinates is in the centre of
gravity of the triangle. By use of (3.7) and (2.11) the specific discharge
components created by a triangle are

$$\left.\begin{aligned}
q_n &= (g/4\pi) \iint (-s_p + s') \, R^{-3} \, ds'dt' \\
q_s &= (g/4\pi) \iint (+n_p) \, R^{-3} \, ds'dt' \\
q_t &= 0
\end{aligned}\right\} \tag{4.2}$$

where integration is over s' and t' throughout the area of the triangle and

$$R = \{n_p^2 + (s_p - s')^2 + (t_p - t')^2\}^{\frac{1}{2}}, \tag{4.3}$$

the distance between the point of integration (0, s', t') and P, the fixed
point of interest, see fig 6.

The components q_n, q_s, q_t are to be converted subsequently in x, y, z
directions by use of transformation relations with the coefficients of (2.2).
Summation of the contribution of all triangles gives the specific discharge
in every desired point P of the aquifer. Especially the specific discharge in
the nodal points are of interest. Dividing these discharges by the porosity
gives the advancing velocity of the interface.

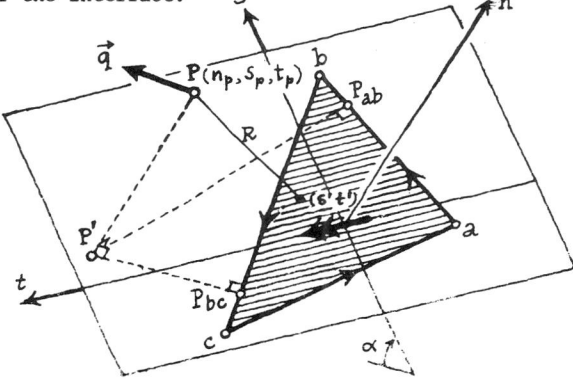

*Fig. 6. Triangle abc on
interface with local n, s, t
coordinates. P is projected
on the s, t plane in point P'.*

When the distance R_o between the centre of gravity of a triangle and the point P is larger than 25 times the smallest triangle side, an error smaller than 0,02% is introduced by concentrating vorticity of the triangle in its centre of gravity. Then the integrals of (4.2) simplify, because $s' = 0$ and $R = R_o$. The result is

$$
\left.
\begin{aligned}
q_n &= (g/4\pi)\ (-s_p)\ R_o^{-3}\ A \\
q_s &= (g/4\pi)\ (+n_p)\ R_o^{-3}\ A \\
q_t &= 0
\end{aligned}
\right\}(4.4)
$$

with $A = \iint ds'\,dt' =$ the area of the triangle. However, if P is closer to the triangle, integration has to be performed with s' and t' variable. This amounts to a cumbersome analysis with the following result.

Result of integration over triangle

Let the cornerpoints a, b, c of the triangle be located such, that viewed from above the triangle is at the left, when traversing the sides from a to b, b to c, c to a, see fig. 6. Let P' be the projection of the point P of consideration on the s, t plane (the plane of the triangle). The distance from P to P' is equal to n_p the coordinate of P normal to the s, t plane; n_p is positive for P above the plane, negative below. The projection of P' on the line ij is P_{ij}.

The specific discharge components q_n q_s created by the triangle abc then are

$$
q_n = \frac{g}{4\pi} \sum \left\{ \frac{(t_i - t_j)}{|ij|}\ \ell n \left(\frac{|iP| + iP_{ij}}{|jP| + jP_{ij}} \right) \right\} \tag{4.5}
$$

$$
q_s = \frac{g}{4\pi} \frac{n_p}{|n_p|} \left[2\pi f + \sum \left\{ -\arctan \left(\frac{|n_p| \cdot iP_{ij}}{|iP| \cdot P'P_{ij}} \right) + \arctan \left(\frac{|n_p| \cdot jP_{ij}}{|jP| \cdot P'P_{ij}} \right) \right\} \right] \tag{4.6}
$$

Summation is over the sides of the triangle, such that ij are subsequently ab, bc, ca. The line pieces between $|\ |$ are always positive. The line iP_{ij} is postive, if the direction of going from P_{ij} to i is in the arrow directions as indicated in fig. 6. In fig. 6 the line aP_{ab} is therefore negative.

The line $P'P_{ij}$ is positve, if P' is located to the left when going from i towards j. In fig. 6, $P'P_{ab}$ is positive, but $P'P_{bc}$ is negative. The factor f equals 1, if all $P'P_{ij}$ are positive. Then P' is located within the triangle abc. The factor f is zero, if one of the lines $P'P_{ij}$ is negative. Then P' is located outside the triangle. The factor f is introduced in order to limit the arc tan $(\ )$ to values between $-\frac{1}{2}\pi$ and $+\frac{1}{2}\pi$.

Singular behaviour in a nodal point

The expression (4.5) becomes infinite, when the point P approaches one of
the sides of the triangle, because either numerator or denominator in the
logarithm becomes zero, when P and P_{ij} conincide. This interferes unacceptably
with the determination of the specific discharge in a nodal point, since all
adjacent triangles then contribute infinite normal components. In order to
establish the influence of these triangles a special procedure has to be
followed, based on a detailed analysis of the specific discharges encountered
in the vicinity of the interface.

At a nodal point several triangles meet, in a diamondshaped interface
schematisation. Therefore the substitution by triangles is not appropriate at
a nodal point, because instead of the top of a pyramid the shape of the
interface is a smoothly rounded surface. A better representation is to
substitute the region, around a nodal point, by a little disc tangent to the
surface and to subdivide the remaining area's of the adjacent triangles into
a number of triangles, that together fit in the rest of the diamondshaped
interface. The remaining triangles contribute acceptable specific discharges,
that can be obtained by use of the equations (4.5)(4.6). Only the vortices
in the little disc have to be considered seperately.

Specific discharge in centre of disc

Consider a circular flat disc of radius
r at an angle α with the horizontal plane.
The vortex strength per unit area is g
from (4.1). The question is to determine
the specific discharge at a point P on the
normal in the centre of the disc, when P
approaches the disc from above or below.
The coordinates of P are taken $n_p = \lambda$, $s_p = 0$,
$t_p = 0$, with λ a small positive or negative
quantity for P approaching the disc centre
from above or below, respectively, see fig 7.

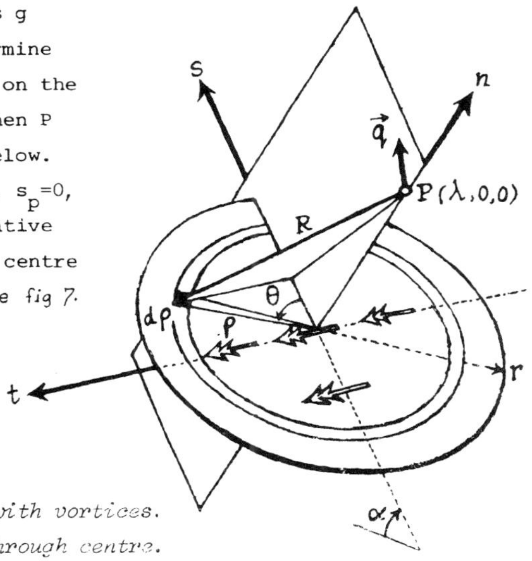

Fig. 7. Circular part of interface with vortices.
Specific discharge in P on normal through centre.

Integration is over a ring shaped area with radius ρ and width $d\rho$. Then in equations (4.2)(4.3) we have $s'=\rho\cos\theta$; $t'=\rho\sin\theta$; $ds'dt'=\rho d\rho d\theta$, $R=(\lambda^2+\rho^2)^{\frac{1}{2}}$ and the specific discharge in P has components

$$q_n = (g/4\pi) \int_o^r \int_o^{2\pi} (\rho \cos \theta)(\lambda^2 + \rho^2)^{-3/2} \rho d\rho d\theta ,$$

$$q_s = (g/4\pi) \int_o^r \int_o^{2\pi} \lambda(\lambda^2 + \rho^2)^{-3/2} \rho d\rho d\theta .$$

$\left.\vphantom{\begin{array}{c}a\\a\\a\\a\end{array}}\right\}$(4.7)

Integration with respect to θ gives

$$q_n = 0$$
$$q_s = \tfrac{1}{2}g \, \lambda \int_o^r (\lambda^2 + \rho^2)^{-3/2} \rho d\rho .$$

$\left.\vphantom{\begin{array}{c}a\\a\\a\end{array}}\right\}$(4.8)

So that finally only the component q_s remaines, giving

$$q_s = \tfrac{1}{2}g \, \frac{-\lambda}{(\lambda^2 + \rho^2)^{\frac{1}{2}}} \Bigg|_o^r = \tfrac{1}{2}g \left\{ \frac{\lambda}{(\lambda^2)^{\frac{1}{2}}} - \frac{\lambda}{(\lambda^2 + r^2)^{\frac{1}{2}}} \right\} . \tag{4.9}$$

For $\lambda \rightarrow \pm 0$ this reduces to

$$q_s = \tfrac{1}{2}g \, (\lambda/|\lambda|) , \tag{4.10}$$

with in the denominator the absolute value of λ, because the root of λ^2 is always positive.

For $\lambda \rightarrow +0$ the point P is above the disc in the lighter fluid with $\gamma = \gamma_1$. For $\lambda \rightarrow -0$ the point is in the fluid with γ_2. Let the discharges in the two fluids be indicated as $\vec{q}^1$ and $\vec{q}^2$ respectively. Then reintroducing (4.1) the result (4.10) can be written as

$$q_s^1 = + \tfrac{1}{2}(\kappa/\mu)(\gamma_2 - \gamma_1)\sin \alpha , \text{for P just above interface in fluid 1}$$
$$q_s^2 = - \tfrac{1}{2}(\kappa/\mu)(\gamma_2 - \gamma_1)\sin \alpha , \text{for P just below interface in fluid 2}$$

$\left.\vphantom{\begin{array}{c}a\\a\end{array}}\right\}$(4.11)

Further according to (4.2) and (4.8)

$$q_t^1 = q_t^2 = 0; \qquad q_n^1 = q_n^2 = 0 \tag{4.12}$$

These relations are identical to those found in the two dimensional case as presented previously (*de Josselin de Jong*, 1977). They show, that the vortices, in a flat, disc shaped area of a sharp interface between two fluids, only create a shear flow parallel to the interface. Above the interface, the flow is in the direction of steepest ascent, and below it in opposite direction, as demonstrated in the adjacent fig. 8.

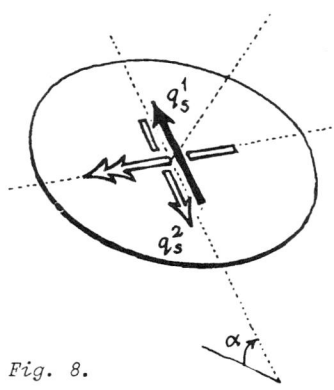

Fig. 8.

Conclusion

An interface between two pore fluids of different specific weight is
according to the three dimensional extension of the vortex theory a vortex
sheet. The vortices have horizontal axes and the contourlines of equal height
on the interface are vortex lines. The ribbons between contourlines are
reentrant vortex tubes. Every vortex on the interface creates a circulating
flow around it and the specific discharge, in an arbitrary point of the
aquifer, is obtained by integration of the influences of all vortices over the
entire interface using equations (3.3), (3.4) and (2.8).

The three dimensional vortex theory was applied to produce a computer
program for determining the movement of an interface with time. This work was
initiated by *Haitjema* as an extension of the computer program he developped
for the two dimensional case and reported in this journal (*Haitjema*, 1977).
The three dimensional elaboration was continued by *Luger*, who elaborated the
use of equations (4.5), (4.6) after they became available.

In the computer program the real, smoothly curved shape of an interface is
replaced by a diamondshaped plane consisting of flat triangles. Each triangle
is a vortex sheet with vortices homogeneously distributed over it. The corner
points of the triangles are the nodal points, whose displacements with course
of time are determined by computing the specific discharge created by all
vortices in all triangles and dividing this discharge by the porosity. Only
the triangles adjacent to the nodal points are exempted because their influence
becomes infinite according to (4.5). These adjacent triangles are replaced
partly by a little circular disc, tangent to the interface at the nodal point,
whose contribution to the specific discharge in its centre is shown in section
4 to consist of a shearflow parallel to the disc, only. This shear flow is
disregarded in the computation, because it does not contribute to the
displacement of the interface.

1. *H.M. Haitjema* (1977),"Numerical application of vortices to multiple fluid
   flow in porous media". Delft Progr.Rep. 2, pp 237-248.
2. *G. de Josselin de Jong* (1969),"Generating functions in the theory of flow
   through porous media" Chapter 9 from "Flow through porous media"
   (R.I.M. de Wiest, editor), Academic Press, New York. pp 377-400
3. *G. de Josselin de Jong* (1977),"Review of vortex theory for multiple fluid
   flow" Delft Progr.Rep. 2, pp. 225-236.
4. *W.C. Knudsen* (1962),"Equations of fluid flow through porous media -
   Incompressible fluids of varying density", J.Geophys.Res. 67, pp 733-737.
5. *H. Lamb* (1932),"Hydrodynamics". 6th ed.Dover Publ., New York.
6. *A. Verruijt* (1970),"Theory of groundwater flow", Macmillan, London.

Proceedings of Euromech 143 / Delft / 2-4 September 1981

The simultaneous flow of fresh and salt water in aquifers of large horizontal extension determined by shear flow and vortex theory

G.DE JOSSELIN DE JONG
Retired from University of Delft, Netherlands

SUMMARY

An interface motion equation is derived, taking into account the complete Edelman shear flow conditions and the Dietz-Dupuit approximations. A solution is verified with a result from exact vortex theory.

1 INTRODUCTION

Shear flow

When two fluids (1,2) of different specific weight γ_1, γ_2 respectively, occupying an aquifer, are separated by a sharp interface there exists a *shearflow* at every point of the interface, where it has a tilt angle α with the horizontal. The first description of this phenomenon is by Edelman (1940), who showed that a difference in q_{s1}, q_{s2}, the specific discharge components parallel to the interface, must occur in order to guarantee that along the interface the pressures in the two fluids at either side is equal. A small error, unfortunately, crept into the end of the derivation, leading to the *incorrect* expression

$$[q_{s1}-(\gamma_2/\gamma_1)q_{s2}]=(\kappa/\mu)(\gamma_2-\gamma_1)\sin\alpha \qquad (1.1)$$

where: κ is intrinsic permeability,
μ is dynamic viscosity.

In a later publication, Edelman (1947, pg 59), removes the error and gives the *correct* equation:

$$(q_{s1}-q_{s2})=(\kappa/\mu)(\gamma_2-\dot\gamma_1)\sin\alpha \qquad (1.2)$$

Also Edelman (1940, 1947) mentions, that at every point of the interface, the normal specific discharge components q_{n1}, q_{n2} in the two fluids at either side must be equal in order to satisfy continuity. In formula

$$(q_{n1}-q_{n2}) = 0 \qquad (1.3)$$

Edelman's work has escaped international attention, because it was written in Dutch. Better known is the work of Hubbert (1940), who also describes shear flow at a sharp interface. His treatise starts with a clarifying description of the mechanism of groundwater flow and its relation to pressures, expressed in terms of equilibrium of forces. From page 842 on, however, the analysis is confused by the presupposition, that groundwater flow is always subject to potentials and the treatment continues with a form of Darcy's law, that is uncapable to describe variable density flow, correctly. Applying an apparently rigourous reasoning an equation (192) is derived for the shearflow, similar to (1.1) here. It would be interesting to know, when this was corrected.

At the time it was not recognised, that density differences create rotation in the flow and that therefore a description with potentials is impossible. Lusczynski (1961) attempted to formulate heads, that could serve as potentials. Here, heads are not recommendable for practical use, because they are pseudo-potentials whose gradients are related to the flow only in particular directions. This is shown by Bear (1975) in his equation (9.5.19).

Using the formulation of Darcy's law in terms of pressures (9.5.6), Bear (1975) derives a correct expression for the shearflow (9.5.7) in the case of combined density and viscosity difference. The same flow relations in terms of pressures served as a basis to the vortex theory for variable density flow, de Josselin de Jong (1960). In this and later papers (1977, 1979) it is shown, that rotation exists proportional to the horizontal component of the density gradient and how the specific discharge can be computed in the entire aquifer by locating vortices in the regions of rotation. By this theory it is possible to treat any distribution of densities, also of the gradual transition zones, that exist between miscible fluids because of diffusion and dispersion. The theory is verified in the papers by reducing a gradual transition zone to a sharp interface and establishing the magnitude of the shearflow. The solutions then become singular and produce discontinuities of magnitudes as given by the equations (1.2) and (1.3), here.

Horizontally extended aquifers

Dietz (1953) treats the combined flow of fluids with different properties in elongated, confined aquifers. The analysis starts with the concept of shearflow created by the equality of pressure on either side of a sharp interface. Then the correct values of the shearflow components, parallel to the aquifer boundaries, are attributed with a Dupuit assumption to the specific discharge components parallel to the aquifer in every plane normal to the aquifer. Finally the partial differential equation (21.a) is obtained for the motion of the interface in course of time.

Bear (1975) pg 535 improves the Dietz-Dupuit analysis by removing a small inaccuracy mentioned by Dietz (1953) pg 88. This leads to Bear's eq.(9.5.64) which is similar to Dietz's (21.a) except for a commutation of a number of terms with the second $\partial/\partial x$ of $\partial^2\eta/\partial x^2$. Comparison of the two papers is somewhat impeded, because in Bear's (9.5.64) the numerator between braces is misprinted. The reader can readily perform the correction by reworking the analysis. Such a correction can be verified with Bear's (9.7.20), which is obtained with a similar analysis.

The work of Dietz and also eq.(9.5.64) of Bear describes an aquifer, that is tilted with respect to the horizon, and the coordinate system is parallel and normal to the aquifer. Other, horizontal coordinates are used in Bear's treatment leading to (9.7.20) and that equation is therefore better suited for comparison with relations developed in this paper.

In the analysis of Dietz (1953) and Bear's (1975) version of it, the discontinuous character of the flow at the interface is only accounted for in a direction parallel to the aquifer. It is an objective of this paper to show, how the analysis is changed by considering relations of the kind (1.2) (1.3), such that the character of the discontinuities in the flow, both parallel and normal to the interface, is taken into account. It appears, that f.i. introduction of aquifer anisotropy changes the interface motion equation only

slightly. As example the solution for the interface motion is considered, when the interface starts as a straight line.

A second objective of this paper is to compare this approximate solution, with an exact solution obtained by vortex theory. For a confined aquifer this theory requires the introduction of an infinite series of image vortices, which is sometimes believed to impede practical application, when the aquifer is elongated and the interface inclination angle is small, because of the integrations required in the analysis. It will be shown here that for the case of parallel, impervious boundaries, the solution is simplified, because the integrals have closed solutions.

The use of vortex theory is required, when the interface inclination angle is not small, as f.i. in the vicinity of wells, and the Dietz-Dupuit approximation is insufficient. In that case a computer is indispensable for establishing the integrations numerically. A description of the application of vortex theory to computer analysis is given by Haitjema (1977). Other programs are in course of development.

The general objective of this paper is to reconcile the Dietz-Dupuit approximation with the complete Edelman shearflow conditions and vortex theory.

2 FLOW CONDITIONS AT AN INTERFACE

The case is considered here of a sharp interface between fresh and salt water, fluid 1,2 respectively. The specific weights are then $\gamma_1<\gamma_2$ and the viscosities are $\mu_1<\mu_2$. The difference in both properties is of the order of several procents. The intrinsic permeabilities of the anisotropic aquifer are κ_h, κ_v in horizontal and vertical directions respectively, and these are considered to be the principal directions. The fluids are miscible and replacement in the pore space is complete such that κ_h and κ_v are the same for both fluids.

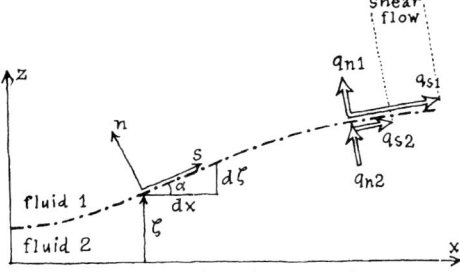

Fig.1. Interface line with coordinates n,s and specific discharges $q_{n1}=q_{n2}$; $q_{s1}>q_{s2}$.

76

The z-coordinate is vertical, the x,y coordinates are horizontal. The height of the interface is called ζ. So the equation for the interface is $z=\zeta(x,y,t)$, where $\zeta(x,y,t)$ is in general a function of x,y and the time t.

First the two dimensional case is considered that the situations in all planes, y=constant, are equal. The interface has then a cylindrical surface, whose intersection line with the arbitrary z,x plane is called the interface line here. This line is represented by $z=\zeta(x,t)$.

The angle of the interface line with the x-coordinate is α (see fig. 1), and so relations exist of the form

$$\tan\alpha=(\partial\zeta/\partial x) \tag{2.1}$$
$$\sin\alpha=(\partial\zeta/\partial x)/[1+(\partial\zeta/\partial x)^2]^{\frac{1}{2}} \tag{2.2}$$
$$\cos\alpha=1/[1+(\partial\zeta/\partial x)^2]^{\frac{1}{2}} \tag{2.3}$$

At the interface local, orthogonal coordinates n,s respectively normal and along the interface line are considered (fig.1). The coordinates n,s are taken positive, when pointing in positive z,x directions respectively. The components q_n, q_s of the specific discharge in the n,s directions are related to the components q_x, q_z in horizontal and vertical directions by

$$q_n=-q_x\sin\alpha+q_z\cos\alpha \tag{2.4}$$
$$q_s=+q_x\cos\alpha+q_z\sin\alpha \tag{2.5}$$

and the inversion

$$q_x=q_s\cos\alpha-q_n\sin\alpha \tag{2.6}$$
$$q_z=q_s\sin\alpha+q_n\cos\alpha \tag{2.7}$$

Continuity

In order to satisfy *continuity*, the normal specific discharge components q_{n1}, q_{n2} on either side of the interface are equal. This is expressed by (1.3), a relation that is not changed by differences in fluid properties or anisotropy of the pore space. So the normal component can be called q_n and with (2.4) can be written as

$$q_n=q_{n1}=-q_{x1}\sin\alpha+q_{z1}\cos\alpha=$$
$$=q_{n2}=-q_{x2}\sin\alpha+q_{z2}\cos\alpha \tag{2.8}$$

Multiplication with μ_1, μ_2 respectively gives

$$(\mu_1-\mu_2)q_n=-(\mu_1q_{x1}-\mu_2q_{x2})\sin\alpha+$$
$$+(\mu_1q_{z1}-\mu_2q_{z2})\cos\alpha \tag{2.9}$$

Equilibrium

In order to satisfy *equilibrium* the fluid pressures p_1, p_2 in the fluids at either side of the interface are equal, if they are miscible in the manner of fresh and salt water. So the relation exists

$$p_1-p_2=0 \tag{2.10}$$

Since this is true for all points along the interface line, also the derivative of (2.10) with respect to s has to be zero or

$$(\partial p_1/\partial s)-(\partial p_2/\partial s)=0 \tag{2.11}$$

Transforming (2.11) in x,z directions gives

$$[(\partial p_1/\partial x)-(\partial p_2/\partial x)]\cos\alpha+$$
$$[(\partial p_1/\partial z)-(\partial p_2/\partial z)]\sin\alpha=0 \tag{2.12}$$

In order to express this relation (2.12) in terms of specific discharge components, the relations between flow and pressure developed from equilibrium of forces are used. These are for the ith fluid:

$$(\mu_i/\kappa_h)q_{xi}=-(\partial p_i/\partial x) \tag{2.13}$$
$$(\mu_i/\kappa_v)q_{zi}=-(\partial p_i/\partial z)-\gamma_i \tag{2.14}$$

Eliminating p_1, p_2 from (2.12),(2.13) and (2.14) gives

$$(\mu_1q_{x1}-\mu_2q_{x2})(\cos\alpha/\kappa_h)+$$
$$(\mu_1q_{z1}-\mu_2q_{z2})(\sin\alpha/\kappa_v)=(\gamma_2-\gamma_1)\sin\alpha \tag{2.15}$$

Discontinuities in the flow components

The relations (2.9),(2.15) can be solved to give

$$(\mu_1q_{x1}-\mu_2q_{x2})[(\cos^2\alpha/\kappa_h)+(\sin^2\alpha/\kappa_v)]=$$
$$=(\gamma_2-\gamma_1)\sin\alpha\cos\alpha-(\mu_1-\mu_2)q_n(\sin\alpha/\kappa_v) \tag{2.16}$$

$$(\mu_1q_{z1}-\mu_2q_{z2})[(\cos^2\alpha/\kappa_h)+(\sin^2\alpha/\kappa_v)]=$$
$$=(\gamma_2-\gamma_1)\sin^2\alpha+(\mu_1-\mu_2)q_n(\cos\alpha/\kappa_h) \tag{2.17}$$

These relations give an impression of the discontinuities in the specific discharge components at the interface.

The normal specific discharge component q_n is somewhat alien in these relations. In section 3 it turns out, that it is convenient to have q_n converted into a term which contains $\partial\zeta/\partial t$, the vertical velocity of the interface, in the following manner.

77

Soil Mechanics and Transport in Porous Media

When the interface moves upwards with a velocity $\partial\zeta/\partial t$, a surface of area A of the interface plane moves during a time interval dt, through a volume of the aquifer of magnitude $\cos\alpha(\partial\zeta/\partial t)dt.A$. (see fig. 2). The amount of fluid in this aquifervolume is ε times the volume, when ε is the porosity of the aquifer. So the amount of fluid displaced is $\varepsilon\cos\alpha(\partial\zeta/\partial t)dt.A$.

The specific discharge q_n is defined as the volume of fluid passing normal to the interface, through a unit area during unit time. So the volume of fluid passing through the area A during a time dt is $q_n Adt$. Comparing the two volumes gives

$$q_n = \varepsilon\cos\alpha(\partial\zeta/\partial t) \qquad (2.18)$$

Using this value for q_n in (2.16),(2.17) and applying (2.2),(2.3) results in

$$(\mu_1 q_{x1} - \mu_2 q_{x2}) =$$
$$[(\gamma_2-\gamma_1)-\varepsilon(\mu_1-\mu_2)\frac{1}{\kappa_v}\frac{\partial\zeta}{\partial t}]\frac{\kappa_h\frac{\partial\zeta}{\partial x}}{1+\frac{\kappa_h}{\kappa_v}(\frac{\partial\zeta}{\partial x})^2} \qquad (2.19)$$

$$(\mu_1 q_{z1} - \mu_2 q_{z2}) =$$
$$[(\gamma_2-\gamma_1)+\varepsilon(\mu_1-\mu_2)\frac{\frac{\partial\zeta}{\partial t}}{\kappa_h(\frac{\partial\zeta}{\partial x})^2}]\frac{\frac{\kappa_h}{\kappa_v}(\frac{\partial\zeta}{\partial x})^2}{1+\frac{\kappa_h}{\kappa_v}(\frac{\partial\zeta}{\partial x})} \qquad (2.20)$$

At this point it may be remarked, that the relations (2.19),(2.20) are valid for any value of α in the interval $-\pi/2\leqslant\alpha\leqslant\pi/2$. Beyond that interval the interface is unstable, because heavier fluid is above lighter.

When the viscosities are equal, such that $\mu_1=\mu_2=\mu$, equations (2.16),(2.17) reduce to

$$(q_{x1}-q_{x2})=\frac{(\gamma_2-\gamma_1)\sin\alpha\cos\alpha}{\mu[(\cos^2\alpha/\kappa_h)+(\sin^2\alpha/\kappa_v)]} \qquad (2.21)$$

$$(q_{z1}-q_{z2})=\frac{(\gamma_2-\gamma_1)\sin^2\alpha}{\mu[(\cos^2\alpha/\kappa_h)+(\sin^2\alpha/\kappa_v)]} \qquad (2.22)$$

Using the tensor character of permeability, it is possible to show that the term $[(\cos^2\alpha/\kappa_h)+(\sin^2\alpha/\kappa_v)]$ represents the reciproval of the intrinsic permeability, κ_s, in s-direction. Using further (2.5) it is found, that

$$(q_{s1}-q_{s2})=(\kappa_s/\mu)(\gamma_2-\gamma_1)\sin\alpha, \qquad (2.23)$$

a relation for the shearflow reminding of (1.2).

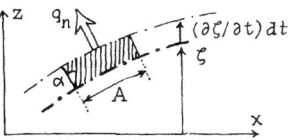

Fig. 2. Interface moving upwards

3. INTERFACE MOTION EQUATION

Continuity in x-direction

Let the aquifer have its upperboundary at $z=Z_1$ and its lower boundary at $z=Z_2$. The height occupied by fluid 1 is $(Z_1-\zeta)$ and for fluid 2 it is $(\zeta-Z_2)$, see fig. 3.

The horizontal specific discharge components q_{x1}, q_{x2} are as an approximation assumed to be constant in a vertical plane. Per unit width the discharges Q_{x1}, Q_{x2} of fluids 1,2 resp. are then

$$Q_{x1}=q_{x1}(Z_1-\zeta) \qquad (3.1)$$
$$Q_{x2}=q_{x2}(\zeta-Z_2). \qquad (3.2)$$

The total discharge of both fluids together over unit width is called, Q_x. Then we have

$$Q_x=Q_{x1}+Q_{x2}=$$
$$=q_{x1}(Z_1-\zeta)+q_{x2}(\zeta-Z_2). \qquad (3.3)$$

Let there be infiltration of fluid 1 through the upperboundary adding a volume S_1, per unit horizontal area and unit time to the discharge Q_{x1} and simirlarly S_2 to Q_{x2} from below. Continuity requires then, that the total discharge Q_x satisfies the relation

$$(\partial Q_x/\partial x)=S_1+S_2 \qquad (3.4)$$

When the interface moves upwards with a velocity $(\partial\zeta/\partial t)$, the volume of fluid 1 added to the discharge Q_{x1} over a unit horizontal area and unit time is $\varepsilon(\partial\zeta/\partial t)$, with ε the aquifer porosity. The same volume is subtracted from Q_{x2}. Continuity of each fluid separately requires

$$(\partial Q_{x1}/\partial x)=S_1+\varepsilon(\partial\zeta/\partial t) \qquad (3.5)$$
$$(\partial Q_{x2}/\partial x)=S_2-\varepsilon(\partial\zeta/\partial t) \qquad (3.6)$$

Equation in ζ alone

When Q_x, S_1, S_2 are known from boundary conditions, a relation in the unknown variable ζ alone is obtained by eliminating q_{xi}, Q_{xi} from equations (2.19) and (3.1) through (3.6). This was the reason to replace q_n by expression (2.18) involving ζ.

78

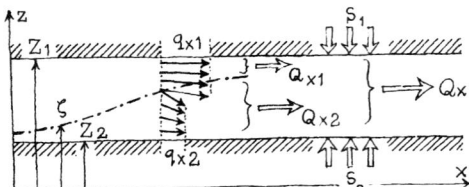

Fig. 3. Aquifer with discharges.

By using the abbreviations

$$\Gamma=(\gamma_2-\gamma_1)\kappa_h/\mu_1 \qquad (3.6)$$

$$\Lambda=\varepsilon[1-(\mu_2/\mu_1)](\kappa_h/\kappa_v)(\partial\zeta/\partial t) \qquad (3.7)$$

$$\Phi=(\kappa_h/\kappa_v)(\partial\zeta/\partial x)^2 \qquad (3.8)$$

equation (2.19) becomes

$$(\mu_1 q_{x1}-\mu_2 q_{x2})=\mu_1(\Gamma-\Lambda)\frac{\partial\zeta}{\partial x}/(1+\Phi) \qquad (3.9)$$

and solving with (3.3) for q_{x1}, q_{x2} gives

$$q_{x1}[\mu_1(\zeta-Z_2)+\mu_2(Z_1-\zeta)]=$$
$$=\mu_2 Q_x+\mu_1(\Gamma-\Lambda)(\zeta-Z_2)\frac{\partial\zeta}{\partial x}/(1+\Phi) \qquad (3.10)$$

$$q_{x2}[\mu_1(\zeta-Z_2)+\mu_2(Z_1-\zeta)]=$$
$$=\mu_1 Q_x-\mu_1(\Gamma-\Lambda)(Z_1-\zeta)\frac{\partial\zeta}{\partial x}/(1+\Phi) \qquad (3.11)$$

It is now possible to obtain a relation in ζ alone, either by eliminating q_{x1} and Q_{x1} by (3.10) in (3.1) and using (3.5), or by eliminating q_{x2} and Q_{x2} by (3.11) in (3.2) and using (3.6). The two expressions are similar but not identical. Averaging them gives

$$\varepsilon(\partial\zeta/\partial t)+\tfrac{1}{2}(S_1-S_2)=$$

$$\frac{\partial}{\partial x}\left\{\tfrac{1}{2}Q_x\frac{\mu_2(Z_1-\zeta)-\mu_1(\zeta-Z_2)}{\mu_2(Z_1-\zeta)+\mu_1(\zeta-Z_2)}+\right.$$
$$\left.+\mu_1(\Gamma-\Lambda)\frac{(Z_1-\zeta)(\zeta-Z_2)(\partial\zeta/\partial x)}{[\mu_2(Z_1-\zeta)+\mu_1(\zeta-Z_2)](1+\Phi)}\right\} \qquad (3.12)$$

This is the *interface motion equation*, comparable with Bear's (9.7.20). When the location of the interface is known at a particular moment, ζ is known as a function of x and the upwards motion $\partial\zeta/\partial t$ of the interface can be determined with (3.12) for every point of it.

In the case of fresh, salt groundwater the difference between μ_1 and μ_2 is small and introducing $\mu_1=\mu_2=\mu$ gives

$$\varepsilon(\partial\zeta/\partial t)+\tfrac{1}{2}(S_1-S_2)=$$

$$\frac{\partial}{\partial x}\left\{Q_x\frac{(Z_1+Z_2-2\zeta)}{2(Z_1-Z_2)}+(\Gamma-\Lambda)\frac{(Z_1-\zeta)(\zeta-Z_2)}{(Z_1-Z_2)(1+\Phi)}\frac{\partial\zeta}{\partial x}\right\} \qquad (3.13)$$

In this last expression Λ is not reduced to zero, since from (3.7) it appears that the difference $[1-(\mu_2/\mu_1)]$ may be overruled by the anisotropy proportion (κ_h/κ_v), which can be quite large in horizontally deposited aquifers. However, since Λ contains $(\partial\zeta/\partial t)$, it is inappropriate to incorporate it in a first determination of $(\partial\zeta/\partial t)$. The procedure is to consider Λ as a correction term, set equal to zero at first and adjusted iteratively afterwards.

The factor Φ in (3.13) also depends on anisotropy, according to (3.8). Here the proportion (κ_h/κ_v) may be large enough to balance the interface inclination factor $(\partial\zeta/\partial x)^2$ and to require that Φ is taken into account from the beginning.

Three dimensional and axial symmetric cases

In the general threedimensional case, the two horizontal coordinates x,y are involved. The total discharge is then a vector $\vec{Q}$ with components Q_x, Q_y. Continuity of the total discharge requires instead of (3.4)

$$\nabla\vec{Q}=S_1+S_2 ,$$

where ∇ stands for $(\partial/\partial x)+(\partial/\partial y)$. The shear flow equations are

$$q_{n1}-q_{n2}=0$$
$$q_{s1}-q_{s2}=\Gamma|\nabla\zeta|(1+|\nabla\zeta|^2)^{\frac{1}{2}}/(1+\frac{\kappa_h}{\kappa_v}|\nabla\zeta|^2)$$
$$q_{t1}-q_{t2}=0 ,$$

where t is the horizontal direction on the interface. The same analysis as above produces the following interface motion equation for $\mu_1=\mu_2=\mu$:

$$\varepsilon(\partial\zeta/\partial t)+\tfrac{1}{2}(S_1-S_2)=$$
$$\nabla[\vec{Q}(Z_1+Z_2-2\zeta)/2(Z_1-Z_2)]+$$
$$+\Gamma\nabla\left\{\frac{(Z_1-\zeta)(\zeta-Z_2)\nabla\zeta}{(Z_1-Z_2)[1+(\kappa_h/\kappa_v)|\nabla\zeta|^2]}\right\} \qquad (3.14)$$

In the axial symmetric case, with r the horizontal radial coordinate, the interface motion equation, with $S_1=S_2=0$, is

$$\varepsilon(\partial\zeta/\partial t)=(Q_r/2\pi r)\frac{\partial}{\partial r}[(Z_1+Z_2-2\zeta)/2(Z_1-Z_2)]+$$
$$+\Gamma\frac{1}{r}\frac{\partial}{\partial r}\left\{r\frac{(Z_1-\zeta)(\zeta-Z_2)(\partial\zeta/\partial r)}{(Z_1-Z_2)[1+(\kappa_h/\kappa_v)(\partial\zeta/\partial r)^2]}\right\} \qquad (3.15)$$

A difference between the equations here and those developed in literature is in the term containing anisotropy: (κ_h/κ_v).

79

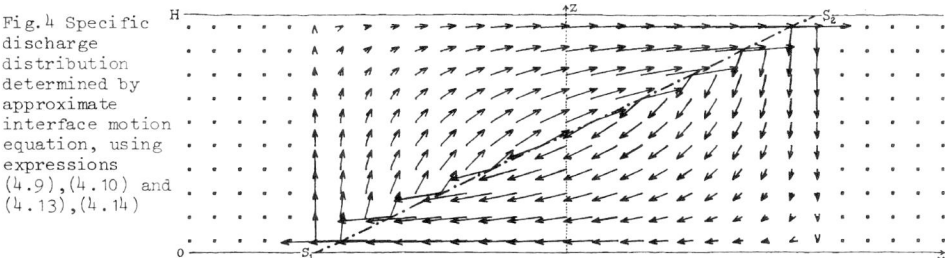

Fig.4 Specific discharge distribution determined by approximate interface motion equation, using expressions (4.9),(4.10) and (4.13),(4.14)

Character of the interface motion equation

The equations (3.12), (3.13), (3.14), (3.15) describing the motion of the interface in different cases, have the character of non-linear diffusion equations with a convection term. The character of solutions satisfying such equations have been studied by Peletier and van Duyn (1977), van Duyn (1979). For a detailed account of the mathematical aspects concerning solutions, the reader is referred to the contribution of van Duyn in these proceedings.

4 EXAMPLE OF A SOLUTION

In order to show a solution of the interface motion equation (3.12), this equation is reduced by considering a simplified case. The aquifer boundaries are at constant heights: $Z_1=H$ and $Z_2=0$. There is neither infiltration, nor convection, so
$$S_1=S_2=Q_x=(\partial Z_1/\partial x)=(\partial Z_2/\partial x)=0 \qquad (4.1)$$
The viscosity differences are small enough to disregard Λ throughout. Then (3.12) is

$$\varepsilon H \frac{\partial \zeta}{\partial t} = \Gamma \frac{\partial}{\partial x}\left\{(H-\zeta)\zeta\frac{\partial \zeta}{\partial x}/[1+\frac{\kappa_h}{\kappa_v}(\frac{\partial \zeta}{\partial x})^2]\right\} \qquad (4.2)$$

The motion is considered of an interface, that at time t is a straight line with inclination angle $\alpha=\text{arc tan } f$, such that it is represented by the relation
$$\zeta(t)=fx+\tfrac{1}{2}H \qquad (4.3)$$
Its intersection points S^1, S^2 with the boundaries have coordinates
$$x_{S1}=-(H/2f); \qquad x_{S2}=+(H/2f) \qquad (4.4)$$
For all $x<x_{S1}$ the aquifer is filled with fresh water, with salt water for $x>x_{S2}$.

Since from (4.3), $\partial\zeta/\partial x=f=$constant with respect to x, relation (4.2) gives

$$\varepsilon H(\partial\zeta/\partial t)=-2\Gamma f^3 x/(1+\frac{\kappa_h}{\kappa_v}f^2) \qquad (4.5)$$

This result shows, that the increase of the interface height ζ is linear in x and therefore the interface apparently turns as a rigid line. This means, that in equation

(4.3) for the interface, the time dependence is only in f, being f(t) a function of t alone. So differentiating (4.3) with respect to t, gives
$$(\partial\zeta/\partial t)=(df/dt)x \qquad (4.6)$$
Combining with (4.5) gives a differential equation in f and t
$$\varepsilon H(df/dt)=-2\Gamma f^3/[1+(\kappa_h/\kappa_v)f^2]. \qquad (4.7)$$

Solving by separation of variables gives

$$\frac{\kappa_h}{2\kappa_v}\ln\frac{f}{f_0}-\frac{1}{4f^2}+\frac{1}{4f_0^2}=-\frac{\Gamma}{\varepsilon H}(t-t_0), \qquad (4.8)$$

where integration constants are added, such that at time t_0, f equals f_0. The result (4.8) shows, that the inclination angle α decreases in course of time.

Using (3.10),(3.11) gives, for $\mu_1=\mu_2=\mu$ and $Q_x=\Lambda=0$, the values of q_{x1}, q_{x2} at the interface
$$q_{x1}H=\Gamma(fx+\tfrac{1}{2}H)f/[1+(\kappa_h/\kappa_v)f^2] \qquad (4.9)$$

$$q_{x2}H=\Gamma(fx-\tfrac{1}{2}H)f/[1+(\kappa_h/\kappa_v)f^2] \qquad (4.10)$$

These values are assumed to be constants over every vertical.
In order to find the components q_{z1}, q_{z2} at the interface, (2.18) is combined with (2.8) to give $\varepsilon(\partial\zeta/\partial t)=-fq_{xi}+q_{zi}$ and use of (4.5) gives with (4.9),(4.10), for $z=\zeta$:

$$q_{\zeta 1}H=\Gamma(-fx+\tfrac{1}{2}H)f^2/[1+(\kappa_h/\kappa_v)f^2]. \qquad (4.11)$$

$$q_{\zeta 2}H=\Gamma(-fx-\tfrac{1}{2}H)f^2/[1+(\kappa_h/\kappa_v)f^2] \qquad (4.12)$$

In order to satisfy continuity in the fresh and salt water regions, the components q_{z1}, q_{z2} are distributed linearly over the heights: $\zeta<z<H$ and $0<z<\zeta$ respectively. Using (4.3) it is found that
$$q_{z1}=\frac{H-z}{H-\zeta}q_{\zeta 1}=\Gamma\frac{H-z}{H}f^2/[1+(\kappa_h/\kappa_v)f^2] \qquad (4.13)$$

$$q_{z2}=\frac{z}{\zeta}q_{\zeta 2}=\Gamma\frac{-z}{H}f^2/[1+(\kappa_h/\kappa_v)f^2] \qquad (4.14)$$

The relations (4.9),(4.10) and (4.13)(4.14) were used to construct fig. 4.

80

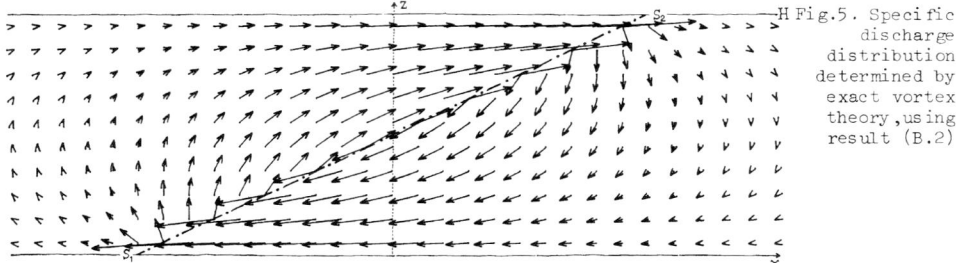

H Fig.5. Specific
    discharge
    distribution
    determined by
    exact vortex
    theory, using
    result (B.2)

EXACT SOLUTION BY VORTEX THEORY

Vortices in confined aquifer

According to vortex theory the specific discharge distribution, associated with a section ds at a point S of an inclined interface, can be determined by locating a vortex of strength Γdz_s in S, with dz_s the vertical height difference spanned by ds. The vortex has a horizontal axis parallel to the interface and creates a rotation, that turns the interface towards the horizontal.

Each vortex contributes to a stream function Ψ, whose derivatives are the specific discharge components

$$q_x = -(\partial\Psi/\partial z) \quad ; \quad q_z = +(\partial\Psi/\partial x) \qquad (5.1)$$

The contribution of the vortex in ds to Ψ is in two dimensions:

$$(\Gamma/2\pi)dz_s \ln(r_{ps}) \qquad (5.2)$$

In this expression r_{ps} is the distance, between P and S, defined by $r_{ps} = [(x_p - x_s)^2 + (z_p - z_s)^2]^{\frac{1}{2}}$, with x_s, z_s the coordinates of the point S, where the vortex is located, and x_p, z_p the coordinates of a point P, where the discharge is considered. The contribution (5.2) to Ψ is reminding of the contribution to a potential Φ by a two-dimensional well of magnitude Γdz_s.

In a confined aquifer, with impermeable boundaries at $z=0$ and $z=H$, an infinite row of image vortices has to be introduced in the manner of image wells. Using a result of Muskat (1937), sect 9.8 concerning an infinite row of wells, each vortex contributes to Ψ by an amount $d\Psi$ given by

$$d\Psi = -\frac{\Gamma}{2\pi} \cdot \frac{1}{2}\ln \frac{\cosh(X_p - X_s) - \cos(Z_p - Z_s)}{\cosh(X_p - X_s) - \cos(Z_p + Z_s)} dz_s \qquad (5.3)$$

where $X_j = \pi x_j/H$; $Z_j = \pi z_j/H$ (5.4)

An interface element between S^1 and S^2 requires integration of (5.3) over z_s from z_{s1} to z_{s2}. Using Euler's relation to modify (5.3), gives for Ψ the value (A.1) in table A. Since x_s is a function of z_s integration of (A.1) is difficult. In order to simplify integration, the discharges in P are combined as follows

$$q_{xp} - iq_{zp} = -(\partial\Psi/\partial z_p) - i(\partial\Psi/\partial x_p) \qquad (5.5)$$

Using (A.1) this gives (A.2). Subtracting 1 from the first term between braces and adding 1 to the second term, (A.3) is produced.

Let the interface interval between S^1 and S^2 have the inclination angle $\alpha = \arctan f$. Then the infinitesimal distance dx_s along the interface equals dz_s/f and using (5.4) it is possible to write dz_s in (A.3) as

$$dz_s = (H/\pi)(dX_s \pm idZ_s)(f/1\pm if) \qquad (5.6)$$

Using this in (A.3) permits integration producing the result (A.4).

The right side of (A.4) is separated in a real and an imaginary part by applying Euler's equation twice. This gives (B.1) in Table B, using abbreviations R_{ps}, θ_{ps}, ξ_{ps}, η_{ps} defined, there. The result (B.1) contains the values for q_x and q_z in a point P, separately and in terms of real quantities. For a straight section of the interface, the value of f is constant and q_x, q_z are found by introducing the integration limit points S^1, S^2, with coordinates

---

TABLE A

$$\Psi = -\frac{\Gamma}{4\pi}\int_{S^1}^{S^2} \ln \frac{[e^{(X_p+iZ_p)} - e^{(X_s+iZ_s)}][e^{(X_p-iZ_p)} - e^{(X_s-iZ_s)}]}{[e^{(X_p+iZ_p)} - e^{(X_s-iZ_s)}][e^{(X_p-iZ_p)} - e^{(X_s+iZ_s)}]} dz_s \qquad (A.1)$$

$$q_{xp} - iq_{zp} = \frac{i\Gamma}{2H}\int_{S^1}^{S^2}\left\{[e^{(X_p+iZ_p)}/(e^{(X_p+iZ_p)} - e^{(X_s+iZ_s)})] - [e^{(X_p+iZ_p)}/(e^{(X_p+iZ_p)} - e^{(X_s-iZ_s)})]\right\} dz_s \qquad (A.2)$$

$$q_{xp} - iq_{zp} = \frac{i\Gamma}{2H}\int_{S^1}^{S^2}\left\{[e^{(X_s+iZ_s)}/(e^{(X_p+iZ_p)} - e^{(X_s+iZ_s)})] - [e^{(X_s-iZ_s)}/(e^{(X_p+iZ_p)} - e^{(X_s-iZ_s)})]\right\} dz_s \qquad (A.3)$$

$$q_{xp} - iq_{zp} = \frac{\Gamma}{2\pi}\left\{-\frac{if}{1+if}\ln[e^{(X_p+iZ_p)} - e^{(X_s+iZ_s)}] + \frac{if}{1-if}\ln[e^{(X_p+iZ_p)} - e^{(X_s-iZ_s)}]\right\}_{S^1}^{S^2} \qquad (A.4)$$

81

$$R_{ps}=[\xi_{ps}^2+\eta_{ps}^2]^{\frac{1}{2}}; \quad R^*_{ps}=[\xi_{ps}^2+\eta_{ps}^{*2}]^{\frac{1}{2}}; \quad \theta_{ps}=\arctan(\eta_{ps}/\xi_{ps}); \quad \theta^*_{ps}=\arctan(\eta^*_{ps}/\xi_{ps})$$

with

$$\xi_{ps}=(e^{X_p}\cos Z_p-e^{X_s}\cos Z_s); \quad \eta_{ps}=(e^{X_p}\sin Z_p-e^{X_s}\sin Z_s); \quad \eta^*_{ps}=(e^{X_p}\sin Z_p+e^{X_s}\sin Z_s)$$

$$q_{xp}-iq_{zp}=\frac{\Gamma}{2\pi}\left(\frac{f}{1+f^2}\right)\left\{-(i+f)\ln R_{ps}+(i-f)\ln R^*_{ps}+(1-if)\theta_{ps}-(1+if)\theta^*_{ps}\right\}_{S1}^{S2} \quad \text{...............} (B.1)$$

$$q_{xp}-iq_{zp}=(\Gamma/\pi)(f/1+f^2)\left\{-f\ln(R_{ps2}/R_{ps1})-if\theta_{ps2}+if\theta_{ps1}+\langle(1-if)\pi\rangle\right\} \quad \text{...............} (B.2)$$

$$q_{xp}-iq_{zp}=(\Gamma/H)(f/1+f^2)\left\{-\tfrac{1}{2}H+fx_p+ifz_p+\langle(1-if)H\rangle\right\} \quad \text{...............} (B.3)$$

x_{s1}, z_{s1} and x_{s2}, z_{s2} the result (B.1) is comparable to Haitjema's (1977) equation (9) derived for a straight interface element AB in an infinite aquifer. The procedure for combining such elements to analyse a curved interface, described by Haitjema, can be applied similarly to (B.1).

In (B.1) angles θ are involved, defined by arcs of tangents. These angles have to be chosen in the interval $0<\theta<\pi$, when θ^*_{ps} is concerned and θ_{ps}, when P is in fluid 2. When P is in fluid 1, and y_p is between y_{s1} and y_{s2}, the value of $(\theta_{ps2}-\theta_{ps1})$ has to be increased with 2π. For a justification see De Josselin de Jong (1960), p3750-3753.

Comparison with example of section 4

In the example the interface is a straight line with limit points S^1S^2 on the boundaries, such that $z_{s1}=0$, $z_{s2}=H$ and x_{s1}, x_{s2} given by (4.4) The boundary values for ξ_{ps} and η_{ps} are then:

$$\xi_{ps1}=e^{X_p}\cos Z_p-e^{-\frac{1}{2}\pi/f}$$
$$\xi_{ps}^2=e^{X_p}\cos Z_p+e^{\frac{1}{2}\pi/f} \quad (5.7)$$
$$\eta_{ps1}=\eta^*_{ps1}=\eta_{ps2}=\eta^*_{ps2}=e^{X_p}\sin Z_p$$

Because of the identity of the η's, (B.1) reduces to (B.2), where the term between $\langle\rangle$ has to be added, if fluid 1 is concerned. The values of (B.2) were used to construct the specific discharge distribution of fig. 5.

When the point of consideration P is located in the region $x_{s1}\langle x_p\langle x_{s2}$ such that

$$e^{-\frac{1}{2}\pi/f}\ll e^{X_p}\ll e^{\frac{1}{2}\pi/f}, \quad (5.8)$$

the variables can be approximated by
$R_{ps}2\approx e^{\frac{1}{2}\pi/f}$; $R_{ps1}\approx e^{X_p}$; $\theta_{ps2}\approx 0$; $\theta_{ps1}\approx Z_p$
and using (5.4), (B.2) reduces to (B.3).

Separation of real and imaginary parts shows, that the result (B.3) gives the same values as (4.9), (4.10) and (4.13),(4.14), when $\kappa_h=\kappa_v$ and the meaning of the brackets $\langle\rangle$ is taken into account. In the case of

anisotropy, such that $\kappa_h\neq\kappa_v$, the vortex theory is similar, only the z-coordinate has to be multiplied by a factor $(\kappa_h/\kappa_v)^{\frac{1}{2}}$ throughout. The only consequence is, that the factor $(1+f^2)$ in (B.3) is replaced by $[1+(\kappa_h/\kappa_v)f^2]$.

REFERENCES

Bear,J. 1975, Dynamics of fluids in porous media, Book 2nd ed.,Am.Elsevier Publ.

Dietz,D.N. 1953, A theoretical approach to the problem of encroaching and bypasing edgewater. Proc.Kon.Ned.Academie van Wetenschappen, series B,56, no.1,83-92.

Van Duyn,C.J. & L.A. Peletier, 1977, A class of similarity solutions of the nonlinear diffusion equation. Nonlinear Analysis, Theory, Methods and Applications, 1, 223-233.

Van Duyn,C.J. 1979, Nonlinear diffusion problems, Dr. Thesis, Leiden.

Edelman,J.H. 1940, Strooming van zoet en zout grondwater, Rapport 1940 inzake de watervoorziening van Amsterdam. Bijlage 2 8-14.

Edelman,J.H. 1947, Over de berekening van grondwaterstromingen, Dr. Thesis, Delft.

Haitjema,H.M., 1977. Numerical application of vortices to multiple fluid flow in porous media. Delft Progr.Rep.2,237-248.

Hubbert,M.K. 1940, The theory of groundwater flow, vol XLVIII, 8, 785-944.

De Josselin de Jong, G. 1960, Singularity distributions for the analysis of multiple fluid flow through porous media, J.Gephys.Res., vol.65, 11, 3739-3758.

De Josselin de Jong, G. 1977, Review of vortex theory for multiple fluid flow, Delft Progr.Rep. 2, 225-236.

de Josselin de Jong, G. 1979, Vortex theory for multiple fluid flow in three dimensions, Delft Progr.Rep. 4, 87-102.

Muskat, M. 1937, The flow of homogeneous fluids through porous media, Book, Mc Graw Hill.

82

Extrait de la Publication n° 41 de l'Association Internationale d'Hydrologie (de l'U.G.G.F.). Symposia Darcy (Dijon, 1956)

L'ENTRAINEMENT DE PARTICULES
PAR LE COURANT INTERSTICIEL

par

Ir. G. de JOSSELIN de JONG
(Laboratoire de la Mécanique des Sols, Delft

Summary

The pore system of a packed bed is schematized to a system of canals in order to permit probability-computations for a strange particle carried by the pore water movement to arrive at a certain place in a certain time.

The computations lead to explicit values for the coefficient of longitudinal and transversal diffusivity.

A test device is described which permits determination of longitudinal diffusivity. Relation between test result and theory is discussed.

1. Introduction

Dans la loi de Darcy tous les phénomènes physiques qui déterminent le mouvement du liquide intersticiel dans un tassement de grains, sont rassemblés d'une façon ingénieusement simple et pratique permettant le calcul rapide de nombreuses applications techniques.

Il existe néanmoins des problèmes relatifs aux mouvements de ces liquides qui demandent à ce que l'on considère d'une façon rigoureuse les phénomènes micro-structurels qui constituent la loi de Darcy.

Un de ces problèmes micro-structurels se pose quand on veut étudier le mouvement d'une particule étrangère entraînée par le courant intersticiel. Par exemple, l'introduction d'eau salée dans un massif saturé d'eau douce peut amener à se poser la question de la répartition de salinité qui s'en suit après que cette quantité salée ait été entraînée pendant un certain temps à travers le tassement de grains. Dans un pays comme la Hollande qui est bordé par la mer et doit protéger l'eau douce nécessaire à la vie de sa population, son cheptel et ses végétaux, contre l'infiltration involontaire d'eau salée, ce problème se rencontre maintes fois.

D'autre part, les problèmes qui se présentent dans les opérations chimiques quand un liquide qui se trouve dans une colonne à remplissage est remplacé par un autre produit, ne se résolvent qu'en considérant les mouvements micro-structurels exécutés par les particules individuelles.

Les expériences ont démontré qu'une quantité de particules étrangères introduite se disperse en toute direction relativement au mouvement moyen du liquide intersticiel par leur mouvement entre les obstacles des grains. Par analogie avec la diffusion moléculaire qui engendre une dispersion semblable l'on a introduit les conceptions de diffusion longitudinale et de diffusion transversale pour indiquer ce phénomène. (Klinkenberg [1], Baron [2]).

Nous voulons démontrer dans cet article comment on peut déterminer les coefficients de diffusion relatifs à ces deux différents modes de dispersion en considérant le mécanisme du mouvement du liquide à travers les pores.

Le calcul complet étant trop long pour être publié ici, nous nous limiterons à la présentation des points de départ, du mode de calcul et des résultats obtenus en réservant la démonstration rigoureuse à une autre publication.

La description d'un essai avec lequel nous avons déterminé le coefficient de diffusion longitudinale, et des résultats obtenus qui démontrent l'applicabilité de la théorie complétera ces considérations.

2. Le mouvement d'une particule étrangère

Nous voulons considérer une particule étrangère qui, entraînée par le courant intersticiel, est assez petite pour n'avoir par sa présence aucune influence sur ce courant.

La question qui se pose est la détermination du déplacement de cette particule après qu'elle ait été entraînée un certain temps. Comme dans un système de grains tassés arbitrairement les interstices ont une forme arbitraire, le parcours que la particule suivra n'est pas à fixer. On ne peut prédire que la probabilité pour la particule d'arriver à un certain endroit.

Le calcul de cette probabilité est le but de notre recherche, car le coefficient de diffusion peut se déduire de la déviation standard de la répartition de la probabilité.

Dans ce calcul la variable stochastique sera déterminée par le passage d'un interstice à un autre, car la longueur du parcours total se compose de la somme des longueurs individuelles de ces passages parcourus et le total du temps passé est la somme des temps de séjour passés dans les divers interstices.

De cette façon des calculs ont été effectués par Danckwerts (³), Klinkenberg (¹) qui, en considérant l'effet de la probabilité, ne se basent pas sur un mécanisme phénoménologique et ont recours à la détermination d'un coefficient numérique qui ne s'obtient que par l'expérience (essais de Kramers (⁴), Klinkenberg (¹).

Les mécanismes introduits par Baron (diff. transv. (²) et Kramers (diff. Longt. (5) donnent déjà une impression de ce coefficient mais sont d'une autre nature trop simple, introduisant une distribution discrète pour le choix d'un déplacement $+ l$ ou $- l$.

Nous voulons traiter ici le cas où la variable stochastique aura une distribution continue, dépendant d'un mécanisme idéalisé.

Il y a trois effets principaux qui déterminent la distribution du variable.

1. Dans la section d'un pore, la vitesse du liquide n'est pas constante mais plus grande vers le centre.

2. Les interstices sont de différentes grandeurs.

3. La direction du courant dans un interstice peut avoir un angle quelconque avec la direction moyenne du courant principal.

ad. 1) Perpendiculairement à la direction du courant dans un pore, les vitesses du iquide ne sont pas les mêmes et il semble nécessaire de considérer à laquelle de ces vitesses la particule prendrait part.

A cause du mouvement Brownien et à grande vitesse à cause de la turbulence, il existe dans le pore une diffusion radiale qui rend la vitesse de la particule égale à la vitesse moyenne du liquide dans certaines conditions.

Le calcul numérique de cet effet est possible avec les résultats de van Deemter (⁶).

Nous avons supposé dans les calculs suivants que cette diffusion radiale soit complète et que le mouvement de la particule soit égale à la vitesse moyenne dans un pore.

ad. 2) En observant la forme d'un interstice qui se produit entre quatre grains de forme à peu près sphéroïdale (fig. 1), on voit que le courant entre par un canal triangulaire assez étroit, arrive dans un ample tétraèdre, se partage en trois et quitte le pore par des canaux également triangulaires.

Ces canaux triangulaires se retrouvent partout dans le tassement de grains et forment la résistance principale dans toutes directions. Si les grains sont tous des sphères et de même grandeur, ces canaux triangulaires ne seront pas égaux à cause du tassement irrégulier.

Nous avons, en première approximation, supposé que la diffusion provenant de l'inégalité de ces résistances serait petite en comparaison du troisième effet et nous avons schématisé notre systeme par des canaux comme indiqués par la fig 2.

ad. 3) Nous voulons supposer que les deux effets précédents sont négligeables en comparaison avec les différences de vitesses qui s'en suivent de la direction des canaux élémentaires.

Plus la vitesse moyenne dans un canal sera grande, plus sa direction correspondra avec la direction principale Z.

Pour la distribution de ces directions, on supposera que le tassement des grains a été tel qu'il n'y a pas de directions de préférence.

140

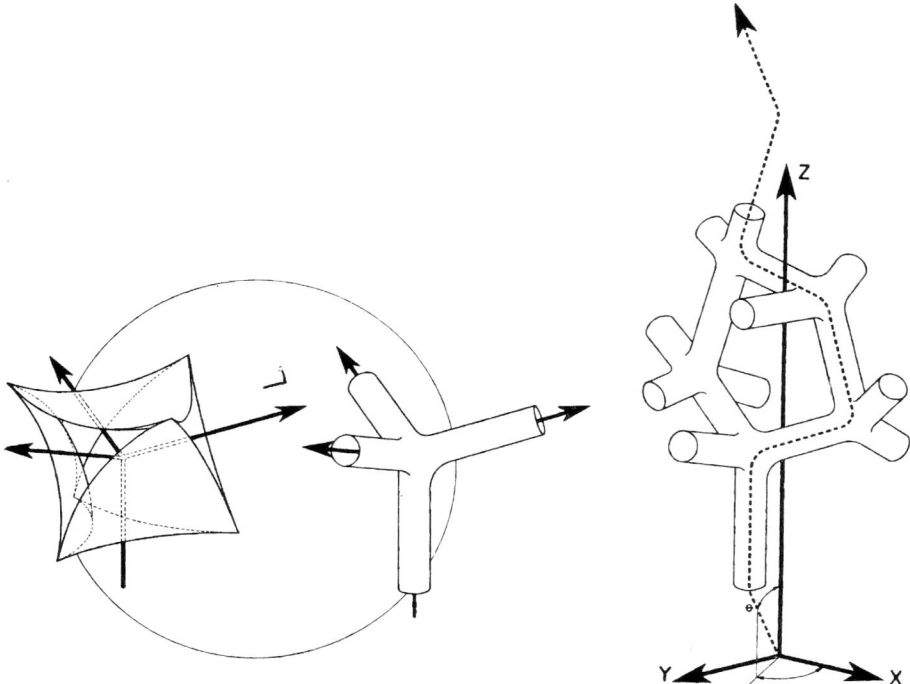

Fig. 1	Fig. 2	Fig. 3
Pore minimal entre quatre sphères.	Pore schematisé par des canaux.	Route arbitraire prise par une particule à travers les canaux.

3. Déduction de la probabilité

Selon ces 3 considérations précédentes, le système de canaux dans lequel la particule est entraînée sera schématisé de la façon indiquée par la figure 3.

Il consiste en des canaux de même longueur — l — de même résistance au courant, orientés uniformément dans toutes directions.

La particule suit une route tortueuse à travers ces divers canaux à une vitesse égale à la vitesse moyenne du liquide dans les canaux différents.

Plaçons la coordonnée Z dans la direction principale du courant intersticiel total, et moyen sur beaucoup de canaux. Nomons l'angle entre cette direction et la direction du canal élémentaire (j) : θ_j et ψ_j l'angle de sa projection avec la coordonnée X (fig. 3).

Supposons que le temps de séjour dans ce canal soit

$$t_j = t_o/\cos\theta_j$$

où t_o est le temps de séjour minimal possible et qui est obtenu dans un canal orienté dans la direction Z.

Quand la particule a passé par N de ces canaux, elle est arrivée à un endroit $X\,Y\,Z$ et dans un temps T donnés par

$$
\begin{aligned}
X &= \Sigma\, x_j = \Sigma\, l\sin\theta_j\cos\psi_j \\
Y &= \Sigma\, y_j = \Sigma\, l\sin\theta_j\sin\psi_j \\
Z &= \Sigma\, z_j = \Sigma\, l\cos\theta_j \\
T &= \Sigma\, t_j = \Sigma\, t_o/\cos\theta_j
\end{aligned}
$$

(1)

où les sommes sont effectuées pour $j = 1$ à N.

141

Dans ces expressions les angles θ_j et ψ_j sont maintenant les variables stochastiques Nous voulons supposer que la probabilité que la particule prenne une direction θ_j, ψ_j au moment, qu'arrivée à une bifurcation, elle doit choisir un nouveau canal à suivre, est donnée par la fraction du liquide total qui coule dans cette direction : θ_j, ψ_j.

Un calcul simple indique que cette probabilité est donnée par

(2)
$$f(\theta_j, \psi_j)\, d\theta\, d\psi = (1/\pi)\sin\theta\,\cos\theta\;d\theta\,d\psi$$

où l'on vérifie que

$$\int_0^{\pi/2} d\theta \int_0^{2\pi} d\psi\, f(\theta, \psi) = 1$$

Pour la détermination de la probabilité, $W_N(X\,Y\,Z\,T)\, dX\,dY\,dZ\,dT$ que la particule arriverait après le passage de N canaux arbitraires dans l'endroit X Y Z et dans un temps T, nous appliquons maintenant le théorème de Markoff (voir p. ex. Chandrasekhar [7]).

En considérant T comme l'un des composants du vecteur stochastique équivalent aux autres composants (X Y Z), on obtient alors

(3)
$$W_N(X\,Y\,Z\,T)\, dX\,dY\,dZ\,dT = \frac{dX\,dY\,dZ\,dT}{(2\pi)^4} \int_{-\infty}^{+\infty} \int_{-\infty}^{+\infty} \int_{-\infty}^{+\infty} \int_{-\infty}^{+\infty} d\xi\,d\eta\,d\zeta\,d\tau \times$$

$$\times \exp\left\{-i(\xi X + \eta Y + \zeta Z + \tau T)\right\} \times A_N$$

avec

$$A_N = [\int_0^{\pi/2} d\theta \int_0^{2\pi} d\psi\, f(\theta, \psi) \exp\left\{+i(\xi x_j + \eta y_j + \zeta z_j + \tau t_j)\right\}]$$

Toutes les intégrations sont faisables par la méthode du col et nous arrivons enfin à une expression pour la probabilité $W_N(R\,Z\,T)$ où R est égale à $(X^2 + Y^2)^{1/2}$, déplacement dans une direction perpendiculaire à la direction principale Z. Toutefois cette probabilité W_N n'est que la probabilité que cet effet soit obtenu en N passages. Mais il est bien possible que la particule arrivera à R, Z, dans un temps T avec différents nombres de passages dépendant de la route qu'elle a prise. Il nous faut donc, pour obtenir la probabilité totale, que la particule arrive dans l'endroit voulu et dans le temps posé, intégrer W_N pour toutes les valeurs possibles de N, c.-à-d.

$$W(R, Z, T)\; dR\,dZ\,dT = dR\,dZ\,dT \int_{N_1}^{N_2} W_N\, dN.$$

Il est possible d'effectuer encore cette dernière intégration par la méthode du col et nous obtenons alors très approximativement

(4)
$$W(R, Z, T)\; dR\,dZ\,dT = \frac{R\,dR \cdot dZ \cdot dT}{8\,f\,[\pi\,(ac - b^2)]^{1/2}} \left[\frac{C}{(A + A')^3}\right]^{1/4}$$

$$\times \exp\left\{1 - 2(A + A')^{1/2}\,C^{1/2} + B\right\}$$

dans laquelle nous avons introduit les notations suivantes :

(5)
$$a = (\lambda - 5/4)\, t^2_0$$
$$b = -1/6\, t_0\, l = 1/2\,(\overline{t_j \cdot z_j} - \overline{t_j} \cdot \overline{z_j})$$
$$c = +1/36\, l^2 = 1/2\,(\overline{z^2_j} - \overline{z_j}^2)$$
$$d = 2\, t_0 = \overline{t_j}$$
$$e = 2/3\, l = \overline{z_j}$$
$$f = 1/8\, l^2 = 1/2\,(\overline{r^2_j} - \overline{r_j}^2) \qquad\qquad \overline{r_j} = 0$$

142

$$A = [aZ^2 — 2bZT + cT^2]/4\ (ac — b^2)$$

$$B = [ae\ Z — b\ (dZ + eT) + cdT]/4\ (ac — b^2)$$

$$C = [ae^2 —\ 2b\ de + cd^2]/4\ (ac — b)^2$$

$$A' = R^2/4\ f.$$

Un trait indique la valeur moyenne d'une variable qui se détermine aisément avec (2).

Par symétrie on attendrait pour a une expression contenant $\overline{t^2}j$, mais parce que cette valeur est infinie elle est évitée dans le résultat du calcul, où l'on rencontre λ qui est défini par

(5') $$\gamma e^\lambda/(\lambda + 1) = 4\ Z/l$$

où γ est la constante de Euler.

L'expression pour $W\ (R, Z, T)$ n'est pas, mais ressemble beaucoup à une répartition normale de Gausz.

Le maximum se trouve à $R = 0$ et $T = dZ/e = 3Z\ t_o/l$.

La répartition autour de ce maximum est définie par les déviations standards suivantes :

$$\sigma_R = \sqrt{\overline{Z\ \overline{r^2}/z}} = l\ \sqrt{3Z/8l}$$

(6) $$\sigma_T = \sqrt{\overline{Z\ [\overline{z^2}\ \overline{t^2} — 2\overline{zt}\ .\ \overline{z}\ .\ \overline{t} + \overline{z^2}\ (2\ \lambda + 3/2)]/z^3}} = t_o\ \sqrt{3Z\ (\lambda — 1/4)/l}$$

$$\sigma_Z = \sigma_T l/3t_o = l\ \sqrt{1/3\ Z\ (\lambda — 1/4)/l}$$

De ce résultat se déduisent les coefficients de diffusion longitudinale (D_Z) et transversale (D_R) qui décrivent la diffusion par rapport au mouvement moyen de vitesse $v_o = Z/T = l/3t_o$, car selon Einstein $D = \sigma^2/2T$. Donc

(7) $$D_Z = \sigma_Z^2/2T = 3\ v_o\ l\ (\lambda — 1/4)$$

$$D_R = \sigma_R^2/2T = 3/16\ v_o\ l.$$

Si les canaux élémentaires ne sont pas égaux, comme nous l'avons supposé au début de cette dérivation, les valeurs de

$$\overline{r^2},\ \overline{z^2},\ \overline{t^2}\ \text{ et }\ \overline{z.t}$$

différeront de ce que nous avons donné plus haut. La répartition de r_j, z_j et t_j connue, il n'est pas difficile d'en tenir compte dans le résultat (6) où l'on remarque que les déviations standard augmenteront.

L'Essai

Un essai de contrôle fut effectué avec un dispositif de percolation (fig. 4).

Un cylindre (ø 6 cm, hauteur 20 cm) fut rempli de boules de verre, sensiblement sphéroïdales et d'un diamètre de 0,2 mm (déviation standard 0,05 mm). Le remplissage contrôlé soigneusement procura une porosité uniforme de 38,5 %.

D'abord le système fut saturé d'une eau distillée dans laquelle on ajoutait une quantité de NaCl correspondant à une concentration de 0,02 normal. Le liquide sous le filtre fut remplacé par de l'eau ayant une concentration de 1,0 normal NaCl et on fit ensuite percoler e système de bas en haut par ce liquide, à une vitesse de 6 cm³/minute. Tenant compte de a porosité, la vitesse moyenne dans les grains fut

$$(6/\ \frac{1}{4}\ \pi.\ 6^2)\ \times\ (1/0{,}385) = 0{,}55\ \text{cm/minute.}$$

A différentes hauteurs 2,5-7,5-12,5-17,5 cm du fond, la paroi était percée par des électrodes qui permettaient de mesurer la concentration de salinité du liquide intersticiel. Ces électrodes consistaient en des fils de platine couverts de noir en platine (diamètre 0,7 mm, 3 mm long). La résistance électrique que ces pointes éprouvèrent en appliquant un courant alternatif (1000 Hz, 50 micro amp.) fut mesurée par le pont de Wheatstone indiqué dans la

143

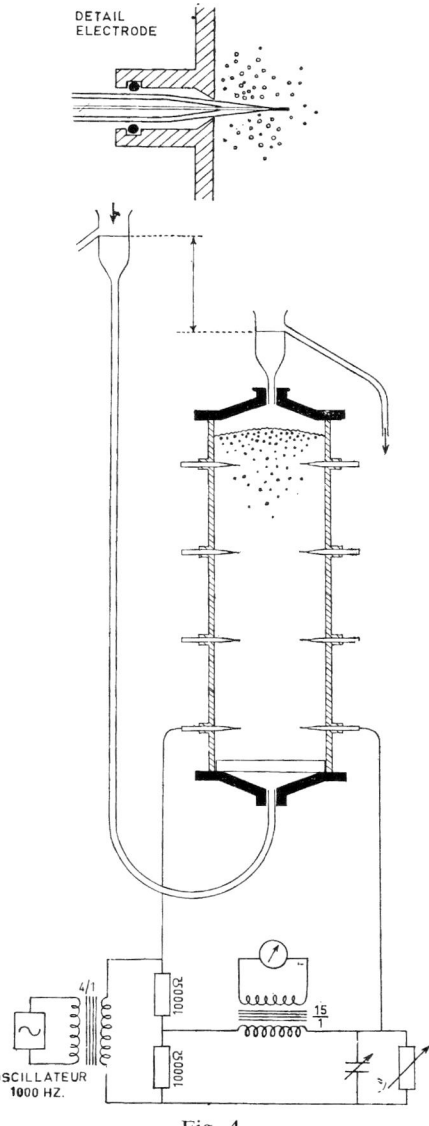

DETAIL
ELECTRODE

OSCILLATEUR
1000 HZ.

Fig. 4
Dispositif de l'essai.

figure 4. Un étalonnage dans les liquides de diverses concentrations de NaCl avait donné la relation entre cette résistance électrique et la salinité.

L'étalonnage fut exécuté aussi bien dans le liquide seul qu'après remplissage avec des boules de verre, donnant des valeurs qui correspondaient avec le calcul théorique.

Au moyen des lectures de ces résistances on pouvait donc suivre les changements de la salinité dans l'entourage de ces électrodes au fur et à mesure que la frontière des concentrations passait de 0,02 normal à 1,0 normal.

144

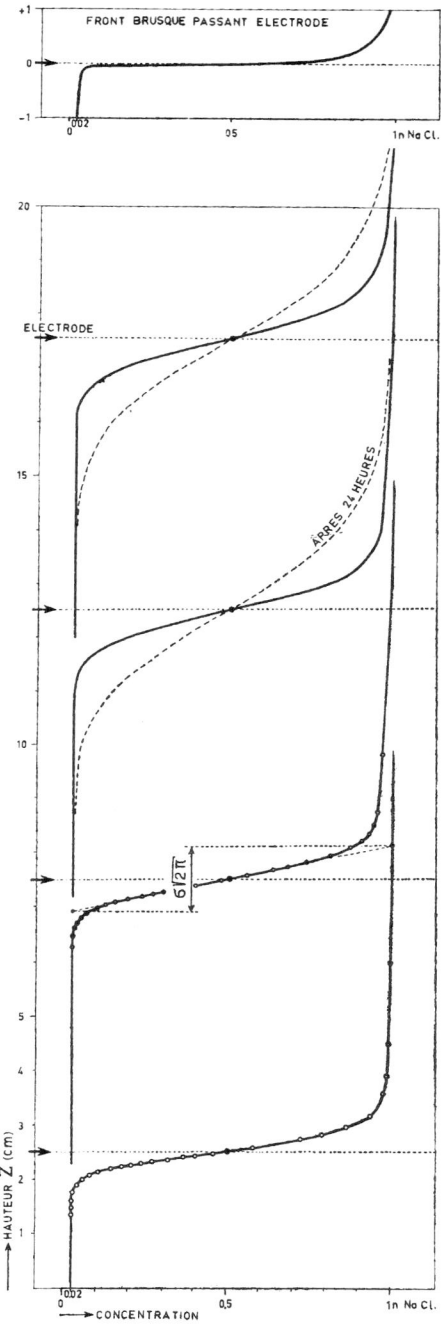

Fig. 5
Salinité enrégistrée par les électrodes.

145

Cette frontière était brusque au début de la percolation, elle se dispersa en montant dans les grains. Dans la figure 5 sont présentés les résultats des observations avec les électrodes.

Interprétations des résultats d'essai

Dans la figure 5 nous avons dessiné la salinité enregistrée par les électrodes au moment où une particule, voyageant à la vitesse moyenne, est arrivée à la hauteur indiquée par l'abscisse.

Pour démontrer l'influence de l'épaisseur de l'électrode, nous avons dessiné au-dessus ce qu'une électrode enregistrerait si une frontière brusque passait. Cette ligne obtenue théoriquement montre que la frontière brusque n'est déformée que très faiblement et que les électrodes sont assez minces pour négliger cette influence.

Les lignes obtenues par les électrodes démontrent sensiblement une distribution normale. On observe que leur pente augmente à mesure que leur emplacement est plus élevé. Cette pente est une mesure pour la déviation standard, comme nous l'avons calculée, et doit donc suivre une loi comme indiquée par σ_z de la formule 6.

Dans la figure 6 est présentée la relation entre $\sqrt{Z}$ la racine de la hauteur des électrodes et la déviation standard σ_z observée.

Pour obtenir la déviation standard σ_z qui ne dépend que de la diffusion longitudinale, il faut encore soustraire l'influence de la diffusion moléculaire. Cette influence a été déterminée par l'expérience. Dans un essai, le front passa d'abord par les électrodes inférieures, puis la percolation fut arrêtée pendant 24 heures et fut rétablie ensuite pour faire passer le front des électrodes restantes.

Les lignes observées, pointillées dans la figure 5, montrent un basculement additif qui correspond à un coefficient de diffusion moléculaire de

$$D_m = 0,7 \times 10^{-5} \text{ cm}^2/\text{sec.}$$

Les essais normaux ne prenant que 30 minutes, l'influence de la diffusion moléculaire fut assez petite comme l'indique la fig. 6 où nous l'avons rapportée.

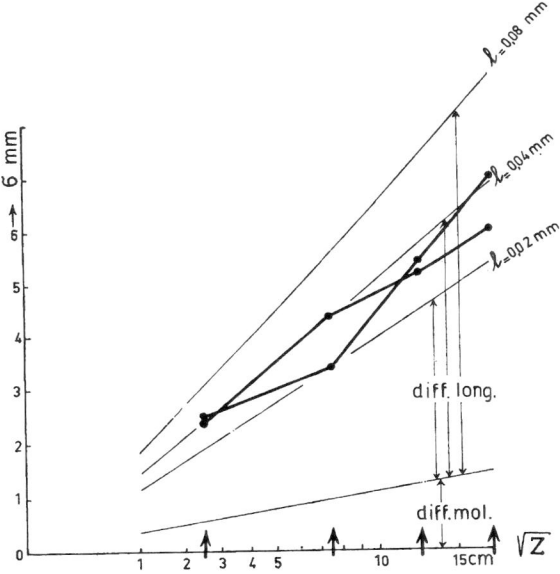

Fig. 6
Déviations standards observées comparées avec la théorie.

146

Dans cette figure sont données également les lignes théoriques obtenues avec la formule 6 pour l étant 0,008, 0,004 et 0,002 cm.

Comme les boules de verre employées ont un diamètre moyen de 0,02 cm, on supposerait que la longueur d'un canal élémentaire ne serait pas plus petite que 0,008 cm étant la distance des centres de deux pores tétraédriques comme dessinés dans la figure 1. Or la valeur de la déviation standard est en réalité plus petite que celle déduite de la théorie. Nous sommes amenés à croire que la suppression de l'effet de la diffusion radiale (ad. 1e) est responsable pour cette anomalie.

Quand nous appliquons les résultats de van Deemter (^{6}) à notre essai avec le sel, nous obtenons une diffusion radiale complète pour un canal de diamètre égal aux pores entre les grains et de longueur égale à la distance des électrodes. Mais pour un canal à longueur égale à la longueur élémentaire — l— cette diffusion radiale n'a pas eu le temps de s'effectuer. Or le sel n'arrive que retardé aux abords et est propagé par préférence au centre des pores. Il s'en suit que dans la route prise par la particule entre les grains les canaux dans la direction Z seront favorisés encore plus que cela a été exprimé par la distribution (formule 2).

Nous ne voulons pas aller plus loin sur ce point mais conclure que les abstractions faites dans la théorie n'affectent son applicabilité qu'à tel point que la prédiction des diffusions longitudinale et transversale en ordre de grandeur est possible.

Nous voulons remercier le professeur Timman pour son concours à surmonter certaines difficultés mathématiques, le professeur Kramers pour ses suggestions dans le domaine physique et M. Mostertman pour sa collaboration.

RÉFÉRENCES

(^{1}) KLINKENBERG, A., SJENITZER, F. Holding time distributions of the Gaussian type. *Chem. Eng. Sci.* Vol. 5, 1956 (paraîtra).
(^{2}) BARON, T. Generalized graphical methods for the design of fixed bed catalytic reactors. *Chem. Eng. Progress*, Vol. 48, 1952, pp. 119.
(^{3}) DANCKWERTS, P. V. Continuous flow systems. Distribution of residence times. *Chem. Eng. Sci.*, 1953, Vol. 2. pp. 1 to 13.
(^{4}) KRAMERS, H., ALBERDA, G. Frequency response analysis of continuous flow systems. *Chem. Eng. Sci.*, 1953, vol. 2, pp. 173 to 181.
(^{5}) KRAMERS, H. Communication privée.
(^{6}) VAN DEEMTER, BROEDER, LAUWERIER. Fluid displacement in capillaries. *Appl. Sci. Res.* Section A, vol. 5, Nr 5, 1955.
(^{7}) CHANDRASEKHAR. Stochastic problems in Physics and Astronomy. *Rev. Mod. Phys.* vol. 15, Nr 1, 1943.

147

Vol. 39, No.1   Transactions, American Geophysical Union   February 1958

Longitudinal and Transverse Diffusion in Granular Deposits

G. de Josselin de Jong

Abstract—The pore system of a packed bed is represented by a system of canals in order to permit probability computations for a foreign particle carried by the pore liquid movement to arrive at a certain place in a certain time. The computations lead to explicit values for the coefficient of longitudinal and transversal diffusion. A test device is described which permits the determination of the longitudinal diffusivity. The relationship between test result and theory is discussed.

Introduction—The flow of liquids through porous media is defined by Darcy's law when bulk movement is considered. In several cases, however, it is of interest to know how elements of volume or discrete particles carried by the liquid will travel. For instance, in the study of ground-water movement, radio active salts are injected into the soil. The salt will travel through different pores and after a given interval of time will arrive at different places, their distance from the starting point being dependent upon how tortuous was the path they followed.

This results in a dispersion of the injected salt which is additional to the molecular diffusion. This dispersion caused by the geometry of the pore canal system also has the character of a diffusion with a greater value in the mean direction of flow than perpendicular to this direction. Therefore the two different concepts longitudinal diffusion [*Klinkenberg* and *Sjenitzer*, 1956] and transverse diffusion [*Baron*, 1952] have been introduced in order to indicate this phenomenon.

The passage of the fluid between the grains consists of the summation of a great number of random phenomena, every single phenomenon being represented by the transition from one pore to another. The randomness is originated by the geometry of the pore canals around the grain, which is determined by the adjacent grains whose positions have a random character. This suggests the application of probability calculus in order to obtain an impression of the diffusion coefficients. Many workers have undertaken such probability computations. *Danckwerts* [1953], *Scheidegger* [1954], *Klinkenberg* and *Sjenitzer* [1956] obtain diffusion coefficients by application of probability, without defining the microstructural mechanism of the particle movement. Therefore their result contains a numerical constant, describing the granulometric properties of the porous medium which can only be determined by experiment (tests of *Kramers* and *Alberda* [1953] and *Klinkenberg* and *Sjenitzer* [1956]). Also *Day* [1956] describes such an approach, using the results of *Scheidegger* [1954]. The latter introduces an equal magnitude of transverse and longitudinal diffusion. Tests, however, pointed out that there is a marked difference (6 to 8 times).

In the determination of the diffusion coefficients presented here the starting point is a schematized pore canal system and the movement of the liquid through the canals is accounted for in detail.

Previously, others have applied mechanisms which introduced the grain dimensions (for transverse diffusion, *Baron* [1952]; for longitudinal diffusion, Kramers in a private communication, and *Rifai* and others [1957]). These mechanisms show a probability distribution of discontinuous character, the traveling particle being able to choose only between movements of +1, 0, or −1.

We preferred an approach where the probability distribution is continuous with respect to the direction of the path and proportional to the discharge of all the canals oriented in the direction considered. This approach enables derivation of an expression, which contains both transverse and longitudinal diffusity combined in one formula, and shows the observed difference of the diffusion coefficients in transverse and longitudinal direction.

This result was published by the author [*de Josselin de Jong*, 1956] in Dijon for brevity's sake without detailed mathematical explanation. Because of a regrettable computational mistake and some printing errors the formulas shown there are incorrect; so this article serves as a rectification and a justification of the mathematical treatment.

Notation—The following notation is used.

ℓ length of elementary canal

x, y, z, r coordinates of the exit of an elementary canal, when the entrance is placed in the origin

t residence time for particle in elementary canal

u residence time for elementary canal in principal direction of flow

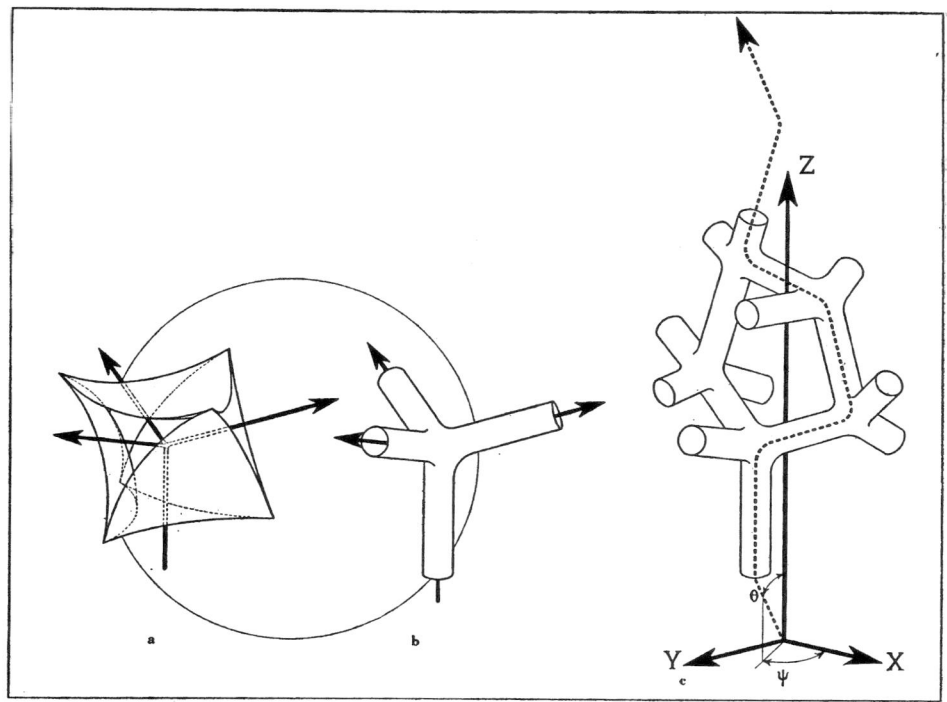

Fig. 1 – Diagram of pore space, (a) tetrahedral pore between four spheres; (b) pore schematized by a canal bifurcation; (c) random path chosen by a foreign particle through the canal system

$\langle x \rangle$, $\langle x^2 \rangle$ $\cdots$ etc., mean values of x, x^2 $\cdots$

θ, ψ angles describing direction of elementary canal

$q_{\theta\psi}$, q_0, Q liquid discharge

v_0 mean velocity of liquid flow

$g_{\theta\psi}$, g_θ, g_ψ distribution functions for choice of paths

P probability function

X, Y, Z, R coordinates of arrival point of a particle, starting in the origin

T time of travel of the particle

N number of canals covered by the particle

X_0, Y_0, Z_0, R_0 coordinates of the maximum number of particles traveling during the time T_0

T_0 arrival time of the maximum number of particles covering a distance Z_0, R_0

N_0 number of canals covered by the maximum number of particles

σ_Z, σ_R, σ_T standard deviations of probability function with respect to Z, R, T

D_Z longitudinal diffusion coefficient

D_R transversal diffusion coefficient

ξ, η, ζ, τ integration variables

a^*, a, b, c, d, e, f functions of the first and second moments of the stochastic variables given by (9) and (12)

A, A', B, C functions given by (15)

λ function of distance Z_0 given by (13) and (23), represented graphically in Fig. 4

$\ln \gamma$ = Euler's constant ≈ 0.577

Schematization of canal system—The pores between grains of spherical form consist of nearly polyhedral cavities connected by triangular shaped canals (Fig. 1a). The resistance to the waterflow is principally determined by the narrow passages, these being oriented at random.

We will schematize this pore system by a network of canals linked together as indicated in Figure 1b. The orientation of the canals is at random, but uniformly distributed in all directions.

We will further suppose that the pressure gradient in each canal is proportional to cos θ, θ being the angle between its direction and the direction of principal flow, Z (Fig. 1c). If for a first approximation the lengths and the conductivity of all elementary canals are taken to be equal, the mean velocity of the liquid in each canal is proportional to cos θ.

The velocity of flow in the canal varies over its section, so that if a foreign particle moves at a velocity equal to that of the fluid in its immediate vicinity, its velocity will be depended on its radial distance from the center. Brownian motion, however, causes a radial diffusion in the canal [see *van Deemter* and others, 1955], so that if the canal section is small enough we may assume that the velocity of the foreign particle is equal to the mean velocity in the canal.

With these assumptions the residence time t_j in the canal j is given by

$$t_j = u/\cos \theta_j \tag{1}$$

where u is the shortest residence time possible, occurring in a canal oriented in the Z direction. If the length of an elementary canal is ℓ, the distance covered in the x, y, z directions by a canal is

$$x_j = \ell \sin \theta_j \cos \psi_j$$
$$y_j = \ell \sin \theta_j \sin \psi_j \qquad z_j = \ell \cos \theta_j \tag{2}$$

Choice of path—A foreign particle carried by the current through the canals has to choose a new direction of motion every time it arrives at a junction. We will assume that the choice of direction between θ, ψ and $\theta + d\theta, \psi + d\psi$ is distributed in proportion to the quantity of water flowing in these directions taken as a fraction of the total discharge of the canals.

This fraction is computed as follows: From a large number A of canals there are by virtue of the uniform distribution with respect to direction

$$A \sin \theta \ (2\pi)^{-1} \ d\theta \ d\psi \text{ canals}$$

oriented in a direction between θ, ψ and $\theta + d\theta$, $\psi + d\psi$. The discharge of these canals is $q_{\theta\psi} = A q_0 (2\pi)^{-1} \cos \theta \sin \theta \ d\theta \ d\psi$ where q_0 is the discharge of an elementary canal oriented in the Z direction. The total discharge of the A canals is

$$Q = (A q_0/2\pi) \int_0^{2\pi} d\psi \int_0^{\pi/2} \sin \theta \cos \theta \ d\theta = \frac{1}{2} A q_0$$

So the fraction of discharge in the $\theta \to \theta + d\theta$, $\psi \to \psi + d\psi$ direction is

$$g_{\theta\psi} d\theta \ d\psi = q_{\theta\psi}/Q = \pi^{-1} \cos \theta \sin \theta \ d\theta \ d\psi \tag{3}$$

The path followed by the particle is determined by the subsequent choices, each being distributed according to the distribution function $g_{\theta\psi}$.

In order to show the scattering effect let us consider how far a particle is deviated when it follows a canal in direction θ, ψ for a time Δt ($< u$). The velocity in this canal is $(\ell/u) \cos \theta$, so that the deviations Δz and Δr are

$$\Delta z = \Delta t \cdot (\ell/u) \cos^2 \theta = \tfrac{1}{2}(\Delta t \cdot \ell/u)(1 + \cos 2\theta)$$

$$\Delta r = \Delta t \cdot (\ell/u) \sin \theta \cos \theta = \tfrac{1}{2}(\Delta t \cdot \ell/u) \sin 2\theta$$

These deviations lie on a sphere with radius $\Delta t \cdot \ell/2u$ and center at $z = \Delta t \cdot \ell/2u$. The distribution function $g_{\theta\psi} = (2\pi)^{-1} \tfrac{1}{2} \sin 2\theta \ d(2\theta) \ d\psi$ also shows that the arrival points on the sphere are uniformly distributed over its surface. This scattering therefore bears a resemblance to Brownian motion super-imposed on a translation, and the problem has become one of random flight. However, there is a difference in that the residence time in the free paths for Brownian motion is independent of the orientation. For the problem considered here, the residence time is inversely proportional to cos θ, so that its value is infinite in the extreme case $\theta = \pi/2$.

For the determination of the diffusion coefficient we can, with some alterations, make use of the probability calculus developed for Brownian motion [*Chandrasekhar*, 1943, pp. 8–16].

Introduction of probability—When the particle has passed through a large number, N, of canals, it has arrived at a point X, Y, Z where

$$\left.\begin{array}{l} X = \sum x_j = \sum \ell \sin \theta_j \cos \psi_j \\[4pt] Y = \sum y_j = \sum \ell \sin \theta_j \sin \psi_j \\[4pt] Z = \sum z_j = \sum \ell \cos \theta_j \end{array}\right\} \tag{4}$$

The summation being from $j = 1$ to N. The time of travel is

$$T = \sum t_j = \sum u/\cos \theta_j \tag{5}$$

With the aid of Markoffs theorem [*Chandrasekhar*, 1943, p. 9] we can determine the probability $P_N(X, Y, Z, T) \ dX \cdot dY \cdot dZ \cdot dT$, that the particle after N passages has arrived in the volume between (X, Y, Z) and $(X + dX, Y + dY, Z + dZ)$ during the time interval T to $T + dT$, as

$$P_N(X, Y, Z, T) \ dX \ dY \ dZ \ dT$$

$$= \frac{dX \ dY \ dZ \ dT}{(2\pi)^4} \int_{-\infty}^{+\infty} d\xi \int_{-\infty}^{+\infty} d\eta \int_{-\infty}^{+\infty} d\zeta \int_{-\infty}^{+\infty} d\tau$$

$$\times \exp \{- i(\xi X + \eta Y + \zeta Z + \tau T)\} \times A_N \tag{6}$$

where

$$A_N = \left[\int_0^{\pi/2} d\theta \int_0^{2\pi} d\psi \ g_{\theta\psi} \right.$$

$$\left. \cdot \exp \{+ i(\xi x_j + \eta y_j + \zeta z_j + \tau t_j)\} \right]^N$$

By introduction of T as the fourth component of the vector whose probability is determined, the dependence of the residence times on the stochastic variables is accounted for. The result P_N, however, is not yet sufficient because the particle may arrive at X, Y, Z in a time T in different ways with different numbers of passages N. For instance, the particle may choose a path which, being oriented in the z direction, brings it quickly very near to XYZ and then take a canal nearly perpendicular to Z with such a long residence time that the rest of T is consumed there. In that case the number of choices, is very small in comparison to the mean number of passages necessary. It is therefore necessary to integrate P_N over all possible values of N to obtain the required probability.

$$P(X, Y, Z, T) = \int_{N_1}^{N_2} P_N(X, Y, Z, T, N) \, dN \quad (7)$$

For points on the Z axis ($X = Y = 0$) the extreme values for N are $N_1 = Z/\ell$ and $N_2 = (ZT/u\ell)^{\frac{1}{2}}$.

Standard deviations—The probability distribution $P(X, Y, Z, T)$ for the arrival of one particle is a measure for the concentration distribution when a great number of particles is injected.

The dispersion may be described by diffusion coefficients which according to Einstein equa$_1$

$$D = \sigma^2/2T \quad (8)$$

where σ is the standard deviation of the probability distribution. As $P(X, Y, Z, T)$ is a function of X, Y, Z, and T we may derive for a given T, the place X, Y, Z where the probability is a maximum, and the standard deviation σ_X, σ_Y, σ_Z in the directions perpendicular and parallel to the principal stream around this maximum. But also we can take a fixed point in space and determine the standard deviation σ_T describing the changes of concentration in course of time at the moment when the maximum concentration passes.

In order to determine these standard deviations from the probability described by (6), several integrations have to be executed.

Integration of the expression for the probability— First the value of A_N has to be determined, which may be effected in a way similar to that shown by *Chandrasekhar* [1943]. We encounter here, however, the difficulty that t_j becomes infinite for $\theta = \pi/2$ according to (1).

We will not enter into the mathematical details, which can be obtained by request to the author, but infer, that to the neglect of small terms for $N \gg 1$, the following result is obtained.

$$A_N = \exp \{N[i\tau d + i\zeta e - (a^*\tau^2 + 2b\tau\zeta + c\zeta^2 + f\xi^2 + f\eta^2)]\} \quad (9)$$

with

$$a^* = [\ln (i/\gamma\tau u) - \tfrac{1}{2}]u^2$$
$$b = \tfrac{1}{2}[\langle tz\rangle - \langle t\rangle\langle z\rangle] = -u\ell/6$$
$$c = \tfrac{1}{2}[\langle z^2\rangle - \langle z\rangle^2] = +\ell^2/36$$
$$d = \langle t\rangle = 2u \qquad e = \langle z\rangle = 2\ell/3$$
$$\langle x\rangle = \langle y\rangle = \langle r\rangle = 0$$
$$f = \tfrac{1}{2}\langle x^2\rangle = \tfrac{1}{2}\langle y^2\rangle = \tfrac{1}{2}\langle r^2\rangle = \ell^2/8$$

where $\langle \, \rangle$ indicates the mean value.

Introduction of this value for A_N into (6) for P_N shows that next integrations versus τ, ζ, ξ, and η have to be executed. With respect to ζ, ξ and η this is readily performed by the use of the well known result

$$\int_{-\infty}^{+\infty} \exp (i\mu\alpha - \mu^2\beta^2) \, d\mu = \pi^{\frac{1}{2}} \beta^{-1} \exp (-\alpha^2/4\beta^2) \quad (10)$$

For the ξ, η part of the integral this gives, using polar coordinates, a result which yields the probability, that the particle should arrive in the ring between R and $R + dR$

$$(2\pi R \, dR/4\pi Nf) \, \exp\{-R^2/4Nf\} \quad (11)$$

From the τ, ζ part of the integral only the integration versus ζ may be treated according to (10) giving

$$(2\pi)^{-2} \, dT \, dZ(\pi/cN)^{\frac{1}{2}} \int_{-\infty}^{+\infty} d\tau \times \exp \{- i\tau[T - dN] + i(b/c)\tau[Z - eN] - ([Z - eN]^2/4cN) - N\tau^2[a^* - (b^2/c)]\}$$

The subsequent integration with respect to τ shows the difficulty that a^* is a function of τ. Here again we cannot extend upon the mathematical details, (which can be obtained by request) by which may be proved that a very good approximation is obtained by substitution of the constant a instead of a^* according to

$$a = (\lambda - \tfrac{1}{2} - \ln \gamma)u^2 \quad (12)$$

with λ the root of

$$N = \tfrac{1}{2}e^{2\lambda}/(\lambda - \tfrac{3}{2} - \ln \gamma) \quad (13)$$

The relationship between λ and N according to (13) is represented in Figure 2. Again the larger N, the better the approximation. With this substitution, the coefficient of τ^2 in the integrant is independent of τ, and (10) may be used. Then finally the following result is obtained for P_N.

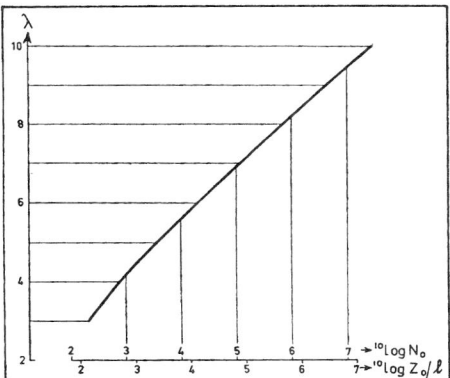

FIG. 2 – Graphical representation of Eq. 13 combined with Eq. 21; λ as a function of Z_0 the distance of the maximum concentration, and ℓ the elementary canal length

$$P_N(T, Z, R)\, dT\, dZ\, dR = (dT\, dZ\, R\, dR/8\pi N^2 f(ac - b^2)^{\frac{1}{2}})$$
$$\times \exp\{-(A + A')N^{-1} + B - CN\} \quad (14)$$

wherein

$$\left.\begin{array}{l} A = [cT^2 - 2bTZ + aZ^2]/4(ac - b^2) \\ B = 2[cdT - b(eT + dZ) + aeZ]/4(ac - b^2) \\ C = [cd^2 - 2b\,de + ae^2]/4(ac - b^2) \\ A' = R^2/4f \end{array}\right\} \quad (15)$$

In order to obtain the diffusion coefficients as the final result of our considerations, the subsequent treatment of (14) for the probability P_N, that a particle after N choices arrives at a point Z, R in a time T, involves an integration with respect to N, and the determination of the standard deviations with respect to Z, R, or T. These operations become easy to perform if it is realized that according to the central-limit theorem the different probabilities involved will approach a normal law for $N \gg 1$. Therefore, if a result is obtained for the probability distribution of a variable μ given by

$$P(\mu)\, d\mu = \exp\{F(\mu)\}\, d\mu$$

we can replace the function F of μ in the exponent by

$$F(\mu) = F(\mu_0) + \tfrac{1}{2}(\mu - \mu_0)^2 F''(\mu_0) \quad (16)$$

where μ_0 is the value of μ for which $F(\mu)$ has its maximum. Because the distribution of $P(\mu)$ is approximately normal the higher order terms in the Taylor expansion of $F(\mu)$ may be neglected.

From (16) it is obvious that the standard deviation of $P(\mu)$ approaches the value

$$\sigma_\mu = [-F''(\mu_0)]^{\frac{1}{2}} \quad (17)$$

The evaluation of an integral of the form

$$\int_{\mu_1}^{\mu_2} P(\mu)\, d\mu = \int_{\mu_1}^{\mu_2} \exp\{F(\mu)\}\, d\mu$$

can by virtue of the substitution (16) be approximated by

$$\int_{-\infty}^{+\infty} \exp\{F(\mu_0) - \tfrac{1}{2}[-F''(\mu_0)](\mu - \mu_0)^2\}\, d(\mu - \mu_0)$$
$$= [2\pi/ - F''(\mu_0)]^{\frac{1}{2}} \exp\{F(\mu_0)\} \quad (18)$$

The error introduced by the additional integrals over the intervals $-\infty \to \mu_1$ and $\mu_2 \to +\infty$ is negligible if the integrand in these intervals consists of an exponential of large negative value. In the case of the integration with respect to N, which has to be performed on (14) according to (7), the values of N_1, N_2, and $F(N)$ are such that in agreement with this condition the extension of the boundaries to $+\infty$ and $-\infty$ is permissible. Therefore the result of the integration versus N can be obtained by determining the value of N_0 for which $F(N)$ in (16) is a maximum and subsequently $F''(N_0)$. These determinations necessitate only differentiations on $F(N)$ being operations which are always feasible. Neglecting terms of small order, there results

$$N_0 = [(A + A')/C]^{\frac{1}{2}} \quad \sigma = [N_0/C]^{\frac{1}{2}} \quad (19)$$

and the probability for the particle to arrive in time T at a point with coordinates Z, R is

$$P(T, Z, R)\, dT\, dZ\, dR = \frac{dT\, dZ\, R dR}{8f[\pi(ac - b^2)]^{\frac{1}{2}}}\left[\frac{C}{(A + A')^3}\right]^{\frac{1}{4}}$$
$$\times \exp\{B - 2[1 + C(A + A')]^{\frac{1}{2}}\} \quad (20)$$

wherein A, A', B, C are given by (15) and a, b, c, d, e, f by (9) and (12).

Character of the probability $P(T, Z, R)$—From (20) for the probability several conclusions may be drawn with respect to standard deviations. The expression is a function of T, Z and R because the factors A, A' and B contain them. The maximum of $P(T, Z, R)$ lies in T_0, Z_0, R_0 given by

$$T_0 = dZ_0/e = 3uZ_0/\ell \quad R_0 = 0 \quad (21)$$

This means that the maximum concentration of an injected substance travels at a velocity v_0 equal to the mean velocity $\langle z \rangle / \langle t \rangle = \tfrac{1}{3}\ell/u$. That the maximum also follows the axis $R = 0$ is evident.

The standard deviations about the maximum may be derived with (17). We obtain then

$$\sigma_R = [2Z_0 f/e]^{\frac{1}{2}} = [Z_0\langle r^2\rangle/\langle z\rangle]^{\frac{1}{2}} = \ell[3Z_0/8\ell]^{\frac{1}{2}}$$

$$\sigma_T = [2Z_0(cd^2 - 2bde + ae^2)/e^3]^{\frac{1}{2}}$$

$$= |Z_0(\langle z^2\rangle\langle t\rangle^2 - 2\langle zt\rangle\langle z\rangle\langle t\rangle$$

$$+ \tfrac{1}{2}(\lambda + \tfrac{3}{2} - \ln \gamma)u^2\langle z\rangle^2)/\langle z\rangle^3]^{\frac{1}{2}}$$

$$\mathfrak{S}_T = u[3Z_0(\lambda + \tfrac{3}{4} - \ln \gamma)/\ell]^{\frac{1}{2}}$$

$$\sigma_Z = (e/d)\sigma_T = \tfrac{1}{3}\ell[3Z_0(\lambda + \tfrac{3}{4} - \ln \gamma)/\ell]^{\frac{1}{2}}$$

(22)

In these expressions Z_0 is the distance along the z axis, where the maximum is at the moment T_0. The maximum number of choices N_0 for the maximum at Z_0 is given according to (19) by

$$N_0 = Z_0/e = 3Z_0/2\ell \qquad (23)$$

The length of ℓ is given by the geometry of the pore canal system, while λ can be read from Figure 2 as a function of N_0 or computed by (13).

The standard deviation σ_T applies to the change of concentration registered at a point on the axis when the maximum passes by. The approximated value of σ_T as given in (22) is applicable for the case of a point injection as well as for the breakthrough curve of a plane front. Exact computation of these two different cases from (20) would yield a slightly different result.

The standard deviations σ_R and σ_Z represent the dispersion of a point injection in longitudinal and transverse directions at a certain moment. From these values the diffusion coefficients D_R and D_Z may be computed by (8) and give

$$D_R = \sigma_R^2/2T_0 = 3\,v_0\ell/16$$
$$D_Z = \sigma_Z^2/2T_0 = v_0\ell(\lambda + \tfrac{3}{4} - \ln \gamma)/6$$

(24)

From this result we deduce that the interrelation between these diffusion coefficients is

$$D_Z/D_R = 8(\lambda + \tfrac{3}{4} - \ln \gamma)/9 \qquad (25)$$

which according to (13) is dependent on the distance, since λ is a function of N_0 and $N_0 = 3Z_0/2\ell$. The order of magnitude of D_Z/D_R would be 6.5 for $N_0 = 10^5$.

In the foregoing computations ℓ the length of the elementary canals was invariable. The introduction of a statistically distributed canal length in the given analysis, entails no essential difficulties. We did not represent its influence because we assumed that for the example given by the experiment this influence was small in comparison to the scattering effect produced by the orientation of the canals.

Experiments—Several experiments have been run with the percolation apparatus represented

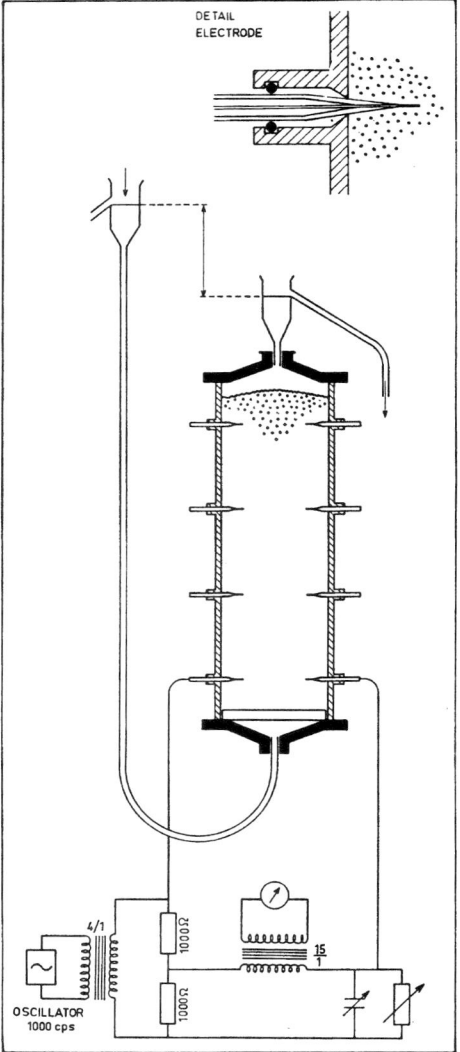

FIG. 3 – Test device

diagrammatically in Figure 3. The cylinder (ϕ 6 cm, height 20 cm) was packed with glass spheres of 0.2 mm diameter (standard deviation 0.05 mm). The packing was carefully controlled to procure a uniform porosity of 38.5 volume pct.

The system was saturated with distilled water with NaCl to a concentration of 0.02 normal. The liquid under the filter was replaced by water with a 1.0 normal NaCl concentration. Next the water-flow was started at a filter velocity upwards of

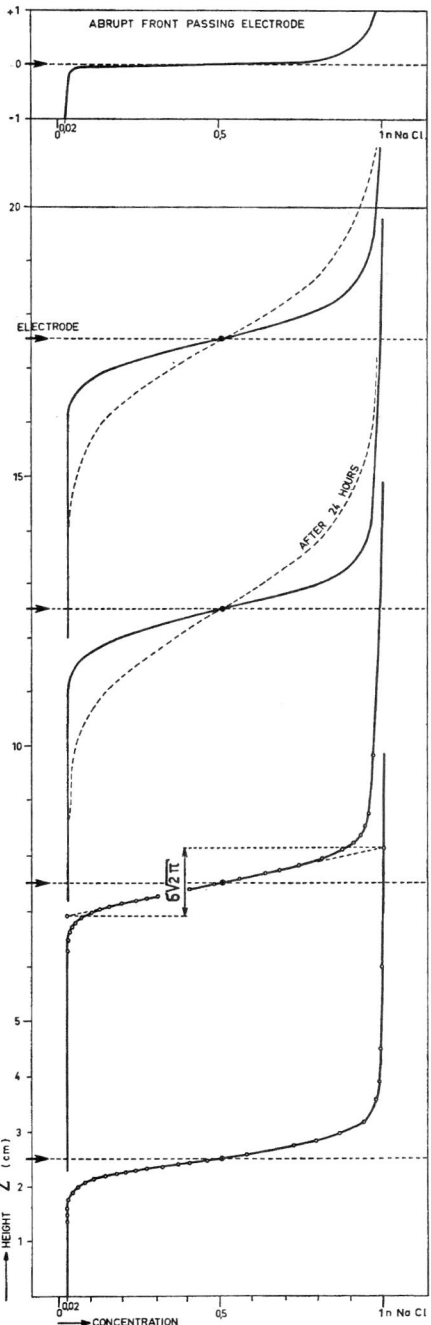

FIG. 4 – Salinity as registered by electrodes

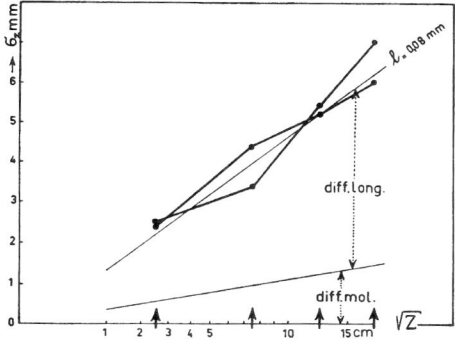

FIG. 5 – Observed standard deviations
as related to theory

6 cm³/minute. The mean velocity through the pores $\langle z \rangle / \langle t \rangle$ was

$$(6 / \tfrac{1}{4} \pi 6^2) \times (1/0.385) = 0.55 \, \text{cm/min}$$

At different distances from the bottom (2.5, 7.5, 12.5, 17.5 cm) electrodes were inserted into the grain material, and permitted salinity measurements of the interstitial liquid. The electrodes consisted of a platinum wire ϕ 0.7 mm, length 3 mm and were covered with platinum-black. The electrical specific resistance of the material around these electrodes was determined with the Wheatestone bridge (1000 cps, 50 μA) represented diagrammatically in Figure 3. Calibration in different NaCl concentrations gave the relation between salinity and electric resistance in the circuit. Calibration in the liquid alone as well as after introduction of the grains gave values which correspond with theoretical computations based on potential distribution around ellipsoids and reduction of conductivity by the grain material. By measuring the resistivity variations with respect to time the break-through curve as the saline front passed by, could be determined.

In Figure 4 is given the salinity as registered by the electrodes at the moment that a particle traveling at the mean velocity would have arrived at the distance indicated by the abscissa. The error introduced by the finite dimension of the electrodes is shown by the curve at the top of Fig. 4 where, based on theoretical calculations of the potential distribution round ellipsoids, the apparent concentration variations measured by an electrode have been drawn for the passing of an abrupt front.

The observed curves show a distribution that is nearly normal. Their inclination increases with the distance from the bottom indicating that also the standard deviations σ_z increase with Z. In Figure 5 the relation between σ_z and $\sqrt{Z}$ is shown. In the same figure we inserted a line for $\ell = 0.08$ mm computed with (22).

For ℓ the value 0.08 mm agrees best. This value is about $\frac{1}{3}$ of the main grain diameter. In a close packing this is the distance of the centers of two neighbouring tetrahedral pores (see Fig. 1a). Here the packing was looser, but the difference in sizes of the spheres may account for the small value of ℓ since the number of smaller spheres present was not negligible.

In Figure 5 the molecular diffusion has also been shown because in the total observed diffusion the effect of Brownian motion is present. Molecular diffusion was determined in the percolation apparatus as follows.

The percolation test was executed as above for the front passing the two lower electrodes. Then the liquid flow was stopped during 24 hours and started again in order to let the front pass the two upper electrodes. The additional inclination of the break-through curves correspond to an acceptable diffusion coefficient of

$$D_m = 0.7 \times 10^{-5} \text{ cm}^2/\text{sec}.$$

The normal runs lasted about 30 minutes, so the influence of molecular diffusion was not sufficient to affect the interpretation of the test results (see Fig. 5).

Conclusions—The introduction of a continuous probability distribution for the random choice of travel direction at a bifurcation in the pore canal system permits the computation of longitudinal and transversal diffusion from a formula which contains both together as a probability expression (20). Both diffusion coefficients are proportional to the mean velocity, as has previously been found by several other workers in this field. Further they are proportional to ℓ which is of the order of the grain size, but depends on the grain-size distributions and packing. In the glass beads packing investigated ℓ amounted to about $\frac{1}{3}$ of the mean diameter of the grains.

The longitudinal diffusion coefficient moreover depends on the distance Z_0, which has been covered by the particle traveling at the mean velocity. This dependence is nearly proportional to $\ln \sqrt{Z_0}$. In experiments, where Z_0 varies over a small range this dependence will remain unob-

served, but the effect may be of importance when scale test results have to be interpreted.

The test results represented by *Rifai* and others [1956] seem not to be in contradiction with this result, that D_z should be a function of Z_0.

Since the transverse diffusivity does not depend on Z_0, the ratio between D_z and D_R increases as the mean distance of travel increases. From the few experimental data available values of D_z/D_R from 6 to 10 have been reported. According to our theory this would correspond to a traveling distance of 10^5 to 10^8 times the elementary canal length.

Acknowledgments—We are indebted to R. Timman for his assistance in the mathematical studies and H. Kramers for his suggestions in physical matters. Miss Hoving carried out the experiments.

REFERENCES

BARON, T., Generalized graphical methods for the design of fixed bed catalytic reactors, *Chem. Eng. Progress*, **48**, 118–124, 1952.

CHANDRASEKHAR, S., Stochastic problems in physics and astronomy, *Rev. Mod. Phys.*, **15**, 1–89, 1943.

DANCKWERTS, P. V., Continuous flow systems; distribution of residence times, *Chem. Eng. Sci.*, **2**, 1–13, 1953.

DAY, P. R., Dispersion of a moving salt water boundary, *Trans. Amer. Geophys. Union*, **37**, 595–601, 1956.

DE JOSSELIN DE JONG, G., L'entertainement de particules par le courant interstiticiel, Publ. 41, Assn. Hydrology, UGGI, *Symposia Darcy*, 139–147, 1956.

JAHNKE, E., AND F. EMDE, *Tables of functions*, Dover Publ., New York, 382 pp., 1945.

KLINKENBERG, A., AND F. SJENITZER, Holding time distributions of the Gaussian type, *Chem. Eng. Sci.*, **5**, 258–270, 1956.

KRAMERS, H., AND G. ALBERDA, Frequency response analysis of continuous flow systems, *Chem. Eng. Sci.*, **2**, 173–181, 1953.

RIFAI, M. N. E., W. J. KAUFMAN, AND D. K. TODD Dispersion phenomena in laminar flow through porous media, *Progress Rep. 2, Canal Seepage Research*, University of California, Berkeley, 157 pp., 1956.

SCHEIDEGGER, A., Statistical hydrodynamics in porous media, *J. Appl. Physics*, **25**, 994–1001, 1954.

VAN DEEMTER, J. J., J. J. BROEDER, AND H. A. LAUWERIER, Fluid displacement in capillaries, *Appl. Sci., Res., Sec. A*, **5**, 5, 374–388, 1955.

Laboratorium Voor Grondmechanica, Postbox 69, Delft, The Netherlands

(Communicated manuscript received March 25, 1957, and, as revised, September 19, 1957; open for formal discussion until July 1, 1958.)

Discussion of "Longitudinal and Transverse Diffusion in Granular Deposits"

BY G. DE JOSSELIN DE JONG

[*Trans.*, **39**, 67–74, 1958]

J. A. Cole (Department of Geodesy and Geophysics, University of Cambridge, Cambridge, England)—The writer finds the paper most valuable, because he [*Cole*, 1957] has made experiments with columns of granular material, basically similar to those described by the author, in order to determine the relative importance of: Effect (a) radial molecular diffusion in each pore combined with a velocity gradient across the pore, and Effect (b) the geometry of the pore system, in producing a longitudinal dispersion of an injected substance.

The author mentions Effect (a), but makes the assumption that it may be neglected if the pore section is small enough. Though valid, such an assumption is not often realized, as the following numerical examples illustrate. *Taylor* [1953, 1954ab] has shown that the diffusion coefficient D_S of Effect (a) with streamline flow in straight cylindrical tubes of radius a is

$$D_S = a^2 v_0^2 / 48\,D_m \tag{26}$$

when

$$4L/a \gg v_0 a/D_m \gg 6.9 \tag{27}$$

L being the distance along the tube in which the concentration of the injected substance has an appreciable gradient, and is approximately equal to $4\,\sigma_Z$. The first condition in (27) can easily be met by using large enough values of Z. Figure 6 is a presentation of D_S versus v_0 on a log-log graph, for two values of a. Where the two lines are broken the second condition in (27) does not hold, and (26) is only an approximation. In applying (26) to granular materials, which do not have cylindrical pores as a rule, one uses a mean pore radius (see $\bar{a}$ below) in computing D_S. The random orientation of the pores results in one's observing not D_S, but $D_S/\xi^2 = a^2(\langle z\rangle/\langle t\rangle)^2/48D_m$, where $\langle z\rangle/\langle t\rangle = v_0/\xi$.

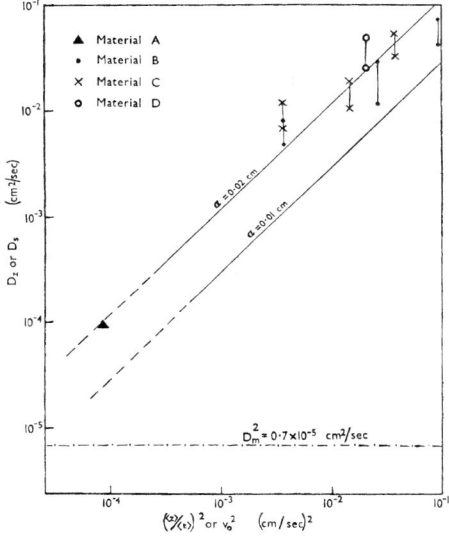

FIG. 6 – Comparison of D_z observed in the granular materials listed in Table 1 with theoretical values of D_s.

TABLE 1 – *Properties of porous columns*

Material	Composition of material	d	p	$\bar{a}$
		cm		cm
A	Glass beads	0.020	0.385	0.0086
B	Glass beads	0.047	0.35	0.018
C	Glass beads	0.027	0.33	0.0105
D	Clean quartz sand	0.034	≈ 0.33	≈ 0.013

Clearly a line corresponding to D_S/ξ^2, for a given value of a, plotted versus $(\langle z\rangle/\langle t\rangle)^2$ is coincident with that of D_S versus v_0 for the same value of a.

The author gives for his measurements in a glass-bead column, $\sigma_Z = 0.6$ cm when $Z = 17.0$ cm and $\langle z \rangle / \langle t \rangle = 0.55$ cm/min $= 0.0092$ cm/sec. Accordingly $D_Z = 0.6^2 \times 0.0092/34$ cm²/sec $= 9.6 \times 10^{-5}$ cm/sec.

This value of D_Z is shown (as the point for Material A) on Figure 6. Also shown are some values of D_Z in columns of glass beads and of clean quartz sand, which were obtained by the writer. Materials A, B, C, and D have the properties specified in Table 1, where $d = $ mean grain diameter, $p = $ coefficient of volume of porosity, and $\bar{a} = (d/2)[p/(1 - p)]^{\frac{1}{3}} = $ the mean radius of the pores, approximating the polyhedra to spheres.

These values of $\bar{a}$ correspond to values of D_S/ξ^2 which are far from negligible. In Material A, the case quoted by the author, it is likely that Effect (a) was appreciable (one tenth of the observed D_Z say). In the other examples Effect (a) was of major importance, amounting to a quarter of the observed D_Z in Materials B and D, and almost all thereof in material C.

In contrast to the above laboratory experiments, the writer [Cole, 1957] also made field experiments with tracers flowing in chalk aquifers. These showed values of D_Z some hundred times greater than would have been expected from Effect (a) alone, and it was presumed that Effect (b) was predominant. Even that was not certain, owing to possible chemisorption effects within the aquifer.

There is thus scope for development of a fuller treatment of longitudinal dispersion in granular materials. Besides Effects (a) and (b), such a treatment would ideally include the work of de Vault [1943], Thomas [1944], and Goldstein [1953] on chemisorption processes in porous columns. Further laboratory work would profit by the determination of the pore radii in the granular material, by measurement of its water content over a range of negative pressures [Foster, 1948].

REFERENCES

COLE, J. A., The flow of ground water, M. S. thesis, University of Cambridge, 204 pp., 1957.

DE VAULT, D., Theory of chromatography, J. Amer. Chem. Soc., 65, 532–540, 1943.

FOSTER, A. G., Pore size and pore distribution, Disc. Faraday Soc., 3, 41–51, 1948.

GOLDSTEIN, S., On the mathematics of exchange processes in fixed columns, pt. 1, Mathematical solutions and asymptotic expansions, Proc. R. Soc., 219A, 151–185, 1953.

TAYLOR, GEOFFREY, Dispersion of soluble matter in solvent flowing slowly through a tube, Proc. R. Soc., 219A, 186–203, 1953.

TAYLOR, GEOFFREY, The dispersion of matter in turbulent flow through a pipe, Proc. R. Soc., 223A, 446–468, 1954a.

TAYLOR, GEOFFREY, Conditions under which dispersion of a solute in a stream of solvent can be used to measure molecular diffusion, Proc. R. Soc., 225A, 473–477, 1954b.

THOMAS, HENRY C., Heterogeneous ion exchange in a flowing system, J. Amer. Chem. Soc., 66, 1664–1666, 1944.

G. de Josselin de Jong (Laboratorium Voor Grondmechanica, Postbox 69, Delft, The Netherlands, Author's closure)—In his discussion Cole gives further information on the deviation from our theory to be expected in experiments for which the basic assumption of piston flow in the elementary canals is unsatisfactory so that the influence of holdback can not be neglected. According to the conception of Van Deemter and others [1955], this holdback amounts to $H = 0.02$ in the test described, because the factor $(D_m t/a^2)^{\frac{1}{2}} = 11$, where $D_m = $ molecular diffusion $= 0.7 \times 10^{-5}$ cm²/sec, $t = $ time to cover the distance $4 \sigma_z$ with mean velocity $= 2.4$ cm/0.0092 cm/sec $= 270$ sec, and $a = $ radius of pores $= 0.004$ cm.

For the interpretation of the test results in the article this amount of holdback was considered as negligible. According to his theory Cole obtains a somewhat larger influence about one tenth of the longitudinal diffusion.

As the principal point of our considerations was to show the interrelation between longitudinal and transverse diffusion and the possibility to describe these two phenomena in one mathematical expression, the influence of holdback and molecular diffusion both were omitted in the theoretical considerations for simplicity's sake. But we do agree that these two effects and the chemisorption as mentioned in the discussion may have an overwhelming influence in certain circumstances.

In order to show the phenomenon of longitudinal and transverse diffusion we add here some photographs of a Christiansen filter (Fig. 7). This is a deposit of crushed optical glass (grain diameter 0.42 to 1.19 mm) saturated with a solution of NH_4J in water of such a concentration that both refraction indices of grains and pore liquid are equal for sodium light.

As the light is not dispersed by deviation of its direction at the interfaces of grains and liquid the system becomes perspicace. (We are indebted to the Shell Laboratory, Amsterdam, for this device.)

The pore liquid runs vertically through the

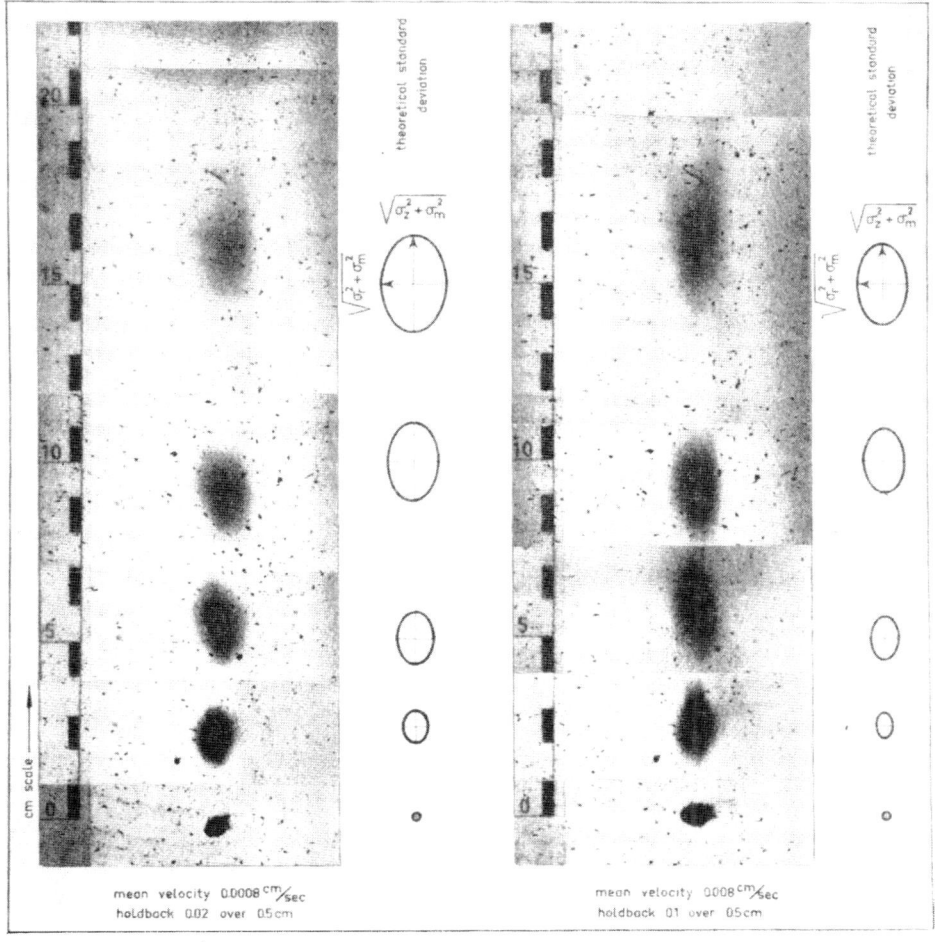

FIG. 7—Christiansen filter showing longitudinal and transverse diffusion

deposit. An analine dye is introduced with an injection needle. The molecular diffusion of the dye is about $D_m = 10^{-5}$ cm² sec. The photographs show the dispersion originated at two velocities, 0.0008 cm/sec and 0.008 cm/sec, which induce a holdback over a distance of 0.5 cm of respectively 0.025 and 0.10. The standard deviations originated by molecular diffusion in combination with longitudinal and transverse diffusion as computed from the theory are added in the figure for comparison.

At the lower velocity the elongation of the standard deviation ellipsoid is principally determined by the longitudinal diffusion, while at the higher velocity the influence of the holdback enters into the observation.

REFERENCE

VAN DEEMTER, J. J., J. J. BROEDER, AND H. A. LAUWERIER, Fluid displacement in capillaries, *Appl. Sci., Res.*, sec. A., **5**, 374–388, 1955.

Journal of Hydrology, 84 (1986) 55—79
Elsevier Science Publishers B.V., Amsterdam — Printed in The Netherlands

[2]

TRANSVERSE DISPERSION FROM AN ORIGINALLY SHARP FRESH— SALT INTERFACE CAUSED BY SHEAR FLOW

G. DE JOSSELIN DE JONG and C.J. VAN DUIJN

Delft University of Technology, P.O. Box 356, 2600 AJ Delft (The Netherlands)

(Received January 7, 1985; revised and accepted October 5, 1985)

ABSTRACT

De Josselin de Jong, G. and Van Duijn, C.J., 1986. Transverse dispersion from an originally sharp fresh—salt interface caused by shear flow. J. Hydrol., 84: 55—79.

In this paper the influence of transversal dispersion and molecular diffusion on the distribution of salt in a plane flow through a homogeneous porous medium is studied. Since the dispersion depends on the velocity and the velocity on the distribution of salt (through the specific weight) this is a nonlinear phenomenon. In particular for the flow situation considered, this leads to a differential equation which has the character of nonlinear diffusion.
    The initial situation (at $t = 0$) is chosen such that the fresh- and salt water are separated by an *interface*, and each fluid has a constant specific weight γ_1 and γ_2, respectively. For this initial situation, the solution of the nonlinear diffusion equation has the form of a similarity solution, depending only on $\zeta/\sqrt{t}$, where ζ denotes the local coordinate normal to the original interface plane and t denotes time.
    Properties of this similarity solution are discussed. In particular it is shown how to obtain this solution numerically. The interpretation of these mathematical results in terms of their hydrological significance is given for a number of worked out examples. These examples describe the distribution of salt, as a function of ζ and t, for various flow conditions at the boundaries $\zeta = \pm\infty$. Also examples are given where the molecular diffusion can be disregarded with respect to the transversal dispersion.

INTRODUCTION

When fluids of different densities are present in an aquifer and the density varies in horizontal directions, the fluid motion contains rotation. In the case of a sharp interface this rotation results in a shearflow, which is proportional to the density difference and the interface inclination. Its magnitude was established by Edelman (1940). The rotations and shearflows resulting in the more general case of gradual density variations were treated by De Josselin de Jong (1960).

Specific discharges in an aquifer are accompanied by dispersion. Therefore it can be expected, that at an inclined interface dispersion occurs, which results in changing the abrupt transition from one density to the other, in a

56

gradual transition zone. Since dispersion can be described in terms of differential equations (see Bear, 1975) it must be possible to express the spreading from an abrupt interface mathematically in terms of the dispersion parameters.

The governing relations form a coupled system, consisting of equations describing the specific discharge rotations, due to the gradual density variations, and the dispersion, caused by the specific discharge distribution. Solving this system analytically in the general case of arbitrary density distributions and additional superimposed specific discharge distributions seems rather tedious. For practical purposes, therefore, Verruijt (1971) proposed to introduce a new parameter to describe the spreading, without specifying its relation to the dispersion parameters and the density variations.

The purpose of this paper is to show, that it is possible to describe the spreading process analytically, when starting from an abrupt interface in certain simple circumstances, such that the discharge remains parallel and a plane flow situation occurs. For that simplified problem the coupled system can be reduced to one ordinary differential equation of diffusion type. The properties of solutions of this equation are known, because they have been studied extensively from the mathematical standpoint, by Gilding and Peletier (1977), Van Duijn and Peletier (1977) and Van Duijn (1986a).

How these solutions are applied to describe the spreading of density variations from an abrupt interface is shown in this paper.

SIMPLIFIED PROBLEM

In this paper the case is considered of a dispersion zone developing from an originally flat, inclined interface, that extends in all directions to infinity. In order to simplify the analysis, the conditions at the boundaries of the infinite aquifer are assumed to be such, that the flow is constant in planes parallel to the original interface plane.

Let coordinates ξ, η be in the original interface with η horizontal and ξ pointing upwards at an angle α with the horizontal, see Fig. 1. The co-ordinate ζ is normal to the original interface plane and points upwards. Flow conditions then are such, that the specific discharge components q_ξ, q_η, q_ζ satisfy:

$$q_\eta = q_\zeta = 0 \qquad \partial q_\xi / \partial \xi = \partial q_\xi / \partial \eta = 0 \qquad (1)$$

indicating that only q_ξ is nonzero and a function of only ζ and the time t, i.e. $q_\xi = q_\xi(\zeta, t)$.

At time $t = 0$ the interface is assumed to be sharp, such that the plane $\zeta = 0$ separates the aquifer into two regions in which the density of the fluids is a constant. Above it is freshwater with density ρ_1 and below it salt water with density ρ_2. In the description below the specific weight $\gamma = \rho g$ is used instead of density ρ. So in formula the situation is initially:

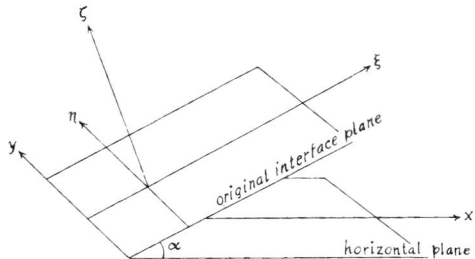

Fig. 1. Interface plane at angle α with horizontal, separating freshwater above it from salt water below it, at time $t = 0$.

$$\begin{aligned} \gamma &= \gamma_1 \quad &\text{for} \quad & \zeta > 0, t = 0 \\ \gamma &= \gamma_2 \quad &\text{for} \quad & \zeta < 0, t = 0 \end{aligned} \tag{2}$$

Under influence of this specific weight difference, initially an Edelman (1940) shear flow $\hat{q}$ exists at the interface of magnitude:

$$\hat{q} = (q_{\xi_1} - q_{\xi_2}) = (\kappa/\mu)(\gamma_2 - \gamma_1) \sin \alpha \tag{3}$$

(see e.g. De Josselin de Jong, 1981). In this expression q_{ξ_1}, q_{ξ_2} are the specific discharge components parallel to the interface in the fresh- and salt water regions, respectively. κ is the intrinsic permeability of the aquifer, considered to be a constant, and μ is the dynamic viscosity of the fluids. For fresh- and salt water the viscosity differs by an amount small enough to disregard its influence on the results of the analysis below: for a justification see e.g. Verruijt (1980).

Superimposed on the shear flow an average specific discharge in ξ direction of magnitude $\beta\hat{q}$ is considered to occur in this paper, with β a number, that remains constant in time. The initial flow conditions are then in accordance with eqn. (3) given by:

$$\begin{aligned} q_{\xi_1} &= (\beta + \tfrac{1}{2})\hat{q} \quad &\text{for} \quad & \zeta > 0, t = 0 \\ q_{\xi_2} &= (\beta - \tfrac{1}{2})\hat{q} \quad &\text{for} \quad & \zeta < 0, t = 0 \end{aligned} \tag{4}$$

The following situations are to be distinguished, see Fig. 2:

$\beta < -\tfrac{1}{2}$ both fresh- and salt water flow downward

$\beta = -\tfrac{1}{2}$ the fresh water is stationary

$-\tfrac{1}{2} < \beta < \tfrac{1}{2}$ fresh flows up, salt flows down

$\beta = \tfrac{1}{2}$ the salt water is stationary

$\beta > \tfrac{1}{2}$ both fresh- and salt water flow upwards.

As time proceeds the transition from γ_1 to γ_2, which originally is sharp at $\zeta = 0$, will spread by hydraulic dispersion and molecular diffusion. Salt water

58

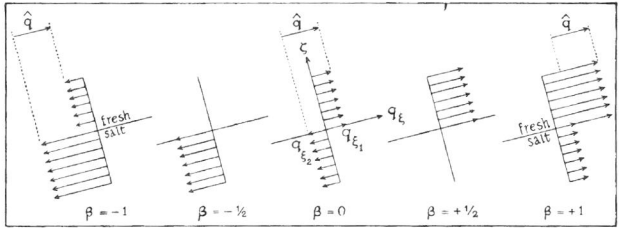

Fig. 2. Distribution of q_ξ at time $t = 0$ for different values of β.

will mix with freshwater, and the specific weight will become a variable function of position and time. Because of the plane character of the case considered, γ will be a function of ζ and t only, i.e. $\gamma = \gamma(\zeta, t)$, which implies that:

$$\partial\gamma/\partial\xi = \partial\gamma/\partial\eta = 0 \tag{6}$$

At infinity the specific weights will tend towards the original values γ_1 at $\zeta \to +\infty$ and γ_2 at $\zeta \to -\infty$.

The mathematical description of the spreading process is developed below from basic equations, describing the changes in specific weight and specific discharge q_ξ in course of time. The boundary conditions at infinity are assumed to be such, that the specific discharges remain constant there and equal to the initial values eqn. (4). Thus:

$$\left.\begin{array}{llll} \gamma = \gamma_1, & q_\xi = (\beta + \tfrac{1}{2})\hat{q} & \text{for} & \zeta \to +\infty, t > 0 \\[2mm] \gamma = \gamma_2, & q_\xi = (\beta - \tfrac{1}{2})\hat{q} & \text{for} & \zeta \to -\infty, t > 0 \end{array}\right| \tag{7}$$

The result of the analysis will be a distribution of specific discharge and specific weight as functions of ζ and t. An example is shown in Fig. 3 for the

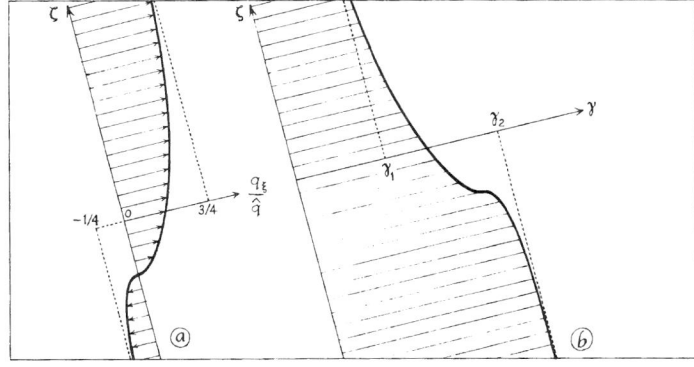

Fig. 3. Distribution of specific discharge (a) and specific weight (b) for the case $m = 0$, $\beta = 1/4$ at some time, $t > 0$.

case $\beta = \frac{1}{4}$ at some later time, $t > 0$. A more detailed description of this figure is given in the sections below.

BASIS EQUATIONS

Flow rule

When an aquifer contains fluids of different specific weight, the flow rule is obtained by considering equilibrium of forces. Let K_q be the force required to act on a unit volume of pore fluid in order to maintain the specific discharge q through a porous medium. According to Darcy's experiments this force equals $K_q = (\mu/\kappa)q$. Let K_p be the force exerted on a unit volume of pore fluid by the gradient in the pressure p and the influence of gravity on the specific weight, γ. This force equals $K_p = -\operatorname{grad} p + \gamma e$, where e is a unit vector in the direction of gravity, i.e. downward. Equating K_q and K_p results in:

$$(\mu/\kappa)q = -\operatorname{grad} p + \gamma e \tag{8}$$

By taking the curl of this relation, the pressure p is eliminated. This leads to:

$$(\mu/\kappa)\operatorname{curl} q = \operatorname{curl}(\gamma e) \tag{9}$$

which, accounting for eqns. (1) and (6), reduces to:

$$(\mu/\kappa)(\partial q_\xi/\partial \zeta) = -(\partial \gamma/\partial \zeta)\sin \alpha \tag{10}$$

Integration of eqn. (10) is possible, because ζ is the only independent variable. This gives:

$$q_\xi = -(\kappa/\mu)\,\gamma \sin \alpha + \text{constant}$$

where the integration constant can be determined by use of eqn. (7). This gives, taking account of eqn. (3):

$$q_\xi = (\kappa/\mu)(\gamma_1 - \gamma)\sin \alpha + (\beta + \tfrac{1}{2})\hat{q}$$

or:

$$q_\xi/\hat{q} = \beta + (\gamma_1 + \gamma_2 - 2\gamma)/2(\gamma_2 - \gamma_1) \tag{11}$$

In this equation q_ξ and γ are the variables, which are both functions of ζ and t. It may be remarked here, that eqn. (11) is a linear relationship between these two variables. This is reflected in Fig. 3, where the curves for q_ξ and γ, respectively, are shown to be each others mirror image, when drawn on appropriate scales.

60

Continuity of fluids

When both fluid and porous medium can be considered to be incompressible, continuity of fluid is satisfied, when div q is zero. Using eqn. (1) it can be verified, that all terms of div q vanish and so continuity of fluid is guaranteed, identically.

Continuity of salt

In stationary groundwater salt is spread by molecular diffusion. In addition, this spreading is enhanced when the fluid moves, because inhomogeneities in the pore space scatter and recombine fluid elements. In the periods of being adjacent, salt is transmitted by molecular diffusion to neighbouring streamlines and carried off in directions deviating from the average flow paths. This process is called mechanical dispersion (see e.g. Bear, 1975).

Averaged over the pore space a salt flux F occurs, which expressed as weight transport per unit time and unit area of the aquifer is:

$$F = - \mathbf{D} \operatorname{grad} \gamma \tag{12}$$

where $\mathbf{D}$ is a second rank tensor. It is the dispersion tensor consisting of terms due to molecular diffusion and mechanical dispersion.

Continuity of salt is satisfied, when the divergence of the exchange flux is balanced by the local rate of change of the specific weight $\partial \gamma / \partial t$ and its convective rate of change $\operatorname{div}(q\gamma)$. When n denotes the porosity of the porous medium, this balance is expressed by:

$$n(\partial \gamma / \partial t) + \operatorname{div}(q\gamma) = - \operatorname{div} F = \operatorname{div}(\mathbf{D} \operatorname{grad} \gamma)$$

Taking account of eqns. (1) and (6) this expression reduces to:

$$n(\partial \gamma / \partial t) = \partial [(nD_{\mathrm{mol}} + \alpha_{\mathrm{T}} |q_\xi|) \, \partial \gamma / \partial \zeta] / \partial \zeta \tag{13}$$

where D_{mol} is the molecular diffusion coefficient and α_{T} is the transverse dispersion length in the direction of ζ (i.e. in the direction perpendicular to q_ξ). A special feature of the dispersion is, that not the specific discharge itself, but its absolute value has to be taken into account.

Equation (13) can be simplified by taking advantage of the linear relationship (11) between γ and q_ξ. This permits to write it in terms of q_ξ only as:

$$n(\partial q_\xi / \partial t) = \alpha_{\mathrm{T}} \, \partial [(\hat{q}m + |q_\xi|) \, \partial q_\xi / \partial \zeta] / \partial \zeta \tag{14}$$

where m is a dimensionless parameter representative for the ratio between the influence of molecular diffusion and mechanical dispersion. It is given by:

$$m = nD_{\mathrm{mol}} / \alpha_{\mathrm{T}} \hat{q} \tag{15}$$

Expressed in these terms the salt flux vector F given by eqn. (12) has only one component F_ζ of magnitude:

$$F_\zeta = + \alpha_T (\gamma_2 - \gamma_1)[(m + |q_\zeta|/\hat{q})\, \partial q_\zeta/\partial \zeta] \tag{16}$$

Equation (14) has to be solved subject to the initial and boundary conditions, eqns. (4) and (7). The mathematical implications of this system of equations is treated in the next sections.

SIMILARITY TRANSFORMATION

The partial differential equation (14) can be converted into an ordinary differential equation with simple boundary conditions, by subjecting it to the Boltzmann (1894) similarity transformation. This means, that the two variables ζ and t are replaced by the independent similarity variable r according to:

$$r = \zeta/(\alpha_T \hat{q} t/n)^{1/2} \tag{17}$$

Introduction of this variable implies that $\partial/\partial \zeta = (\partial r/\partial \zeta)\mathrm{d}/\mathrm{d}r \ldots$ etc., so that eqn. (14) is transformed into the following ordinary differential equation:

$$-\tfrac{1}{2} r(\mathrm{d}q_\zeta/\mathrm{d}r) = \mathrm{d}[(m + |q_\zeta|/\hat{q})\, \mathrm{d}q_\zeta/\mathrm{d}r]/\mathrm{d}r \tag{18}$$

A solution of eqn. (18), which only depends on r is called a similarity solution of the original equation (14). Let the new variable $w = w(r)$ be defined by:

$$w = q_\zeta/\hat{q} \tag{19}$$

Then eqn. (18) reduces to:

$$\tfrac{1}{2} r(\mathrm{d}w/\mathrm{d}r) + \mathrm{d}[(m + |w|)\, \mathrm{d}w/\mathrm{d}r]/\mathrm{d}r = 0 \tag{20}$$

This is a nonlinear, ordinary differential equation in which the relevant coefficient has the form $(m + |w|)$. By its dependence on the absolute value of w, this coefficient creates a special nonlinear character of the problem.

The boundary conditions (4) and (7) reduce by introduction of eqns. (17) and (19) to:

$$\left.\begin{array}{llll} w = \beta + \tfrac{1}{2} & \text{for} & r \to +\infty \\ w = \beta - \tfrac{1}{2} & \text{for} & r \to -\infty \end{array}\right| \tag{21}$$

Problem P(m, β)

The problem, defined by the differential equation (20) with the boundary conditions (21), will be referred to as $P(m, \beta)$, indicating that m and β are the essential parameters. Since these parameters are related to the molecular diffusion (see eqn. 15) and to the specific discharge at infinity (see eqn. 7), it is justified physically to assume that they are real numbers, satisfying:

$$0 \leqslant m < \infty \qquad \text{and} \qquad -\infty < \beta < \infty \tag{22}$$

Using the substitutions (17) and (19) it is possible to write the salt flux from eqn. (16) in the form:

$$F_\zeta = (\gamma_2 - \gamma_1)(\alpha_T \hat{q} n/t)^{1/2} [(m + |w|)dw/dr] \tag{23}$$

Application of similarity solution

A solution of $P(m, \beta)$ is called a similarity solution of eqns. (14), (4) and (7). In Fig. 4 it is shown, how such a solution w as a function of r is related to the curves for q_ζ as a function of ζ and t. The heavy curve in Fig. 4a is the solution $w(r)$ of $P(0, \frac{1}{4})$, determined in a manner that is explained in example 3-ii in the section on practical application.

This curve intersects the axis $r = 0$ in $w_0 = 0.372$. This means that according to eqns. (17) and (19) $q_\zeta = 0.372\,\hat{q}$ at $\zeta = 0$ for every time, $t > 0$. Thus, all the curves in Fig. 4b, which represent the distribution of q_ζ over the height ζ at different time $t > 0$, have the intersection with the axis $\zeta = 0$ in common.

The heavy curve in Fig. 4a intersects the axis $w = 0$ in $r_0 = -0.503$. This means according to eqns. (17) and (19) that the plane where the discharge q_ζ is zero, is located in $\zeta_0 = -0.503\,(\alpha_T \hat{q}t/n)^{1/2}$. As a consequence this plane descends with time proportional to $t^{1/2}$ as shown in Fig. 4c. The curves in Fig. 4b all have the same shape as the heavy curve in Fig. 4a but are stretched with a factor, that is proportional to $t^{1/2}$.

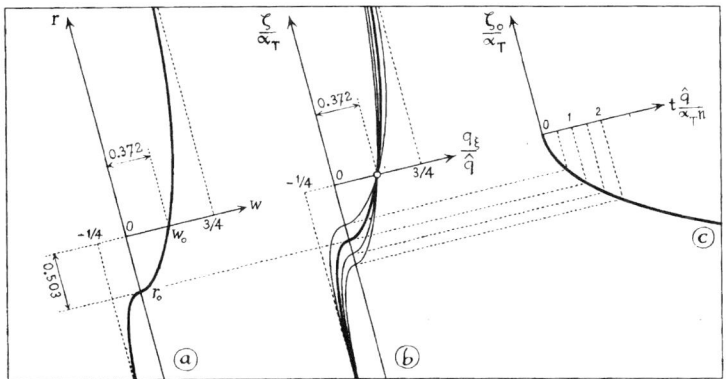

Fig. 4. Relation between a solution $w(r)$ in (a) and the specific discharge q_ζ (ζ, t) in (b). The plane, where $q_\zeta = 0$ descends in course of time according to $\sqrt{t}$, see (c). The curves are for the case $m = 0$, $\beta = 1/4$.

AUXILIARY PROBLEM Q

The problem $P(m, \beta)$ is not explicitly solvable, because of its nonlinear character, although much is known about its solutions. A special difficulty mentioned already is the occurrence of the absolute value $|w|$, which causes solutions to consist of combinations of parts, where $w > 0$ and $w < 0$.

This difficulty is solved by considering first the auxiliary problem Q, which is defined by the differential equation:

$$\tfrac{1}{2} s(du/ds) + d[u(du/ds)]/ds = 0 \tag{24}$$

with boundary condition:

$$u = 1 \quad \text{for} \quad s \to +\infty \tag{25}$$

and the additional condition:

$$0 \leqslant u(s) \leqslant 1 \quad \text{for} \quad -\infty < s < +\infty \tag{26}$$

In this section the solution set of this problem Q is considered. In subsequent sections it is shown, that with these solutions it is possible to produce the solutions of $P(m, \beta)$ for all m and β by application of an appropriate rescaling procedure.

Solutions of Q

The solutions required in this paper are given by the family of curves represented in Fig. 5.

Solutions of equations similar to eqn. (24) were studied extensively by Gilding and Peletier (1977), Van Duijn and Peletier (1977), Gilding (1980) and Van Duijn (1986a). They gave rigorous mathematical proofs about existence and uniqueness of solutions and they studied their behaviour. Using ideas developed in these papers, the following basic facts about solutions of Q can be established, see Appendix A.

All the curves from Fig. 5 are strictly increasing, with $du/ds > 0$, at points where $u > 0$. Different curves cannot intersect. Further, all curves approach the upper boundary $u = 1$ as a complementary error function such that:

$$1 - u = O[\text{erfc}\,(+\tfrac{1}{2} s)] \quad \text{for} \quad s \to +\infty \tag{27}$$

Expressions for the curves in Fig. 5 cannot be given in closed form. They were constructed with a shooting method, which was executed by using a finite difference approximation of eqn. (24), described in Appendix B.

The curves are indicated by type numbers I and II, such that type I curves are located above the heavy line in Fig. 5, and type II curves below that line. The separation line is called *separatrix*. In the description below a few remarks are inserted on the construction of the curves. These are included for those readers who wish to dispose of more accurate values than can be inferred from Fig. 5.

64

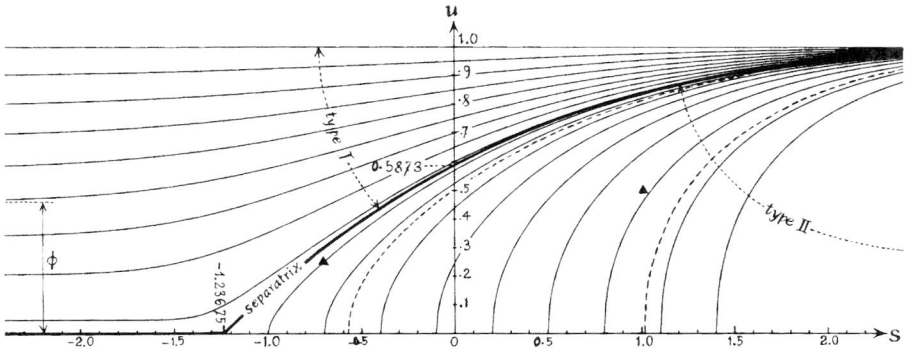

Fig. 5. Family of curves representing solutions of problem Q. Triangles are for example 3-i, where $m = 1/4$, $\beta = 1/4$. Dashed lines are for example 3-ii, where $m = 0$, $\beta = 1/4$.

Type I curves

These remain strictly positive in the entire interval for s, such that u tends to a positive value ϕ as a lower boundary, i.e.:

$$u = \phi \quad \text{with} \quad 0 < \phi < 1 \qquad \text{for} \qquad s \to -\infty \tag{28}$$

This lower boundary is also approached as a complementary error function, in this case such that:

$$u - \phi = O[\text{erfc}\,(-\tfrac{1}{2}s/\phi^{1/2})] \qquad \text{for} \qquad s \to -\infty \tag{29}$$

The numerical procedure for establishing the type I curves is to start from different values u_0 on the vertical axis, $s = 0$. Next u'_0, the value of $(du/ds)_0$ in each starting point, is chosen in such a manner that constructing the curve to the right, it reaches the value $u = +1$ asymptotically as a complementary error function according to eqn. (27). Subsequently starting with these values u_0 and u'_0 the curves are constructed to the left. The asymptotic value ϕ mentioned in eqn. (28) is established using again the complementary error function approximation, now with eqn. (29).

For every value of ϕ, the corresponding starting values are assembled in Fig. 6, u_0 on the left-hand scale and u'_0 on the right-hand scale. For example, the value $\phi = 0.580$ corresponds to $u_0 = 0.8$, $u'_0 = 0.132$.

Type II curves

These remain zero for all values of s below a value s_0, specific for each curve, i.e.:

$$u = 0 \qquad \text{for} \qquad -\infty < s \leqslant s_0 \tag{30}$$

At s_0 the curves start on the base line, $u = 0$, with a vertical tangent $u' \to +\infty$, such that:

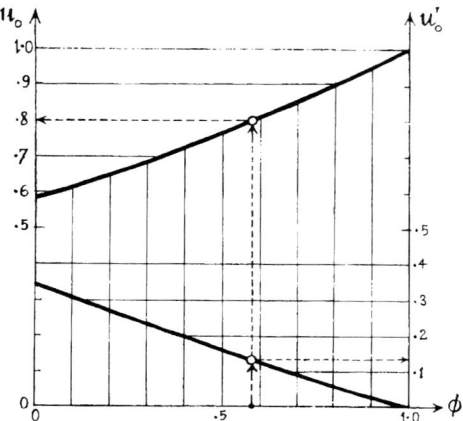

Fig. 6. Startvalues u_0 and $u_0' = (du/ds)$ in $s = 0$ corresponding to different endvalues ϕ for type-I curves.

$$2u(du/ds) = \lambda \qquad u = 0 \qquad \text{in} \qquad s = s_0 \tag{31}$$

with λ a value specific for each curve.

The numerical procedure for establishing the type II curves is to start from different s_0 values on the base line, to choose a λ and to use the series expansion in terms of λ mentioned in Appendix C for small values of $(s - s_0)$. Subsequently, the curves are constructed towards the right, using the finite difference scheme of Appendix B. The asymptotic end value of u is established by use of the complementary error function approximation (eqn. 27). The value of λ is finally chosen in such a manner that the end value equals $u = 1$. The λ values corresponding to different start values s_0 are assembled in Fig. 7.

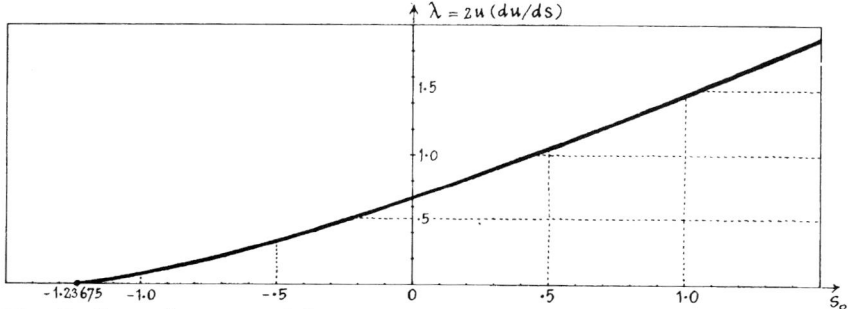

Fig. 7. Startvalues s_0 and $\lambda = 2u(du/ds)$ on $u = 0$ for type-II curves ending in $u = 1$, $s = +\infty$.

66

Separatrix

The separatrix is the limiting case between the two types of curves. It is a type I curve with ϕ reduced to zero, and a type II curve with λ reduced to zero. The series expansion for small values of $(s - s_0)$ of the separatrix differs from the series applicable to the type II curves (see Appendix C).

PRACTICAL APPLICATION

In this section it is shown how curves of Fig. 5, representing the solutions of problem Q, can be used for solving problem $P(m, \beta)$ in practical situations. It is assumed here, that the relevant hydraulic parameters m and β, defined by eqns. (15), (4) and (7) are known. The situations where $|\beta| \geqslant \frac{1}{2}$ and $|\beta| < \frac{1}{2}$ are distinguished.

When $|\beta| \geqslant \frac{1}{2}$ the boundary conditions (eqn. 21) show that either $w(\infty) > 0$ and $w(-\infty) \geqslant 0$ (case 1) or $w(\infty) \leqslant 0$ and $w(-\infty) < 0$ (case 2). In both cases the corresponding solution of $P(m, \beta)$ does not change sign on the entire interval $(-\infty, \infty)$. It will be shown that in both cases a solution of $P(m, \beta)$ can be obtained from a solution of problem Q (i.e. a curve from Fig. 5 for that matter) by applying an elementary transformation involving rescaling and displacing the curve in u-direction.

When $|\beta| < \frac{1}{2}$ the boundary conditions (eqn. 21) have opposite sign such that $w(-\infty) < 0$ and $w(-\infty) > 0$ (case 3). Consequently, the solution w changes sign on the interval $(-\infty, \infty)$. This introduces an additional difficulty caused by the absolute value of w in the differential equation (20). Let r_0 be the value of r where the solution vanishes: i.e. $w(r_0) = 0$, then $w(r) < 0$ for $r < r_0$ and $w(r) > 0$ for $r > r_0$. The solution thus consists of two parts ($w > 0$ and $w < 0$). Each part is an appropriate transformed solution of problem Q. They are joined together at r_0 using the continuity of the concentration (or velocity q_ξ) and the continuity of the salt flux.

The treatment here is aimed to describe the required procedure of rescaling and combining the curves appropriately, without too much mathematical details. For more detailed information the reader is referred to Van Duijn (1986b).

Case 1: $\beta \geqslant \frac{1}{2}$

When β is larger than $\frac{1}{2}$, both fresh- and salt water flow upwards. When β equals $\frac{1}{2}$, only the freshwater flows upwards while the salt water is stagnant. Both situations are shown in Fig. 2. Now let $u(s)$ be a solution of problem Q and consider for $\sigma > 0$ the transformation:

$$r = \sigma s \tag{32}$$

$$w(r) = \sigma^2 u(r/\sigma) - m \tag{33}$$

This transformation consists of a rescaling (caused by σ) and a displacement (over m) of a relevant curve from Fig. 5. By the transformation the differential equation (24) becomes:

$$\tfrac{1}{2} r \, dw/dr + d[(m + w) \, dw/dr]/dr = 0 \tag{34}$$

because the σ cancels.

Next let $u(s)$ be a solution of problem Q of type I or the separatrix. Then $u(\infty) = 1$ and $u(-\infty) = \phi \geqslant 0$. Using this in eqn. (33) gives:

$$w(\infty) = \sigma^2 - m \tag{35}$$

and:

$$w(-\infty) = \sigma^2 \phi - m \tag{36}$$

Thus when choosing σ such that:

$$\beta + \tfrac{1}{2} = \sigma^2 - m \qquad \text{or} \qquad \sigma = (\beta + \tfrac{1}{2} + m)^{1/2} \tag{37}$$

and ϕ such that:

$$\beta - \tfrac{1}{2} = \sigma^2 \phi - m \qquad \text{or} \qquad \phi = (\beta - \tfrac{1}{2} + m)/(\beta + \tfrac{1}{2} + m) \tag{38}$$

it follows that the function $w(r)$ satisfies the boundary conditions (eqn. 21). Moreover $w(r) \geqslant 0$ for all $-\infty < r < \infty$, implies $w(r) = |w(r)|$. Therefore eqn. (34) is identical to eqn. (20). Thus the function $w(r)$, defined according to eqns. (32), (33) and (37), (38) is a solution of $P(m, \beta)$.

Example 1-i: $m = 1$ and $\beta = 0.881$

Then from eqns. (37) and (38) there results $\sigma = 1.543$ and $\phi = 0.580$. Using Fig. 6, the value of ϕ indicates that the relevant curve is a type I curve of Fig. 5 passing through the point $u_0 = 0.8$, $s = 0$ with inclination $u_0' = 0.132$. Transformed with eqns. (32) and (33), the u, s values of this type I curve produce the following w, r values:

$$w = \sigma^2 u - m = 2.381 \, u - 1$$

and:

$$r = 1.54 \, s$$

These values form the curve of Fig. 8, which is readily verified to satisfy the boundary conditions (eqn. 21). The curve in Fig. 8 resembles a complementary error function. Indeed this kind of function is to be expected as a solution of problem $P(m, \beta)$ in the limit-case where m is large with respect to one.

Example 1-ii: $m = 0$ and $\beta = \tfrac{1}{2}$

Then the molecular diffusion can be disregarded with respect to the effect of the lateral dispersion and the salt water is stagnant. This situation can be considered as the limit of the case where $\beta > \tfrac{1}{2}$ and $m > 0$. From eqns. (37)

68

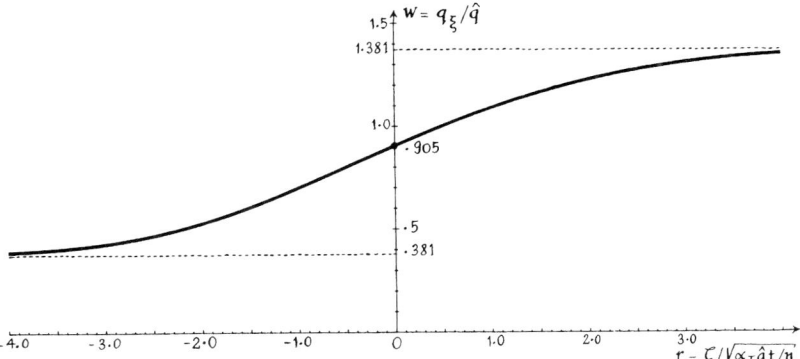

Fig. 8. Similarity solution for $m = 1$, $\beta = 0.881$.

and (38) it follows that:

$$\sigma = 1 \qquad \text{and} \qquad \phi = 0 \tag{39}$$

As mentioned above, this value of ϕ indicates that the relevant curve is the separatrix of Fig. 5 and the value of σ shows, that the similarity solution is the undeformed separatrix. Since this simple result produces a clarifying example, the corresponding specific weight and specific discharge distributions are shown in Fig. 9. This figure is to be interpreted in the manner as explained for Figs. 3 and 4.

In Fig. 5 the separatrix is specified by the intersections with the co-ordinate axes. These are:

$$\begin{aligned} s &= 0 & u_0 &= 0.5873 \\ s_0 &= -1.23675 & u &= 0 \end{aligned} \tag{40}$$

These values are reencountered in Fig. 9 in the following manner. Since the scale factor in this case is $\sigma = 1$, see eqn. (39), the s, u values are directly the r, w values, which are related to physical quantities by eqns. (17) and (19).

Using these relations it follows that $s = 0$ corresponds to $\zeta = 0$ for all time t which is the original height of the fresh—salt interface. The first line of eqn. (40) therefore indicates that the specific discharge at the original interface height is constant for all t and equal to $q_\xi = 0.5873\,\hat{q}$, see Fig. 9b.

The second line of eqn. (40) indicates that the depth ζ_0, below which the groundwater is still stationary and the specific weight is not yet reduced, equals $-1.23675\ (\alpha_T\hat{q}t/n)^{1/2}$. This means that this depth increases proportional to root time, see Fig. 9c.

Case 2: $\beta \leqslant -\tfrac{1}{2}$

When β is smaller than $-\tfrac{1}{2}$, both fresh- and salt water move downwards. When β equals $-\tfrac{1}{2}$, only the salt water flows downward while the freshwater

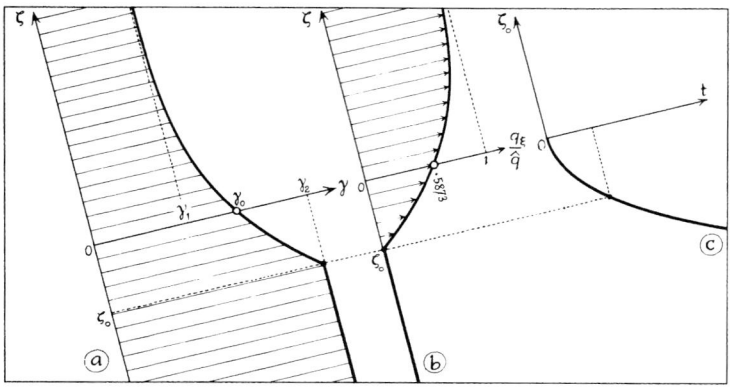

Fig. 9. Example 1-ii, case $m = 0$, $\beta = 1/2$. Development of brackish zone, when molecular diffusion can be disregarded and the salt water is stationary.

is stagnant. Both situations are shown in Fig. 2. Now the appropriate transformation is:

$$r = -\sigma s \tag{41}$$

$$-w(r) = \sigma^2 u(-r/\sigma) - m \tag{42}$$

Applying this transformation, eqn. (24) becomes:

$$\tfrac{1}{2} r \, dw/dr + d[(m-w) \, dw/dr]/dr = 0 \tag{43}$$

Let again $u(s)$ be of type I or the separatrix. Then eqns. (41) and (42) imply that:

$$-w(-\infty) = \sigma^2 - m \tag{44}$$

and:

$$-w(+\infty) = \sigma^2 \phi - m \tag{45}$$

Now σ must be chosen such that:

$$-\beta + \tfrac{1}{2} = \sigma^2 - m \qquad \text{or} \qquad \sigma = (-\beta + \tfrac{1}{2} + m)^{1/2} \tag{46}$$

and ϕ such that:

$$-\beta - \tfrac{1}{2} = \sigma^2 \phi - m \qquad \text{or} \qquad \phi = (-\beta - \tfrac{1}{2} + m)/(-\beta + \tfrac{1}{2} + m) \tag{47}$$

in order that $w(r)$ satisfies eqn. (21). Moreover, $w(r) \leqslant 0$, implies that $-w(r) = |w(r)|$. Thus eqn. (43) is identical to eqn. (20), showing that $w(r)$ in this case is in fact a solution of $P(m, \beta)$.

Summarizing, for this case the solution of $P(m, \beta)$ again consists of a type I or separatrix curve from Fig. 5. Since both eqns. (41) and (42) contain a minus sign the relevant curve from Fig. 5 is rotated over $180°$ to produce the similarity solution in the r, w plane.

70

Case 3: $-\frac{1}{2} < \beta < +\frac{1}{2}$

In this case fresh- and salt water flow in opposite directions (see Fig. 2) and so q_ξ and therefore also w from eqn. (19) change sign in the region of integration. The positive part of the solution is denoted by w^+ and its negative part by w^-. Then w^+ resembles case 1 and w^- resembles case 2.

Before showing the practical elaboration of two examples in the subsection "use of Fig. 10" below, a few concepts required in the procedure are mentioned here first. The positive part of $w(r)$ is defined according to:

$$w^+(r) = \sigma^{+2}u^+(r/\sigma^+) - m \qquad \text{with} \qquad \sigma^+ = (m + \beta + \tfrac{1}{2})^{1/2} \qquad (48)$$

and the negative part by:

$$-w^-(r) = \sigma^{-2}u^-(-r/\sigma^-) - m \qquad \text{with} \qquad \sigma^- = -(m - \beta + \tfrac{1}{2})^{1/2} \qquad (49)$$

In eqns. (48) and (49), u^+ and u^- are parts of two different curves from Fig. 5. Because of the minus signs in eqn. (49), the part u^- is rotated over $180°$ in the rescaling process. By choosing σ according to eqns. (48) and (49), it follows that $w^+(\infty) = \beta + \tfrac{1}{2}$ and $w^-(-\infty) = \beta - \tfrac{1}{2}$.

It remains to organize the solution in such a manner that the two parts w^+ and w^- match together at the point where $w = 0$. More precisely, it remains to select r_0 and curves $u^+(s)$ and $u^-(s)$ from Fig. 5 so that the composite function:

$$w(r) = \left.\begin{array}{lll} w^+(r) & \text{for} & r > r_0 \\[2mm] w^-(r) & \text{for} & r < r_0 \end{array}\right| \qquad (50)$$

satisfies certain continuity properties.

In a study of the more general nonlinear partial differential equation $\partial u/\partial t = \partial^2 (|u|^{k-1}u)/\partial x^2$ with $k > 1$, it is shown by Van Duijn (1986a) how to join the solutions on the base line, $u = 0$ in his case. He points out, that the fitting conditions are to be deduced from additional physical considerations. In the present problem the condition to be satisfied is that the salt flux should be continuous. Using expression (23) this condition can be written as:

$$2(m + w^+)dw^+/dr = 2(m - w^-)dw^-/dr = \Lambda \qquad \text{for}$$

$$w^+ = u^- = 0 \quad \text{at} \quad r = r_0 \qquad (51)$$

Elaboration of this condition is rather involved, since the relation between Λ and r_0 cannot be determined directly. In order, however, to provide a first approximation of the solution, Fig. 10 is added here, in which Λ and r_0 values are assembled for a relevant region of m, β values.

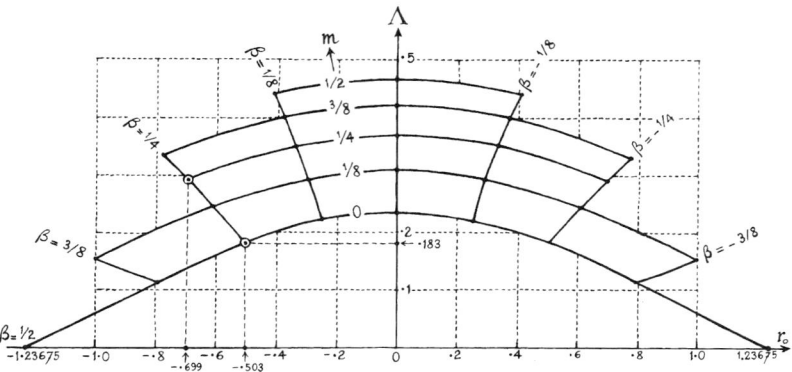

Fig. 10. Values for r_0 and Λ for different m, β combinations.

Use of Figure 10

Example 3-i: $m = \frac{1}{4}$ and $\beta = \frac{1}{4}$

The similarity solution in r, w coordinates is the dashed line in Fig. 11. This curve is obtained as follows. From Fig. 10 the appropriate values for this example are found to be $\Lambda = 0.292$, $r_0 = -0.699$. The value of r_0 means that the dashed curve in Fig. 11 intersects the axis $w = 0$ in the point $r = r_0 = -0.699$.

The value of Λ indicates with eqn. (51) that the inclination of the curve is:

$$dw/dr = \Lambda/2m = 0.585 \qquad \text{in} \qquad w = 0, \qquad r_0 = -0.699$$

This information is sufficient to construct the two parts of the curve with the finite difference scheme of Appendix B, by extending them from the starting point $r = r_0$, $w = 0$ in both directions up to infinity.

It is also possible to obtain the curve by transforming two curves from Fig. 5. From eqns. (48) and (49) it follows that the scale factors are $\sigma^+ = 1$ for the positive part and $\sigma^- = -(\frac{1}{2})^{1/2} = -0.707$ for the negative part, respectively. Again from eqns. (48) and (49) it can be deduced that the starting points of the corresponding curves in Fig. 5 are:

$$u^+ = \frac{1}{4} \qquad \text{and} \qquad s_0^+ = r_0/\sigma^+ = -0.699$$

and:

$$u^- = \frac{1}{2} \qquad \text{and} \qquad s_0^- = r_0/\sigma^- = +0.989$$

The points are indicated by two triangles in Fig. 5. It is readily verified, that the upper part of the dashed curve in Fig. 11 is identical to the undeformed curve starting in the left triangle in Fig. 5, undeformed because $\sigma^+ = 1$. The other one is rescaled by $\sigma^- = -0.707$.

72

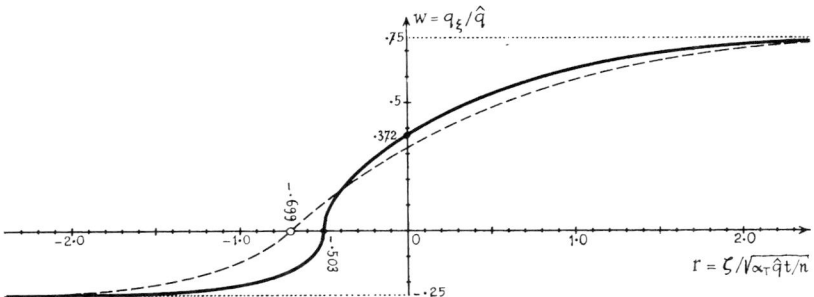

Fig. 11. Similarity solutions for the example 3-i, case $m = 1/4$, $\beta = 1/4$ (dashed line) and example 3-ii, case $m = 0$, $\beta = 1/4$ (full line).

Example 3-ii: $m = 0$ and $\beta = \frac{1}{4}$

The solution in r, w coordinates is the full line in Fig. 11. This curve is obtained as follows. From Fig. 10 the appropriate values for this example are found to be $\Lambda = 0.183$ and $r_0 = -0.503$. The value of r_0 indicates that the full line in Fig. 11 intersects the $w = 0$ axis in the point $r = r_0 = -0.503$.

The value of Λ being nonzero, it follows from eqn. (51) in this case where $m = 0$, that dw^+/dr and dw^-/dr are infinite. The two parts of the curve have a vertical tangent in the intersection point with the axis $w = 0$. They can be constructed, however, by starting on either side of the intersection point with the series expansion of Appendix C because $2w^+dw^+/dr = -2w^-(r)$ $dw^-/dr = \Lambda$ is finite and known in this case. Using the finite difference scheme of Appendix B the curve can be extended towards $r \to \pm \infty$.

It is also possible to obtain the curve by transforming two appropriate curves of Fig. 5. From eqns. (48) and (49) it follows that the starting points of the curves are:

$$u^+ = 0 \qquad \text{and} \qquad s_0^+ = r_0/\sigma^+ = -0.503/(3/4)^{1/2} = -0.581$$

and:

$$u^- = 0 \qquad \text{and} \qquad s_0^- = r_0/\sigma^- = -0.503/-(1/4)^{1/2} = +1.006$$

The corresponding curves are indicated by dashed lines in Fig. 5.

It may be remarked, that u^+ intersects the axis $s = 0$ in $u^+ = 0.496$. Rescaled with eqn. (48) this point becomes $r = 0$, $w_0 = 0.372$. This value and $r_0 = -0.503$ mentioned above are reencountered in Fig. 4, which was dicussed in the subsection "Application of Similarity Solution".

ANISOTROPY

From the practical standpoint the case of anisotropy is of importance. The results of the analysis presented here are still valid in that case. Only the coefficients κ for the intrinsic permeability and α_T for the transverse dispersion, have to be adjusted. It was shown (De Josselin de Jong, 1981) that for the case of a soil with a horizontal permeability κ_h deviating from κ_v the vertical permeability, the relation (3) for the shear flow, $\hat{q}$, remains valid. Only κ has to be replaced by κ_s, the intrinsic permeability in the direction of the interface, such that $(1/\kappa_s) = (\cos^2\alpha/\kappa_h) + (\sin^2\alpha/\kappa_v)$.

With respect to the transverse dispersion it may be remarked here, that the custom to attribute α_T a value of one tenth of the longitudinal dispersion length α_L, which is common in these days, is not justified in general. It is certainly an overestimation in the case of anisotropy created by lenses of more permeable material, that are elongated in horizontal direction.

The longitudinal dispersion length in the direction of the lenses may be of the order of the lens lengths and/or their mutual distance. The longitudinal dispersion length perpendicular to the lenses, may be of the size of the lens thickness and/or spacing. But the transverse dispersion may be much smaller because that effect is due to the possibility for the groundwater to exchange salt with fluid elements in neighbouring streamlines. Since the elaboration of streamline patterns and the ensuing exchange possibilities is rather involved, this point is not pursued in detail here. For practical use it may suffice to mention, though, that it is more realistic to envisage a value much smaller than one tenth for the ratio between α_T and α_L.

RECAPITULATION OF THE RESULTS

In the preceding sections, the mixing process of fresh- and salt groundwater due to molecular diffusion and transverse dispersion was discussed. This was done for plane flow under several different conditions. In all cases, the initial distribution (at time $t = 0$) of the specific weight is that the fresh and salt fluids have constant specific weights, γ_1 and γ_2 respectively, and are separated by a sharp interface. The difference is in the specific discharges q_ξ of the two unmixed fluids. These are considered to have a constant value in each of the two regions above and below the interface at $t = 0$, and to keep that same magnitude at infinity both above and below, $\zeta = \pm \infty$, for all later times $t \geqslant 0$. For $t > 0$ the fluids become more or less mixed and the specific weight γ becomes a function of the local height ζ and the time t. The specific discharge q_ξ changes accordingly because it depends linearly on γ, see eqn. (11).

Because of the plane flow and other simplifying assumptions, the system of partial differential equations, that describes the spreading process can be reduced to the single differential equation (20). This is achieved by introducing the similarity variables w and r. From these, w is related to the specific discharge q_ξ by eqn. (19) and to the specific weight γ by using eqn. (11). The variable r is related to the height ζ and the time t by eqn. (17).

The governing equation (20) has as relevant solutions the family of curves shown in Fig. 5. By rescaling, displacing and combining in various manners the appropriately chosen curves of this family, it is possible to construct the solution for various values of molecular diffusion and flow conditions at infinity. Molecular diffusion in comparison to dispersion is described by the parameter m, see eqn. (15), the flow conditions originally and at infinity by the parameter β, see eqns. (4) and (7).

The differential equation (20) is a nonlinear diffusion equation with $(m + |w|)$ as diffusion coefficient. The absolute value of w in this coefficient is unusual and requires a special treatment when w changes sign in the integration interval. By eqn. (19), this occurs when the fluids flow in opposite directions.

In this paper three cases are considered that differ in the way the two unmixed fluids flow. In the cases 1 and 2 both fluids flow initially and at infinity in the same direction. In case 1, the choice $\beta \geqslant \frac{1}{2}$ guarantees that both fluids flow upwards or only the freshwater flows upwards and the salt water is stagnant ($q_\xi \geqslant 0$). In case 2 ($\beta \leqslant -\frac{1}{2}$), both flow downwards or only the salt water flows downwards and the freshwater is stagnant ($q_\xi \leqslant 0$). In case 3 ($-\frac{1}{2} < \beta < \frac{1}{2}$), the two fluids flow in opposite directions.

When $\beta \geqslant \frac{1}{2}$ or $\beta \leqslant -\frac{1}{2}$, it follows from eqn. (21) that the function w is nonnegative ($w \geqslant 0$) or nonpositive ($w \leqslant 0$), respectively. In these two cases the solution consists of one rescaled and displaced curve of Fig. 5 from the subfamily called type I curves. In the section "Practical Application" two numerical examples (1-i and 1-ii) are elaborated to demonstrate the procedure.

Mathematically of more interest is case 3 where w changes sign according to eqn. (21) and $-\frac{1}{2} < \beta < \frac{1}{2}$. The solution then consists of two rescaled and displaced curves of Fig. 5. The procedure is now more involved because the curves have to be selected in such a manner that they fit together correctly in the point, where $w = 0$. The transition condition is derived from the requirement of continuity of salt flux. In case 3 both curves of type I and type II from Fig. 5 can be required to produce the end result. Examples 3-i and 3-ii show these results explicitly.

In practical situations the flow conditions are in general not as simple as assumed in this study. Plane flow is an exception, the interface is generally not flat and flow is not necessarily parallel to the original interface. However, being an exact solution of a simplified situation, the results of this analysis may be useful for verifying numerical procedures that describe variable density flow with dispersion of a more general purpose character.

ACKNOWLEDGEMENT

The authors are indebted to J. van Kan for a number of clarifying discussions about the numerical approach.

APPENDIX A

In this appendix, some elementary properties of solutions of problem Q are discussed. For convenience, the notation $u' = du/ds$ is being used here.

Uniqueness

Consider eqn. (24). Suppose that at some point s_0 the values:

$$u(s_0) = C_1 \quad \text{and} \quad u'(s_0) = C_2 \tag{A1}$$

are prescribed, where C_1 and C_2 are given constants. Taking $C_1 > 0$, it was shown by Atkinson and Peletier (1971) that for any $-\infty < C_2 < \infty$, there exists a unique solution of eqn. (24), which satisfies the conditions (A1) and which exists on the largest possible interval, where it is positive. A consequence of the uniqueness is monotonicity of solutions of Q.

Monotonicity

Let u be a solution of problem Q and suppose that there exists a point s_0 where:

$$u(s_0) = C_1 \quad \text{with} \quad 0 < C_1 < 1 \quad \text{and} \quad u'(s_0) = 0 \tag{A2}$$

Now observe, that the constant function $\hat{u}(s) = C_1$ for $-\infty < s < \infty$ also satisfies eqns. (24) and (A2). Then the uniqueness requires, that the solutions u and $\hat{u}$ must be identical, which implies that $u(s) = C_1$ for all $-\infty < s < \infty$. This contradicts the boundary condition (25). Therefore the only solution of Q, which is not strictly increasing is the constant $u = 1$. All other solutions with $0 < u < 1$ must satisfy $u' > 0$ in order to satisfy the boundary condition (25).

Intersection

Next it is shown, that two solutions of Q cannot intersect. Let u_1 and u_2 be two solutions of Q and suppose, that there exists an intersection point s_0, where $u_1(s_0) = u_2(s_0)$. By the uniqueness, $u'_1(s_0) \neq u'_2(s_0)$, because otherwise u_1 and u_2 would be identical. Without loss of generality it is assumed here, that $u'_1(s_0) > u'_2(s_0)$. Then two situations can arise:

(1) There exists an other intersection point s_1, such that:

$$u_1(s_1) = u_2(s_1) \quad \text{and} \quad u_1(s) > u_2(s) \quad \text{for} \quad s_0 < s < s_1$$

(2) $u_1(s) > u_2(s)$ for all $s > s_0$ and both solutions satisfy eqn. (25).

Ad (1)

Integration of eqn. (24) with respect to s from s_0 to s_1 gives for u_1 and u_2, respectively:

$$u_1(s_1)u'_1(s_1) - u_1(s_0)u'_1(s_0) + \tfrac{1}{2}s_1 u_1(s_1) - \tfrac{1}{2}s_0 u_1(s_0) - \tfrac{1}{2}\int_{s_0}^{s_1} u_1(s)ds = 0 \tag{A3}$$

and:

$$u_2(s_1)u'_2(s_1) - u_2(s_0)u'_2(s_0) + \tfrac{1}{2}s_1 u_2(s_1) - \tfrac{1}{2}s_0 u_2(s_0) - \tfrac{1}{2}\int_{s_0}^{s_1} u_2(s)ds = 0 \tag{A4}$$

Subtracting these equations gives:

$$u_1(s_1)[u'_1 - u'_2](s_1) - u_1(s_0)[u'_1 - u'_2](s_0) - \tfrac{1}{2}\int_{s_0}^{s_1} [u_1(s) - u_2(s)]ds = 0 \tag{A5}$$

However, by the above assumptions $[u'_1 - u'_2](s_1) < 0$, $[u'_1 - u'_2](s_0) > 0$ and $\int_{s_0}^{s_1} [u_1(s) - u_2(s)]ds > 0$. This contradicts the equal sign in eqn. (A5). Therefore case (1) cannot arise.

Ad (2)

A similar argument gives for this case also a contradiction. Thus both (1) and (2) cannot arise and thus no intersection point exists.

Asymptotic behaviour

Next an argument due to Peletier (1970) is used to obtain eqns. (27) and (29). The starting point is the following observation. At points where $u > 0$, eqn. (24) can be written as:

$$\tfrac{1}{2}(s/u)(u^2)' + (u^2)'' = 0 \qquad (A6)$$

which can be integrated to give:

$$(u^2)'(s) = (u^2)'(s_0) \exp\left\{-\tfrac{1}{2}\int_{s_0}^{s} [z/u(z)]\,dz\right\} \qquad (A7)$$

Here s_0 is an arbitrary chosen point, where $u(s_0) > 0$.

Since $u(s) < 1$, it follows from eqn. (A7), that:

$$(u^2)'(s) \leqslant (u^2)'(s_0) \exp[-\tfrac{1}{4}(s^2 - s_0^2)] \qquad \text{for} \qquad s \geqslant s_0 \qquad (A8)$$

Using $u'(s) > 0$ and thus $u(s) > u(s_0)$ for $s > s_0$ in eqn. (A8) gives:

$$0 < u'(s) \leqslant u'(s_0) \exp[-\tfrac{1}{4}(s^2 - s_0^2)] \qquad \text{for} \qquad s \geqslant s_0 \qquad (A9)$$

Integration of eqn. (A9) with respect to s from a point $\bar{s} \geqslant s_0$ to ∞ gives:

$$1 - u(\bar{s}) \leqslant u'(s_0) \exp(\tfrac{1}{4}s_0^2) \int_{\bar{s}}^{\infty} \exp(-\tfrac{1}{4}s^2)\,ds = u'(s_0) \exp(\tfrac{1}{4}s_0^2)\sqrt{\pi}\,\text{erfc}\,(\tfrac{1}{2}\bar{s}) \qquad (A10)$$

where $\text{erfc}\,(x) = (2/\sqrt{\pi}) \int_{x}^{\infty} \exp(-z^2)\,dz$.

From eqn. (A10), condition (27) follows when $\bar{s} \to \infty$. A similar argument gives eqn. (29), whenever $\phi > 0$.

APPENDIX B

A finite difference method to obtain the solutions of problem Q as shown in Fig. 5, is discussed here.

The solutions of problem Q were distinguished in type-I curves, type-II curves and a separatrix. In all three cases, first a value of u and du/ds was chosen at a point where u is positive. For curves of type I this was done at $s = 0$ and the appropriate values for $u(0)$ and $du/ds(0)$ were found by a try and error method. For curves of type II and for the separatrix a series expansion was used to approximate the value of u and du/ds at a point s_1 which was chosen sufficiently close to the point s_0 where the solution vanishes, see Appendix C.

Thus for the finite-difference approximation one has to solve an initial value problem of the form:

$$(u^2)'' + su' = 0 \qquad s > \hat{s} \qquad (B1)$$

$$u(\hat{s}) = \hat{u} \qquad (B2)$$

$$u'(\hat{s}) = \hat{u}' \qquad (B3)$$

where eqn. (B1) is a rewritten version of eqn. (24) and where $\hat{s}$, $\hat{u}$ and $\hat{u}'$ are three given numbers such that $\hat{u} > 0$ and $\hat{u}' > 0$.

Let $s_i = \hat{s} + i\Delta s$, where Δs denotes the discretization interval. Integration of eqn. (B1) from s_i to s_{i+1} gives:

$$(u^2)'(s_{i+1}) - (u^2)'(s_i) + \int_{s_i}^{s_{i+1}} su'(s)\,\mathrm{d}s = 0 \tag{B4}$$

Integrating the third term in eqn. (B4) by parts and setting $u' = p$ gives:

$$2u_{i+1}p_{i+1} = 2u_i p_i + s_i u_i - s_{i+1}u_{i+1} + \int_{s_i}^{s_{i+1}} u(s)\,\mathrm{d}s \tag{B5}$$

where $u_i = u(s_i)$ and $p_i = u'(s_i)$.

A second equation is needed to solve eqn. (B5) and this can be:

$$u_{i+1} = u_i + \int_{s_i}^{s_{i+1}} p(s)\,\mathrm{d}s \tag{B6}$$

The integrals in eqns. (B5) and (B6) are approximated by a third-degree Hermite polynomial, see Ralston (1965, p. 60). Using the notation $u'' = q$ this leads to the finite-difference scheme:

$$2u_{i+1}p_{i+1} = 2u_i p_i + s_i u_i - s_{i+1}u_{i+1} + \frac{\Delta s}{2}(u_i + u_{i+1}) + \frac{\Delta s^2}{12}(p_i - p_{i+1}) \tag{B7}$$

and:

$$u_{i+1} = u_i + \frac{\Delta s}{2}(p_i + p_{i+1}) + \frac{\Delta s^2}{12}(q_i - q_{i+1}) \tag{B8}$$

This scheme has a local truncation error of $O(\Delta s^5)$, see Ralston (1965, p. 212), and it is therefore expected that the global accuracy is of $O(\Delta s^4)$.

The values of q at s_i and s_{i+1} are obtained by using the differential equation (B1) at $s = s_i$ and $s = s_{i+1}$. For q_i this gives:

$$q_i = -\frac{p_i^2}{u_i} - s_i \frac{p_i}{2u_i} \tag{B9}$$

Substituting this expression and the corresponding expression for q_{i+1} into eqn. (B8) yields:

$$u_{i+1} = u_i + \frac{\Delta s}{2}(p_i + p_{i+1}) - \frac{\Delta s^2}{12}\left(\frac{p_i^2}{u_i} + s_i \frac{p_i}{2u_i} - \frac{p_{i+1}^2}{u_{i+1}} - s_{i+1}\frac{p_{i+1}}{2u_{i+1}}\right) \tag{B10}$$

Thus for given values of s_i, s_{i+1}, u_i and p_i, the two nonlinear equations (B7) and (B10) have to be solved to obtain u_{i+1} and p_{i+1}. However, because of the nonlinearity this cannot be done directly and the following iteration process was used to obtain approximate values for u_{i+1} and p_{i+1}. Let $u_{i+1}^{(0)} = u_i$. Substituting this value into eqn. (B7) and solving the resulting linear equation for p_{i+1}, gives the approximate value $p_{i+1}^{(0)}$.

This value in turn is substituted into eqn. (B10). The resulting quadratic equation in u_{i+1} can be solved directly to obtain the value $u_{i+1}^{(1)}$ as a next approximation for u_{i+1}. Since this process converges rapidly only a few of these iteration steps were needed at each value of $i = 0, 1, 2, \ldots$

APPENDIX C

In this appendix an approximation is considered of curves of type II and the separatrix from Fig. 5 for small values of u. This approximation has the form of a series expansion which solves the initial value problem defined by the differential equation (24) for $s > s_0$ and the initial conditions (eqn. 31).

Type-II curves

For type-II curves, $\lambda > 0$ and a series expansion of the form:

$$\bar{u}(s) = [\lambda(s - s_0)]^{1/2} + a(s - s_0) + b(s - s_0)^{3/2} + c(s - s_0)^2 \tag{C1}$$

is chosen as an approximation in a sufficiently small neighbourhood of s_0. Clearly, any approximation of the form (C1) satisfies eqn. (31). The values of the constants a, b and c are obtained by substituting eqn. (C1) into the differential equation (24). This gives the following set of equations:

from the term with $(s - s_0)^{-1/2}$: $2a\sqrt{\lambda} = -\frac{2}{3}s_0\sqrt{\lambda}$

from the term with $(s - s_0)^{0}$: $a^2 + 2b\sqrt{\lambda} = -\frac{1}{2}as_0$

from the term with $(s - s_0)^{1/2}$: $2ab + 2c\sqrt{\lambda} = \frac{2}{3}\lambda - \frac{2}{5}bs_0 - \frac{4}{5}\sqrt{\lambda}$

from the term with $(s - s_0)$: $b^2 + 2ac = \frac{1}{3}c - \frac{2}{3}a$

from the term with $(s - s_0)^{3/2}$: $2bc = -\frac{4}{7}b$

from the term with $(s - s_0)^2$: $c^2 = -\frac{1}{2}c$

Thus for $\lambda > 0$, the choice:

$$a = -\tfrac{1}{3}s_0 \qquad b = s_0^2/36\sqrt{\lambda} \qquad c = (s_0^3/270\lambda) - (1/15) \tag{C2}$$

guarantees, that the approximation (C1) satisfies equation (24) up to terms with $(s - s_0)^{1/2}$.

The first point on the curve (u_1, s_1) is chosen in such a manner, that $(s_1 - s_0)$ is a sufficiently small number for approaching the value of u_1 by the series expansion. The value of u'_1, which represents (du/ds) in the first point, is taken from the derivative of the series expansion. The values u_1, u'_1, s_1 are the starting values for establishing the rest of the curve upwards and to the right by applying the finite-difference procedure of Appendix B. The parameter to adjust in this case is λ from eqn. (31).

Separatrix

For the separatrix, an approximation of the form (C1) is chosen with $\lambda = 0$ and $s_0 < 0$. Then the following two sets of values for a, b and c occur:

$$a = 0 \qquad\qquad b = 0 \qquad c = 0 \tag{C3}$$

and:

$$a = -\tfrac{1}{2}s_0 \qquad\qquad b = 0 \qquad c = -s_0/(3s_0 + 1) \tag{C4}$$

The set of values (C3) leads to an approximation, which is identically zero. This solution corresponds to the part of the separatrix in the region $s \leqslant s_0$. In the region $s > s_0$, the set (C4) gives:

$$\bar{u}(s) = -\tfrac{1}{2}s_0(s - s_0) - s_0(s - s_0)^2/(3s_0 + 1) \tag{C5}$$

and this approximation satisfies eqn. (24) up to terms with $(s - s_0)^{3/2}$. To obtain numerically the solution, that starts at this value of s_0 an identical procedure as for the case of type-II curves was used. The parameter to adjust in this case is s_0. It is found, that for $s_0 = -1.23675$ the corresponding solution reaches the value $u = +1$ asymptotically, when $s \to +\infty$.

REFERENCES

Atkinson, F.V. and Peletier, L.A., 1971. Similarity profiles of flow through porous media. Arch. Ration. Mech. Anal., 42: 363—379.

Bear, J., 1975. Dynamics of Fluids in Porous Media. Elsevier, New York, N.Y., 2nd ed.

Boltzmann, L., 1894. Zur Integration der Diffusionsgleichung bei variabeln Diffusions-coefficienten. Ann. Phys. Chem., 53: 959—964.

De Josselin de Jong, G., 1960. Singularity distributions for the analysis of multiple fluid flow through porous media. J. Geophys. Res., 65 (11): 3739—3758.

De Josselin de Jong, G., 1981. The simultaneous flow of fresh and salt water in aquifers of large horizontal extension determined by shear flow and vortex theory. In: A. Verruijt and F.B.J. Barends (Editors), Proc. Euromech. 143, Balkema, Rotterdam, pp. 75—82.

Edelman, J.H., 1940. Strooming van zoet en zout grondwater. Rapport 1940 inzake de watervoorziening van Amsterdam, Bijl. 2, pp. 8—14.

Gilding, B.H., 1980. On a class of similarity solutions of the porous media equation III. J. Math. Anal. Appl., 77: 381—402.

Gilding, B.H. and Peletier, L.A., 1977. On a class of similarity solutions of the porous media equation II. J. Math. Anal. Appl., 57: 522—538.

Peletier, L.A., 1970. Asymptotic behaviour of temperature profiles of a class of nonlinear heat conduction problems. Q. J. Mech. Appl. Math., 23: 441—447.

Ralston, A., 1965. A First Course in Numerical Analysis. McGraw-Hill, New York, N.Y.

Van Duijn, C.J., 1986a. On a class of similarity solutions of the porous media equation. (In prep.)

Van Duijn, C.J., 1986b. A mathematical analysis of density dependent dispersion in fresh—salt groundwater flow. Proc. 9th Salt Water Intrusion Meeting, Delft. (In prep.)

Van Duijn, C.J. and Peletier, L.A., 1977. A class of similarity solutions of the nonlinear diffusion equation. Nonlinear Anal., Theory, Methods, Appl., 1: 223—233.

Verruijt, A., 1971. Steady dispersion across an interface in a porous medium. J. Hydrol., 14: 337—347.

Verruijt, A., 1980. The rotation of a vertical interface in a porous medium. Water Resour. Res., 16: 239—240.

THE TENSOR CHARACTER OF THE DISPERSION COEFFICIENT IN ANISOTROPIC POROUS MEDIA

G. De Josselin de Jong

INTRODUCTION

In their attempt to describe dispersion phenomena in mathematical form several authors arrived at the conclusion, that a porous medium possesses a coefficient of dispersion, which has the character of a tensor. This tensor is formulated in such a manner, that it represents the geometrical aspects of the porous medium responsible for the scatter of tracer particles, when carried by a fluid flowing through it. It is therefore a property of the porous medium alone, and can be considered to materialize the tortuosity of the particle trails caused by the random arrangement and the interconnectivity of the channels constituing the pore space.

Nikolaevskii (1959) originated the tensor. He constructed it with a dimension of length and predicted the magnitude of this length to be in the order of particle size. He obtained a tensor of the fourth rank for an isotropic porous medium by postulating, that the random modification of the velocity vector, from mean velocity to local velocity, is a tensor of second rank.

This postulate requires that the velocity vectors considered possess the property of linear superposition. It will be shown subsequently that this requirement is not always fulfilled, and that, for instance, Nikolaevskii's result for the isotropic medium cannot be extended to the general case of an anisotropic medium. Nikolaevskii states that the rank of the tensor must be even, but includes the possibility of any number for that rank.

Bear (1961) started with the results of the dispersion computations proposed by the author (De Josselin de Jong, 1958). These computations were executed for an isotropic model material exhibiting a particular manner of scattering tracer particles. The basic scattering mechanism adopted was apparently realistic enough to reproduce mathematically the dispersivity phenomena observed in tests and especially the difference in longitudinal and transverse dispersion.

The specific problem treated was the determination of concentration distribution developing from a point injection of tracer particles if these were carried away by the fluid flow through the porous medium. The resulting concentration distribution turns out to be almost normal (Gaussian), in all directions, with the points of standard deviation lying on an ellipsoid.

259

This concentration distribution can be uniquely defined by its mean point and the variances or second central moments with respect to the three space coordinates. By superposition of point injections and rotation of coordinates, Bear (1961) showed that the variances can be considered as the components of a second rank tensor, and he developed the relation between this tensor and Nikolaevskii's fourth rank tensor.

Inspired by the work of Nikolaevskii and Bear, Scheidegger (1961) suggested that the dispersion constant is also a fourth rank tensor in the general anisotropic case. This assertion was not substantiated by considering the physics of the scattering mechanism underlying the dispersion phenomenon. Therefore his suggestion is not imperative.

It is the purpose of this paper to reestablish the dispersion tensor for the anisotropic medium starting from the basic scattering mechanism. Since the mechanism used by the author, (De Josselin de Jong, 1958) proved realistic enough for the isotropic case it was considered to be acceptable for the anisotropic case as well.

SCATTERING MECHANISM

The scattering mechanism adopted previously consists of assigning a choice to every tracer particle when arriving at a junction point in the pore channel system. It was suggested that the probability for a particle to choose a certain direction is proportional to the discharge in that direction considered as a fraction of the total discharge through all junction points and in all directions.

Once a particular channel has been chosen in junction A the particle travels a certain distance in a certain direction to the next junction point B. This distance is described by its space coordinates x, y, z with origin in A. The

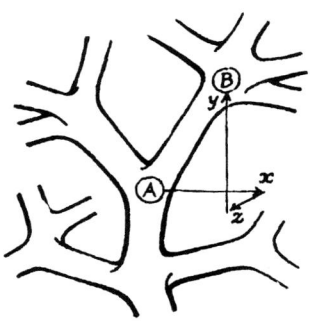

Fig. 1.

travel time required for the particle to travel from A to B is called the residence time, t.

The components x, y, z, t are stochastic variables with a probability distribution determined by the adopted flow mechanism in the channels and in the junction points. In this work use will be made of the mean values of these stochastic variables and their squares. These mean values are indicated by a bar and combinations by a circonflex.

(1)
$$\bar{x} = \text{mean value of } x$$
$$\overline{x^2} = \text{mean value of } x^2$$
$$\overline{xy} = \text{mean value of } x \cdot y$$
$$\widehat{xx} = \overline{x^2} - \bar{x}^2 = \text{mean value of square minus square of mean value}$$
$$\widehat{xy} = \overline{xy} - \bar{x} \cdot \bar{y}.$$

The summation of the stochastic components x, y, z, t of each individual channel over the many channels covered by a tracer particle form its total travel path and travel time. The probability for a particle to arrive at a certain point X, Y, Z, after time T_0 is the combined probability for that particle to travel through a certain combination of channels. The probability distribution $P(X, Y, Z, T_0)$ in space of the arrival points is the tracer concentration distribution at time T_0 after the moment that an amount of particles was injected in the origin of the X, Y, Z coordinate system.

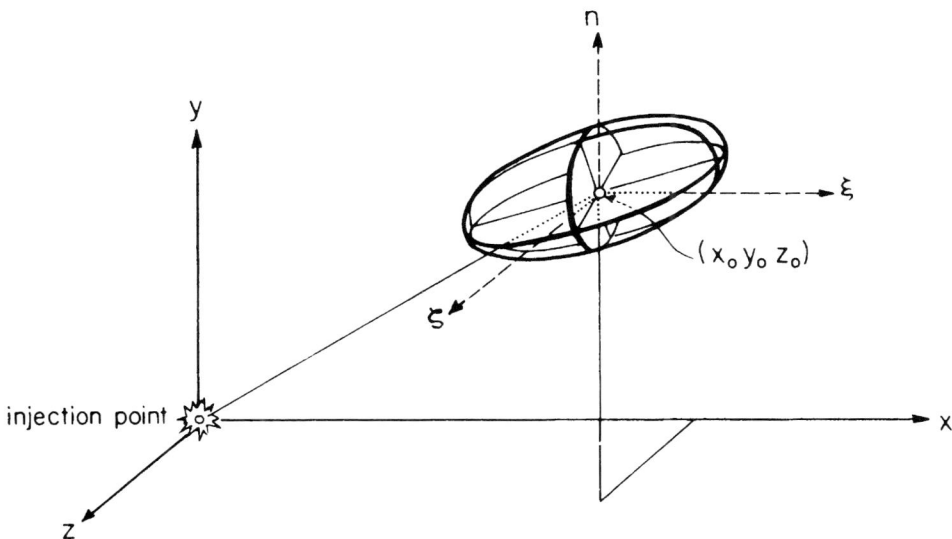

Fig. 2. Ellipsoid of standard deviation

By use of Chandrasekhar's analysis of Markoff processes and a special form of saddle point analysis this probability turns out to be given in good approximation by

(2) $$P(X, Y, Z, T_0) = \frac{\sqrt{|f|}}{(2\pi)^{3/2}} \exp\left[-\tfrac{1}{2} f_{ij} \xi_i \xi_j\right].$$

In this expression ξ_i and ξ_j are space coordinates with their origin in the point

$$X_0 = T_0 \frac{\bar{x}}{\bar{t}}; \quad Y_0 = T_0 \frac{\bar{y}}{\bar{t}}; \quad Z_0 = T_0 \frac{\bar{z}}{\bar{t}} \quad \text{(see Fig. 2)}.$$

This is the point of average displacement of the tracer particles during travel time T_0. Throughout this paper subscripts can have three values 1, 2, 3 corresponding respectively to x, y, z, if they are Roman, and if they are Greek they run over 4 values corresponding to x, y, z and t.

$|f|$ is the determinant of a matrix f with the components f_{ij}. These 9 components f_{ij} are related to the mean squares and square means of the stochastic variables, by the following expressions given here without proof.

(3) $$f_{ij} = \frac{\bar{t}}{T_0} \frac{(b_{ij} b_{\mu\nu} - b_{i\mu} b_{j\nu}) \bar{x}_\mu \bar{x}_\nu}{b_{\mu\nu} \bar{x}_\mu \bar{x}_\nu}.$$

In this formula $b_{\mu\nu}$ are components of matrix b, which is the inverse of a matrix α given by

(4) $$b^{-1} = \alpha = \begin{pmatrix} \widehat{xx} & \widehat{xy} & \widehat{xz} & \widehat{xt} \\ \widehat{xy} & \widehat{yy} & \widehat{yz} & \widehat{yt} \\ \widehat{xz} & \widehat{yz} & \widehat{zz} & \widehat{zt} \\ \widehat{xt} & \widehat{yt} & \widehat{zt} & \widehat{tt} \end{pmatrix}.$$

By the inverse of α, it is meant that the components of b and α are related by the expressions

(5) $$b_{\mu\nu} \alpha_{\nu\rho} = \delta_{\mu\rho}.$$

The form (2) for the probability distribution is Gaussian in all directions, and the ellipsoid of standard deviation centered in the mean point is given by

(6) $$f_{ij} \xi_i \xi_j = 1.$$

Since the distribution is Gaussian it is uniquely defined by the second moments of the concentration around the mean point. These second moments can be combined as the components of a matrix a such that

(7)
$$a_{pq} = \frac{\displaystyle\int\int\int_{-\infty}^{+\infty} \xi_p \xi_q P(X, Y, Z, T_0) d\xi_1 d\xi_2 d\xi_3}{\displaystyle\int\int\int_{-\infty}^{+\infty} P(X, Y, Z, T_0) d\xi_1 d\xi_2 d\xi_3}.$$

By taking the partial derivative of (2) with respect to f_{pq} it is possible to establish, that the matrix a is the inverse of the matrix f, so that their components are related by

(8)
$$a_{pq} f_{qr} = \delta_{pr}.$$

By use of (3) and (4) elaboration of a_{pq} gives the following expression

$$a_{pq} = \frac{T_0}{\hat{t}^3} \left[\hat{t}\hat{t}\bar{x}_p \bar{x}_q - \hat{x_p} t \bar{x}_q \hat{t} - \hat{x_q} t \bar{x}_p \hat{t} + \hat{x_p} x_q \hat{t}^2 \right].$$

By introduction of (1) this can be reduced to

(9)
$$a_{pq} = \frac{T_0}{\bar{t}^3} \left[\overline{t^2} \bar{x}_p \bar{x}_q - \overline{x_p t}.\bar{x}_q \bar{t} - \overline{x_q t}.\bar{x}_p \bar{t} + \overline{x_p x_q} \bar{t}^2 \right].$$

The variances a_{pq} are the components of a second rank tensor. It was shown by BEAR (1961) that the relation between the variances and the components $\partial\phi/\partial x_r$ of the hydraulic gradient $\nabla\phi$ is

$$a_{pq} = D_{pqrs} \ L \ \{ \frac{\partial\phi}{\partial x_r} \frac{\partial\phi}{\partial x_s} \ / \ |\nabla\phi|^2 \}$$

Since the expression between braces is a second rank tensor, the constant D_{pqrs} is a fourth rank tensor, representing the medium's dispersivity. L is a scalar equal to the distance travelled by the centre of gravity of the injected particles.

In the anisotropic case the dependance of the variances on the hydraulic gradient vector is not that simple, because products of the gradient components appear of the order $(2n+2)$, where n covers the range from zero to infinity. These products form tensors of even but infinite rank.

Since the constant of dispersivity relates these tensors of infinite rank to the variances a_{pq} which are the components of a second rank tensor, the medium's dispersivity can only be expressed in terms of a tensor of infinite rank in the anisotropic case.

In order to verify the dependence of a_{pq} on the hydraulic gradient vector it is necessary......

necessary to elaborate the mean values of the stochastic variables as occurring in expression (9). For the two-dimensional anisotropic case this will be done in the next section with $\bar{x}$, $\bar{y}$, $\bar{t}$ and $\overline{x^2}$, providing the possibility to consider the form of a_{11}.

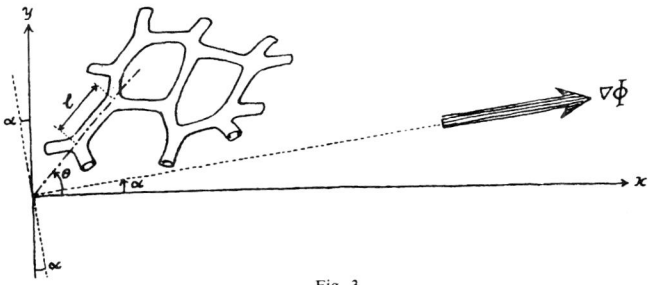

Fig. 3.

ELABORATION OF MEAN VALUES

Let $n(\theta)d\theta$ be the amount of channels having an angle between θ and $\theta + d\theta$ with the positive x-direction, such that the total amount of channels considered: $N = \int_{-1/2\pi}^{+1/2\pi} n(\theta)d\theta$ is a large number. The integral only covers half the circle, because every channel has two opposite directions.

Let $\lambda(\theta)$ be the conductivity of the channels $n(\theta)d\theta$. Anisotropy is obtained, if $n(\theta)\lambda(\theta)$ varies with θ.

Let $l(\theta)$ be the length of the channels $n(\theta)d\theta$. We will take for $l(\theta)$ a constant length l here, because it simplifies the computations without effecting the result essentially.

Let $c(\theta)$ be the cross sectional area of the channels $n(\theta)d\theta$.

In the fluid a gradient of head, $\nabla\Phi$, is assumed to exist which is everywhere equal and makes an angle α with the positive x-direction. This means that the gradient in each of the channels is equal to $|\nabla\Phi| \cdot \cos(\theta - \alpha)$. The discharge of such a channel is then

$$(10) \qquad q(\theta) = \lambda(\theta)|\nabla\Phi| \cdot \cos(\theta - \alpha).$$

According to the assumption, that the probability for a particle to choose a certain direction is proportional to the amount of fluid flowing in that direction, that probability is for a direction between θ and $\theta + d\theta$ equal to

$$(11) \qquad P(\theta \to \theta + d\theta) = \frac{n(\theta)q(\theta)d\theta}{Q}.$$

In this expression Q is the total amount of fluid passing through all the N-channels. Using (10) this becomes

$$(12) \quad Q = \int_{-1/2\pi+\alpha}^{+1/2\pi+\alpha} n(\theta)q(\theta) \cdot d\theta = |\nabla\Phi| \int_{-1/2\pi+\alpha}^{+1/2\pi+\alpha} n(\theta)\lambda(\theta)\cos(\theta - \alpha)d\theta.$$

Integration is performed over the range $-\frac{1}{2}\pi + \alpha < \theta < +\frac{1}{2}\pi + \alpha$, because only channels in those directions carry water away from the junction points. That only discharge departing from the junction points is counted, is a requirement for the probability calculus. The choice is made forward in time and not backward.

The x-component of one step in the Markoff process is the projection of the channel in x-direction. For each of the $n(\theta)d\theta$ channels, this is $l\cos\theta$. The mean value $\bar{x}$ of these x-components is this value $l\cos\theta$ multiplied by its probability of occurrence $P(\theta \to \theta + d\theta)$ integrated over all possible directions. This gives with (10), (11), (12)

$$(13) \qquad \bar{x} = \frac{\displaystyle\int_{-1/2\pi+\alpha}^{1/2\pi+\alpha} n(\theta)l\cos\theta q(\theta)d\theta}{Q} = \frac{l\displaystyle\int n(\theta)\lambda(\theta)\cos\theta\cos(\theta - \alpha)d\theta}{\displaystyle\int n(\theta)\lambda(\theta)\cos(\theta - \alpha)d\theta}.$$

For the y-component, which is $l \sin \theta$, this gives

$$(14) \quad \bar{y} = \frac{\displaystyle\int_{-1/2\pi+\alpha}^{1/2\pi+\alpha} n(\theta)l \sin \theta \, q(\theta)d\theta}{Q} = \frac{l \displaystyle\int n(\theta)\lambda(\theta) \sin \theta \cos(\theta - \alpha)d\theta}{\displaystyle\int n(\theta)\lambda(\theta) \cos(\theta - \alpha)d\theta}.$$

The residence time t of one step in the Markoff process, is the time a particle stays in one channel. If the discharge is $q(\theta)$ and the volume of the channel is $l \cdot c(\theta)$ then this residence time is $t = (l \cdot c(\theta)/q(\theta))$. The mean value $\bar{t}$ of the residence time is this value multiplied by its probability of occurrence, $P(\theta \to \theta + d\theta)$, integrated over all possible directions. This gives with (10) (11), (12),

$$(15) \quad \bar{t} = \frac{\displaystyle\int_{-1/2\pi+\alpha}^{1/2\pi+\alpha} n(\theta)l \cdot c(\theta)d\theta}{Q} = \frac{l \displaystyle\int n(\theta)c(\theta)d\theta}{|\nabla\Phi| \displaystyle\int n(\theta)\lambda(\theta) \cos(\theta - \alpha)d\theta}.$$

In the same manner we find

$$(16) \quad \overline{x^2} = \frac{l^2 \int n(\theta)\lambda(\theta) \cos^2\theta \cos(\theta - \alpha)d\theta}{\int n(\theta)\lambda(\theta) \cos(\theta - \alpha)d\theta}.$$

The general case of anisotropy is obtained by assigning to the $n(\theta)d(\theta$ channels an arbitrary distribution with respect to θ of the combined conductivities $n(\theta)\lambda(\theta)d\theta$ and the combined cross sectional areas $n(\theta)c(\theta)d\theta$. Every possible distribution can be expressed as a Fourier series. Let these be

$$(17) \quad \begin{cases} \dfrac{n(\theta)}{N} \lambda(\theta) = \displaystyle\sum_{n=0}^{\infty} A_{2n} \cos 2n(\theta - \alpha_{2n}) \\[2mm] \dfrac{n(\theta)}{N} c(\theta) = \displaystyle\sum_{n=0}^{\infty} B_{2n} \cos 2n(\theta - \beta_{2n}). \end{cases}$$

Only the even terms appear, because every channel figures in two oppposite directions.

Execution of the integrals in (13), (14), (15) and (16), all between the limits $-\tfrac{1}{2}\pi + \alpha$ and $\tfrac{1}{2}\pi + \alpha$ gives finally

$$(18) \quad \begin{aligned} \bar{x} &= \frac{l}{R} \frac{\pi}{2}[A_0 \cos \alpha + \tfrac{1}{2} A_2 \cos(2\alpha_2 - \alpha)] \\[2mm] \bar{y} &= \frac{l}{R} \frac{\pi}{2}[A_0 \sin \alpha + \tfrac{1}{2} A_2 \sin(2\alpha_2 - \alpha)] \\[2mm] \bar{t} &= \frac{l}{R} \frac{\pi}{|\nabla\Phi|} B_0 \end{aligned}$$

$$\text{with } R = \int_{-1/2\pi+\alpha}^{1/2\pi+\alpha} \frac{n(\theta)}{N} \lambda(\theta) \cos(\theta - \alpha)d\theta.$$

The real mean velocity v has x- and y-components given by $v_x = \bar{x}/\bar{t}$ and $v_y = \bar{y}/\bar{t}$. With (18) this is

(19)
$$
\begin{cases}
v_x = \dfrac{\bar{x}}{\bar{t}} = \left[\dfrac{A_0}{2B_0}\cos\alpha + \dfrac{A_2}{4B_0}\cos(2\alpha_2 - \alpha)\right] \cdot |\nabla\Phi| \\[2mm]
v_y = \dfrac{\bar{y}}{\bar{t}} = \left[\dfrac{A_0}{2B_0}\sin\alpha + \dfrac{A_2}{4B_0}\sin(2\alpha_2 - \alpha)\right] \cdot |\nabla\Phi|.
\end{cases}
$$

Since the gradient of Φ has the direction α, the partial derivatives of Φ can be introduced as $\partial\Phi/\partial x = |\nabla\Phi| \cdot \cos\alpha$, $\partial\Phi/\partial y = |\nabla\Phi|\sin\alpha$. Then (19) becomes

(20)
$$
\begin{cases}
v_x = \left(\dfrac{A_0}{2B_0} + \dfrac{A_2}{4B_0}\cos 2\alpha_2\right)\dfrac{\partial\Phi}{\partial x} + \dfrac{A_2}{4B_0}\sin 2\alpha_2\,\dfrac{\partial\Phi}{\partial y} \\[2mm]
v_y = \dfrac{A_2}{4B_0}\sin 2\alpha_2\,\dfrac{\partial\Phi}{\partial x} + \left(\dfrac{A_0}{2B_0} - \dfrac{A_2}{4B_0}\cos 2\alpha_2\right)\dfrac{\partial\Phi}{\partial y}.
\end{cases}
$$

These expressions form the relation between the real mean velocity vector, v and the gradient vector, $\nabla\Phi$. Because the vector components are obtained by linear superposition of the four coefficients at the right of (20), these coefficients form the components of a second rank tensor.

This tensor is the anisotropic permeability tensor divided by the porosity, because specific discharge is equal to real mean velocity multiplied by the porosity. This result is in agreement with the concept developed by Ferrandon (1948).

In the permeability tensor expressed by (20) only A_0, B_0 and A_2 appear, showing that the other terms of the infinite series (17) are of no importance to the permeability. In contrast to this simple result the dispersion phenomenon cannot be described without using all the terms in the infinite series. This will be seen by elaborating a_{pq} for the case that the conductivity distribution is expressed in an infinite series as given by (17).

Introduction of v_x and v_y in (9) gives for the coefficient $a_{11} = a_{xx}$, corresponding to $x_p = x_q = x$, the form

(21)
$$
a_{11} = a_{xx} = \frac{T_0}{\bar{t}}\left[\overline{\bar{t}^2 v_x^2} - 2\overline{x\bar{t}\,v_x} + \overline{x^2}\right].
$$

Consider as an example the last term between brackets. Elaboration of the integral (16) with (17) gives the value of $\overline{x^2}$. Because of $T_0/\bar{t}$ before brackets, this has to be multiplied by $(RT_0|\nabla\Phi|/\pi l B_0)$, to give finally

(22)
$$
\frac{T_0}{\bar{t}}\overline{x^2} = \frac{T_0 l}{\pi B_0}|\nabla\Phi| \cdot
$$

$$
\cdot \sum_{n=0}^{\infty}(-1)^n A_{2n}\frac{[(4n^2 + 3)\cos 2\alpha + (8n^3 - 10n)\sin 2\alpha - 4n^2 + 9]}{(4n^2 - 9)(4n^2 - 1)} \cdot
$$

$$
\cdot \cos 2n(\alpha - \alpha_{2n})
$$

Considered as a polynome in $(\cos\alpha)$ and $(\sin\alpha)$, this term of a_{xx} contains $(\cos\alpha)^{2n+2}$ and $(\sin\alpha)^{2n+2}$. The contribution of the other two terms in (21) does not reduce the order of the polynome.

This means, that a_{pq} can be considered as the components of an even tensor of rank $(2n + 2)$. Since n can run up to infinity it is an even tensor of infinite rank. └ the dependence of a_{pq} on the hydraulic gradient

CONCLUSION

dependence of the ... on the hydraulic gradient

Since the variance of dispersion for the general anisotropic case is a tensor of infinite rank the dispersion constant is also a tensor of infinite rank. As a consequence of this fact the differential equation for dispersion developed for the isotropic case can only be generalised to the anisotropic case by introducing a dispersion coefficient, whose dependence on the direction of flow is expressed by a series of infinite terms.

REFERENCES

1. Bear, J. (1961), On the tensor form of dispersion in porous media, *J. Geoph. Res.*, *66*, 1185–1197.

2. De Josselin de Jong, G. (1958), Longitudinal and transverse diffusion in granular deposits, *Trans. Amer. Geoph. Union*, *39*, 67–74.

3. Nikolaevskii, V. N. (1959), Convective diffusion in porous media, *Prikl. Mat. Mekh.*, *23*, 1042–1050.

3. Scheidegger, A. E. (1961), General theory of dispersion in porous media, *J. Geoph. Res.*, *66*, 3273–3278.

5. Ferrandon, J. (1948), Les lois de l'écoulement de filtration, *Le Génie Civil, 125*, 24–28.

Civil Engineering Department,
  Delft University of Technology,
    Oostplantsoen 25, Delft, Netherlands

DISPERSION IN FISSURED ROCK

by
G de Josselin de Jong
and
Shao-Chih Way

Geoscience Department
New Mexico Institute of Mining
and Technology, Socorro,
New Mexico 87801
July 1972

Introduction

It is the purpose of this study to show that the dispersion of particles, transported by a fluid flowing through a system of fissures in rock, can be treated as a special case of the general theory for dispersion, developed by use of the probability theory (DE JOSSELIN DE JONG, 1969).

A case study of the dispersion of a large amount of locally injected particles will first be presented computation wise. It will be computed how the particles are partitioned at each intersection of the fissures, how the subgroups of particles are transported through the fissures and where the subgroups will be located at successive time intervals. The result consists of a distribution of particles at a certain time, T , after injection, with discrete amounts of particles at discrete points.

In the second place the general theory developed on the basis of probability concepts will be applied to the same case. This theory gives as a result a continuous distribution of particles, which is essentially Gaussian.

In order to show that the discrete distribution obtained from the computation and the continuous distribution obtained from the probability have the same general form, a figure (Fig.4^a) will be used. Numerically, the two results will be compared by considering the centre of gravity of the dispersed particles and the second moments of the distribution around that centre.

The case study is simplified by considering a system of two intersecting families of fissures. Let the fissure planes be parallel to Z (Fig.1), and let the conditions in all planes perpendicular to Z be identical. Thus the problem to be treated is two-dimensional and only the x,y coordinates are essential. Fluid flow through the fissures proceeds parallel to the xy plane. Discharges are per unit depth in the Z-direction.

Sections 1, 2, 3, 4 deal with the computational experiment. Sections 5, 6, 7, 8 deal with the theoretical prediction. The two resulting particle distributions will be compared in Section 9.

1. Particle Behaviour in Fissures and their Intersections

The behaviour of particles in passing through the individual fissures and their intersections is partly deterministic, partly random.

Let us first consider, what happens within a fissure

In the following the discharges through the fissures are divided in lamellae. Each lamella carries the same amount of discharge per unit depth. Since the fluid velocity in a fissure cross section is not uniform, the velocity in the different lamellae is not the same.

It is assumed here, that the length, ℓ_μ , of the individual fissure segments between intersections, and the average fluid velocity, v_μ , through them are known. The index, μ , refers to the particular family of fissures (characterized by their direction, length, conductivity,...etc.).

In the case considered here, there are two different kinds of fissures, so μ has the value I , $I\!I$, see Fig.1.

Furthermore it is assumed that the width, c_μ , of the fissure is small with respect to its length, ℓ_μ , and molecular diffusion, D_{mol} , large enough for the quantity $c_\mu^2 v_\mu / D_{mol} \ell_\mu$ to be small with respect to one. Then Brownian motion will force a particle to visit all fluid stream lamellae within a fissure, such that the overall particle velocity corresponds to the average fluid velocity. The time, t_μ , that a particle will reside in a fissure segment is then equal to

$$(1.1) \qquad\qquad t_\mu = \ell_\mu / v_\mu$$

This influence of Brownian motion on the particle behaviour creates a deterministic effect, because the residence time, t_μ , has a known value for every particle travelling through the fissure.

Brownian motion, however, also has a second probabilistic influence on the particle behaviour. Because a particle is scattered over all stream lamellae it has an equal probability to be found at any point of the cross section at the exit of a fissure, irrespective of its location in the cross section at the entrance. The choice of stream lamellae that carries the particle at the exit is probabilistic and not correlated to the entrance stream lamellae.

Let us now consider, what happens at an intersection.

Redistribution of particles at an intersection of fissures is again deterministic, if the fluid and the particles behave in the manner proposed by SNOW (19_?_). He assumes, that there is no turbulence and that the time a particle spends within an intersection is too small to permit a particle to switch stream lamellae.

According to SNOW the redistribution of stream lamellae at an intersection can be determined from the discharges in different directions. Since the particles stay with the stream lamellae and they are equally distributed over the stream lamellae by Brownian motion in their preceding travel through the fissure, the redistribution of particles at an intersection is the same as the redistribution of the fluid. And since the fluid distribution is determinable, also the particle distribution is deterministic.

2. Regular Array of Fissures

The example considered is a system of cracks, consisting of two intersecting families of fissures indicated by subcripts I and II (see Fig.1).

The angles of the fissures with respect to the x-axis are respectively: θ_I and θ_{II}. Their lengths between intersections are: ℓ_I and ℓ_{II}. The projections of the fissure segments on the x- and y-coordinate axes are

$$
\begin{array}{ll}
X_I = \ell_I \cos \theta_I & X_{II} = \ell_{II} \cos \theta_{II} \\
y_I = \ell_I \sin \theta_I & y_{II} = \ell_{II} \sin \theta_{II}
\end{array}
\tag{2.1}
$$

The case considered here consists of a system of parallel and equidistant fissures, such that for each family the fissure length and direction angle are constant throughout the field.

A uniform hydraulic gradient J at an angle α with respect to the x-axis is acting throughout the flow field (Fig.1). Because of the regularity of the crack system, the gradients in the two families of cracks are respectively

$$
(2.2) \qquad J_I = J \cos(\alpha - \theta_I) \quad \text{and} \quad J_{II} = J \cos(\alpha - \theta_{II})
$$

If the hydraulic conductivity per unit depth of the fissures is λ_I and λ_{II}, the discharges q_I and q_{II} per unit depth of the fissures are

$$
\begin{array}{l}
q_I = \lambda_I J_I = \lambda_I J \cos(\alpha - \theta_I) \\
q_{II} = \lambda_{II} J_{II} = \lambda_{II} J \cos(\alpha - \theta_{II})
\end{array}
\tag{2.3}
$$

If the width of the two kinds of fissures are respectively C_I and C_{II}, then the average velocities will be

$$
(2.4) \qquad v_I = \frac{q_I}{C_I} \quad \text{and} \quad v_{II} = \frac{q_{II}}{C_{II}}
$$

The corresponding residence times are using (1.1)

(2.5) $t_I = \dfrac{\ell_I}{v_I} = \dfrac{\ell_I \, C_I}{q_I}$ and $t_{II} = \dfrac{\ell_{II}}{v_{II}} = \dfrac{\ell_{II} \, C_I}{q_{II}}$

Computations will be performed for the special case, where (see Fig.3)

(2.6) $\theta_I = 30°$; $\theta_{II} = -30°$; $\alpha = 30°$

Furthermore all fissures are equal, such that

(2.7) $\ell_I = \ell_{II} = \ell$; $C_I = C_{II} = c$; $\lambda_I = \lambda_{II} = \lambda$

then formula (2.3) becomes

(2.8) $q_I = \lambda J \cos(0°) = \lambda J$; $q_{II} = \lambda J \cos(60°) = \tfrac{1}{2}\lambda J$

and formula (2.5) becomes

(2.9) $t_I = \dfrac{\ell c}{\lambda J}$; $t_{II} = 2\,\dfrac{\ell c}{\lambda J}$

3. Continuation Probability

In the case considered, the ratio of q_I to q_{II} is 2 to 1, see (2.8). In Fig.2, the discharges are represented in a ratio 5 to 2, in order to avoid ambiguity resulting from the equal subdivisions, $\frac{1}{2}:\frac{1}{2}$. The discharges are represented by stream lamellae: a, b, c, d, e, f, g, each of which carries the same amount of fluid per unit depth.

Since there is no mixing or turbulence at an intersection, the lamellae do not interchange position. Let us determine first the probability $f^R_{p \to q}$, that a particle carried through a fissure of direction θ_p will continue after the intersection R in direction θ_q. This probability will be called the continuation probability.

The lamellae a, b, carrying the discharge q_{II} through the fissure AC in θ_{II}-direction, continue their path beyond C in the θ_I-direction. At C, therefore, certainty exists, that a particle arriving along θ_{II} will continue along θ_I, and not along θ_{II}. As a consequence the continuation probabilities $f^c_{II \to I}$ and $f^c_{II \to II}$ have the values

$$(3.1) \qquad f^c_{II \to I} = 1 \qquad ; \qquad f^c_{II \to II} = 0$$

The five lamellae c, d, e, f, g, carrying discharge q_I through fissure BC are divided into two groups. The lamellae c, d, e, continue along θ_I, the lamellae f, g, continue along θ_{II}. As can be seen from the figure this subdivision is such that

$$c + d + e = q_I - q_{II} \quad ; \qquad f + g = q_{II}$$

Particles entering fissure BC at B are evenly distributed over the lamellae by Brownian motion when arriving in C. This is independent of their entrance position at B. Therefore, the continuation probability,

$f_{I \to I}^{c}$, for a particle started at B along θ_I to continue in the θ_I-direction is proportional to c+d+e, and $f_{I \to \pi}^{c}$ to continue in the θ_π-direction is proportional to f+g.

Since the particles started at B in the θ_I-direction are certain to continue in some direction, it is necessary, that

$$f_{I \to I}^{c} + f_{I \to \pi}^{c} = 1$$

Combining these results, the following values for the continuation probabilities are obtained

$$(3.2) \qquad f_{I \to I}^{c} = \frac{q_I - q_\pi}{q_I} \qquad ; \qquad f_{I \to \pi}^{c} = \frac{q_\pi}{q_I}$$

In this case where all intersections have the same characterestics, the letter C in the expressions for the continuation probabilities $f_{I \to I}^{c}$, ...etc can be replaced by the letter indicating any other intersection.

$$f_{\pi \to I}^{c} = 1 \qquad\qquad f_{\pi \to \pi}^{c} = 0$$

4. Computational Experiment

The computational experiment consists in pursuing the travel paths of a large amount of particles injected along a line in Z-direction. This line intersects the xy plane in a point P, which will be called the injection point.

As shown in Section 3, the amount of particles passing an intersection is subdivided into subgroups that continue their path in different directions. The probability to continue in a certain direction was expressed as a ratio given by the formulae (3.1) and (3.2). Using for the specific discharges q_I and q_{II} the values (2.8),

$$q_I = \lambda J \qquad\qquad q_{II} = \tfrac{1}{2}\lambda J$$

the continuation probabilities become

(4.1)
$$f^c_{I \to I} = \tfrac{1}{2} \qquad\qquad f^c_{I \to II} = \tfrac{1}{2}$$

$$f^c_{II \to I} = 1 \qquad\qquad f^c_{II \to II} = 0$$

The ratios of subdivision of the injected amount are represented in Fig 3, where the location of the subgroups is shown after time intervals of magnitude Δt, $2\Delta t$, $3\Delta t$, $4\Delta t$, $5\Delta t$. The interval Δt is equal to the residence time in fissures I, such that according to (2.9)

(4.2)
$$\Delta t = t_I = \frac{\ell c}{\lambda J}$$

Since according to (2.9) the residence time in fissures II is twice t_I, subdivision of the particle groups only occurs at time intervals, which are integer multiples of Δt.

After continuing the procedure of subdivision, (shown for the first 5 timesteps in Fig.3) over a time interval

T , consisting of 40 timesteps, such that

(4.3) $$T = 40\,\Delta t = 40\,\frac{\ell c}{\lambda J}$$

the result of Fig.4 is obtained. The particle groups are
indicated by solid circles, whose areas correspond approximately
to the sizes of the subgroups. The numbers accompanying
the circles are the ratios of subdivision, from total
amount of particles to subgroups. The circles do not represent
a spatial distribution, because all particles are considered
to be located at the centre of each circle. All centres
are located on the dash dot line at an angle $\psi^e = 60°$,
see Fig.4.

In Fig.4[a] the distribution of particles along the
line of spread is shown by the stepwise solid line, which
represents the cumulation curve.

Numerically the results of this experiment are expressed
by the coordinates of the centre of gravity of the particles
and the second moment around that centre. The origin of
the coordinates X , Y used for the centre of gravity
is located in the injection point, P, see Fig.4.

Using a *superscript* e for the experimental values, it
is found that after $T = 40\,\frac{\ell c}{\lambda J}$ the centre of gravity has
the coordinates

(4.4) $$X^e = 25.93\,\ell \qquad ; \qquad y^e = 4.91\,\ell$$

The second moment is taken in the direction of the
line of spread at 60°, the result will be called a_{11}^e .
The spread perpendicular to the line at 60° will be called
a_{22}^e . Since there is no spread in that direction, the
second moment, a_{22}^e is zero. So we find

(4.5) $$a_{11}^e = 3.90\,\ell^2 \qquad ; \qquad a_{22}^e = o$$

5. General Theory of Dispersion

The general theory of dispersion based on probability theory using CHANDRASEKHAR's generalisation of MARKOFF's method (DE JOSSELIN DE JONG, 1969), predicts that a point injection of particles, after being transported during a time interval T, is spread into a spatial Gaussian distribution with the points corresponding to the standard deviation in any direction located on an ellipsoid. If the flow system is two-dimensional, points of standard deviation are located on an ellipse. We present here the two dimensional case only, because extension to 3 dimensions is obtained by adding terms of a similar character as those presented here.

The centre of the ellipse has coordinates X, y, (with origin in the injection point) given by

$$(5.1) \qquad X = T \frac{\langle x \rangle}{\langle t \rangle} \qquad\qquad y = T \frac{\langle y \rangle}{\langle t \rangle}$$

The quantities $\langle x \rangle$, $\langle y \rangle$ are the averages of the displacement coordinates x_m, y_m corresponding to the passage through elementary conveyor units (defined in Section 6) and $\langle t \rangle$ is the average of the residence time t_m spent in an elementary conveyor unit. The quantities x_m, y_m, t_m are the stochastic variables, subject to a probabilistic choice, because the elementary conveyor units have different magnitude and orientation and particles choose to enter them according to probabilistic rules.

The size, shape and orientation of the ellipsoid can be computed from the variances or second moments a_{xx}, a_{xy}, a_{yy}, whose magnitude according to the general theory is given by

$$(5.2) \quad \begin{cases} a_{xx} = \dfrac{T}{\langle t \rangle} \left[\langle xx \rangle - 2 \langle xt \rangle \dfrac{\langle x \rangle}{\langle t \rangle} + \langle tt \rangle \dfrac{\langle x \rangle^2}{\langle t \rangle^2} \right] \\[2mm] a_{xy} = \dfrac{T}{\langle t \rangle} \left[\langle xy \rangle - \langle xt \rangle \dfrac{\langle y \rangle}{\langle t \rangle} - \langle yt \rangle \dfrac{\langle x \rangle}{\langle t \rangle} + \langle tt \rangle \dfrac{\langle x \rangle \langle y \rangle}{\langle t \rangle^2} \right] \\[2mm] a_{yy} = \dfrac{T}{\langle t \rangle} \left[\langle yy \rangle - 2 \langle yt \rangle \dfrac{\langle y \rangle}{\langle t \rangle} + \langle tt \rangle \dfrac{\langle y \rangle^2}{\langle t \rangle^2} \right] \end{cases}$$

The quantities $\langle xx \rangle$, $\langle xy \rangle$, $\langle yy \rangle$, $\langle xt \rangle$, $\langle yt \rangle$, $\langle tt \rangle$ are the averages of the products of the stochastic variables x_m, y_m, t_m as combined between the brackets, $\langle \ \rangle$. By introducing the second moments of the stochastic variables and indicating them

(5.3)
$$\hat{xx} = \langle xx \rangle - \langle x \rangle^2$$
$$\hat{xt} = \langle xt \rangle - \langle x \rangle \langle t \rangle \ \ldots \ etc$$

the expressions for the variances obtain the form

(5.4)
$$\begin{cases} a_{xx} = \dfrac{T}{\langle t \rangle} \left[\hat{xx} - 2\hat{xt} \dfrac{\langle x \rangle}{\langle t \rangle} + \hat{tt} \dfrac{\langle x \rangle^2}{\langle t \rangle^2} \right] \\[2mm] a_{xy} = \dfrac{T}{\langle t \rangle} \left[\hat{xy} - \hat{xt} \dfrac{\langle y \rangle}{\langle t \rangle} - \hat{yt} \dfrac{\langle x \rangle}{\langle t \rangle} + \hat{tt} \dfrac{\langle x \rangle \langle y \rangle}{\langle t \rangle^2} \right] \\[2mm] a_{yy} = \dfrac{T}{\langle t \rangle} \left[\hat{yy} - 2\hat{yt} \dfrac{\langle y \rangle}{\langle t \rangle} + \hat{tt} \dfrac{\langle y \rangle^2}{\langle t \rangle^2} \right] \end{cases}$$

The relation between the variances and the size, the shape and the direction of the ellipse is as follows (see BEAR, 1961, by 1191). The quantities a_{xx}, a_{xy}, a_{yy} are the components of a second rank tensor whose principal values are a_{11}, a_{22}, given by

(5.5)
$$\begin{cases} a_{11} = \frac{1}{2}(a_{xx} + a_{yy}) + \frac{1}{2}\sqrt{(a_{xx} - a_{yy})^2 + 4a_{xy}^2} \\[2mm] a_{22} = \frac{1}{2}(a_{xx} + a_{yy}) - \frac{1}{2}\sqrt{(a_{xx} - a_{yy})^2 + 4a_{xy}^2} \end{cases}$$

The roots of these principal values are equal to the magnitudes of the principal axes of the ellipse, σ_1 and σ_2 according to

(5.6)
$$\sigma_1 = \sqrt{a_{11}} \qquad ; \qquad \sigma_2 = \sqrt{a_{22}}$$

The angle, ψ, between the major principal axis, σ_1, and the x-axis is given by

(5.7)
$$\tan \psi = \frac{2a_{xy}}{(a_{xx} - a_{yy}) + \sqrt{(a_{xx} - a_{yy})^2 + 4a_{xy}^2}}$$

6. Elementary Conveyor Units

The results mentioned in Section 5 are theoretically obtained as a distribution of the probability to find a particle in a certain location at a certain time. A particle can arrive there following different paths. The analysis consists in deriving the probability for a particle to choose a certain path and to select from all possible paths only those, that end up in the desired location at the desired time. Mathematically this produces an integral expression with a closed solution.

It does not fit in the framework of this presentation to expose all the details of this analysis (a separate report deals with this aspect). It is, however, necessary to mention here that the analysis requires, that the particle path be subdivided into elementary steps of a special nature.

In the first place the steps must be such that it is possible to determine the magnitude of the probability for a particle to choose a step of a particular kind.

In the second place it is necessary that the steps are not correlated, the reason for noncorrelation is that the individual steps must be combined into a product. It is known that according to probability theory the probability for the subsequent occurence of several events consists of the multiplication of the probabilities of the seperate occurence of the individual events, only if the events are not correlated. In order to be able to apply the mentioned product, it is therefore necessary that the steps are not correlated.

When particles are transported by a fluid through a porous system, the above mentioned analysis can be applied to predict their dispersion if the paths followed by

the particles can be subdivided into steps that satisfy
the requirements imposed by the theory. We will use the
form « probabilistic step » , if the steps do indeed satisfy
these requirements.

Physically the particle path consists of a number of
traverses through pores, cavities or fissures, depending
on the nature of the porous system. In general there is
not necessarily a coincidence between the probabilistic
steps of the particle path and the traverses through
individual pores, cavities or fissures. It can be any
combination of them.

In order to indicate the physical counterpart of
the mathematical steps, the term «elementary conveyor unit »
is proposed here. An elementary conveyor unit can be any
part of the porous medium traversed by transported particles.
It can be one channel between grains or any combination
of channels or fissures up to complete lenses or layers
of soil. The combinations of conduits constituting
elementary conveyor units can be different for different
directions and magnitudes of the hydraulic gradient.

In order to constitute an elementary conveyor unit
the combination of conduits must satisfy the four
requirements listed below. These requirements guarantee
that the passage through the elementary conveyor units
from their entrances to their exits, coincide with the
probabilistic steps. In the following the parameter, m,
will be used to indicate the properties of an elementary
conveyor unit of the mth kind.

The requirements are

1   a particle, travelling from entrance to exit through
an elementary conveyor unit of the mth kind, covers a
distance, whose coordinate components x_m , y_m , z_m , have
a known magnitude.

[2]   a particle, travelling from entrance to exit through an elementary conveyor unit of the mth kind, remains within the unit during a residence time, t_m , with a known magnitude.

[3]   the choice being made by a particle to enter the next unit after having completed its travel through a previous unit has a probability distribution, g_m , which is a known function of theparameter, m, of the unit it will enter.

[4]   the probability distribution of choice, g_m , does not depend on the parameter, m, of the previous unit. This independence guarantees that subsequent steps are uncorrelated.

·Because the values of x_m , y_m , z_m , t_m as chosen successively by the particle during its travel through the porous system are subject to a probabilistic choice, these quantities are called the stochastic variables.

In the case considered here, there are two kinds of elementary conveyor units. Therefore m has the values 1, 2. In order that the distribution of choice represents a correctly mormalized probability it has to satisfy the condition

$$(6.1) \qquad \sum^{m} g_m = 1$$

The averages $\langle x \rangle$, $\langle y \rangle$, $\langle t \rangle$, $\langle xx \rangle$, $\langle xy \rangle$,..........etc mentioned in the results obtained from the general theory are defined as follows by summations of the stochastic variables multiplied by the corresponding probability of choice

$$(6.2) \quad \begin{cases} \langle x \rangle = \sum^{m} x_m g_m & \langle xx \rangle = \sum^{m} x_m^2 \, g_m \\[2mm] \langle y \rangle = \sum^{m} y_m g_m & \langle xy \rangle = \sum^{m} x_m y_m g_m \\[2mm] \langle t \rangle = \sum^{m} t_m g_m & \dots\dots etc \end{cases}$$

If m, instead of consisting of a finite number of discrete values, is continuously distributed, then the requirement on g_m is

(6.3)
$$\int_{-\infty}^{+\infty} g_m \, dm = 1$$

In this case the averages are defined as follows by integrals over all possible values of m

(6.4)
$$\begin{cases} \langle x \rangle = \int_m x_m \, g_m \, dm & \langle xx \rangle = \int_m x_m^2 \, g_m \, dm \\[2mm] \langle y \rangle = \int_m y_m \, g_m \, dm & \langle xy \rangle = \int_m x_m \, y_m \, g_m \, dm \\[2mm] \langle t \rangle = \int_m t_m \, g_m \, dm & \quad \dots \dots \, etc \end{cases}$$

7. Combination of Fissures into
Elementary Conveyor Units

In order to assign the status of elementary conveyor unit to fissure segments or combinations of fissure segments, the 4th requirement provides the decisive criterion. Therefore, we have to examine first, where in the particle transportation system the correlation disappears between local decision of subsequent path direction and the preceding path direction. This can be done by considering the stream lamellae of Fig.2 again.

Particles entering the fissure CE at C are carried by either of the two lamellae, f or g. Both those lamellae were previously in the fissure BC. Therefore fissures CE and BC are correlated and cannot be considered as two seperate elementary conveyor units, but have to be combined. Furthermore at E both stream lamellae, f and g, will continue in the θ_I-direction as can be deduced from the behaviour of the stream lamellae, a and b, as they pass intersection C.

Particles entering the fissure BC at B are transported by one of the lamellae c, d, e, f or g. If entering with c or d the particle had previously the θ_{II}-direction, if entering with e, f or g the particle had previously the θ_I-direction. Molecular diffusion scatters particles randomly over the lamellae such that arriving in C, the location of a particle is no more influenced by which lamellae carried them at B. One could say that the particle has forgotten its past history. At C, part of the particles continues along θ_I (those carried by c, d, e) and part continues along θ_{II}. The continuation probabilities at C are, because of molecular diffusion, independent of the particle location at B and therefore not correlated to the travel direction of the particle before arriving at B.

Since the rupture in correlation is produced by
molecular diffusion within the θ_I fissure (such as BC)
the transition points between elementary conveyor units,
where exits and entrances of subsequent units meet, must
be located halfway along these fissure segments.

The discharge of particles is subdivided in C. The
lamellae c, d, e continue through CD in the θ_I-direction
and reach an exit of an elementary conveyor unit halfway
along CD. The lamellae f and g continue first along θ_{II}
through CE and then, in E, procede along θ_I through EF,
reaching the exit of an elementaty conveyor unit halfway
between E and F.

This is represented by the shaded fluid flow in figure
5a. The elementary conveyor units are represented in figure
5b by the solid arrows overlying the dotted fissures.
There are two kinds of elementary conveyor units labeled
m=1 and m=2.

The conveyor unit, m=1, consists of two halves of
fissure segments in the θ_I-direction, such as going from
a point halfway between B and C to a point halfway between
C and D. This distance has a length ℓ_I and a direction
θ_I .

The conveyor unit, m=2, consists of one half of a
fissure segment in the θ_I-direction, a full length of
fissure segment in the θ_{II}-direction, and again a half
in the θ_I- direction, such as going from halfway between
BC over C to E upto halfway between E and F.

The first two requirements on elementary conveyor
units are satisfied, because it is possible to compute
the displacements and residence times. These are, using
(2.1) and (2.5)

For m=1

$$(7.1) \quad \begin{cases} x_1 = \frac{1}{2} x_I + \frac{1}{2} x_I = \ell_I \cos \theta_I \\ y_1 = \frac{1}{2} y_I + \frac{1}{2} y_I = \ell_I \sin \theta_I \\ t_1 = \frac{1}{2} t_I + \frac{1}{2} t_I = \frac{\ell_I C_I}{q_I} \end{cases}$$

For m=2

$$(7.2)\quad\begin{cases}x_2 = \frac{1}{2}x_I + x_{II} + \frac{1}{2}x_I = \ell_I \cos\theta_I + \ell_{II}\cos\theta_{II}\\[4pt] y_2 = \frac{1}{2}y_I + y_{II} + \frac{1}{2}y_I = \ell_I \sin\theta_I + \ell_{II}\sin\theta_{II}\\[4pt] t_2 = \frac{1}{2}t_I + t_{II} + \frac{1}{2}t_I = \dfrac{\ell_I\,C_I}{q_I} + \dfrac{\ell_{II}\,C_{II}}{q_{II}}\end{cases}$$

In order to show that finally also the third requirement is satisfied and q_1, q_2 can be computed, it is necessary to determine the probability that a particle, starting halfway between B and C, will end up either halfway between C and D or halfway between E and F. These probabilities can be obtained by considering the continuation probabilities in C and E.

A particle can be considered to traverse an elementary conveyor unit, m=1, if arriving in C from the θ_I-direction it will continue in the θ_I- direction. The probability for a particle to do so is $f^{\,c}_{I\to I}$ as defined in Section 3, and therefore with (3.2) we obtain

$$(7.3)\qquad q_1 = f^{\,c}_{I\to I} = \frac{q_I - q_{II}}{q_I}$$

A particle can be considered to traverse an elementary conveyor unit, m=2, if arriving in C from the θ_I-direction, it will continue along θ_{II} and subsequently, arriving in E from the θ_{II}-direction it will continue along θ_I . The probability for a particle to do so consists of two consecutive probabilities, $f^{\,c}_{I\to II}$ combined with $f^{\,E}_{II\to I}$. According to the rules of probability theory these two probabilities have to be multiplied. This gives for the probability, q_2 , that a particle chooses an elementary conveyor unit, m=2, with (3.2) and (3.1)

$$(7.4)\qquad\begin{aligned}q_2 &= f^{\,c}_{I\to II} * f^{\,E}_{II\to I} = f^{\,c}_{I\to II} * f^{\,c}_{II\to I}\\[6pt] q_2 &= \frac{q_{II}}{q_I} * 1 = \frac{q_{II}}{q_I}\end{aligned}$$

The two probabilities of choice g_1 and g_2 satisfy the requirement (6.1), because

$$(7.5) \qquad g_1 + g_2 = \frac{g_I - g_{II}}{g_I} + \frac{g_{II}}{g_I} = 1$$

Because the combination of fissure segments as combined above into the units, m=1 and m=2, satisfy the four requirements imposed by probability calculus, these units represent the elementary conveyor units in this case. Computation of the averages $\langle x \rangle$, $\langle y \rangle$, $\langle t \rangle$, $\langle xx \rangle$, $\langle yy \rangle$,...etc. amounts to the use of (7.3) and (7.4), (7.5) in the expressions (6.2). This gives the following result. for the averages

$$(7.6) \begin{cases}
\langle x \rangle = x_1 g_1 + x_2 g_2 = (\ell_I \cos\theta_I \, g_I + \ell_{II} \cos\theta_{II} \, g_{II})/g_I \\[4pt]
\langle y \rangle = y_1 g_1 + y_2 g_2 = (\ell_I \sin\theta_I \, g_I + \ell_{II} \sin\theta_{II} \, g_{II})/g_I \\[4pt]
\langle t \rangle = t_1 g_1 + t_2 g_2 = (\ell_I c_I + \ell_{II} c_{II})/g_I \\[4pt]
\langle xx \rangle = x_1^2 g_1 + x_2^2 g_2 = (\ell_I^2 \cos^2\theta_I \, g_I + 2\ell_I \ell_{II} \cos\theta_I \cos\theta_{II} \, g_{II} \\[2pt]
\qquad\qquad + \ell_{II}^2 \cos^2\theta_{II} \, g_{II})/g_I \\[4pt]
\langle xy \rangle = x_1 y_1 g_1 + x_2 y_2 g_2 = (\ell_I^2 \cos\theta_I \sin\theta_I \, g_I + \ell_I \ell_{II} \cos\theta_I \sin\theta_{II} \, g_{II} \\[2pt]
\qquad\qquad + \ell_I \ell_{II} \cos\theta_{II} \sin\theta_I \, g_{II} + \ell_{II}^2 \cos\theta_{II} \sin\theta_{II} \, g_{II})/g_I \\[4pt]
\langle yy \rangle = y_1^2 g_1 + y_2^2 g_2 = (\ell_I^2 \sin^2\theta_I \, g_I + 2\ell_I \ell_{II} \sin\theta_I \sin\theta_{II} \, g_{II} \\[2pt]
\qquad\qquad + \ell_{II}^2 \sin^2\theta_{II} \, g_{II})/g_I \\[4pt]
\langle xt \rangle = x_1 t_1 g_1 + x_2 t_2 g_2 = \ell_I \cos\theta_I \frac{(\ell_I c_I + \ell_{II} c_{II})}{g_I} + \ell_{II} \cos\theta_{II} \frac{(\ell_I c_I \, g_I + \ell_{II} c_{II} \, g_{II})}{g_I^2} \\[4pt]
\langle yt \rangle = y_1 t_1 g_1 + y_2 t_2 g_2 = \ell_I \sin\theta_I \frac{(\ell_I c_I + \ell_{II} c_{II})}{g_I} + \ell_{II} \sin\theta_{II} \frac{(\ell_I c_I \, g_I + \ell_{II} c_{II} \, g_{II})}{g_I^2} \\[4pt]
\langle tt \rangle = t_1^2 g_1 + t_2^2 g_2 = \frac{\ell_I^2 c_I^2}{g_I^2} + 2\frac{\ell_I \ell_{II} c_I c_{II}}{g_I^2} + \frac{\ell_{II}^2 c_{II}^2}{g_I g_{II}}
\end{cases}$$

for the second moments, as defined by (5.3)

$$
(7.7) \quad
\begin{cases}
\widehat{xx} = \langle xx \rangle - \langle x \rangle^2 = \ell_{II}^2 \cos^2\theta_{II} \; \dfrac{q_I(q_I - q_{II})}{q_I^2} \\[3mm]
\widehat{xy} = \langle xy \rangle - \langle x \rangle\langle y \rangle = \ell_{II}^2 \cos\theta_{II} \sin\theta_{II} \; \dfrac{q_I(q_I - q_{II})}{q_I^2} \\[3mm]
\widehat{yy} = \langle yy \rangle - \langle y \rangle^2 = \ell_{II}^2 \sin^2\theta_{II} \; \dfrac{q_{II}(q_I - q_{II})}{q_I^2} \\[3mm]
\widehat{xt} = \langle xt \rangle - \langle x \rangle\langle t \rangle = \ell_{II}^2 \cos\theta_{II} \, C_{II} \; \dfrac{(q_I - q_{II})}{q_I^2} \\[3mm]
\widehat{yt} = \langle yt \rangle - \langle y \rangle\langle t \rangle = \ell_{II}^2 \sin\theta_{II} \, C_{II} \; \dfrac{(q_I - q_{II})}{q_I^2} \\[3mm]
\widehat{tt} = \langle tt \rangle - \langle t \rangle^2 = \ell_{II}^2 C_{II}^2 \; \dfrac{(q_I - q_{II})}{q_I^2 \, q_{II}}
\end{cases}
$$

8. Application to the Example

The example, considered for the experiment, had the following values for the direction of the fissures and the direction of the overall hydraulic gradient (Fig.3)

$$(8.1) \qquad \theta_I = +30° \quad , \qquad \theta_{II} = -30° \quad , \qquad \alpha = +30°$$

All lengths were equal, and all cross sections were equal, so that

$$(8.2) \qquad \ell_I = \ell_{II} = \ell \quad , \qquad C_I = C_{II} = c \quad , \qquad \lambda_I = \lambda_{II} = \lambda$$

The discharges are according to (2.8)

$$(8.3) \qquad q_I = \lambda J \cos(0°) = \lambda J \quad , \qquad q_{II} = \lambda J \cos(60°) = \tfrac{1}{2}\lambda J$$

Using these values in (7.6) and (7.7) gives

$$(8.4) \quad
\begin{cases}
\langle x \rangle = \dfrac{3\sqrt{3}}{4}\, \ell \\[4pt]
\langle y \rangle = \dfrac{1}{4}\, \ell \\[4pt]
\langle t \rangle = 2\, \dfrac{c\ell}{\lambda J} \\[4pt]
\langle xx \rangle = \dfrac{15}{8}\, \ell^2 \\[4pt]
\langle xy \rangle = \dfrac{\sqrt{3}}{8}\, \ell^2 \\[4pt]
\langle yy \rangle = \dfrac{1}{8}\, \ell^2 \\[4pt]
\langle xt \rangle = \dfrac{7\sqrt{3}}{4}\, \dfrac{c\ell^2}{\lambda J} \\[4pt]
\langle yt \rangle = \dfrac{1}{4}\, \dfrac{c\ell^2}{\lambda J} \\[4pt]
\langle tt \rangle = 5\, \dfrac{c^2\ell^2}{\lambda^2 J^2}
\end{cases}$$

The angle $\mathcal{V}$ between x-axis and path followed by the centre of gravity is given by

(8.5) $\qquad \mathcal{V} = \tan^{-1} \dfrac{\langle y \rangle}{\langle x \rangle} = 10.9°$

Applying the results (8.4) to the formulae of Section 5 developed from the general dispersion theory gives:

for the coordinates of the midpoint displacement

(8.6) $\qquad X = \dfrac{3\sqrt{3}}{8} \dfrac{\lambda J}{c} T \qquad\qquad y = \dfrac{1}{8} \dfrac{\lambda J}{c} T$

for the second moments

(8.7) $\qquad \begin{cases} a_{xx} = \dfrac{3}{128} T \ell^2 \dfrac{\lambda J}{\ell c} \\[2mm] a_{xy} = \dfrac{3\sqrt{3}}{128} T \ell^2 \dfrac{\lambda J}{\ell c} \\[2mm] a_{yy} = \dfrac{9}{128} T \ell^2 \dfrac{\lambda J}{\ell c} \end{cases}$

With T , the time after injection equal to

(8.8) $\qquad T = 40 \dfrac{\ell c}{\lambda J}$

this gives, with a superscript th for theoretical:

(8.9) $\qquad \begin{cases} a_{xx}^{th} = 0.94 \; \ell^2 \\[2mm] a_{xy}^{th} = 1.62 \; \ell^2 \\[2mm] a_{yy}^{th} = 2.81 \; \ell^2 \end{cases} \qquad \begin{aligned} X^{\ell} &= 25.98 \; \ell \\[2mm] y^{th} &= 5.00 \; \ell \end{aligned}$

Using these values in (5.5), (5.6) and (5.7) gives

(8.10) $\qquad \begin{cases} a_{11}^{th} = 3.75 \; \ell^2 \\[2mm] 6_1^{th} = 1.94 \; \ell \end{cases} \qquad \begin{aligned} a_{22}^{th} &= 0 \qquad\quad \mathcal{Y}^{th} = 60° \\[2mm] 6_2^{th} &= 0 \end{aligned}$

The values (8.9) and (8.10) are used to produce the dotted continuous cumulation curve in Fig.4$^{\alpha}$, which according to the theory has the form of a Gaussian distribution.

9. Discussion of Results, Conclusions and Further Development

The results of the computational experiment and of the theory, expressed in the values of the coordinates of the center of gravity and the second moments as obtained in (4.5), (4.6) and (8.8), (8.10), are summarized as follows

$$(9.1) \quad \begin{cases} x^e = 25.93 \, \ell & x^{th} = 25.98 \, \ell \\ y^e = 4.91 \, \ell & y^{th} = 5.00 \, \ell \\ a_{11}^e = 3.90 \, \ell^2 & a_{11}^{th} = 3.75 \, \ell^2 \\ a_{22}^e = 0 & a_{22}^{th} = 0 \\ \Upsilon^e = 60° & \Upsilon^{th} = 60° \\ \delta_1^e = 1.975 \, \ell & \delta_1^{th} = 1.94 \, \ell \\ \delta_2^e = 0 & \delta_2^{th} = 0 \end{cases}$$

Both experimental computation and theory predict a dispersion ellipse, whose major principal axis has a direction of 60° with the X-axis and a minor principal axis equal to zero. This means, therefore, that particles are spread in a line at 30° with respect to the hydraulic gradient. (see Fig.4).

The values of the centre of gravity coordinates and the standard deviation disagree by less than 2%. The experimental centre of gravity lies below the theoretical prediction.

These errors are largely due to the arbitrariness of the condition of injection. If the injection point had been chosen at a point Q along θ_I, halfway between

the fissure intersections, rather than at P (Fig.6), the
experiment would have given the same spread of particles
with the same experimental values for the second moments
as injection at P.

By displacing the injection point, also the origin of
the XY coordinates used for locating the centre of gravity
is desplaced. Since the displacement consists of components
$PR = \frac{1}{4}\ell\sqrt{3}$ and $RQ = \frac{1}{4}\ell$, the coordinates X, Y have to be
increased by these amounts respectively. The result of
the experiment starting from Q therefore is:

$$(9.2) \quad \begin{cases} X_Q^\ell \doteq 25.93 + 0.433 = 26.36\,\ell & \sigma_{1_Q}^\ell = 1.975\,\ell \\ \\ Y_Q^\ell = 4.91 + 0.25 = 5.16\,\ell & \sigma_{2_Q}^\ell = 0 \end{cases}$$

The travel time to obtain the values given by Fig.4 starting
from Q is longer than starting from P by half the residence
time in a θ_I fissure. So T must be taken to be

$$(9.3) \qquad T_Q = T + \tfrac{1}{2} t_I = 40\tfrac{1}{2} \frac{\ell c}{\lambda J}$$

Using this value T_Q in (8.6) and (8.7) gives the
theoretical prediction

$$(9.4) \quad \begin{aligned} X_Q^{th} &= 26.3\,\ell & \sigma_{1_Q}^{th} &= 1.95\,\ell \\ \\ Y_Q^{th} &= 5.06\,\ell & \sigma_{2_Q}^{th} &= 0 \end{aligned}$$

The error in centre of gravity location is again less than
2%, but the experimental centre of gravity now lies above
the theoretical prediction. The error in standard deviation
σ_1 is now about 1%.

These errors can still be reduced by starting the
injection in a more balanced way. This means that the
distribution of particles over the two kinds of fissures

Soil Mechanics and Transport in Porous Media

at the injection point is chosen in such a manner that
this distribution remains unaltered, although after every
time interval Δt , particles are redistributed. Then
the centre of gravity coordinates of the experiment will
not follow the tortuous path as shown in Fig.6 but a
straight line that matches the theoretical values exactly
and the error in standard deviation reduces to less than
1%. We will not elaborate this refinement further because
explanation of a balanced injection may confuse more than
its contribution warrants.

Besides this numerical comparison, Fig.4[a] permits
a visual comparison of the cumulation curves of the particle
distributions from computational experiment (solid, stepwise
curve) and theory (dashed, continuous curve). The two
curves fit everywhere within the mesh width of the fissure
pattern.

Since experiment and theory agree numerically and
visually within values that are smaller than the mesh width
of the fissure pattern, the conclusion may be drawn that
the theory is capable of predicting the outcome of the
experiment within acceptable limits.

The theory is expressed in terms of averages of the
stochastic variables. The prediction formulae can therefore
be used in every case where the stochastic variables and
their probability distribution is known. As an example
of further development it will be shown here how to apply
the theory if the lengths of the fissure segments are not
equal (as in the previous treatment), but are randomly
distributed . Then the distribution of choice g_m must be
changed to include the variability of length, ℓ .

Let the distribution of the lengths for both fissure
families be Gaussian with the same mean value ℓ_μ and the
same standard deviation, ℓ_σ . The probability of choice
will then contain a term

(9.5)
$$\frac{1}{\ell_\sigma\sqrt{2\pi}} \; exp\left\{-(\ell-\ell_\mu)^2/2\,\ell_\sigma^2\right\} d\ell$$

which represents the probability that a fissure segment
has a length of magnitude between ℓ and $\ell+d\ell$.

The previously obtained values for g_m given by (7.3)
and (7.4) then become

(9.6)
$$\begin{cases} g_1 = \frac{q_I - q_{I\!I}}{q_I} \; \frac{1}{\ell_\sigma\sqrt{2\pi}} \; exp\left\{-(\ell-\ell_\mu)^2/2\,\ell_\sigma^2\right\} d\ell \\[2mm] g_{I\!I} = \frac{q_{I\!I}}{q_I} \; \frac{1}{\ell_\sigma\sqrt{2\pi}} \; exp\left\{-(\ell-\ell_\mu)^2/2\,\ell_\sigma^2\right\} d\ell \end{cases}$$

The factor $\frac{1}{\ell_\sigma\sqrt{2\pi}}$ is added in order to normalize the
probability of choice properly, satisfying the requirement

$$\int_{-\infty}^{\infty} \sum_m g_m \, d\ell = \frac{1}{\ell_\sigma\sqrt{2\pi}} \int_{-\infty}^{\infty} exp\left\{-(\ell-\ell_\mu)^2/2\,\ell_\sigma^2\right\} \left[\frac{q_I - q_{I\!I}}{q_I} + \frac{q_{I\!I}}{q_I}\right] d\ell = 1$$

Besides a summation of the form (6.2), this requirement
contains an integration over all possible lengths. In order
to obtain simple results from the integration, it is common
practice to integrate ℓ over the range from $-\infty$ to ∞ .
This means, that unreleastic negative values of ℓ are
included. However, if ℓ_σ is small with respect to ℓ_μ ,
the error introduced is small with respect to the convenience
of a simpler expression.

The averages of the stochastic variables are obtained
by formalae similar to (6.2) and (6.4). For the average
of the x-component of the displacement this gives with
(7.1) and (7.2)

(9.7)
$$\langle x \rangle = \frac{1}{\ell_\sigma\sqrt{2\pi}} \int_{-\infty}^{\infty} exp\left\{-(\ell-\ell_\mu)^2/2\,\ell_\sigma^2\right\} *$$

$$* \left\{[\ell \cos\theta_I (q_I - q_{I\!I}) + (\ell\cos\theta_I + \ell\cos\theta_{I\!I}) q_{I\!I}]/q_I\right\} d\ell =$$

$$= \ell_\mu (\cos\theta_I q_I + \cos\theta_{I\!I} q_{I\!I})/q_I$$

and for the average of the x-component squared

$$(9.8) \qquad \langle xx \rangle = \frac{1}{\ell_\sigma \sqrt{2\pi}} \int_{-\infty}^{\infty} \exp \left\{ - (\ell - \ell_\mu)^2 / 2\ell_\sigma^2 \right\} \times$$

$$\times \left\{ [\ell^2 \cos^2 \theta_I (q_I - q_{I\!I}) + (\ell \cos \theta_I + \ell \cos \theta_{I\!I})^2 q_{I\!I}] / q_I \right\} d\ell =$$

$$= (\ell_\mu^2 + \ell_\sigma^2) [\cos^2 \theta_I \, q_I + 2 \cos \theta_I \cos \theta_{I\!I} \, q_{I\!I} + \cos^2 \theta_{I\!I} \, q_{I\!I}] / q_I$$

and similar expressions for the other averages.

Considering the example of Section 8 and introducing the following values for the mean length and its standard deviation

$$(9.9) \qquad \ell_\mu = \ell \qquad\qquad \ell_\sigma = \ell \sqrt{0.1}$$

the averages become (to be compared with (8.4))

$$(9.10)$$

$$\langle x \rangle = \frac{3\sqrt{3}}{4} \ell \qquad \langle xx \rangle = \boxed{1.06} \frac{15}{8} \ell^2 \qquad \langle xt \rangle = \boxed{1.06} \frac{7\sqrt{3}}{4} \ell^2 \frac{c}{\lambda J}$$

$$\langle y \rangle = \frac{1}{4} \ell \qquad \langle xy \rangle = \boxed{1.1} \frac{\sqrt{3}}{8} \ell^2 \qquad \langle yt \rangle = \boxed{\frac{1}{4} \ell^2 \frac{c}{\lambda J}}$$

$$\langle t \rangle = 2 \frac{c\ell}{\lambda J} \qquad \langle yy \rangle = \boxed{1.3} \frac{1}{8} \ell^2 \qquad \langle tt \rangle = \boxed{1.06} \times 5 \frac{c^2 \ell^2}{\lambda^2 J^2}$$

From this the second moments become $\text{moet allerverd 1,1 zijn} \atop \text{al } \ell_\mu^2 + \ell_\sigma^2 = \ell^2(1+0,1)$

$$(9.11)$$

$$a_{xx} = 1.3 \frac{3}{128} T \ell^2 \frac{\lambda J}{\ell c}$$

$$a_{xy} = 1.1 \frac{3\sqrt{3}}{128} T \ell^2 \frac{\lambda J}{\ell c}$$

$$a_{yy} = 1.3 \frac{9}{128} T \ell^2 \frac{\lambda J}{\ell c}$$

and this gives for $T = 40 \frac{\ell c}{\lambda J}$ a dispersion ellipse with $\quad$ *als us Serefodi moet dit zijn*

major principal axis: $\qquad \sigma_1 = 2.13 \ell \quad\longrightarrow \qquad \sigma_1 = 2.021$

minor principal axis: $\qquad \sigma_2 = 0.527 \ell \quad\longrightarrow \qquad \sigma_2 = 0$

direction of major principal axis: $\qquad \psi = 62.2°$

Dispersion in this case spreads the particles in two directions because $\sigma_2 \neq 0$, instead of along a line as was the case when all fissure segments had the same length.

REFERENCES

J. Bear,
《 On the Tensor Form of Dispersion in Porous Media. 》
J. Geoph. Research, Vol. 66, No. 4, Pages 1185-1197, 1961.

S. Chandrasekhar,
《 Stochastic Problems in Physics and Astronomy. 》
Reviews of Modern Physics, Vol. 15, No. 1, Pages 1-89, 1943.

G. de Josselin de Jong,
《 Longitudinal and Transverse Diffusion in Granular Deposits. 》
Trans. Amer. Geoph. Union, 39, Pages 67-74, 1958.

G. de Josselin de Jong,
《 The Tensor Character of the Dispersion Coefficient in Anisotropic Porous 》
Proc. IAHR Symposium, Haifa, 1969, Pages 259-267.

D. T. Snow,
? ?

Soil Mechanics and Transport in Porous Media

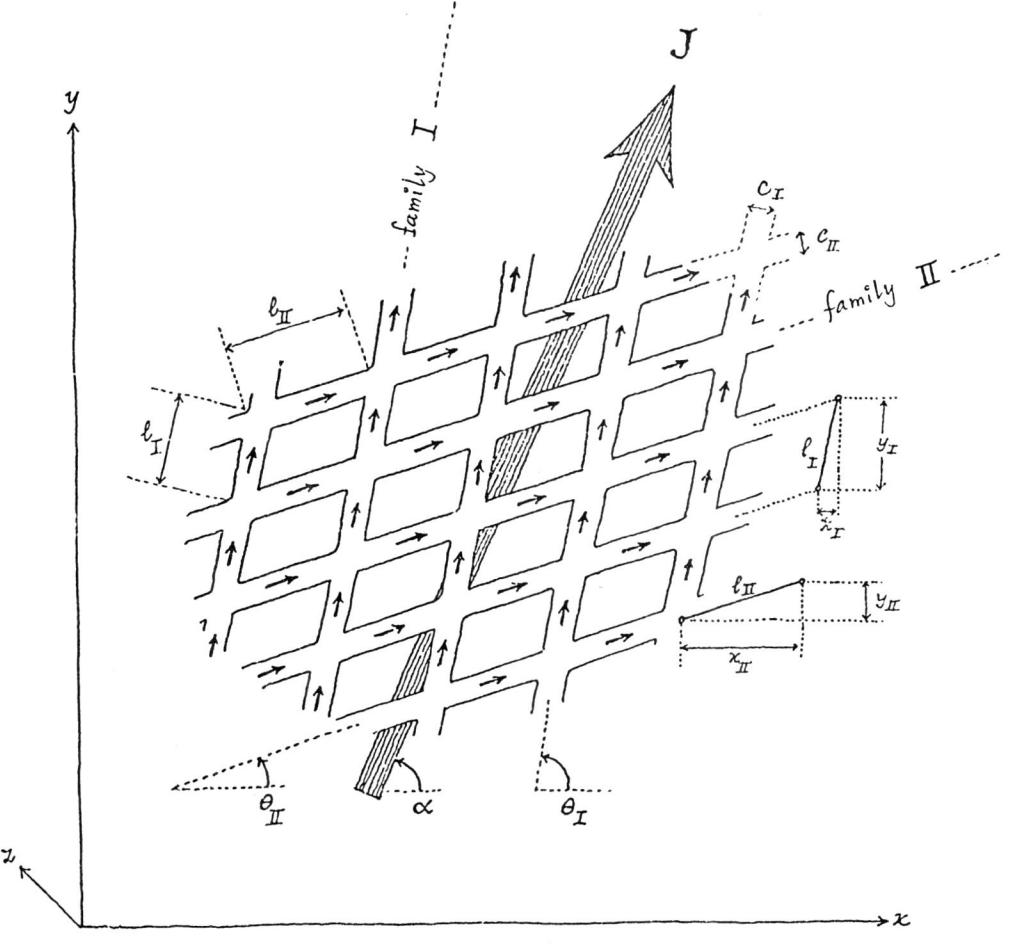

Orientation of the two families of fissures and
of the hydraulic gradient, J .

Fig. 1

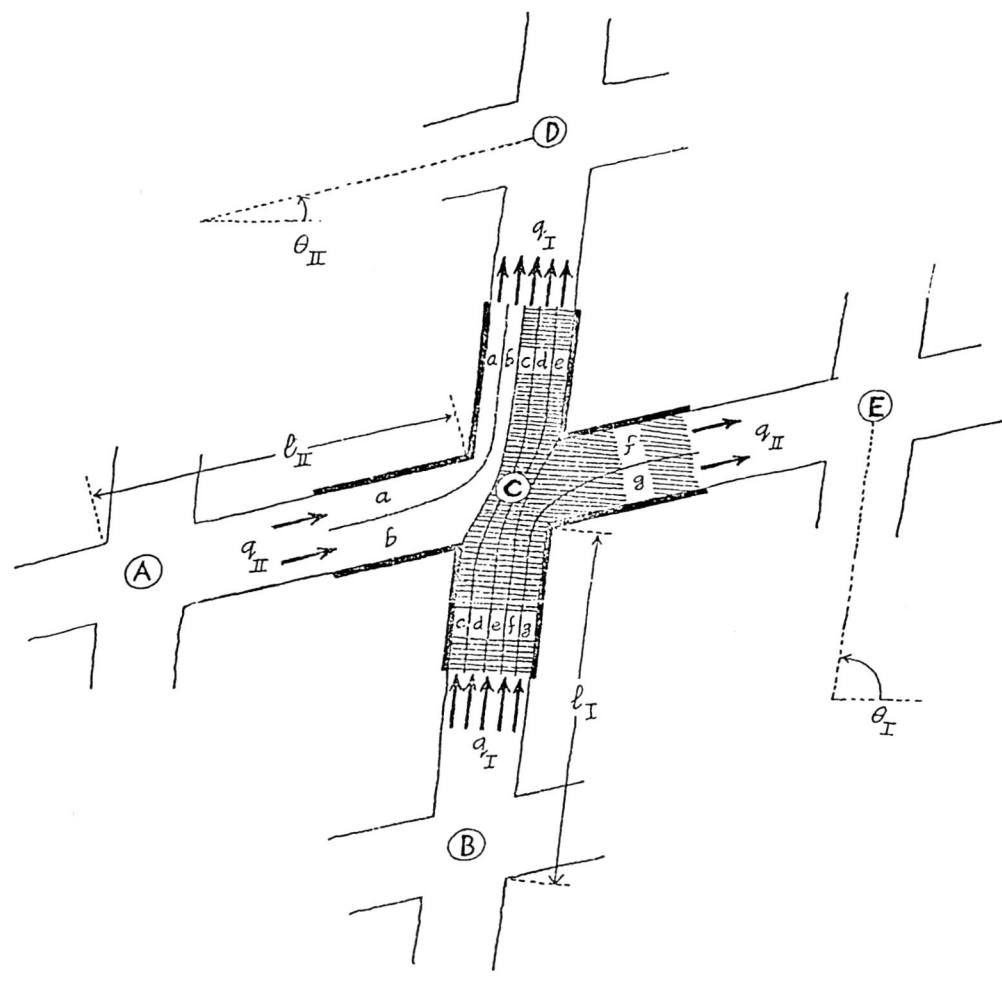

Subdivision of the discharges at intersections of fissures.

Fig. 2

Soil Mechanics and Transport in Porous Media

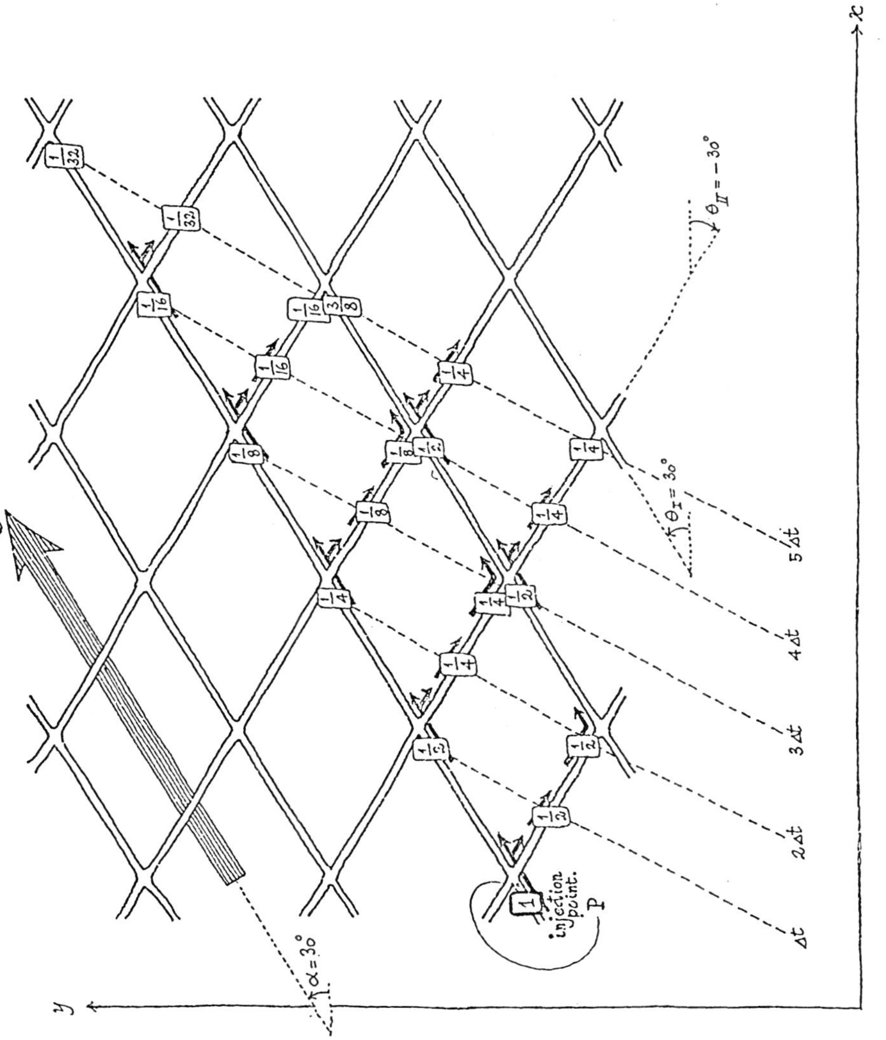

Division of point injection into subgroups.

Fig. 3

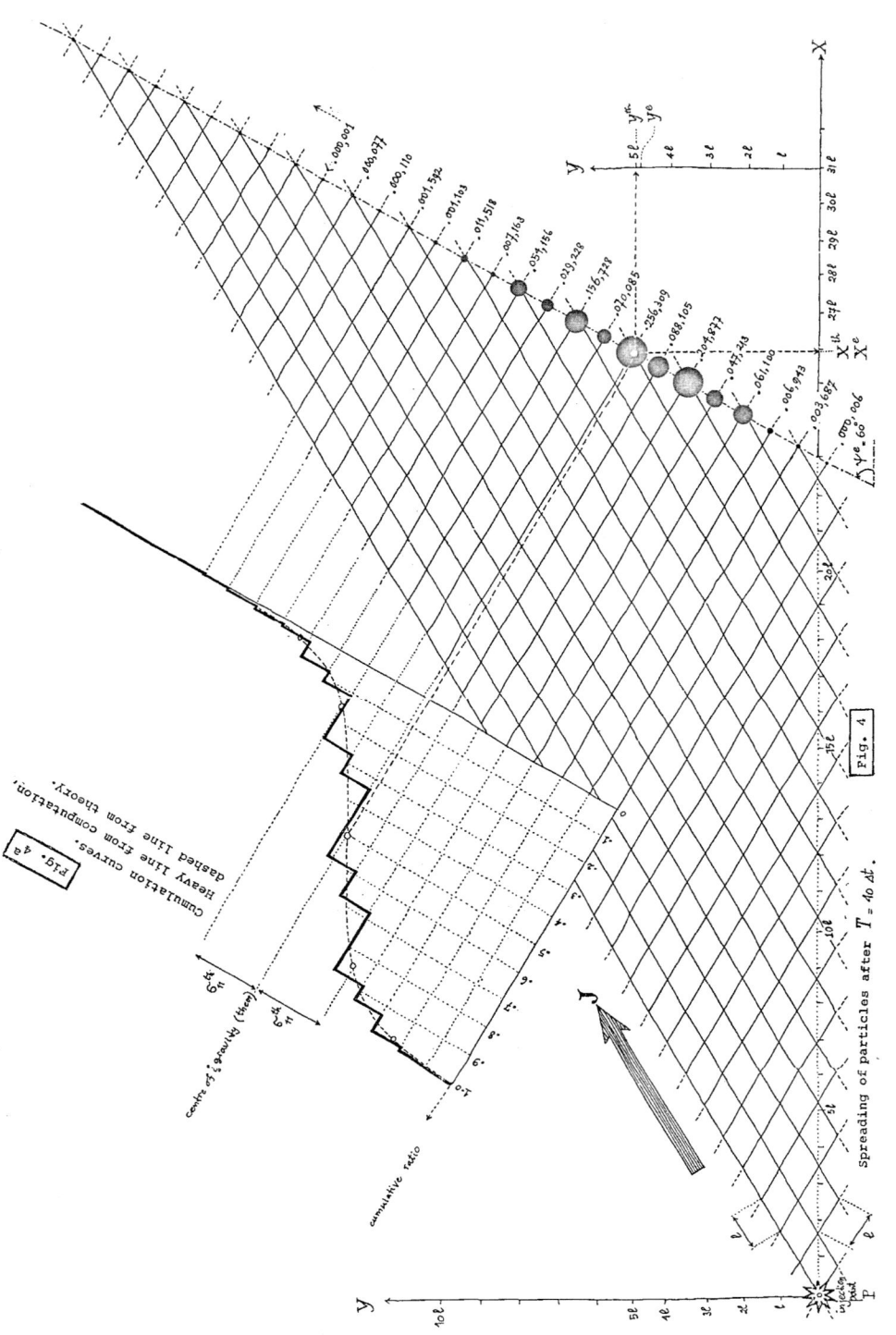

Fig. 4

Spreading of particles after $T = 40 \, \Delta t$.

Soil Mechanics and Transport in Porous Media

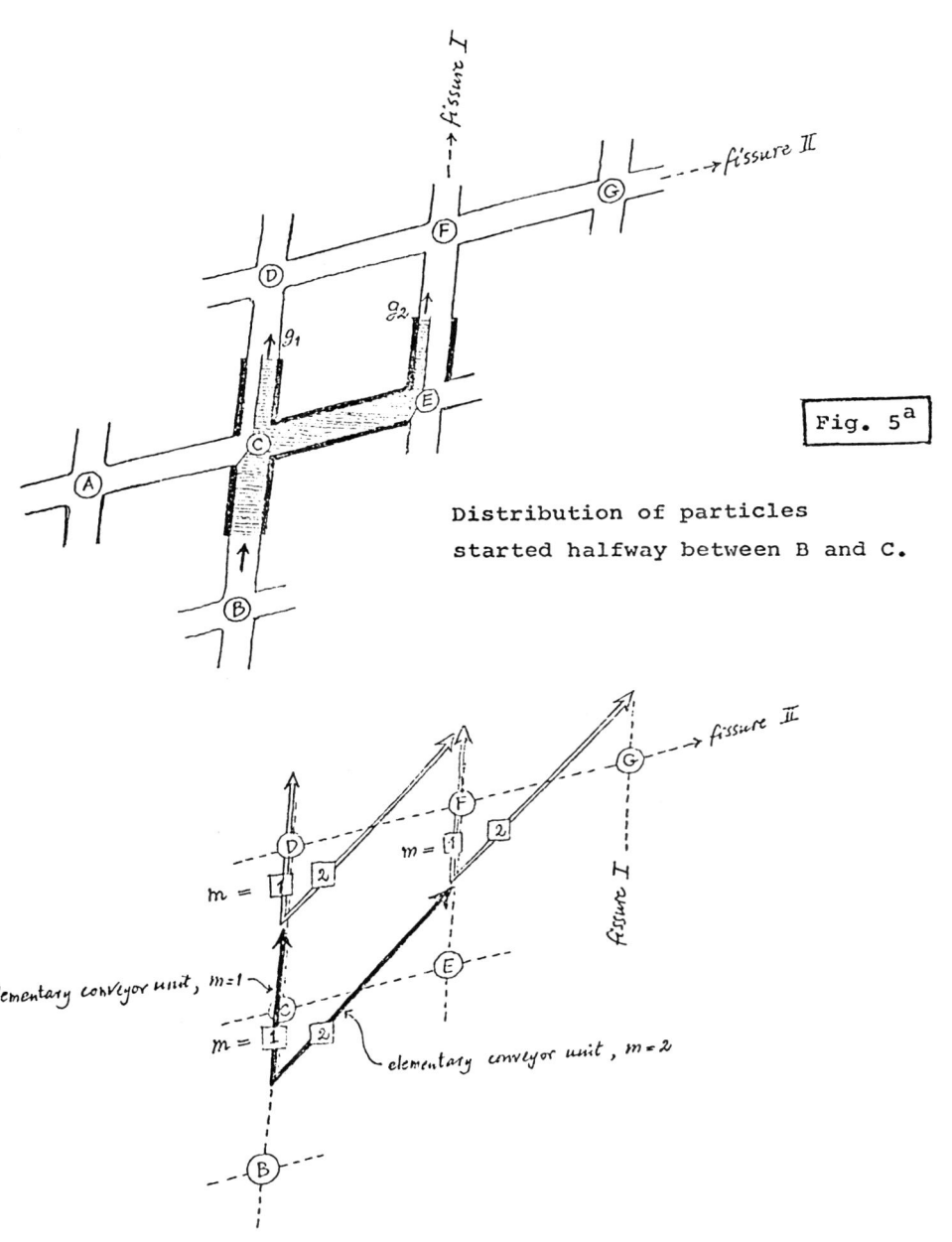

Distribution of particles
started halfway between B and C.

Fig. 5^a

Elementary Conveyor Units, m=1 and m=2.

Fig. 5^b

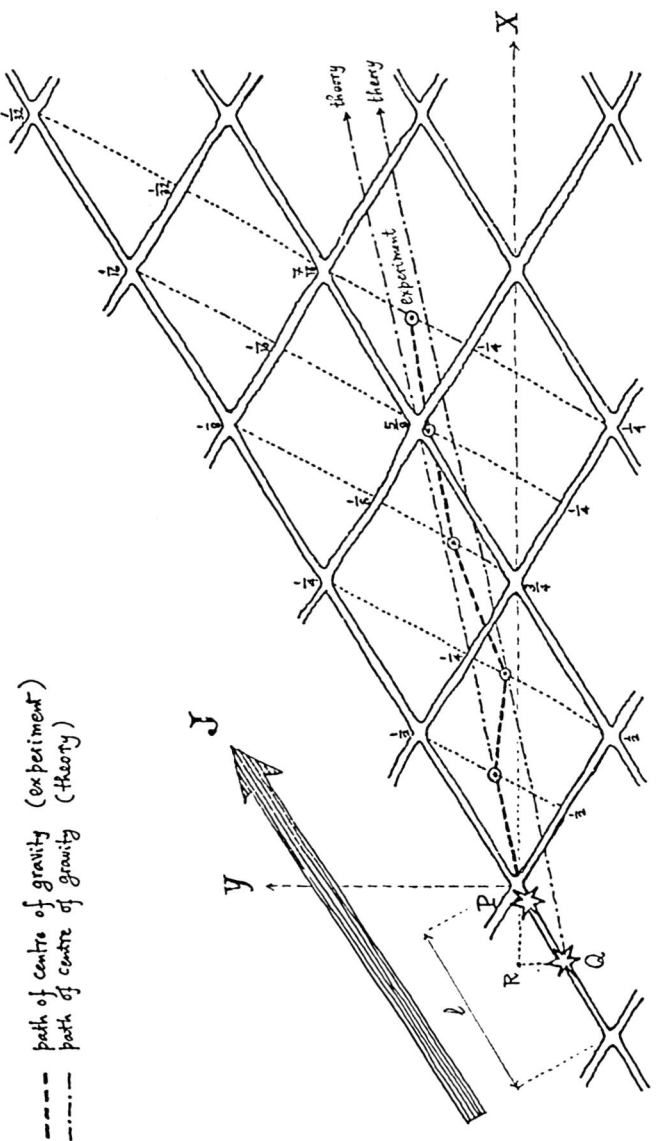

Paths followed by centre of gravity of particles.

---- path of centre of gravity (experiment)
—·— path of centre of gravity (theory)

Fig. 6

NIEUW ARCHIEF VOOR WISKUNDE (4), Vol.3, (1985) 207-208

Cube in Tessaract

An Introduction to the Following Article

J.H. de Boer & J. van de Craats

(Section Editors)

Tessaract (or four-dimensional measure polytope) is the four-dimensional analogue of cube. In the November 1966 issue of "Scientific American", Martin Gardner devoted his column "Mathematical Games" to higher dimensional polytopes ("polytope" is the general term in the sequence: point, segment, polygon, polyhedron, ...). He raised the question of "finding the largest cube, that can be fitted in a unit tessaract". This problem was inspired by the three-dimensional phenomenon, that inside a unit cube a square may be constructed with sides larger than unity, leading to the surprising result, that it is possible to cut a hole in a unit cube such that a cube with edges *larger* than 1 can pass through it (the hole must be cut in a direction perpendicular to the square). Pieter Nieuwland (1764 - 1794) found the maximum edge length for such a cube (cf. Section 3 of the following article). The weaker result that a cube can be perforated in such a way that the second cube of the *same* size may pass through the hole, is ascribed by John Wallis (1616 - 1703) to Prince Rupert, Count Palatine of the Rhine (1619 - 1683) (cf. D.J. Schrek: Prince Rupert's problem and its extension by Pieter Nieuwland, Scripta Mathematica 16 (1950) pp. 73-80 and pp. 261-267).

In 1971, G. de Josselin de Jong (professor of Soil Mechanics, now retired from Delft University of Technology) sent a letter to Gardner containing a description of the construction of a cube with edges b in a tessaract with edges a, where $b > a$. He also expressed his opinion that this cube might be the *largest* cube that can be fitted in the tessaract. He did not have a proof of this conjecture, which he based on visual intuition only.

Martin Gardner answered, that he did not feel competent to judge the merits of this result, and he recommended to send it to a mathematical journal. To find out wether the result was worth publishing, de Josselin de Jong asked the advice of H.S.M. Coxeter who proposed to formulate it as a problem in the American Mathematical Monthly. It appeared as problem 5886 in 1972 (vol. 79, p. 1140). No reactions were received. However, in 1974 (vol. 81, p. 294) the result of Section 6 of the following article was given, but without any explanation how it was obtained.

Nine years later, de Josselin de Jong, still welcoming an adequate opportunity for showing the geometrical aspects of the result, came in contact with one of the editors of our section "Recreational Mathematics". Upon seeing the material we asked him to publish his construction in our journal. The result is the following article.

We expect that many will enjoy reading it, and, of course, we also hope that some will find it a challenge to improve upon the results by constructing a larger cube, or to supply a proof that in fact the cube is the largest possible. Also the general n-dimensional case might be worth considering.

NIEUW ARCHIEF VOOR WISKUNDE (4), Vol.3, (1985) 209-217

Cube with Edges Larger than those of

the Enclosing Tessaract

G. de Josselin de Jong

Ary Schefferstraat 227
2597 VT Den Haag
The Netherlands

1. INTRODUCTION

The purpose of this paper is to show, how inside a *tessaract (four-dimensional measure polytope, or hyper-cube)* a three-dimensional cube can be constructed with edges larger than those of the enclosing tessaract. As an example of a tessaract consider in euclidian four-space $\mathbb{R}^4$ the polytope

$$T = \{(x_1,x_2,x_3,x_4)\in\mathbb{R}^4 \mid -\tfrac{1}{2}a \leqslant x_1,x_2,x_3,x_4 \leqslant +\tfrac{1}{2}a\}.$$

T has 16 vertices (in coordinates $(\pm\tfrac{1}{2}a, \pm\tfrac{1}{2}a, \pm\tfrac{1}{2}a, \pm\tfrac{1}{2}a)$), 32 edges (of length a), 24 two-dimensional faces and 8 three-dimensional cells.

To show that T contains a cube with edges larger than a, it may suffice to give the coordinates of the eight vertices of the cube (see (10) in Section 7) and leave it as an exercise to the reader to verify the claim by linear algebra. However, it might be of interest to show, how the result was found by rotating a rectangular parallelepiped around an axis, adjusting the lengths of its edges in such a manner, that its vertices are on the bounding surface of the tessaract and its edges are of equal length. The procedure results in producing the largest cube, that can be constructed by this particular rotation. Since all other rotations the author could imagine, did not produce a larger cube, it is conjectured that the maximal cube is found. This conjecture is based on visual intuition only, and still requires a rigourous proof.

345

2. A NOTE ON THE FIGURES

Let it first be shown how a tessaract can be represented visually as in Fig. 2. The construction of this figure is understood by starting with the oblique parallel projection of a three-dimensional cube, as in Fig. 1. Two faces of the cube are undistorted and presented as two congruent squares. The four other square faces appear as oblique parallelograms.

In an analogous way an oblique parallel projection of a (four-dimensional) tessaract in three-space may be produced by presenting two boundary cells undistorted as two congruent cubes. From these the front cube is drawn with heavy lines in Fig. 2. The back cube is displaced with respect to the front cube in an arbitrary direction backwards. Then corresponding vertices are connected by 8 parallel edges. The six remaining boundary cubes appear as six oblique parallelopipeds.

Figure 2 gives a plane picture of the three-dimensional projection of a tessaract. This figure should be visualized as a three-dimensional model. It is instructive to locate on this model all 8 boundary cells, which in $\mathbf{R}^4$ are cubes.

Before treating the problem of finding a cube with edges larger than those of the enclosing tessaract, it is helpful to consider first the problem of constructing the largest square in a cube. This problem is treated here in a circumstantial manner, unnecessary for such a trivial case. This is done in order to introduce the elements of the reasoning which, being obvious in three dimensions, can be extended by analogy to the less obvious four-dimensional case.

3. SQUARE IN CUBE

Visual insight suggests, that the most favourable position for the square is obtained by choosing its plane through the centre M of the cube. This centre lies (Fig. 3) halfway on the line connecting the midpoints n and n' of two opposite vertical edges of the cube. Consider this line to be the axis of rotation for a plane α, that will contain the square. Note that in a cube two edges are opposite when, although being parallel, they are not in a common boundary face.

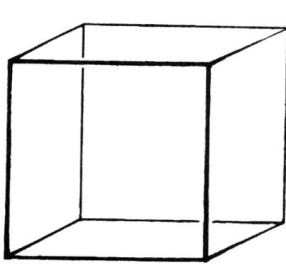

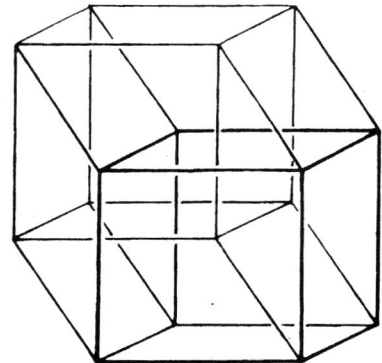

Fig. 1. Cube projected in
        plane of paper

Fig. 2. Projection of tessaract in $\mathbf{R}^3$,
        projected in plane of paper

The plane α intersects the upper and lower faces of the cube by lines parallel to the diagonals in these faces. The lengths of these lines is limited by the edges of the cube, giving symmetrically located vertices for the square.

Let the edges of the cube have lengths a and those of the square b. Introducing a parameter λ according to Fig. 3, gives the relation

$$mm' = b = a\sqrt{2} - 2\lambda. \tag{1}$$

The upright edges of the square are located in a plane normal to the axis of rotation. This plane intersects the cube in the dotted rectangle with sides 2λ and a, such that with Pythagoras' theorem

$$b^2 = (2\lambda)^2 + a^2. \tag{2}$$

Elimination of λ gives the relation

$$a\sqrt{2} - b = (b^2 - a^2)^{\frac{1}{2}} \tag{3}$$

with the root $b/a = 3/2\sqrt{2} \cong 1.0607$, showing that b can be larger than a. This is Pieter Nieuwland's solution mentioned in the editor's introduction. Visual insight in this case clearly indicates that it is impossible to manoeuvre the square in a position which produces a higher value for b/a.

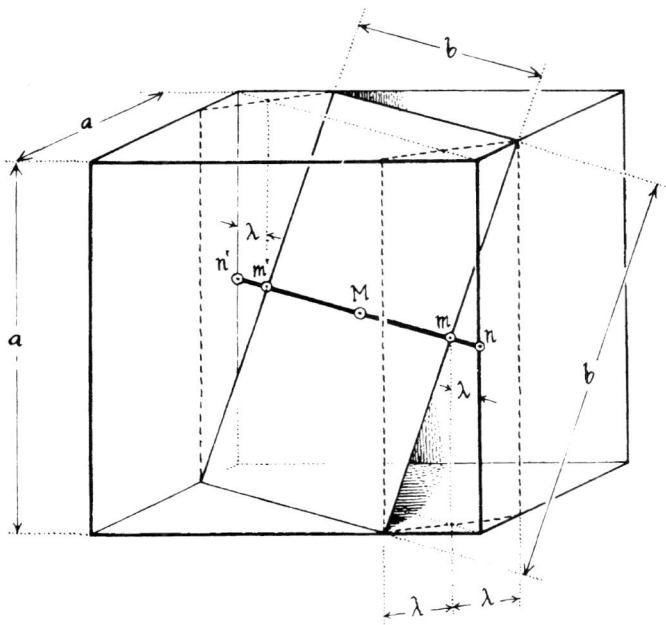

Fig. 3.   Largest flat square in cube

4. CUBE IN TESSARACT

A similar procedure is now applied to the construction of a cube with edge length b, that can be inserted in a tessaract with edge length a, such that $b > a$. The tessaract is the body represented in Figure 4 by

$$A \; B \; C \; D \; E \; F \; G \; HA' \; B' \; C' \; D'E' \; F' \; G' \; H'$$

The cube, that has to fit in it, is shown in Fig. 6 as $P \; Q \; R \; S \; P' \; Q' \; R' \; S'$.

Visual intuition suggests that the most favourable position for the cube is obtained, when its centre coincides with the centre of the tessaract. Let m and m' be the intersection points of the diagonals of the opposite faces $P \; Q \; R \; S$ and $P' \; Q' \; R' \; S'$ of the cube (Fig. 6). The centre M of the cube lies halfway mm'. For the tessaract its centre M lies halfway the centres n and n' of the two opposite faces $B \; F \; F' \; B'$ and $D \; H \; H' \; D'$ (Fig. 4). These faces are opposite, because, although being parallel, they are not in a common cubic boundary cell.

In order to position the cube in the tessaract the choice is now made to superimpose the lines mMm' and nMn' so that the points M coincide. Further, the line $nmMm'n'$ is considered to be the axis of rotation for a plane β, that will contain the diagonal plane $PRR'P'$ of the cube.

The plane β intersects the upper and lower faces of the tessaract ($E'F'G'H'$ and $A \; B \; C \; D$ in Fig. 4) in lines parallel to the diagonals in those faces. The segment $R \; R'$ in face $E' \; F' \; G' \; H'$ is shown separately in Fig. 5. Its position is specified by the parameters λ and μ. Its length b then satisfies the relation

$$RR' = PP' = b = a\sqrt{2} - 2\lambda \tag{4}$$

The symmetric positions of PP' and $R \; R'$ in the upper and lower faces guarantee that $P \; R \; R' \; P'$ is a rectangle. The lengths of the upright edges of this rectangle are $b\sqrt{2}$ and with Pythagoras' theorem it follows from Fig. 4 that

$$(PR)^2 = (P'R')^2 = 2b^2 = 2a^2 + 4\mu^2. \tag{5}$$

(Note that the distance between the faces $ABCD$ and $E' \; F' \; G' \; H'$ is equal to $a\sqrt{2}$.)

Rotation of the plane β around the axis $n \; m \; M \; m' \; n'$ varies the parameter μ. In Fig. 5 the situation $\mu < \lambda$ is assumed. It might be expected that in the most favourable position, i.e. the position for which b is maximal, μ would be equal to λ. Then the points R and R' would be on the edges $F' \; G'$ and $G' \; H'$ of the upperface $E' \; F' \; G' \; H'$ (as in Fig. 3). However, it is shown below (eqs (9)) that the solution requires μ to be smaller than λ. This prevents the other vertices Q, Q', S', S of the cube to exceed the limits imposed by the tessaract.

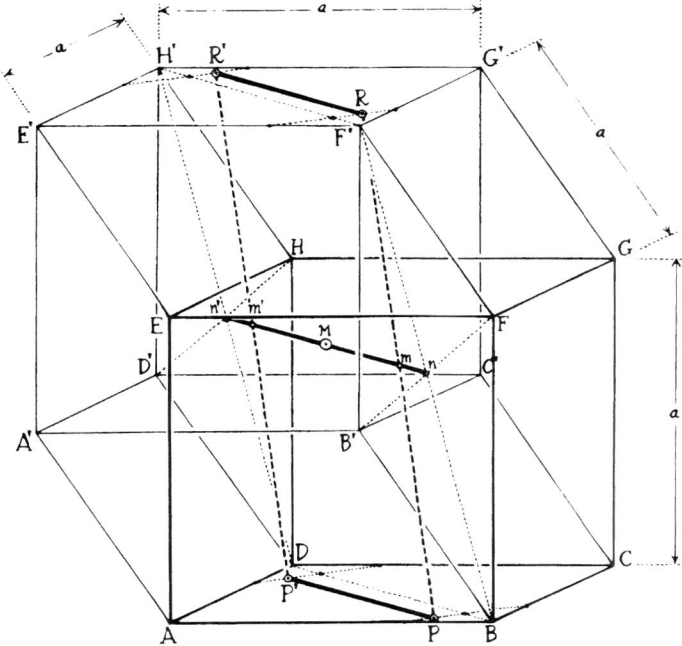

Fig. 4. Tessaract with axis of rotation $n'm'Mmn$ and
vertices $PRR'P'$ of the cube

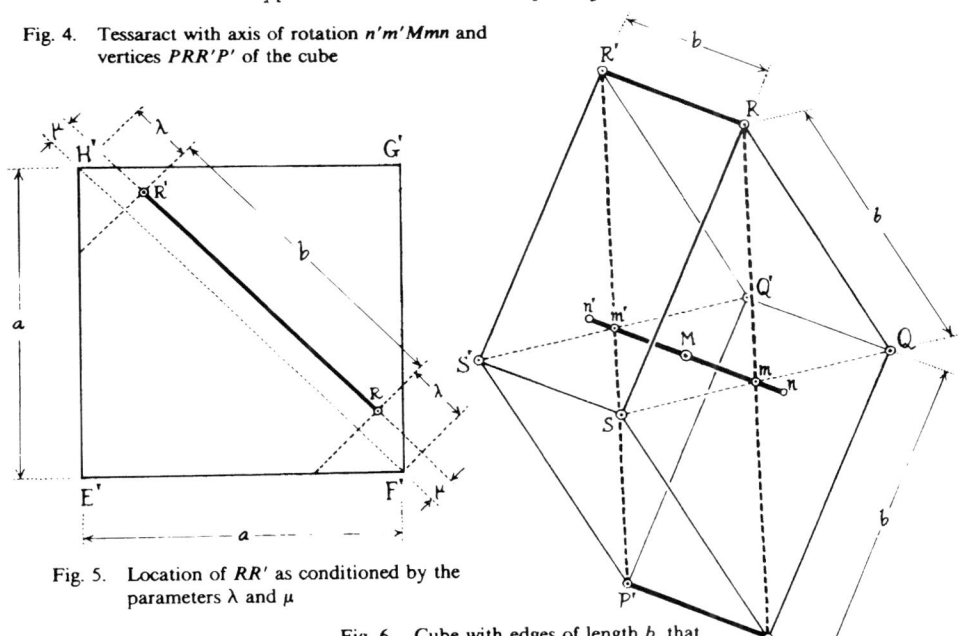

Fig. 5. Location of RR' as conditioned by the
parameters λ and μ

Fig. 6. Cube with edges of length b, that

5. THE REMAINING FOUR VERTICES

The plane $P\,R\,R'\,P'$ is kept fixed in the tessaract during the procedure to locate the other diagonal plane $S\,Q\,Q'\,S'$ of the cube. Unlike in three-space, where the position of a cube is completely determined by one diagonal plane, in four-space it is possible to rotate the other diagonal plane: $S\,Q\,Q'\,S'$ around the axis $n\,m\,M\,m'\,n'$, while keeping it perpendicular to the plane $P\,R\,R'\,P'$.

In order to study the possibilities offered by this second kind of rotation, it is helpful to construct the intersection of the tessaract with the three-dimensional space through m normal to $n\,m\,M\,m'\,n'$. The line PR is located in this space and $S\,Q$ can rotate freely in the plane normal to $P\,R$ in this space. After choosing the most favourable position for $S\,Q$ the cube is determined completely.

In Fig. 7 the dotted orthogonal parallelepiped is the intersection body of the tessaract with the space through m perpendicular to $n\,m\,M\,m'\,n'$. This assertion can be verified by noting that the dotted lines, being mutually orthogonal, are all normal to the diagonals $F'\,H'$ or BD, which in turn are parallel to $n\,m\,M\,m'\,n'$. Moreover, all dotted lines are situated in faces of the tessaract and therefore on its boundary.

The dotted intersection body is shown separately in Fig. 8. It has eight edges of length a, one set of four being parallel to $B\,F$, the other set parallel to $F\,F'$. The remaining four edges, from which two carry P and R, have lengths 2λ. This is verified in Fig. 5, where the dotted line through R (perpendicular to $R\,R'$) is the upper edge of the intersection solid in Fig. 8.

The diagonal $S\,Q$ of the inserted cube is situated in the dotted intersection body of Fig. 8, and more specifically in a plane γ through m normal to $P\,R$. This plane γ intersects the dotted body in the heavy lined hexagon in Fig. 8. The segment SQ has to remain within this hexagon which is presented undistortedly and with its dimensions in Fig. 9.

The most favourable position for $S\,Q$ is as shown in Fig. 9. $S\,Q$ is a diagonal in the cube and therefore its length is $b\sqrt{2}$. Using the values shown in Fig. 9 gives

$$(SQ)^2 = 2b^2 = (a\sqrt{2} - \frac{2\lambda\mu\sqrt{2}}{a})^2 + 4\lambda^2(1+2\frac{\mu^2}{a^2}). \qquad (6)$$

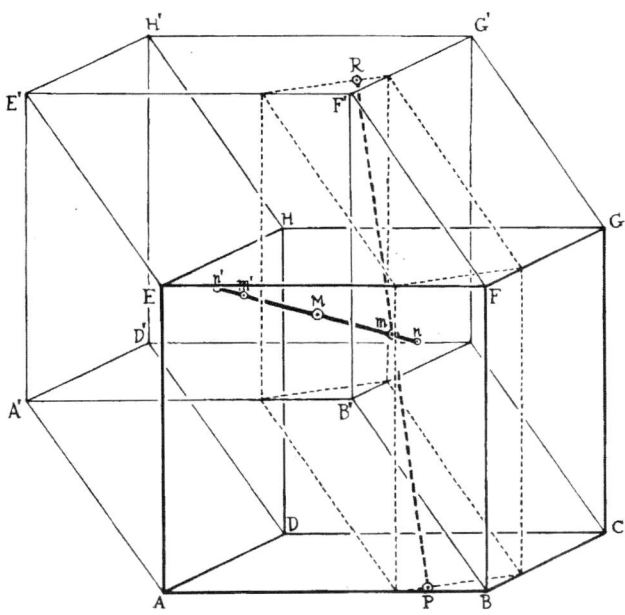

Fig. 7. Tessaract intersected by three-space through m, perpendicular to the rotation axis

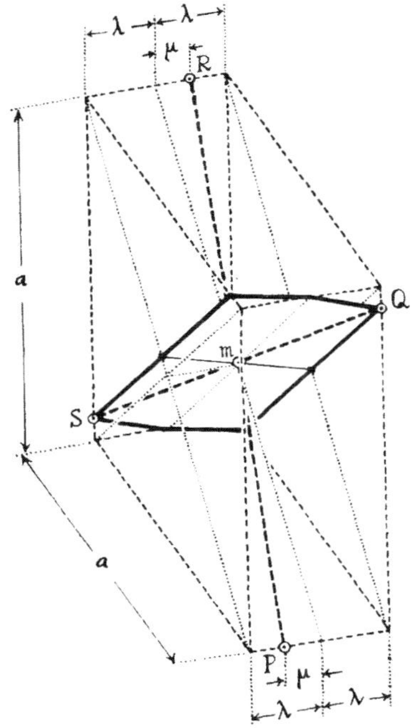

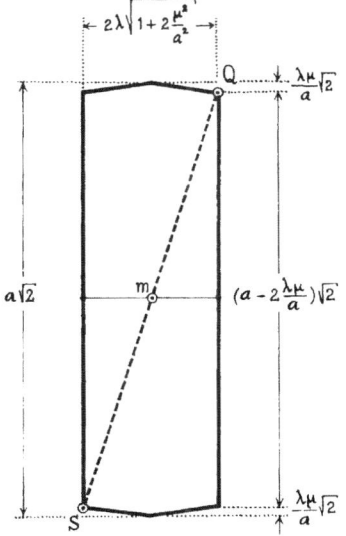

Fig. 8. Shape of intersection body with plane normal to PR

Fig. 9. Hexagon, limiting the size of SQ in plane γ through m normal to PR

6. RESULT

Solving (6) for λ, after μ has been eliminated by means of (5), gives

$$2\lambda = \pm a \sqrt{2}(b^2 - a^2)^{\frac{1}{2}} / (b\sqrt{2} \mp a).$$

Only the upper sign is applicable, since λ cannot be negative. Use of (4) then gives

$$(a\sqrt{2} - b)(b\sqrt{2} - a) = a\sqrt{2}(b^2 - a^2)^{\frac{1}{2}}. \tag{7}$$

There is only one root, viz.

$$b/a \simeq 1.007435. \tag{8}$$

This result shows, that there exists indeed a cube with edges larger than those of the tessaract. From (4) and (5) it follows that

$$\lambda/a \simeq 0.2034 \quad , \quad \mu/a \simeq 0.0864. \tag{9}$$

In the solution of the problem in the Am. Math. Monthly, mentioned in the editor ~~prime~~'s introduction, the ratio (8) was given as a root of an equation of degree eight, instead of the more simple equation (7). This was due to a less adequate elimination of λ and μ from (4), (5) and (6).

Fig. 10 shows how the cube fits into the tessaract. It may be noted that the point Q of the cube is located in the face $B\,C\,G\,F$ of the boundary cell $A\,B\,C\,D\,E\,F\,G\,H$. Since both P and P' are situated in the face $A\,B\,C\,D$ of that boundary cell the entire face $P\,Q\,Q'\,P'$ is located in that cell. Similarly, the face $S\,R\,R'\,S'$ is a subset of the opposite boundary cell $A'\,B'\,C'\,D'\,E'\,F'\,G'\,H'$.

Only the two faces $P\,Q\,Q'\,P'$ and $S\,R\,R'\,S'$ of the cube are located in the boundary space of the tessaract. All other planes of the cube are in the interior. The vertices, however, are all situated on the boundary of the tessaract.

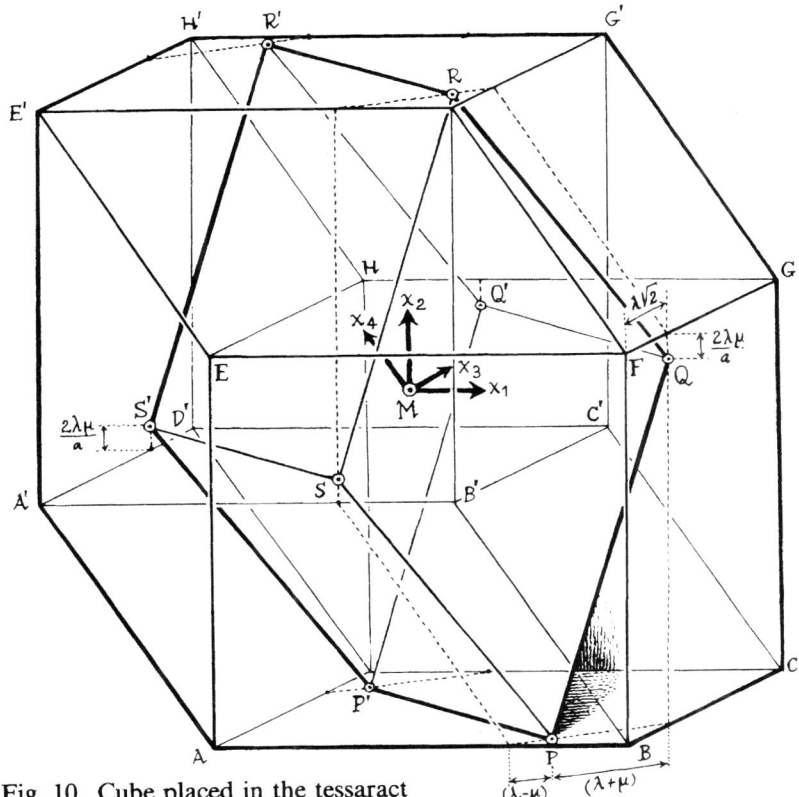

Fig. 10. Cube placed in the tessaract

7. COORDINATES

In order to facilitate verification of the result, the coordinates of the vertices of the cube are given below as located in the tessaract of Section 1 with vertices $(\pm\frac{1}{2}a, \pm\frac{1}{2}a, \pm\frac{1}{2}a, \pm\frac{1}{2}a)$. The lower signs in these expressions relate to the coordinates of the points R', S', P', Q'.

$$P, R' = (\pm(\tfrac{1}{2}a - \tfrac{1}{2}(\lambda + \mu)\sqrt{2}), \mp\tfrac{1}{2}a, \mp(\tfrac{1}{2}a - \tfrac{1}{2}(\lambda - \mu)\sqrt{2}), \mp\tfrac{1}{2}a)$$

$$Q, S' = (\pm\tfrac{1}{2}a, \pm(\tfrac{1}{2}a - \frac{2\lambda\mu}{a}), \mp(\tfrac{1}{2}a - \lambda\sqrt{2}), \mp\tfrac{1}{2}a)$$

$$R, P' = (\pm(\tfrac{1}{2}a - \tfrac{1}{2}(\lambda - \mu)\sqrt{2}), \pm\tfrac{1}{2}a, \mp(\tfrac{1}{2}a - \tfrac{1}{2}(\lambda + \mu)\sqrt{2}), \pm\tfrac{1}{2}a)$$

$$S, Q' = (\pm(\tfrac{1}{2}a - \lambda\sqrt{2}), \mp(\tfrac{1}{2}a - \frac{2\lambda\mu}{a}), \mp\tfrac{1}{2}a, \pm\tfrac{1}{2}a).$$

It is possible to verify by linear algebra that the conditions guaranteeing P Q R S P' Q' R' S' to be a cube, lead to equations (4), (5) and (6).

Complete Bibliography of Scientific Publications

6.1 List of publications

The numbering of the papers is identical to the original numbering made by Gerard de Josselin de Jong, and therefore still completely in line with the actual hardcopy archive of the papers.

[1] De spanningverdeling rondom vertikale in zandige dekterreinen geboorde holten in stand gehouden door zware vloeistof. (The stress distribution around a vertical hole in cohesionless soil.), G. de Josselin de Jong & J. Geertsema, *De Ingenieur*, 1953, pp M1-M5

[2] Consolidation around pore-pressure meters, G. de Josselin de Jong, *J. Appl. Phys.*, Vol. 24, no. 7, 1953, pp 922-928, American Institute of Physics

[3] Wat gebeurt er in de grond tijdens het heien? (What happens in the soil during piled-driving?), *De Ingenieur*, 1956, pp B77-B88

[4] Discussion, *Proc. 3rd Int. Conf. Soil Mech. & Found. Eng.*, Zurich, 1953, p 161

[5] L'entrainement de particules par le courant intersticiel, *Proc. Symposia Darcy*, Dijon, 1956, Publ. nr 41 of the UGGI, pp 139-147, UGGI

[6] Spoorbaanafschuiving (Slide in a railway embankment), *LGM Mededelingen*, Vol. 1, 1956, pp 7-40

[7] Verification of the use of peak area for the quantitative differential thermal analysis, G. de Josselin de Jong, *J. Am. Ceram. Soc.*, Vol. 40, no. 2, 1957, pp 42-49, American Ceramic Society

[8] A capacitative cell apparatus , G. de Josselin de Jong & E.C.W.A. Geuze, *Proc. 4th Int. Conf. Soil Mech. & Found. Eng.*, London, 1957, Vol. 1, pp 52-55, Butterworth Scientific Publications

[9] Application of stress functions to consolidation problems, G. de Josselin de Jong, *Proc. 4th Int. Conf. Soil Mech. & Found. Eng.*, London, 1957, Vol. 1, pp 320-323

[10] Discussion, *Proc. 4th Int. Conf. Soil Mech. & Found. Eng.*, London, 1957, Vol. 13, pp 148-149

[11] Grafische bepaling van glijlijn-patronen in de grondmechanica (Graphical method for the determination of slipline fields in soil mechanics) , Hodograaf zonder rotatie is fout., *De Ingenieur*, 1957, pp B61-B65

[12] Grafische methode ter bepaling van glijlijn-patronen in grond met weinig samenhang (Graphical method for the determination of slip-line fields in cohesionless soil), *LGM Mededelingen*, Vol. 2., 1957, pp 61-78

[13] Longitudinal and transverse diffusion in granular deposits, G. de Josselin de Jong, *Trans. A. Geophys. Union*, Vol. 39, No.1, 1958, pp 67-74, AGU 13 a. Discussion of "Longitudinal and transverse diffusion in granular deposits", G. de Josselin de Jong, *Trans. A. Geophys. Union*, Vol. 39, No. 6, 1958, pp 1160-1162, AGU

[14] The indefiniteness in kinematics for friction materials, *Proc. Brussels Conf. on Earth Pressure Problems*, Vol. 1, 1958, pp 55-70

[15] Discussion, *Proc. Brussels Conf. on Earth Pressure Problems*, Vol. 3, 1958, pp 57-58, pp 64-65

[16] Statics and kinematics in the failable zone of granular material, *Doctor's Thesis Delft University of Technology*, Waltman Delft, 1959

[17] Vortex theory for multiple phase flow through porous media, *Water Resources Center Contribution*, 23, Hydr. Lab, Univ. of California, 1959 (intern report)

[18] Foto-elastisch onderzoek van korrelstapelingen, *LGM Mededelingen*, Vol. 4, 1960, pp 119-134

[19] Singularity distributions for the analysis of multiple-fluid flow through porous media, G. de Josselin de Jong, *J. Geophys. Res.*, Vol. 65, no. 11, 1960, pp 3739-3758, American Geophysical Union

[20] Moiré patterns of the membrane analogy for ground-water movement applied to multiple fluid flow, *J. Geophys. Res.*, Vol. 66, no. 10, 1961, pp 3625-3628, American Geophysical Union

[21] Discussion, *Proc. 5the Int. Conf. Soil Mech. & Found. Eng.*, Paris, 1961, Vol. 3, pp 326-327
21 a. Discussion of paper by Jacob Bear on the Tensor Form of Dispersion in Porous Media, G. de Josselin de Jong & M.J. Bossen, *J. Geophys. Res.*, Vol. 66, No. 10, 1961, pp 3623-3624
21 b. Refraction moiré analysis of curved surfaces, *Proc. Symposium Shell Research*, Delft, 1961, North-Holland Publishing Comp., pp 302-308

[22] Vacuum voorgespannen zand als constructiemateriaal, I (Vacuum pre-stressed sand as construction material, I), *De Ingenieur*, 1962, B127-B134

[23] Vacuum voorgespannen zand als constructiemateriaal , II (Vacuum pre-stressed sand as construction material, II), *De Ingenieur*, 1963, B113-B121

[24] Discussion, *Proc. Eur. Conf. Soil Mech. & Found. Eng.*, Wiesbaden, 1963, p 35

[25] Discussion, *Proc. Eur. Conf. Soil Mech. & Found. Eng.*, Wiesbaden, 1963, p 76

[26] Consolidatie in drie dimensies, I, *LGM Medelingen*, 7, 1963, pp 57-73

[27] Consolidatie in drie dimensies, II, *LGM Medelingen*, 8, 1963, pp 25-38

[28] Consolidatie in drie dimensies, III, *LGM Medelingen*, 8, 1964, pp 53-68

[29] A many-valued hodograph in an interface problem, G. de Josselin de Jong *Water Resour. Res.*, Vol. 1, no. 4, 1965, pp 543-555, American Geophysical Union

[30] Primary and secondary consolidation of a spherical sample, G. de Josselin de Jong & A. Verruijt, *Proc. 6th Int. Conf. Soil Mech. & Found. Eng.*, Montreal, 1965, Vol. 1, pp 254-258

[31] Lower bound collapse theorem and lack of normality of strainrate to yield surface for soils, G. de Josselin de Jong, *Proc. IUTAM Symp. on Rheology and Soil Mechanics*, edited by G. Kravtchenko and PM Sirieys Grenoble, 1964, pp 69-78, Springer-Verlag

[32] Discussion, G. de Josselin de Jong, *Proc. Eur. Conf. Soil Mech. & Found. Eng.*, Oslo, 1968, pp 199-200, Norwegian Institute of Technology

[33] Consolidation models consisting of an assembly of viscous elements or a cavity channel network, G. de Josselin de Jong, *Géotechnique*, Vol. 18, 1968, pp 195-228, The Institute of Civil Engineers

[34] Generating functions in the theory of flow through porous media, G. de Josselin de Jong, Chapter 9, *Flow Through Porous Media*, Academic Press, 1969, pp 377-400

[35] Etude photo-élastique d'un empilement de disques, G. de Josselin de Jong & A. Verruijt, *Cahier Groupe Français de Rhéologie*, Vol. 2, No. 1, 1969, pp 73-86

[36] The double sliding, free rotating model for granular assemblies, G. de Josselin de Jong, *Géotechnique*, 1971, pp 155-163, The Institution of Civil Engineers

[37] Discussion, *Proceedings Roscoe Memorial Symposium*, Cambridge, 1971, pp 258-262

[38] Dispersion of a point injection in an anisotropic porous medium, Socorro Report, New Mexico 87801, 1972
38 a. Dispersion in fissured rock, G. de Josselin de Jong & Shao Chih Way, New Mexico 87801, 1972

[39] Dispersion described by differential equation developed with Lagrangian Correlation Functions, *Rapport Geotechniek Delft*, 1973

[40] The tensor character of the dispersion coefficient in anisotropic porous media, G. de Josselin de Jong, *Proc. JAHR Congress*, Haifa Israël, 1972, pp 259-267

[41] Photoelastic verification of a mechanical model for the flow of a granular material, A. Drescher & G. de Josselin de Jong, *J. Mech. Phys. Solids*, 1972, Vol. 20, pp 337-351, Pergamon Press

[42] A limit theorem for material with internal friction, *Proc. Symp. Plasticity*, Cambridge, 1973, pp 12-21

[43] Aelotropie van gestructureerde materialen, *Cement XXVI*, No. 4, 1974, pp 166-176

[44] Rowe's stress dilatancy relation based on friction, *Géotechnique*, 178, pp 527-534

[45] Grondmechanische aspecten van korrelstapelingen, *Procestechniek*, Jaargang 31, No. 12, 1976
45 a. Evaluation of principle stress difference in photo-elastic granular media by use of a compensator in pure bedding, *Rapport Laboratorium voor Geotechniek Delft*, June 1976

[46] Mathematical elaboration of the double sliding free rotating model, *Archiwun Mechaniki Stosowanej*, 29, No. 4, Warszawa, 1977
46 a. Model for the behaviour of granular materials in progressive flow, *Contribution Conf. in Jablona*, Poland, 1976, lecture notes

[47] Review of vortex theory for multiple fluid flow, *Delft Progress Report*, 2, 1977, pp 225-236

[48] Constitutive relations for the flow of a granular assembly in the limit state of stress, *Speciality Session 9, Proc. Int. Conf. Soil Mech. & Found. Eng.*, Tokyo, 1977, pp 87-95

[49] Lecture notes Int. Center for Mech. Sciences (CISM), Udine, 1974, Model for the behaviour of granular material in progressive flow

[50] Improvement of the lowerbound solution for the vertical cut off in a cohesive, frictionless soil, G. de Josselin de Jong, *Géotechnique*, Technical notes June 1978, pp 197-201, The Institution of Civil Engineers

[51] Vortex theory for multiple fluid in three dimensions, G. de Josselin de Jong, *Delft Progress Report*, 4, 1979, pp 87-102

[52] Diskontinuitäten in Grenzspannungsfeldern, *Geotechniek*, Jahrgang 2, Heft 1, 125-129 (translated by Smoltczyk)

[53] Het verloop van een drukstoot door een paal, *KIVI Publicatie*, Vreedenburghdag 1977, 19 oktober 1977

[54] Application of the calculus of variations to the vertical cut off in cohesive frictionless soil, G. de Josselin de Jong, *Géotechnique*, 30, No. 1, 1980, pp 1-16, The Institution of Civil Engineers

[55] Illustraties (Illustrations), Uitgegeven door de vakgroep Geotechniek ter gelegenheid van het afscheid van prof.dr.ir. G. de Josselin de Jong, June 25, 1980

[56] Tribute to Professor de Josselin de Jong, *LGM Mededelingen*, Part XXI, No. 2, June 1980

[57] A limit theorem for soils, *Delft Progress Report*, 5, 1980, pp 280-291

[58] A variational fallacy, G. de Josselin de Jong, *Géotechnique*, June 1981, Vol. 31, no. 2, pp 289-290, The Institution of Civil Engineers

[59] The simultaneous flow of fresh and salt water in aquifers of large horizontal extension determined by shear flow and vortex theory, G. de Josselin de Jong, *Proc. Euromech. Colleg.*, Edited by A. Verruijt & F. B. J. Barends, Sept. 1981, pp 75-82, A. A. Balkema

[60] Behavior of an elasto-plastic double-sliding free-rotating material if subjected to an ideal simple shear test, *Proc. IUTAM Conf. on Deformation and Failure of Granular Materials*, Edited by Vermeer & Luger, Delft, 1982, pp 555-562

[61] Cube with edges larger than those of the enclosing tessaract. G. de Josselin de Jong, *Nieuw Archief voor de Wiskunde*, 4, Vol. 3, 1985, pp 209-217 (+ introduction: pp 207-208), Wiskundig Genootschap

[62] Transverse dispersion from an originally sharp fresh-salt interface caused by shear flow. G. de Josselin de Jong and C.J. Van Duijn, *J. Hydrology*, 84, 1986, pp 55-59, Elsevier

[63] Elasto-plastic version of the double sliding model in undrained simple shear tests. G. de Josselin de Jong, *Géotechnique*, 38, No. 4, 1988, pp 533-555 (+ 63a Discussion G. de Josselin de Jong *Géotechnique*, 39, No. 3, 1989, pp 565-566). The Institution of Civil Engineers

[64] Keverling Buisman Lezing, 1989

[65] Co-rotational solution in simple shear tests, *Wroth Memorial Symposium*, Oxford, 1992. Proceedings Predictive Soil Mechanics, Thomas Telford, 1993

7

Acknowledgements

The following articles have been reproduced with the kind permission of the copyright holder. The numbers in brackets refer to the numbering of the articles in the complete bibliography (Chapter 6).

[2] Consolidation around pore pressure meters, G. de Josselin de Jong, *J. Appl. Phys.*, Vol. 24, no. 7, 1953, pp 922-928, American Institute of Physics. Reprinted with the permission of the American Institute of Physics

[5] L'entrainement de particules par le courant intersticiel, G. de Josselin de Jong, *Proc. Symposia Darcy,* Dyon, 1956, Publ. nr 41 of the UGGI, pp 139-147, UGGI. Reprinted with the kind permission of UGGI

[7] Verification of the use of peak area for the quantitative differential thermal analysis, G. de Josselin de Jong, *J. Am. Ceram. Soc.,* Vol. 40, no. 2, 1957, pp 42-49, American Ceramic Society. Reprinted with the permission of the American Ceramic Society, Post Office Box 6136, Westerville, Ohio 43086-6136

[8] A capacitative cell apparatus, G. de Josselin de Jong & E.C.W.A. Geuze, *Proc. 4th Int. Conf. Soil Mech. & Found. Eng.,* London, 1957, Vol. 1, pp 52-55, Butterworth Scientific Publications. Reprinted with kind permission of the copyright holder Prof. J. de Josselin de Jong

[9] Application of stress functions to consolidation problems, G. de Josselin de Jong, *Proc. 4th Int. Conf. Soil Mech. & Found. Eng.,* London, 1957, Vol. 1, pp 320-323. Reprinted with kind permission of the copyright holder Prof. J. de Josselin de Jong

[13] Longitudinal and transverse diffusion in granular deposits, G. de Josselin de Jong, *Trans. A. Geophys, Union,* Vol. 39, 1958, pp 67-74, AGU. Reprinted with the kind permission of the American Geophysical Union (AGU)

[13a] Discussion of "Longitudinal and transverse diffusion in granular deposits", G. de Josselin de Jong, *Trans. A. Geophys. Union,* Vol. 39, No. 6, 1958, pp 1160-1162, AGU. Reprinted with the kind permission of the American Geophysical Union (AGU)

[19] Singularity distributions for the analysis of multiple-fluid flow through porous media, G. de Josselin de Jong, *J. Geophys. Res.,* Vol. 65, no. 11, 1960, pp 3739-3758, American Geophysical Union. Reprinted with the kind permission of the American Geophysical Union (AGU)

[20] Moiré patterns of the membrane analogy for groundwater movement applied to multiple fluid flow, G. de Josselin de Jong, *J. Geophys. Res.,* Vol. 66, no. 10, 1961, pp 3625-3628, American Geophysical Union. Reprinted with the kind permission of the American Geophysical Union (AGU)

[29] A many-valued hodograph in an interface problem, G. de Josselin de Jong, *Water Resources Research,* Vol. 1, no. 4, 1965, pp 543-555, American Geophysical Union. Reprinted with the kind permission of the American Geophysical Union (AGU)

[31] Lower bound collapse theorem and lack of normality of strainrate to yield surface for soils, G. de Josselin de Jong, *Proc. IUTAM Symp. on Rheology and Soil Mechanics,* edited by G. Kravtchenko and PM Sirieys Grenoble, 1964, pp 69-78, Springer-Verlag. Copyright Springer-Verlag

[32] Discussion, G. de Josselin de Jong, *Proc. Eur. Conf. Soil Mech. & Found. Eng.,* Oslo, 1968, pp 199-200, Norwegian Institute of Technology. Reprinted by kind permission of the Norwegian Geotechnical Institute

[33] Consolidation models consisting of an assembly of viscous elements or a cavity channel network, G. de Josselin de Jong, *Géotechnique,* Vol. 18, 1968, pp 195-228, The Institute of Civil Engineers. Reprinted by kind permission of the Institution of Civil Engineers

[34] Generating functions in the theory of flow through porous media, G. de Josselin de Jong, Chapter 9, *Flow Through Porous Media,* Academic Press, 1969. Reprinted with kind permission of Academic Press-Elsevier

[35] Etude photo-élastique d'un empilement de disques, G. de Josselin de Jong & A. Verruijt, *Cahier Groupe Francais de Rhéologie,* No. 1, 1969, pp 73-86. Reprinted with the kind permission of the Cahier Groupe Francais de Rhéologie

[36] The double sliding, free rotating model for granular assemblies, G. de Josselin de Jong, *Géotechnique,* 1971, pp 155-163, The Institution of Civil Engineers. Reprinted by kind permission of the Institution of Civil Engineers

[38a] Dispersion in fissured rock, G. de Josselin de Jong & Shao-Chih Way, New Mexico 87801, 1972. Unpublished material. Reprinted with kind permission of the copyright holder Prof. J. de Josselin de Jong

[40] The tensor character of the dispersion coefficient in anisotropic porous media, G. de Josselin de Jong, *Proc. JAHR Congress*, Haifa Israël, 1972, pp 259-267. Reprinted with kind permission of the copyright holder Prof. J. de Josselin de Jong

[41] Photoelastic verification of a mechanical model for the flow of a granular material, A. Drescher & G. de Josselin de Jong, *J. Mech. Phys. Solids,* 1972, Vol. 20, pp 337-351, Pergamon Press. Reprinted with kind permission of Elsevier

[50] Improvement of the lowerbound solution for the vertical cut off in a cohesive, frictionless soil, G. de Josselin de Jong, *Géotechnique*, Technical notes June 1978, pp 197-201, The Institution of Civil Engineers. Reprinted by kind permission of the Institution of Civil Engineers

[51] Vortex theory for multiple fluid flow in three dimensions, G. de Josselin de Jong, *Delft Progress Report,* 4, 1979, pp 87-102. Reprinted by kind permission of Delft University Press

[54] Application of the calculus of variations to the vertical cut off in cohesive frictionless soil, G. de Josselin de Jong, *Géotechnique*, 30, No. 1, 1980, pp 1-16, The Institution of Civil Engineers. Reprinted by kind permission of the Institution of Civil Engineers

[58] A variational fallacy, G. de Josselin de Jong, *Géotechnique*, June 1981, Vol. 31, no. 2, pp 289-290, The Institution of Civil Engineers. Reprinted by kind permission of the Institution of Civil Engineers

[59] The simultaneous flow of fresh and salt water in aquifers of large horizontal extension determined by shear flow and vortex theory, G. de Josselin de Jong, *Proc. Euromech. Colleg.*, Edited by A. Verruijt & F. B. J. Barends, Sept. 1981, pp 75-82, A. A. Balkema. Reprinted by kind permission of AA Balkema Publishers-Taylor and Francis The Netherlands

[61] Cube with edges larger than those of the enclosing tessaract, G. de Josselin de Jong, *Nieuw Archief voor de Wiskunde*, 4, Vol. 3, 1985, pp 209-217 (+ introduction: pp 207-208), Wiskundig Genootschap. Reprinted by kind permission of "Nieuw Archief voor Wiskunde"

[62] Transverse dispersion from an originally sharp fresh-salt Interface caused by shear flow, G. de Josselin de Jong and C. J. van Duijn, *J. of Hydrology,* 84, 1986, pp 55-79, Elsevier. Reprinted with kind permission of Elsevier

[63] Elasto-plastic version of the double sliding model in undrained simple shear tests, G. de Josselin de Jong, *Géotechnique*, 38, No. 4, 1988, pp 533-555 (+ 63a discussion G. de Josselin de Jong *Géot.*, 39, No. 3, 1989, pp 565-566). The Institution of Civil Engineers. Reprinted by kind permission of the Institution of Civil Engineers

Theory and Applications of Transport in Porous Media

Series Editor:
Jacob Bear, *Technion – Israel Institute of Technology, Haifa, Israel*

18. R. de Boer: *Trends in Continuum Mechanics of Porous Media.* 2005
ISBN 1-4020-3143-2

19. R.J.Schotting, H.(C.J.) van Duijn and A.Verruijt (eds.): *Soil Mechanics and Transport in Porous Media.* Selected Works of G. de Josselin de Jong. 2006
ISBN 1-4020-3536-5